Genesis, Pathophysiology and Management of Venous and Lymphatic Disorders

Genesis, Pathophysiology and Management of Venous and Lymphatic Disorders

N. RADHAKRISHNAN
Medical Director, St. Thomas Institute of Research on Venous Diseases, Kerala, India

Academic Press is an imprint of Elsevier
125 London Wall, London EC2Y 5AS, United Kingdom
525 B Street, Suite 1650, San Diego, CA 92101, United States
50 Hampshire Street, 5th Floor, Cambridge, MA 02139, United States
The Boulevard, Langford Lane, Kidlington, Oxford OX5 1GB, United Kingdom

British Library Cataloguing-in-Publication Data
A catalogue record for this book is available from the British Library

Library of Congress Cataloging-in-Publication Data
A catalog record for this book is available from the Library of Congress

ISBN: 978-0-323-88433-4

For Information on all Academic Press publications
visit our website at https://www.elsevier.com/books-and-journals

Publisher: Stacy Masucci
Acquisitions Editor: Katie Chan
Editorial Project Manager: Pat Gonzalez
Production Project Manager: Punithavathy Govindaradjane
Cover Designer: Miles Hitchen

Typeset by MPS Limited, Chennai, India

This book is dedicated to the great teacher,

My spiritual master, **Swami Vivekananda**, who reminds me always

"Arise! Awake!

and stop not

till the goal is reached."

Epigraph

"I shall be telling this with a sigh
Somewhere ages and ages hence:
Two roads diverged in a wood, and I—
I took the one less traveled by,
And that has made all the difference"

. . .Robert Frost

Contents

About the author

Dr. N. Radhakrishnan (born in 1945) is a Senior Consultant Vascular Surgeon and Medical Administrator at St. Thomas Hospital, Chethipuzha, Changanassery, in Kerala, India. He is also the Medical Director of St. Thomas Institute of Research on Venous Diseases. He graduated from Government Medical College, Thiruvananthapuram, University of Kerala, India, and later took a postgraduate degree in General Surgery from the same institution. He has had a professional career spanning over 52 years and has focused on the management of patients with chronic venous disease (CVD) over the past 25 years. He has treated more than 40,000 patients with advanced stages of CVD, with minimally invasive procedure (modified microfoam sclerotherapy) based on new concepts and innovative techniques. He has collaborated with the Rajiv Gandhi Center for Biotechnology, Thiruvananthapuram, to enhance the understanding of the molecular and genetic basis for varicose veins and with the Mahatma Gandhi University, Kerala, in developing nanotechnology-based strategies for healing chronic wounds.

He has authored a reference book *A Treatise on Venous Disease* (Jaypee, 2014) and published several articles in peer-reviewed international journals. He has also delivered lectures in many national and international professional meetings. He has been awarded several fellowships including FRCS (England), FRCS (Glasgow), FRSM (England), FICS (Phlebolymphology), FAIS (Founder Fellow of Association of Surgeons of India), Fellow of the IMA Academy of Medical Specialties, MRCP (United Kingdom) and FACS (USA & Canada). He is also a member of the American Venous Forum, European Venous Forum, and American College of Phlebology. He is a life member of the Venous Association of India, Indian Medical Association, Indian Academy of Medical Specialties, and Association of Surgeons of India.

He has been bestowed with several national and international accolades and is the patron of several social and philanthropic organizations. The Dr. N. Radhakrishnan Foundation for Research on Venous Diseases (2012) offers scholarships for postdoctoral fellows in the Union of India to advance scientific research into venous diseases.

Foreword

The second book of Dr. Radhakrishnan, *Genesis, Pathophysiology and Management of Venous and Lymphatic Disorders*, is virtually new. Far from a text on venous disease, it covers a wide range including history, developmental biology, vascular physiology, pharmacology, and much more. It goes into such details as the transcriptional control of endothelial differentiation, angiogenesis, and vascular network development in the embryo. The lymphatic system receives serious attention to address its poorly recognized role in the onset of complications in venous disease. The discussion on the immunological markers of lymphatic endothelium and migration of lymphatic endothelial cells in the context of the genesis of lymphovenous valves is an example of the depth to which the author goes in explaining histological observations in the venous system. Similarly, the surgical anatomy of the venous system in the lower limb is explained lucidly, with emphasis on the peripheral and truncal veins, perforators, and the all-important venous valves. Many figures of high quality are presented to facilitate the reader's understanding of the centrifugal flow of blood in the veins and the consequences of its reversal, including varicose veins (VV) and their complications. The clinical manifestations of VV and their staging as per international protocols are also well described.

Dr. Radhakrishnan comes into his own when he takes up the treatment of VV. He would rank high among the few surgeons who love the history of medicine and salute the master surgeons of long ago. He lists a large number of distant European surgeons who made pioneering contributions to venous studies, such as Ambroise Pare, Brodie, and Trendelenburg. Who would not be delighted to learn that Trendelenburg wrote his dissertation during training on "Surgery in ancient India"? Though prolific in paying homage to pioneers, Dr. Radhakrishnan concludes his historical narrative by affirming that ancient techniques, such as stripping and ligation of VV, must yield place to new methods which, according to him, include, for example, "chemical bombing" of the network of diseased superficial veins around the ankle and foot. Dr. Radhakrishnan does full justice to contemporary techniques for managing VV by describing in detail endovascular laser treatment, endovascular radiofrequency ablation, and Clarivein, which consists of an intraluminal catheter with a rotating tip that releases a vein-occluding chemical at the targeted area. All these techniques, postprocedure management, and complications are discussed with photographs of the equipment in current use.

Fully conversant with alternate approaches to the treatment of VV, Dr. Radhakrishnan's forte is "modified microfoam sclerotherapy"(MMFST). He devotes a separate chapter (Chapter 14: Sclerotherapy) to MMFST after a graphic description of the history, historical use of sclerosants, methods of administration, postprocedure care, and complications of conventional sclerotherapy (Chapter 13: Radiofrequency Ablation). In contrast to conventional sclerotherapy which relies on the slow and steady injection of a sclerosant into large veins near junctions and perforators and into the tributaries, MMFST is directed at the location of the tiny venous network tributaries in the tissue [Radhakrishnan N, Jayakrishnan R, Deepu G. Microfoam sclerotherapy for varicose veins: a retrospective analysis of a modifies technique. Indian J Surg (2015);77(Suppl 3):S816–21].

This technique had been used in 14,707 patients by Dr. Radhakrishnan at the St. Thomas Institute of Research on Venous Diseases, Changanassery, Kerala, India, by May 2013, when the paper was written. In a follow-up of 6350 patients undergoing MMFT, 5397 were available for review. Complications included 12% recurrence and other minor events. The recurrences were managed successfully by repeat procedures. In an encyclopedic approach to venous diseases, Dr. Radhakrishnan also deals with important subjects such as deep vein thrombosis, anticoagulation for the prevention and treatment of thromboembolism, and chronic venous ulcer with the same attention to detail. The intellectual feast concludes with a dessert of 191 exquisite figures of veins and venous conditions in disease.

While commending Dr. Radhakrishnan's innovative approach to the treatment of VV by MMFST, the sheer volume of surgical procedures done in a personal series and setting up the Institute for Research in Venous Disease at St. Thomas Hospital, I would also like to compliment him for reaching out to Professor Kartha's distinguished group at the Rajiv Gandhi Centre for Biotechnology for collaborative studies on the genomic basis of venous diseases. As already mentioned, this joint effort has already given new insights such as the association of small VVs with severe venous insufficiency in the absence of truncal saphenous reflux and the role of altered fox C2 - D114 signalling the structural alterations of saphenous vein in VV. These studies have opened a new frontier for research into vascular biology.

I am confident that this new book will enjoy a wide readership among vascular surgeons, general surgeons and physicians, medical students, and biological scientists and researchers.

M.S. Valiathan
National Research Professor, Padma Vibhushan Awardee (2005), Former Vice-Chancellor, Manipal Academy of Higher Education, Former President, Indian National Science Academy, Former Director, Sree Chitra Tirunal Institute for Medical Sciences & Technology, Thiruvananthapuram, India

June 2, 2021

Preface

Varicose Vein (VV) is not a disease. It is the normal physiological phenomenon of all the veins. The pathological process is called Venous Disease (VD). Venous Disease along with lymphatic system involvement at capillary level is called Chronic Venous Disease (CVD).

Venous disease consists of four difficult issues: first, the life-threatening bleeding due to unpredictable rupture of the vein; second, the disfiguring, dreadful congenital venous anomalies; third, extremely painful chronic venous ulceration of foot; and finally, the silent killer, deep vein thrombosis. Of these four problems, chronic venous disease due to varicose veins is prominent in causing long-term difficulties, not only mentally and physically but also socially and economically. Sir Benjamin Brodie (1814), while lecturing at the beside of a patient suffering from varicose ulceration, remarked: "This is a case in which there is no question of patient's life or death, and I think it is probable that many among you may pass by the bedside of such a patient without thinking worthy of attention. But I am not disposed to regard it in this manner. Although the patient will not die of the disease, yet, without great care, it may render her very miserable. The disease may be very much relieved by art, and it is one of very common occurrence ... such a case as may meet you at every turn of your practice; and your reputation in early life will depend more upon your understanding a case of this kind, than upon your knowledge of one of more rare occurrence." Even after centuries of investigation, and with exceptional advances in medicine and surgery, we are unable to offer any substantial treatment for varicose vein sufferers. The core problem of recurrences and the illness due to ulcerations following repeated surgeries has never been analyzed in depth. The principles behind the surgical intervention have been always overlooked, instead of focusing only on the modifications of instruments so that the interventions qualified as "minimally invasive," which is the current passion of many. "The Future is only the Past again entered through another gate" (Sir Arthur Pinero; The Second Mrs. Tanqueray, Act IV), still holds well. However, it is high time to revalidate the reentry of the newer technique based on a newer outlook of CVD. An analysis of the causes of surgical failures based on a deep understanding of the causes of the disease, and the treatment designed on that basis, is what is needed to solve this enigma.

While going into the various historical landmarks of the treatment of varicose veins to pay tribute to the great men who made their challenging contributions, I was surprised to find that the surgery we do today is only a modification of the procedure practiced in the early 1st century AD, revised in the early 19th century by Sir Benjamin Collins Brodie, who stated "I always observed that, if I cured one cluster of veins surgically, two smaller ones appeared, one on each side, and that, ultimately, I left the patient no better than I found him." Even the internationally celebrated surgeon, Friederich Trendelenburg, after whom the procedure for varicose vein surgery was named and still considered the "gold standard," found that the procedure failed over a period of 4 years due to the high recurrence rate.

I was inspired by Sir Benjamin Brodie and Trendelenburg, as I found the same problems in my early days of varicose vein management (in 1995). One patient, who was referred to a reputed surgeon, very well known for treatment of varicose veins, was refused surgery due to poor surgical fitness attributed to old age and other coexisting problems, returned to me and asked, "Should I die of this misery of nonhealing eroding ulceration of foot?" I was terribly moved by his desperate situation and started thinking of a way to help. I carried out sclerotherapy for the first time on him and even today, he is perfectly healthy with no venous problems. He started referring patients to the point that now I have developed an institute of research into venous diseases supplying surgical services to more than 3000 patients a year from various places around the world with an outpatient attendance of an average of 100 patients per day. We carry out genetic studies also, in collaboration with Rajiv Gandhi Centre for Biotechnology, Government of India.

My work is very different than that of others in the same discipline. I have been privileged to treat around 40,000 patients using sclerotherapy. I analyzed the basics for the evolution of varicose veins, and structural, etiological, and pathophysiological evaluations were done. The biophysics behind the pathology was well evaluated and the principles of treatment restructured to suit the natural functioning of the body. The response of patients with chronic venous disease, especially after venous surgery, was so encouraging that I was able to develop a major research institute on venous diseases.

"Treat the cause with a physiological basis" is the lesson I learned. The veins of the legs are flowing against gravity. The pathology is the failure of the uplifting mechanism, and not a reverse flow of blood. Flooding in a river cannot be controlled either by making dams across the river (multiple ligation or excision of truncal veins) or at the estuary (junctional blocks) or at any exit points (perforator block), or by filling the whole river with sand (truncal ablation, like stripping, endovenous laser treatment, radiofrequency ablation, or truncal sclerotherapy). The tributaries are most important. The tributaries should be diverted, especially at the lowermost level where the venous pressure is at its maximum. Remember, no instrument can be passed through these small tributaries. Also remember, that *no fluid will flow from a lower pressure gradient to a higher pressure gradient* compartment. Therefore the exit points should never be blocked. Make a physiological atmosphere. *Flooding in a river is never due to a high tide from the sea; the tributaries are the most important, hence never try to make a dam across a river or at the estuary to prevent flooding in the river.*

> *History is often shaped by small groups of forward-looking innovators rather than by the backward-looking masses (Yuval Noah Harari, Homo Deus: A Brief History of Tomorrow)*

My first book, *A Treatise on Venous Disease*, published by Jaypee Brothers in early 2014, received excellent support from readers and this response prompted me to have a reference book published. The last 5 years have seen a huge increase in research works on venous diseases. Phlebolymphology has consequently developed from its infancy to great heights and is now an independent branch of medical science, thus conquering new areas in the understanding of the disease, both in the outlook and the management.

While the first edition was intended for students and practitioners to gain a basic knowledge of venous diseases, and this new edition elaborates on this to assist researchers to acquaint themselves with current advances in research by referring to original articles in different international journals published until late 2019, with an emphasis on basic sciences, embryology, genetics, macroscopic and microscopic anatomy, biophysics, and pathophysiology.

In the previous book, I was concerned with the genomic predominance of the disease. However, having done more than 40,000 procedures for different classes of chronic venous disease (CVD) many of my patients were treated by me after repeated surgical interventions (namely Trendelenburg's operation with multiple ligations, microphlebectomy, and different types of endovenous procedures), and I was able to develop a different outlook on the disease. Based on this new outlook, the treatment modality was modified, bringing excellent results.

Universally, the large veins, from minor tributaries to truncal veins, are considered the morbid group. The popular belief is that as the size of the vein and level of the involved veins increase, the morbidity is considered progressively increase. The smallest visible veins measuring less than 2 mm are considered nonmorbid or even only causing cosmetic disfigurement. *However, this concept is wrong.* The veins visible to the naked eye constitute just 5%–10% of the total venous system. The rest are the microscopic veins concealed in the tissues as intercellular venous plexus. CVD is due to the microcirculatory impediment arising from the increased venous pressure transmitted to the capillaries due to the weakness (loss of integrity) of the venous wall or valves in the region of venulets (the portion of venules closer to the capillary and having the same dimensions as the capillary and differentiated by the presence of microscopic valves), venules, and minor tributaries. The incompetence of valves in the truncal and major tributaries alone does not carry much significance in the pathogenesis of CVD. Such incompetence of major veins, along with the incompetence of microscopic veins, has significance by indirectly increasing the capillary venous pressure. The hydrostatic pressure in the capillaries depends on two factors, namely, gravitational force and the weight of the blood column. *This outlook has revolutionized the management of varicose veins of the lower limb.*

The sequential changes in CVD are due to the inefficiency of the microscopic valves in the lower two-thirds of the leg and foot, particularly centered around the ankle region. Varicosity of the lower limb affects only the superficial venous system, which drains only the skin and fat outside the deep fascia of the lower limb. The superficial venous system is only a bypass in case of any venous block in the deep venous system.

As such, the management should be based on the new outlook and not on the ancient concepts. In order to reduce the capillary venous pressure, ideally, the microscopic venous plexuses have to be destroyed. The elimination of macroscopic veins alone in the presence of small veins in the lower two-thirds of the leg and foot involvement will only aggravate the condition. The details of the modified microfoam sclerotherapy procedure are given in detail in this book, with an emphasis on the anatomy and vein mapping. It is worth remembering that veins and lymphatics are highly significant body systems and have tremendous capacity for regeneration and do not have a definite course, unlike the arteries and nerves.

In order to obtain a clear idea of the lymphovenous system, this book gives elaborate details from angiogenesis in the embryo to recent management. I have also endeavored to facilitate your task by appending my journal articles at the end of this book. I believe this book will help the reader to understand all about the veins and lymphatics and will be a very useful source of knowledge for researchers also.

Before I conclude, let me quote my spiritual master, Swami Vivekananda:

Take up one idea

Make that one idea your life — think of it, dream of it, live on idea.

Let the brain, muscles, nerves, every part of your body,

be full of that idea,

and just leave every other idea alone.

This is the way to success.

I will be blessed if the new generation is enlightened by these newer thoughts.

N. Radhakrishnan

MRCP (London), MS, FICS (Phlebolymphology), FAIS, FIAMS, FRCS (Glasgow), FRCS (London), FACS (USA & Canada), FRSM

Acknowledgments

"Very often, events in one's life are done through *him rather than* by *him."*

This book was brought about through encouragement from my colleagues, my teachers, and well-wishers. Padmavibhooshan Prof. M.S. Valiathan, National Research Professor, Govt. of India was instrumental in supporting and guiding me throughout this new venture, and the frequent suggestions and advice received from Emeritus Prof. Mathew Varghese MS, FRCS, were inspiring.

I would also like to express my gratitude to Swami Golokanandaji of the Sri Ramakrishna Order for introducing me to Prof. M.S. Valiathan; this marked a turning point in my life and paved the way to advanced research and professional excellence.

I am very grateful to Archbishop Mar Joseph Perumthottam, the patron of St. Thomas Hospital, Emeritus Bishop Mar Joseph Powathil, and Nuncio Mar George Kochery for their blessings, prayers, and moral support through all my endeavors.

The great support offered from the Rajiv Gandhi Centre for Biotechnology, Thiruvananthapuram, Kerala, is greatly appreciated. Prof. Dr. M. Radhakrishna Pillai, Director; Dr. C.C. Kartha, Professor Eminence of Cardiac Molecular Biology Division; and Dr. S. Sumi PhD, Senior Research Fellow, require special mention. Our joint research in molecular analysis has given me great enthusiasm to dive into the depths of venous diseases and prompted me to write this book.

Writing this book has been very time-consuming; collating all the available data from leading journals and books has been extremely complex and taken more than 2 years to complete. It was only possible to achieve, despite my busy professional schedule, because of the sincere and dedicated work of my research fellow Dr. Aloy Antony, MSc (Biomedical Science, UEL, United Kingdom), who has worked tirelessly day and night. The technical support offered by my secretary Mr. Tony Chacko, in shaping the book, has been indispensable.

My son, Dr. R. Jayakrishnan H. Dip Surgery (SA), FCS (SA), FICS, Head of Surgery, Cecilia Makiwane Hospital, East London, South Africa, has contributed with much important data from various sources, and his suggestions have been very helpful.

I am particularly thankful to the management of St. Thomas Hospital, Changanassery, Kerala, India, for having offered me all the assistance I required in creating the Research Institute for Venous Diseases.

Writing this book has been something of an emotional rollercoaster ride, with many ups and downs. I am sure that at times this made me a very difficult partner to live with, and so I am sincerely thankful to my wife, Jayasree, for putting up with my mood swings during these times. Though far away from our homeland, my daughter, Priya, was always a source of moral support and deserves special mention.

I would be failing in my duty if I did not give particular mention to my coworkers, who have worked tirelessly in assisting with this project.

Last, but not least, my special thanks to those at Elsevier who assisted throughout the writing and publication of this book.

N. Radhakrishnan

Abbreviations

CTGF connective tissue growth factor
CVD chronic venous disease
ECM extracellular matrix
EVF European venous forum
MMPs matrix metalloproteinases
SMD standard mean difference
TGF-β transforming growth factor beta
TIMP tissue inhibitor of metalloproteinases
VAD venoactive drugs

CHAPTER 1

The lymphovenous system

1.1 Embryological considerations

1.1.1 Early embryonic development

The cardiovascular system is the initial functional organ system formed in mammalian embryos. Furthermore, maintenance and stabilization of this complex cardiovascular network must occur throughout adulthood, as impairments in vessel integrity and hemodynamic function can result in poor health or early death [1]. The vascular system forms as a branching network of endothelial cells that inherit their identity as arterial, venous, hemogenic, or lymphatic vessels.

The vascular system initially erupts in the embryo as a highly branched network of structurally primitive vessels composed of endothelial cells and their basement membranes. Embryonic endothelial cells (EEC) express their identity as arterial, venous, lymphatic, or hemogenic components, and then further specialize in an organotypic manner [2] (Fig. 1.1).

In addition to EEC, erythromyeloid progenitors in the yolk sac give rise to tissue-resident fetal macrophages that play a vital role in the morphogenesis and function of developing angiogenic blood vessels [1].

The primitive erythroid cells circulate within the embryonic vascular network and arise from hemogenic endothelial cells in the yolk sac blood islands around E7.5 [3]. Gradually, these primitive erythroid cells are replaced by definitive erythroid cells. The morphogenesis of the embryonic vascularate commences with the accumulation of presumptive endothelial cells (PECs) into loosely associated cords following their segregation from the mesoderm [4]. In 1915, Raegen showed that blood vessels of the embryo originate within the body proper, not by invasion from the highly vascular extraembryonic yolk sac.

The dorsal aortae and posterior cardinal veins are the first major blood vessels that form in situ by segregation of mesenchymal cells from the mesoderm. The dorsal aortae soon form a continuous cord at the ventrolateral edge of the somites and are continuous into the head to fuse with the ventral aortae forming the first aortic arch by the six-somite stage. The primitive endocardium fuses at the midline above the anterior

Genesis, Pathophysiology and Management of Venous and Lymphatic Disorders
DOI: https://doi.org/10.1016/B978-0-323-88433-4.00006-1

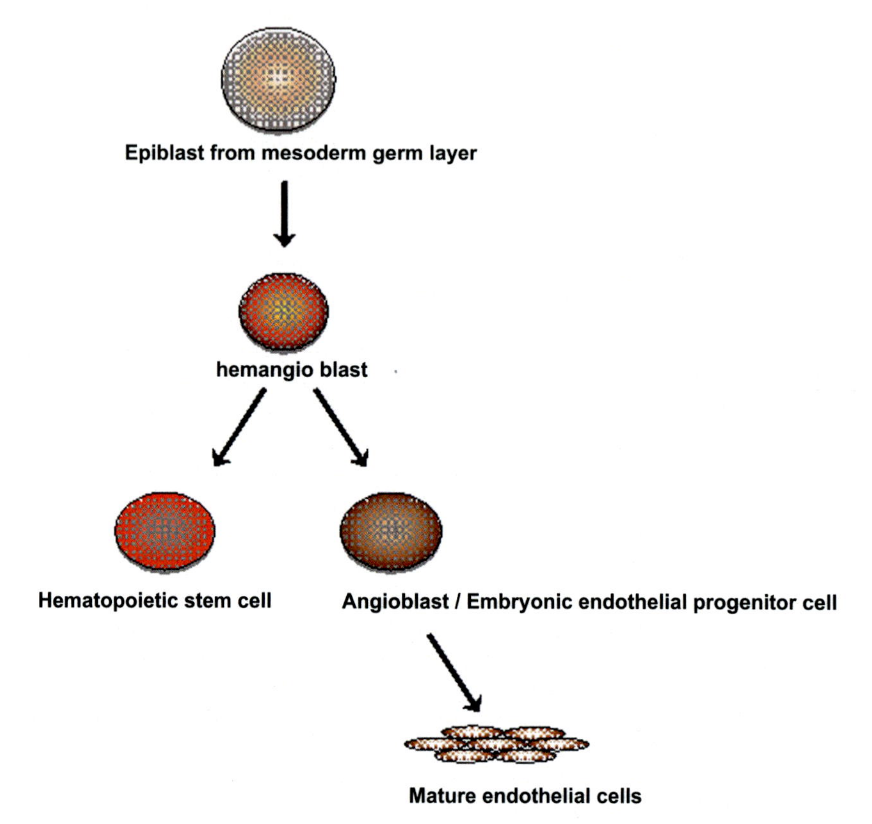

Figure 1.1 Transformation of epiblast from mesoderm to mature endothelila cells.

intestinal portal by the three-somite stage and the ventral aorta extends craniad. Intersomitic arteries begin to sprout from the dorsal aorta at the seven-somite stage. The posterior cardinal veins form from single cells which segregate from the somatic mesoderm at the seven-somite stage to form a loose plexus which moves mediad and wraps around the developing Wolffian duct in later stages [5].

These studies suggest two models of origin of embryonic blood vessels. The dorsal aortae and cardinal veins evidently arise in situ by the local segregation of PECs from the mesoderm. The early vessel rudiments lead to the formation of intersomitic arteries, vertebral arteries, and cephalic vascularate by the process of sprouting. The importance of cell migration in the morphogenesis of endocardium, ventral aorta, and aortic arches is also essential in the development of the vascular network. Highlighting the physical and molecular cues requisite to activate the morphogenetic events that anticipate during vascular development has been a remarkable challenge to cancer biologists, physiologists, scientists, and developmental biologists of many other disciplines for centuries. Elucidating these mechanisms is indispensable for determining how a normal vascular network develops and how aberrant vascular development can contribute to disease conditions [6].

The sprouting form of angiogenesis has been much more comprehensively studied recently as it is the mechanism by which cancer cells recruit a new vascular supply [7]. The angiogenic factors have also been discovered in developing systems. The kidneys and brain produce angiogenic factors which resemble tumor angiogenic factors [8].

1.1.2 Transcriptional control of endothelial differentiation

Two well-defined mechanisms of blood vessel formation in the early embryo have been described. The de novo formation of endothelial cells by differentiation from angioblasts followed by their self-assembly into vascular structures is called vasculogenesis. The process of vasculogenesis is highly dependent on specification of endothelial cell fate from progenitors in the mesoderm via activity of the E26 transformation-specific domain transcription factor (Etv2—also called Ets variant gene 2 or ETSRP71) [2]. The requirement of Etv2 in embryonic angiogenesis, either alone or in combination with other endothelial transcription factors, was addressed by Craig et al. [9] using a zebrafish model of vascular development. The above-mentioned study examined functional interactions between Etv2 and the Ets domain-containing transcription factors Fli 1a and Fli 1b in early vascular development. Embryos double-deficient for Etv2 and Fli 1b failed to form angiogenic sprouts and exhibited greatly increased endothelial apoptosis throughout the developing vasculature. Endothelial specification depends on gene targets transcribed by Ets domain-containing factors, including Etv2, together with the activity of chromatin-remodeling complexes containing Brahma related gene-1 (BRG1) [10].

Once nominative and structured into vessels, mechanisms restricting lumen diameter and axial growth ensure that the structure of the branching vascular network matches the need for perfusion of target tissues. In addition to this mechanism, the important morphogenic cues provided by the blood vessels will guide or alter the development of organs forming around them. Eventually, as the embryo grows and the diameter of the lumen increases, the smooth muscle cells (SMCs) wrap around the nascent vessel walls to provide mechanical strength and vasomotor control of the circulation [11]. Finally, the SMC differentiation via coupling of actin cytoskeleton remodeling to myocardin and serum response factor-dependent transcription is promoted by increased mechanical strength and wall strain.

One essential target for Etv2 in endothelial cells is an Ets factor-binding intronic enhancer element in the gene encoding vascular endothelial growth factor receptor (Flk1/VEGFR2) [2]. This enhancer sequence also contains an essential Gata factor-binding motif that likely interacts with Gata2 in endothelial cells. Etv2 is highly requisite for angiogenesis in adult tissues. Overexpression of Etv2 has been used to direct vascular progenitor cells in the postnatal arterial adventitia toward an endothelial cell fate. The chromatin-remodeling enzyme BRG1 is vital for early vascular development and primitive hematopoiesis [10].

1.1.3 Vascular lumen diameter controlled by endothelial cells

In small arteries and arterioles, the lumen diameter is restricted mainly by myogenic and neurogenic control of vasomotor activity in circumferentially arranged vascular SMCs [12]. In larger conduit arteries, the lumen diameter is primarily restricted by wall remodeling and is blood flow responsive and endothelial dependent [2]. The work by the Zhang et al. group demonstrated that intermedin promotes increases in blood flow through a neovascular network by increasing the size of the vascular lumen and reducing the number of excessive vascular sprouts. To generate intermedin-deficient mice using CRISP/Cas9, Wang et al. were able to exhibit that intermedin promotes increased vascular lumen size by stimulating the proliferation of confluent endothelial cells while preserving organized cell—cell contacts with no epochal changes in overall cell shape.

The vascular SMC responsiveness to mechanical stress and wall strain is a major morphogenic pathway in the vascular development, an important adaptive pathway in hypertension, and a pathogenic pathway in aneurysm formation. However, maladaptive responses to prolonged or aggravated stretch can also occur. A study conducted by Rodriguez et al. investigated the role of nicotinamide adenine dinucleotide phosphate (NADPH) oxidase isoform-1 (Nox1) in maladaptive stretch-induced SMC phenotypes. They discovered that Nox1 expression in a myocyte enhancer factor 2B (Mef2B)-dependent manner was increased by cyclic stretch (10% at 1 Hz). Increases in Nox1 activity led to an increase in osteopontin expression and downregulated contractile phenotype markers calponin-1, smoothelin-B, and total actin fiber density [13]. Marfan syndrome is correlated with mutations in FBN1 (fibrillin 1), an extracellular matrix protein that plays a vital role as a component of microfibrils that help to configure Eln filaments in the walls of elastic arteries [14]. Fbn1 also directly binds the large latent transforming growth factor-beta (TGF-β) complex and bone morphogenetic proteins and impounds the growth factors in an inactive form [15] (Fig. 1.2).

Genetic fate-mapping approaches have shown that arterial SMCs arise from multiple embryonic origins in vertebrate development [16]. Various SMC origins usually map on to different axial domains that underpin spatially to the anterior—posterior organization of the early embryo. The aortic root and ascending aorta are composed of SMCs that originate either from cardiac neural crest SMCs or from the second heart field (SHF—SMCs) [17]. A well-characterized Mef2c SHF enhancer-Cre mouse was used to reexamine the distribution of SHF—SMCs in the outflow tract and ascending aorta from 3 to 25 weeks of age [17]. This confirmed the previous findings and expanded them to show that the SHF-derived SMCs actually distribute from the aortic root along the outside of the developing aortic wall to form the outer medial layers of the ascending aorta. The overall organization and function of the epigenome are determined by the embryonic origin and lineage history of a cell.

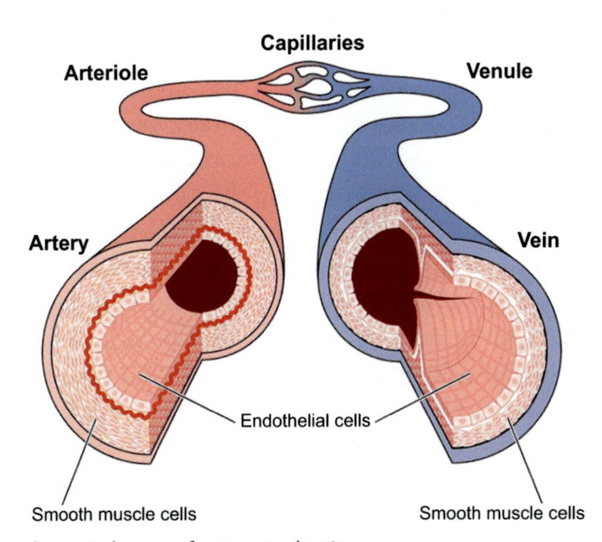

Figure 1.2 Embryonic layers of artery and vein.

The completion of aortic media formation around E15.5 marks the initiation of development of the adventitia [2]. The adventitia is the most complex layer of the artery wall and is composed of different types of fibroblasts, inflammatory cells, nerves, microvessels, and resident progenitor cells [18]. The adventitia is a sensor of biomechanical wall strain, which contains a microvascular network called the vasa vasorum, plays a critical role in immune cell trafficking, and provides a progenitor niche environment for the artery wall [2]. Lineage-tracking studies using Wnt1-cre mice showed that aortic adventitial cells in the ascending aorta were not derived from progenitors in the cardiac neural crest. Adventitial fibroblasts also are active innate immune regulators producing proinflammatory cytokine interleukin-6 and monocyte chemoattractants such as monocyte chemoattractant protein 1 in response to an inflammation-provoking stimulus [2]. In addition to innate immune functions, adventitial cells are also major mediators of artery wall fibrosis that accompanies vascular remodeling in atherosclerosis, restenosis, and aortic aneurysm formation [2].

1.2 Vascular network development in animal models

1.2.1 Mouse model

To learn the physical and molecular regulation of vascular development, the mouse model is used extensively for conducting experiments. The molecular pathways involved in the establishment, remodeling, and maintenance of the cardiovascular network are well understood by the availability of genetic manipulation in the mouse model. The initial establishment of blood vessels occurs in the absence of hemodynamic force in mouse embryo [19]. The first primitive vessel fused by the cells located in the yolk sac begins to form during gastrulation, and occurs on mouse embryonic day 6.5 (E6.5). During this time, the epiblast, which is a single epithelial cell layer, undergoes an epithelial-to-mesenchymal transition to form the primitive streak, which gives rise to the germ layers of the embryo proper—the definitive endoderm, embryonic ectoderm, and mesoderm [20]. The de novo formation of a network of blood vessels is the initial phase of vessel development, where blood and endothelial cell precursors (angioblast) migrate from the mesoderm, aggregate, and form the blood island located in the proximal extraembryonic yolk sac by E7.5.

After the blood island has formed, the angioblasts situated at the outer edges divide and differentiate into endothelial cells and migrate distally into the yolk sac, forming an unsophisticated capillary network that encompasses the yolk sac. This network of interconnected rudimentary small vessels, homogeneous in shape and size, is the primitive capillary plexus [1]. During vasculogenesis, the initial establishment of the capillary plexus is interdependent on activation of the vascular endothelial growth factor (VEGF) signaling pathway. Various genetics-related studies in the mouse model have distinguished VEGFR2/Flk1 as a modulator of de novo vasculogenesis, that is also crucial for the development of both blood cells and the primitive capillary plexus during embryogenesis [1]. VEGF signaling is moderately balanced to permit the migration and proliferation of endothelial cells into the primitive vascular network during the organization of the capillary plexus of the yolk sac, as well as the early development of the vascularate within the embryo.

Furthermore, the Scl (stem cell leukemia)/T-cell acute lymphoblastic leukemia transcription factor, a member of the basic-helix—loop-helix (Bhlh) transcription factor family, has been identified in the specification of Flk1-positive multipotent progenitors, which give rise to both blood and endothelial cells [21]. In fact, putative Scl-binding sites have been found on the Flk1 promoter, suggesting that Scl directly acts as a regulator of Flk1 expression and multipotent properties of early endothelial and hematopoietic progenitor cells, critical for the establishment of the initial capillary plexus during vasculogenesis [1]. The sequence of highly complex dynamic morphogenetic events which occurs during early vascular development is the establishment of initial capillary network in the yolk sac.

1.2.2 Mouse yolk sac: vascular remodeling

The primitive capillary plexus of the yolk sac embryo is achieved by around E8.5. This meshwork of vessels goes through rapid morphological changes for the convenience of expanding yolk sac and growing embryos. This is an angiogenic process, where the established vessels remodel to expand and increase in diameter, and smaller vessels fuse to form the large major vessels of the yolk sac and embryo [22]. Meanwhile, other vessels undergo pruning to exclude excessive branches, intussusception to create branch points in large vessels, and nascent vessels sprout from existing larger vessels to form small capillary beds. This morphological process of vascular remodeling occurs between E8.5 and E9.5, and results in a hierarchically branched, highly organized vascular network that allows vessels to become low-resistance conduits for the transport of blood, oxygen, nutrients, and waste to and from the yolk sac and embryo proper. The process of remodeling is considered as a crucial time point necessary for the ability of the cardiovascular system to function and support the growing embryo. The viability of the embryo is no longer sustainable if the vessels are not remodeled correctly, and this process is regulated by both molecular and biomechanical signaling. The most studied and earliest identified molecular signaling pathways that are vital for angiogenesis are the Tie family of receptor tyrosine kinases [23].

Expression analysis of Tie-1 and Tie-2 receptors show localization within vascularate from the onset of angioblast specification (E7.0), continuing on through development and into adulthood [24]. Furthermore, targeted mutations of Tie receptors in mice revealed significant impairments on vascular development during angiogenic processes, including yolk sac vascular remodeling [1]. Although Tie receptors were initially thought to be orphan receptors, extensive studies have identified four ligands that can bind to Tie receptors: angiopoietins 1−4 [25]. These ligands have also been shown to be necessary for the regulation of yolk sac vascular remodeling in the mouse embryo [1]. These results confirm the Tie family of receptors and ligands as critical molecular regulators of vascular remodeling during angiogenesis, after the initial establishment of the capillary plexus during vasculogenesis (Fig. 1.3).

Another example of the molecular regulation of yolk sac angiogenesis and vascular remodeling is the TGF-β superfamily, which encompasses transforming growth factors and receptors, bone morphogenic proteins, and activins [26]. Homozygous deletion of TGF-β family members result in embryonic lethality by E10.5−E11.5, at which point the mice are shown to have an unremodeled yolk sac vascularate, defects in endothelial and hematopoietic differentiation, delayed angiogenesis, and weak vessels [1]. The genetic experiment mentioned here provides examples of the individual contribution of members of the TGF-β pathway in remodeling the vasculature of the yolk sac and embryo.

Once the complex process of vascular remodeling is done, the vasculature of yolk sac and embryo are subdivided into a highly organized network of structurally and functionally distinguishable arteries and veins, each with its own specialized purpose for delivery of oxygen and nutrients to the developing embryo. It has been hypothesized that the differential effects of hemodynamic force of blood flow on developing vessels is a major factor for the molecular distinction between arteries and veins. The earliest identified and most studied factors which regulate arterial/venous

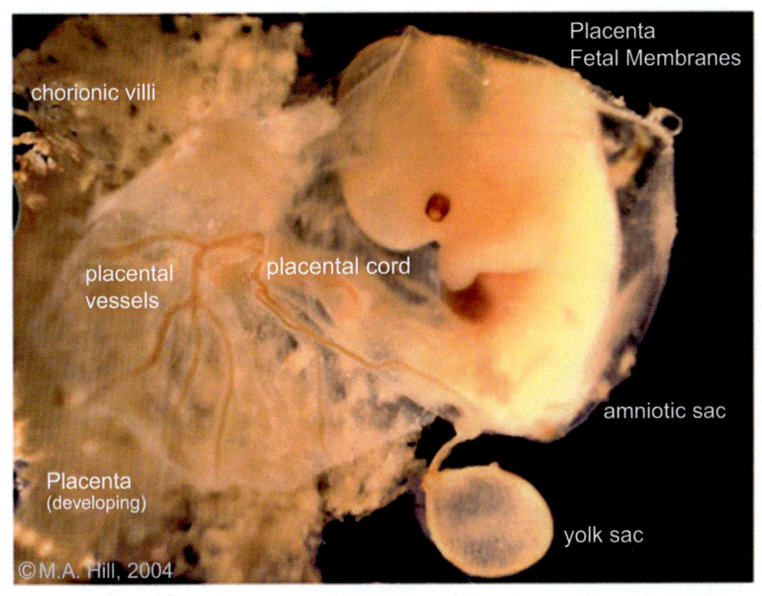

Figure 1.3 Embryological development of placenta.

identity are the Eph receptors, a large family of receptor tyrosine kinases and their membrane-bound ligands, and ephrins [27]. Analysis of EphrinB2 and its receptor EphB4 at the earliest stages of vascular development revealed compartmentalized expression of EphrinB2 within the endothelial cells of developing arteries, whereas expression of EphB4 is localized to venous endothelial cells [28], indicating that the differences between arteries and veins are genetically predetermined and are crucial for proper vascular function during development. Researchers around the world have outlined the interaction between EphrinB2 and EphrinB4 as reciprocal, in which signaling between the arterial and venous cell fate, downstream of pathways that are necessary for the establishment of arterial and venous progenitors [29].

1.2.3 Signaling pathways

The notch signaling pathway has been implicated as a modulator of arterial and venous cell fate upstream of ephrin signaling. This family consists of the Notch receptors (Notch1–4), which interact with the ligands Delta-like (DLL)-1, -3, -4, Jagged1, and Jagged2. Upon interaction between the ligand and its receptor, the intercellular domain of the receptor is cleaved, where it can translocate to the nucleus and active transcription of the target genes [30]. Analysis of Notch ligands and their receptors has revealed expression in the endothelium, with many family members restricted to the arterial compartment [31]. Targeted mutation of Notch pathway members in mice has provided further evidence that this pathway is key in regulating arterial endothelial cell identity and yolk sac vascular morphology [32]. The specification of arterial identity is regulated by various different transcription factor and signaling molecules, whereas the specification of venous cell fate is primarily mediated through the orphan nuclear receptor chicken ovalbumin upstream promoter-transcription factor 2 (CoupTF2). Targeted deletion of CoupTF2 in mice results in an unremodeled yolk sac vasculature and defects in cardiac morphogenesis [33].

CoupTF2 is expressed solely in the venous compartment of the endothelium, and upon conditional deletion of CoupTF2 in endothelial cells, expression of CoupTF2 results in the fusion of arteries and veins. It has been concluded that CoupTF2 functions cell autonomously to suppress the activation of arterial regulators to maintain venous cell fate [34]. The confinement and sustenance of arterial and venous identity are critical for viability as the embryo continues to grow. The circulation can become impaired in the absence of structured and organized functional arteries and veins. After the process of remodeling, the freshly formed vessels of the yolk sac and embryo begin to stabilize through the recruitment of mural cells to the endothelium, including SMCs and pericytes. This recruitment is initiated by the secretion of platelet-derived growth factor B (PDGFB) from endothelial cells, which in turn binds to PDGF receptors on the mural cell surface and allows them to contribute to the stabilization and formation of the vessel wall [35] (Fig. 1.4).

1.2.4 Endothelial cell regulation by hemodynamic force during vascular remodeling

Physical forces imparted by blood flow are known to regulate vascular remodeling within the yolk sac and embryo proper [36]. In vitro analysis of how endothelial cells respond to mechanical force has been insightful in identifying mechanosensitive signal transduction pathways that can modulate endothelial cell morphology and function [37]. For example, the application of shear stress, the frictional force of flowing blood dragging along the vessel wall, to cultured endothelial cells, has been shown to activate a signal transduction complex composed of three receptors, VEGFR2 (Flk1), vascular endothelial cell cadherin (VE-cadherin), and

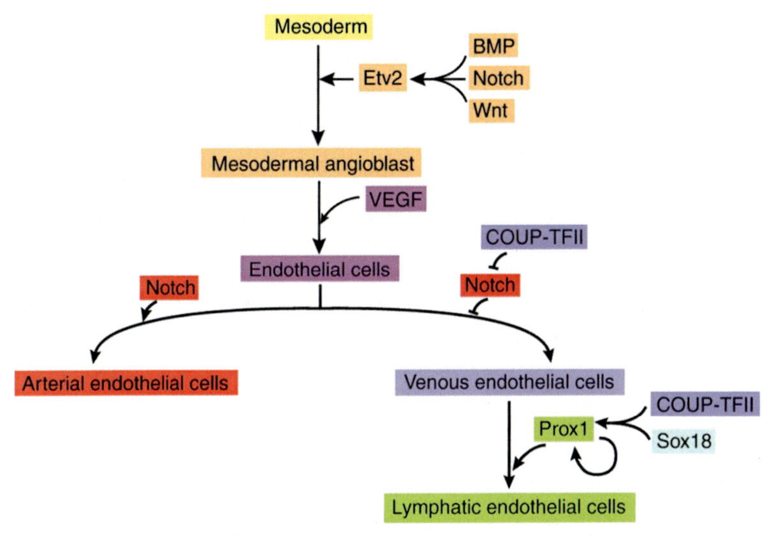

Figure 1.4 Mesodermal differentiation.

platelet endothelial cell adhesion molecule1 (PECAM1) in vitro [38]. Upon overexpression of Flk1, PECAM1, and VE-cadherin in cells which do not respond to mechanical shear stress normally, cells realigned parallel to the direction of flow, becoming mechanosensitive [1].

These data rendered evidence of a molecular complex that is triggered solely by mechanical force, which serves as an obligatory cue for the ability of endothelial cells to change morphology and polarity within a cultured monolayer. The mechanism for activation of this mechanosensory complex is through direct transmission of mechanical force via PECAM1 to VE-cadherin, which acts as an adapter protein activating FLK1 and catalyzing the activation of the phosphoinositide 3-kinase (PI3K) signaling cascade. Flk1 is critical for the earliest stages of both vasculogenesis and hematopoiesis [1], and VE-cadherin is necessary for vascular morphogenesis. However, while PECAM1 is expressed in the endothelium during embryonic vascular development, genetic deletion of this gene shows no deficiencies in vasculogenesis, angiogenesis, or vascular remodeling [39].

Furthermore, several in vitro experiments performed to determine the mechanism behind the mechanosensitivity of endothelial cells have identified a wide array of proteins, including ion-channels, G-protein-coupled receptors, nitric oxide release molecules, and signaling transduction pathways, all of which may play a potential role in the regulation of vascular remodeling in vivo [40]. The initiation of the heartbeat in the mouse at

E8.25 along with the entry of blood cells into the circulation leads to the development of hemodynamic force, which plays a critical role in the regulation of endothelial cell function and vascular remodeling, forming a low-resistance, highly organized yolk sac vascularate. The direct role of the onset of blood flow at this early stage of vascular development has been studied for several years and remains highly debated; some authors have suggested that blood flow is necessary for the transport of oxygen to the growing embryonic tissue, while others have suspected blood flow in the nascent vascularate is required for the transport of soluble nutrients and growth factors essential for proper endothelial cell function [41]. In vitro experiments revealed that shear stress and hemodynamic force are considered to be the main physical factor in regulating endothelial cell function.

Indeed, a number of mice with mutations causing deficiencies in early cardiac development and function are embryonic lethal due to impairments in vascular remodeling [42]. The mutation of the gene atrial myosin light chain 2a, a component of the atrial myofibrillar apparatus, confers significantly reduced atrial contraction and reduced hemodynamic force (low, oscillatory flow), which leads to lethality by E10.5–11.5. Not only do embryos display cardiac morphogenesis abnormalities, but also a completely unremodeled yolk sac vasculature at E9.5 [2]. Similarly, deletion of the Na/Ca^{2+} exchanger causes a complex lack of heartbeat and shows deficiencies in vascular remodeling [39]. The deletion of the homeobox transcription factor Nkx2.5, which is a modulator of dorsal mesoderm specification, results in impaired cardiac development, defects in the vascular remodeling in the yolk sac and embryo, and also impairments in hematopoiesis. Finally, embryos with mutations in the cardiomyocyte protein Titin, which acts as a spring during muscle contraction, have impaired blood flow, and endothelial cells of the yolk sac and embryo proper display defects in cell morphology, including irregular cell–cell contact and abnormal spatial distribution during vasculogenesis and angiogenesis [43]. The above data display the importance of cardiac output and its contribution to hemodynamic force during vascular development. The vessels of the yolk sac fail to remodel when the movement of blood flow within the capillary plexus decreases. This process is interrelated to the absence of a strong heartbeat and to the corresponding pulsatile nature, where the embryo can no longer sustain growth, resulting in lethality.

Early studies revealed that vascular development is a sequential and highly orchestrated process molded by cell–cell interactions and biophysical forces. Vascular remodeling, in which the primitive capillary plexus of the yolk sac is remodeled to form a branched hierarchical network of arteries, veins, and capillary beds, is a crucial step in early cardiovascular development. Through high requisiteness, blood vessels progressively extend to all parts of the growing embryo and play different roles in the patterning and organogenesis of diverse tissues. The blood vessels are phenomenal mosaics whose walls are assembled from progenitor cells that arise from various embryonic origins and thus confer origin-specific properties on the final structures developed. Simultaneously, blood vessels and the cells within their walls are extremely responsive to changes in the local environment and possess remarkable and unpredicted plasticity. Advance studies conducted to investigate the mechanism of vascular development can deliver surprising hints about vascular disease as cells respond to tissue damage and emphasize developmental programs to repair injury and remodel vessel walls [44].

Research studies conducted in vasculogenesis and angiogenesis have been supported by their clinical relevance, in which what is learnt can be adapted to disease states such as cancer, atherosclerosis, and osteoporosis. Although, extensive knowledge has been acquired in the process of signal transduction pathways that control both vasculogenesis and angiogenesis, new genes are being introduced almost daily, which can contribute to the regulation of endothelial cell function, vascular morphogenesis, and vascular homeostasis. Continuing studies that take advantage of animal disease models and genetic manipulations will be necessary to highlight the complexities of signal transduction and gene regulation during embryonic vascular development, to provide new gene targets and therapeutics for health interventions [45].

1.3 Angiogenesis

Angiogenesis is very conspicuous in developing and establishing organisms. In the adult, it is only active in specific situations and tissues, such as in wound healing, in the cyclic ovary, and in the life cycle of the female mammary glands and uterus. It is also associated with diverse pathological conditions, in which it can have a positive or negative impact on the pathological process [46]. Angiogenesis can repair damage inflicted by ischemia or cardiac failure, but in certain situations the activation of angiogenesis may aggravate the pathology [47]. Examples of the latter are tumor growth and cardiovascular diseases such as atherosclerosis, chronic inflammation, diabetic retinopathy, psoriasis, endometriosis, and rheumatoid arthritis [48]. The mechanism of angiogenesis has been recognized as a promising therapeutic target that has to be enhanced or inhibited, depending on the pathology [49]. Until recently, the process of angiogenesis was thought to mainly proceed through the so-called sprouting of blood vessels from the preexisting vasculature. However, different mechanisms of angiogenesis have been discovered, that is, sprouting angiogenesis (SA) and intussusceptive angiogenesis (IA). SA was initially thought to be the only process of angiogenesis, but in the 1990s IA was discovered as an alternate process [50]. SA is purely responsible for vascular growth, whereas IA can also involve vascular remodeling through pruning of excessive blood vessels. In addition to these mechanisms, two alternative forms of angiogenesis have been described, vascular cooption and vascular mimicry, however these mechanisms are assumed to be restricted to pathological situations [50]. The heterogeneity of angiogenic processes should be taken into account in future antiangiogenic therapies since recent data have revealed that the angiogenic phenotype may switch to a different form of angiogenesis after angiogenic treatments [51]. In addition, a better knowledge of these forms of angiogenesis can lead to further fine-tuning of proangiogenic therapy to reestablish a functional vascular network in ischemic tissue [52].

1.3.1 Intussusceptive angiogenesis

IA determines the process in which transluminal tissue pillars develop within capillaries, small arteries, and veins and subsequently coalesce, thus delineating new vascular entities or resulting in vessel remodeling. The concept of IA was first floated by Caduff et al. [53], when they discovered numerous tiny holes in the vascular casts of developing pulmonary vessels. They postulated the holes to be spaces for tissue posts that had been inserted into the vascular lumina and that were digested away during tissue corrosion. These authors coined the name "intussusceptional angiogenesis" to describe the process of "in-itself" vascular growth. This terminology was modified to IA by Burri and Tarek who demonstrated the tissue posts to be pillars in the vascular lumina of developing vessels by serial sectioning. Such pillars were likened to slipping a piece of tissue into another one, and the presence of such pillars is, therefore, the quintessence of IA [54]. IA was discovered in the late 1980s and remains poorly investigated [51]. Nevertheless, IA seems to play a major role in the growth and remodeling of most vascular beds, including the tumor vascular beds [55].

1.3.2 Mechanism of intussusceptive angiogenesis

The hallmark of IA, which is pillar formation, follows distinct phases that result in the development of a tissue post across the vascular lumen. The establishment of interendothelial cell contacts will lead to the protrusion of opposite sides of the culprit vascular walls of the same capillary. Reorganization of the endothelial cell junctions preceded perforation of the bilayer by invading interstitial tissue, pericytes, and myofibroblasts. Three forms of IA are recognized, depending on the outcome or phenotype of these forms, that is, intussusceptive microvascular growth (IMG), intussusceptive arborization (IAR), and intussusceptive branching remodeling (IBR) [56]. IMG leads to the quick expansion of a preexisting vascular network through the continuous new formation and expansion of pillars into the network. This results in a simple network of similarly sized capillaries. IMG is characterized by the diffuse appearance of numerous pillars. In a mechanism that is most likely driven by blood flow, these pillars fuse and split the vessels, expanding the capillary network and forming the organ-specific architecture [57]. The other form of IA, IBR, can be defined as the mechanism that optimizes the number of vessels to efficiently supply a tissue with blood by either changing the branching pattern of blood vessels or pruning the vascular network from superfluous vessels [58]. IBR is fairly well distinguished by the occurrence of a tissue pillar close to the bifurcation of two blood vessels. These pillars enlarge and gradually merge with the perivascular connective tissue. As such, the bifurcation is further narrowed and the bifurcation point is relocated more proximally. In the case of pruning, pillars also appear close to a bifurcation, but they are situated more eccentrically and asymmetrically [57].

1.3.3 Occurrence of intussusceptive angiogenesis

In the majority of developing vascular beds, the initial network is formed through SA, and IA gradually takes over [50]. IA is a quicker process than SA and can occur without interfering with the local physiological conditions because no blind-ending capillary segments are formed. The basement membrane remains intact during the course of IA, preventing the blood vessels from becoming leaky. In addition, endothelial migration and proliferation are kept to a minimum as the endothelial cells do not necessarily proliferate but rather increase in size and flatten [50]. This results in a relatively lower metabolic cost of IA in comparison with SA [50].

Since its discovery in developing lungs in the late 1980s, evidence for ongoing IA has been detected in a wide range of developing tissues such as bone, retina, muscle, kidney, ovary, mammary gland, and many more [50]. Interestingly, investigations into the different developmental stages in the mammary gland in developing glomeruli and in the cyclic ovary revealed that SA is predominant in the early stages of angiogenesis, while IA predominates in the later stages of vascular growth and remodeling [59]. IA is also active in diverse pathologies and during tumor growth. IA has been observed in different murine disease models, including models of liver cirrhosis, models of inflammation, and the hypoxic mouse retina [60] (Fig. 1.5).

In addition, IA was observed to start glomerular repair after induced Thy-1.1 nephritis in rats [61]. IA has recently been observed to participate in tumor growth, such as in B-cell nonHodgkin's lymphoma, different types of gliomas, mammary tumors, or renal hepatocellular and colon carcinoma [62]. Observations in mammary tumors of mice revealed that SA is active in small tumors (less than 8 mm in diameter), whereas IA is predominant in large tumors [55]. Recent evidence suggests that tumors respond with an angiogenic switch from SA to IA following ionizing radiation or antiangiogenic drug therapy using PTK787/ZK222584, a wide spectrum tyrosine kinase inhibitor. These treatments are mainly designed to stop SA but lose their effect because the process of IA enables the vascular network to keep growing [63]. IA might also be exploited in proangiogenic therapy to reestablish a functional vascular plexus in ischemic tissues. This form of angiogenesis results in quick and better structured vascular plexus in comparison to SA. Current proangiogenic therapies

Intussusceptive Angiogenesis

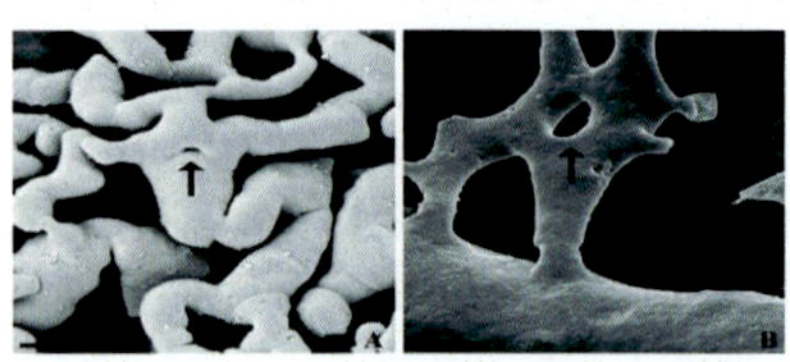

Intussusception is important because it is a reorganization of existing cells. It allows a vast increase in the number of capillaries without a corresponding increase in the number of endothelial cells. This is especially important in embryonic development as there are not enough resources to create a rich microvasculature with new cells every time a new vessel develops.

Figure 1.5 Intussusceptive angiogenesis.

mainly aim to induce SA, leading to the accumulation of fluid and transient edema in tissues [64]. This mechanism could be the reason for the leaky nature of recently developed vascular sprouts during SA. The specific induction of IA after an initial colonization of angiogenic sprouts into the ischemic tissue could enhance vascular normalization, improving the therapeutic outcome. These proangiogenic therapies could be used in various forms of ischemic disease including ischemic heart diseases, large wound healing, and atherosclerosis [65].

1.3.4 Regulation of intussusceptive angiogenesis

Hemodynamic forces play an important role during the process of IA [47]. Blood flow has been observed to establish a hierarchical vasculature with venules and arterioles from a preexisting capillary plexus in the chick chorioallantoic membrane (CAM) [66]. Experiments using chick CAM revealed an increase in IA in blood vessels in which the blood flow is enhanced by clamping the side branches [67]. Interestingly, in 1939 the remodeling of a vascular plexus after an experimentally induced increased blood flow was already observed to occur through the formation of small vessel loops, but this network remodeling was not recognized as IA [68]. It is still unclear whether the effect of hemodynamics on IA is mainly induced by change in hydrostatic pressure, cyclic stretch, shear stress, or a combination of all of these factors [69]. Shear stress seems to play a major role, since in silico models revealed that tissue pillars occur in places with low shear stress caused by turbulence due to increased blood flow [70]. Apart from hemodynamic forces, IA is also regulated by molecular factors. Due to the lack of appropriate experimental assays, only scarce data are available on the molecular control. Initial work in developing chick glomeruli and in the chicken embryonic lung has indicated that VEGF mainly affects SA and that it is downregulated in vascular beds that are growing by IA [50]. However, recent investigations on chick CAM and observations in human gliomas suggest that VEGF and VEGF isoforms play an important but unclear role during IA [71]. A decrease in VEGF levels after depletion of VEGF-releasing hydrogels on growing chick CAM is associated with a reduction of the vascular network through intussusceptive vascular pruning (IPR). A study of the gene expression in developing glomeruli revealed upregulation of VEGFR2 in early glomerular development but a significant downregulation of this receptor in later development when glomeruli were still growing [72].

Apart from VEGF, the angiopoietins are also important angiogenic growth factors, and there are clear indications that these factors play an important role during IA [73]. Targeted deletion of TIE2 expression in mice leads to deficient pillar formation [50]. Both ANGPT1 and ANGPT2 are continuously expressed in growing kidney glomeruli. Studies on the developing chick vasculature also indicated that fibroblast growth factor (FGF2) and PDGFB might play a regulatory role during IA [74]. PDGF is important for pericyte recruitment and, since pericytes play a role in the formation of tissue pillars, it might be hypothesized that FGF2 regulates IA by inducing PDGFB responsiveness in pericytes through upregulation of the PDGF receptors [42,44,75]. Recently, IA was observed to be upregulated in response to hypoxia with a chronic hypoxia model in mice [76]. Hypoxia can trigger the expression of several angiogenic genes through upregulation of hypoxia-inducible factors. As an example, hypoxia-inducible factor 2 alpha upregulates erythropoietin (EPO) expression [77]. EPO has been described as a possible regulator of IA, as addition of EPO to chick CAM enhanced IMG [78]. Because of the current scarcity of experimental and descriptive studies on IA, it can be expected that additional angiogenic factors might also influence the process of IA. These factors play an important role in regulating EC—pericyte interactions and cell—cell junctions. Among the factors regulating cell—cell interactions, VE-cadherin might be an interesting candidate since deletion of this factor in mice leads to disturbed morphogenesis of developing blood vessels [79]. In addition, the ephrins and eph-B receptor or monocyte chemotactic protein 1 have also been proposed as other potential regulators of IA [80].

1.3.5 Phases and phenotypes of intussusceptive angiogenesis

IA may be divided into three major phases depending on the outcomes or phenotypes accomplished at the end of the processes [50]. The fundamental function to all of them is the formation of tissue pillars and also the differences being inherent in the direction and arrangement of such pillars. The three phases include IMG, IAR, and IBR. In normally developing organs, these three phases have been seen to initially establish in tandem, later becoming synchronous in development.

IMG inaugurates the primordial capillary network expansion, while IAR adapts the vasculature into the typical vascular tree pattern [50], IBR finally remodels the vasculature to optimum local perfusion requirements.

Normally, IA supplants sprouting angiogenesis after the establishment of the basic plexus but remodeling of the vasculature is achieved through IBR [81] (Fig. 1.6).

IMG includes the process of initiation of pillars and their subsequent expansion with the result that the capillary surface area is greatly increased [51]. The deft remodeling of the blood vessels establishes specific organ angioarchitecture. In the embryonic avian metanephric kidneys, initial large blood vessels that form the primitive glomerular tufts are split in the middle through the process of intussusception [82]. These entail arrangement of tissue pillars in a line along the longitudinal axis of the vessel, their subsequent fusion, and thus delineation of two branches with the result that two daughter glomeruli are formed [51]. In the developing vasculature of the retina, lung, muscle, etc., different pillar arrangements and fusion patterns determine the organ-specific angioarchitecture. Numerous reports in the literature indicate that IA occurs in many developing organs, but in some situations it has not been out-rightly reorganized and

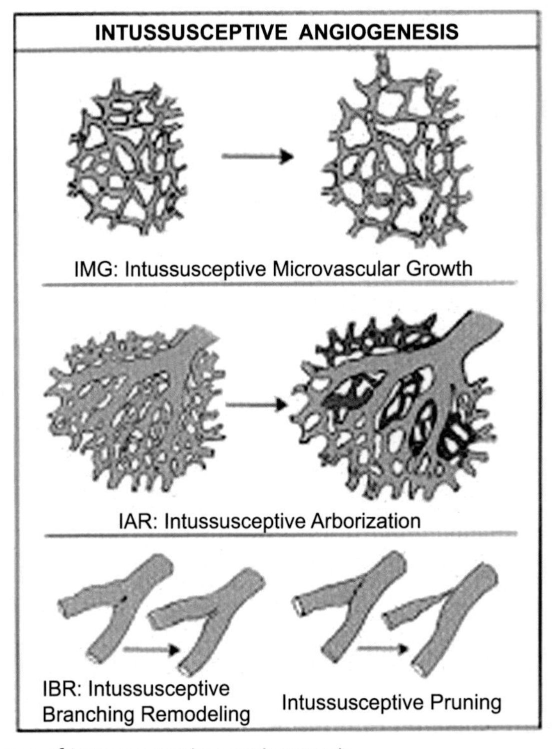

Figure 1.6 Stages of intussusceptive angiogenesis.

has been given different names such as longitudinal splitting, longitudinal division, intraluminal splitting angiogenesis, or internal division [51]. Despite some reports to the contrary, there is no evidence that longitudinal vessel division is any different from IA. In the capillaries of skeletal muscle, pillars appear more elongated than in other tissues due to the parallel arrangement of myofibers so that the spaces left for the capillaries have a general longitudinal disposition [83] (Fig. 1.7).

IAR defines the process which represents smaller generations of future feeding and draining vessels and contributes immensely to the formation and expansion of the vascular tree. The original pattern of blood vessels formed either through vasculogenesis or sprouting angiogenesis is a disorganized meshwork and hardly resembles the tree-like arrangement of the mature vasculature [51]. The adaptation of the organ vascular tree is achieved through IAR, which entails formation of serried "vertical pillars" that demarcate lower generations of vessels. Numerous circular pillars are formed in rows, thus demarcating future vessels; pillar reshaping and pillar fusions result in the formation of narrow tissue septa. There is delineation, segregation, growth, and extraction of the new vascular entities by merging of septa (Fig. 1.8).

IBR describes the process that results in adaptation of the architecture and number of vascular branches to optimum local requirements. Accomplishments of IBR occur mainly via transluminal pillars that are formed close to arterial or venous bifurcation sites. Enlargement of such pillars, their subsequent approximation and fusion, and also fusion with the connective tissue at the bifurcation narrows the bifurcation angle by relocating the branching point proximally [84]. Symorphosis is a concept that postulates a match between function and structural design, while

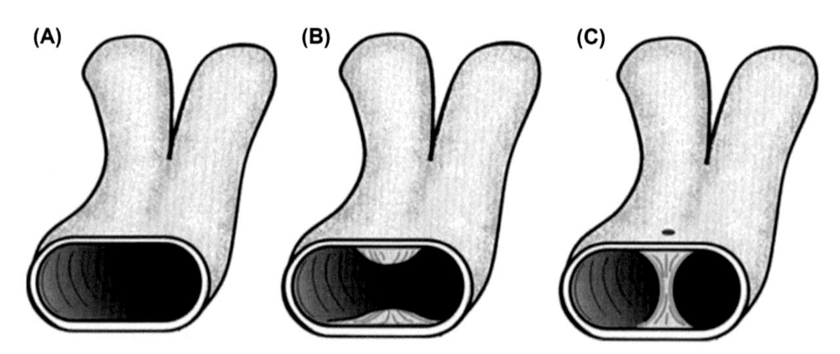

Figure 1.7 Sprouting angiogenesis.

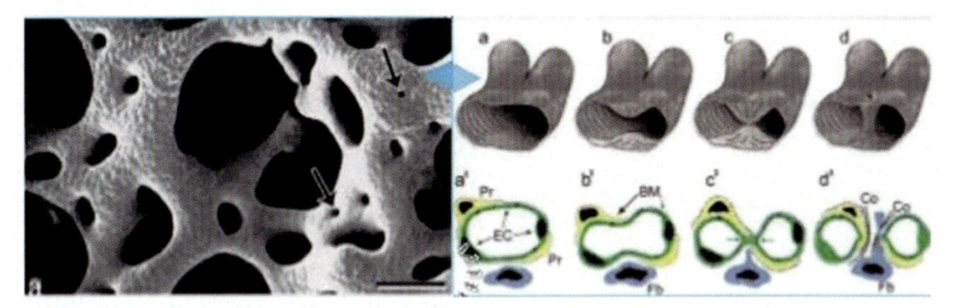

Figure 1.8 Intussusceptive angiogenesis.

Murray's law pictured an ideal situation of minimum power consumption and constant shear stress in blood flow.

In remodeling mature vessels, IBR severs superfluous vessels, a process known as IPR. This process is achieved through unconventional establishment of pillars in rows across the breadth of the target vessel at bifurcation sites. Expansion and subsequent fusion of pillars results in reduced blood flow, the consequences being regression, retraction, and atrophy of the affected vessel. The phenomenon of vascular pruning was first reported in retinal vessels long before IA was described, and a classical description of the process and mechanisms involved was provided by Djonov et al. and has been elucidated in several reviews [85]. It has also been demonstrated in the remodeling vasculature of the developing metanephros and the maturing CAM vasculature. IPR is related to reductions in blood flow and oxygen tension. In the absence of blood flow, pruning was seen to be accelerated while hypoxia resulted in downregulation of VEGF and subsequent vascular pruning [51]. Vessel pruning was demonstrated many decades ago by Clarke and Clarke when they documented that the vasculature of an inflamed rabbit ear reverted to a simple architecture after relief from inflammation. Another type of pruning achieved through leukocytes has been demonstrated in the retina [53]. Leukocytes adhere to the vessels and this leads to Fas-ligand-mediated endothelial cell apoptosis. Whether this has any relation to pillar formation is not clear. Generally, a reduction of vascular branches is referred to as vascular pruning, but it is only on a few occasions that IPR per second has been demonstrated [86].

1.3.6 Temporospatial distribution of intussusceptive and sprouting angiogenesis

IA occurs only on preexisting vasculature, formed either through sprouting angiogenesis or vasculogenesis. IA is antedated by SA and forms an important part of the vascular remodeling. In the much studied CAM, blood vessels grow by SA in its first phase of development (E5–E7), and in the second phase (E8–E12) they grow mainly by IMG, while the final phase grows without a substantial increase in vascular complexity [82]. It was further shown that expansion of the CAM vasculature during the second phase was via IAR while IBR remodeled the vessels during the maturation phase. In the growing rat, the mammary vasculature during the pubertal, adult virgin, and early-pregnancy stages grows by sprouting angiogenesis, after which this process gives way to massive IA. In the ephemeral ovarian follicles, sprouting angiogenesis characterizes the initial phase of development and this is soon supplanted by IA, which expands and remodels the vasculature toward the time of follicular maturation [87]. The effect of IA does not become apparent in the embryonic avian lung until about the 15th day of incubation, where there is an angiogenic switch with an upsurge in the number of transluminal pillars. In addition, IA is responsible for remodeling the fine capillaries that interlace with air capillaries to form the thin blood–gas barrier characteristic of the avian lung.

In pathological conditions that are characterized by excessive angiogenesis such as psoriasis, rheumatic disease, retinopathy, and tumorigenesis, IA is suspected to be one of the participating processes, but this far overwhelming evidence has only been adduced for tumorigenesis [88].

1.3.7 Control of intussusceptive angiogenesis

IA occurs during embryonic development, in physiological adaptations as may be seen in exercised muscles, and also in pathological situations such as tumorigenesis. The stimuli for this type of angiogenesis are either physiological, such as alterations in blood flow, as occur during exercise, or may be local biochemical alterations, as may be created in increased expression of genes that transcribe molecules specific for endothelial cell growth [89].

1.3.8 Hemodynamic control

Blood flow within vessels results in stress, which is referred to as shear stress. Shear stress may be laminar, thus acting tangentially or parallel to the endothelial surface, or otherwise is oscillatory, also regarded as turbulent [90]. Enhanced laminar shear has been shown to preserve the integrity of the walls of vessels, in part through Ets-1-dependent induction of protease inhibitors. In addition, laminar shear stress is known to inhibit tubule formation and migration of endothelial cells by an angiopoietin-2 (Ang-2)-dependent mechanism, which entails downregulation of Ang-2 and is associated with IA. Laminar shear stress also increases endothelial actin filament bundles [91], but whether these fibers are important in the endothelial cell movements during pillar formation is unclear. In contrast, turbulent shear stress results in angiogenesis and remodeling of the vessels, with an increase in cell proliferation and migration, a process characteristic of sprouting angiogenesis.

Oscillatory shear stress results in the production of Ang-2 in endothelial cells and plays a critical role in migration and tubule formation and may be a culprit in diseases with disturbed flow and angiogenesis. The role of hemodynamics in control of IA was demonstrated by clamping of one of the dichotomous branches of an artery in the developing CAM microvasculature. Increases in blood flow and pressure in cognate artery resulted in an almost immediate effect on branching morphology with pillars beginning to appear within 15–30 min of clamping and a concomitant reduction in branching angles by about 20% after 40 min. This indicates that a hemodynamic alteration results in an immediate vascular adaptation [50]. Egginton et al. reported that capillary growth in muscles with increased blood flow occurred through intraluminal splitting, with no sprouting, a mechanism typical of IA. During increased blood flow, the mechanical stretch of vessels may result in expansion, providing the medium for intussusceptive microvascular augmentation. Indeed, capillary network remodeling occurs in response to the mechanical forces of increased shear stress and cell stretch [54].

Changes in shear stress are sensed by the endothelial cells and the signal is transduced by molecules such as PECAM/CD31. This proceeds through a cascade of events that result in transcription of many molecules involved in angiogenesis such as endothelial nitric oxide synthase (eNos) and growth factors. Shear stress is known in vitro to activate the VEGFR-2 pathway, and can indirectly influence the mitogenic effect of VEGF via nitric oxide [92]. During development, sensitivity to the local hemodynamic environment facilitates assembly and remodeling of appropriate microvascular network structures and may continue to be critical in maintaining and remodeling capillary networks in adulthood [93]. The essential role played by hemodynamics was demonstrated many decades ago by Clarke and Clarke when they illustrated an increase in vessel complexity as a result of enhanced blood flow subsequent to inflammation. The same authors had earlier demonstrated that a vessel branch became separated completely from the main vessel within 3 days, and subsequently became obliterated [54].

1.3.9 Molecular control

The mechanism that entails pillar formation in the process of IA is highly mediated by molecular control. The primordial indicators of pillar initiation are intraluminal endothelial protrusions, distinguished as minute shallow depressions on intravascular casts or projections into the vessel lumen in serial sections. Such protrusions are followed by endothelial cell contacts, reorganization of endothelial cell junctions, and invasions of the pillar core by myofibroblasts and pericytes, which lay down collagen fibrils [94]. The most important growth factor in angiogenesis is VEGF, which is known to support both sprouting and IA [50]. Other endothelial growth factors implicated in IA include bFGF and PDGF-B. It has been shown that bFGF upregulates PDGFR-α and -β expression levels in the newly formed blood vessels and PDGF-AB and BB act through PDGFR-β to enhance vessel stability. The factors that initiate transluminal pillars are unknown; however, PDGF and Ang-2 are both known to be important in pericyte recruitment and may therefore play a role in pillar formation. In knockout mice lacking angiopoietin-1 (Ang-1) and Tie-2, vessel growth is arrested at an early stage of development and further remodeling does not occur [95]. On injection of a monoclonal antibody against PDGFR-β, the receptor for PDGF-B, in murine neonates completely blocks mural cell recruitment in the developing retinal vessels. PDGF-B promotes pericyte recruitment by stimulating both proliferation and migration of such cells. Ang-1 cannot initiate angiogenesis; it is constitutively expressed throughout the body and promotes vascular remodeling, maturation, and stabilization of vessels via its TIE-2 receptor. Molecules specifically associated with cell migration such as neuropilin, restin, and midkine are downregulated during the phase of IA. To date, no direct evidence linking a specific molecule to IA has been cited. However, a synergism between VEGF and Ang-1 is highly suspected. Transgenic mice with overexpression of VEGF in the skin has numerous tortuous and leaky capillaries, whereas those with Ang-1 overexpression have enlarged but less leaky vessels. In contrast, Ang-1 is necessary for later stages of vessel development, and mice with deficiency die of problems associated with vessel remodeling and maturation. Future investigations into the molecular control of IA are creating a focus on the interactions between angiogenic growth factors (especially VEGF), angiopoietins, and their cognate receptors [96].

1.3.10 Experimental models to study intussusceptive angiogenesis

Experimental models are crucial to achieve evidence in the process of IA. Detailed study of in vitro and in vivo angiogenesis models has been described over the past decades. However, most models are only optimized to study SA or do not enable discernment of sprouting from IA [85]. The ideal experimental model for angiogenesis should provide a good mimic of the process under study. In addition, it should be highly reproducible, provide the possibility to monitor the process of angiogenesis in real time without inducing extra environmental bias, and enable the possibility to carefully monitor and adjust the concentration of angiogenic factors in a spatial and temporal fashion. The process of angiogenesis in an in vivo model provides an enthusiastic substitute which occurs in natural physiological conditions. Due to the highly complex structure of in vivo models, the concentrations of angiogenic factors and other environmental factors cannot be easily monitored and adjusted. Moreover, real-time monitoring of the angiogenic process at high resolution is almost impossible in in vivo models [50].

1.3.11 In vivo study

The vascular corrosion casting technique is extensively used in most in vivo angiogenesis models to study the occurrence of IA. Unfortunately, vascular corrosion casting is not regularly used in most angiogenic investigations, and thorough screening for the phenotype of the angiogenic mechanism is scarcely performed. Recently, IA was reported to occur in the retina of mice that were raised in hypoxic conditions. This chronic hypoxia model is often used to study retinal angiogenesis and might also form an interesting model for studying IA. Another secondary interesting candidate to conduct the study of the mechanism of angiogenesis is the kidney glomeruli. These vascular beds are already used as models of angiogenic repair through IA. During their development, growing glomeruli display a remarkable switch of the angiogenic phenotype from SA to IA. In the rat arteriovenous loop model, an arteriovenous fistula (loop) is constructed from the femoral neurovascular bundle and embedded in a fibrin matrix within a Teflon isolation chamber [50]. This procedure, along with corrosion casting and SEM, allows a quantitative investigation of the influence of angiogenic growth factors on the angiogenic phenotype including intussusceptive pillar formation. The above-mentioned in vivo models are guaranteed tools to investigate the molecular pathways regulating IA. However, they only permit the evaluation of the capillaries after sacrificing the animals and do not allow evaluation of the process of IA in real time.

The chick CAM assay is the only model that was recently developed to study IA in real time and ranks among the most frequently used angiogenesis assays [1]. This remodeling two-dimensional capillary bed can be monitored in real time with a fluorescent microscope after injection of a fluorescent dye into the circulation of the growing embryo. In this way, the process of pillar formation at the bifurcation point of splitting vessels can be demonstrated, and the influence of various pro- and antiangiogenic substances can be evaluated.

1.4 Sprouting angiogenesis

This chapter provides a brief overview of some relevant topics explaining sprouting angiogenesis which includes (1) the concept of functional specialization of endothelial cells during different phases of this process, involving the specification of endothelial cells into tip cells, stalk cells, and phalanx cells bearing different morphologies and functional properties; (2) the interplay between numerous signaling pathways, including Notch and Notch ligands, VEGF and VEGFRs, semaphorins, and netrins, in the regulation and modulation of the phenotypic characteristic of these cells; and (3) some fundamental and consecutive morphological processes, including lumen formation and perfusion, network formation, remodeling, pruning, leading to the final vessel maturation and stabilization.

1.4.1 Endothelial tip, stalk, and phalanx cells

Initiation of sprouting demands the specification of endothelial cells into tip and stalk cells bearing different morphologies and functional properties. Endothelial tip cells initially migrate but proliferate minimally, in contrast to endothelial stalk cells, which do proliferate. In 2003, the concept of "tip" and "stalk" cell phenotype was described for emerging sprouts, even though filopodia-studded cells at the front were already described before this work by Kurz et al. and, later, by Ruhrberg et al. [97]. According to this new model, two principal cells are involved in sprouting angiogenesis, namely the "tip cell" and "stalk cell." The tip cell is migratory and polarized, while the stalk cell proliferates during sprout extension and forms the nascent vascular lumen cell. The phenotypic specialization of endothelial cells as tip or stalk cells is very transient and reversible, depending on the balance between proangiogenic factors, such as VEGF and Jagged-1 (JAG-1), and suppressors of endothelial cell proliferation, such as DLL4-Notch activity [98] (Fig. 1.9).

Tip cells exhibit high levels of DLL-4, PDGF-b, unc-5 homolog b, VEGFR-2, and VEGFR-3/Flt-4, and have low levels of Notch signaling activity [98]. The zebrafish tip cell is a highly branched structure, while in mouse angiogenesis models, such as brain and retina, the tip cell extends numerous filopodia that serve to guide the new blood vessel in a certain direction toward an angiogenic stimulus, proliferate minimally, and adapt a highly branched shape while moving [99]. Bentley et al. developed a new physics-based model tightly coupled to in vivo data and proposed a novel filopodia adhesion-driven migration mechanism, displaying that cell—cell junction size is a key factor in establishing a stable tip/stalk pattern. The stalk cells produce fewer filopodia, are more proliferative, and form tubes, branches, and a vascular lumen. Stalk cells also constitute

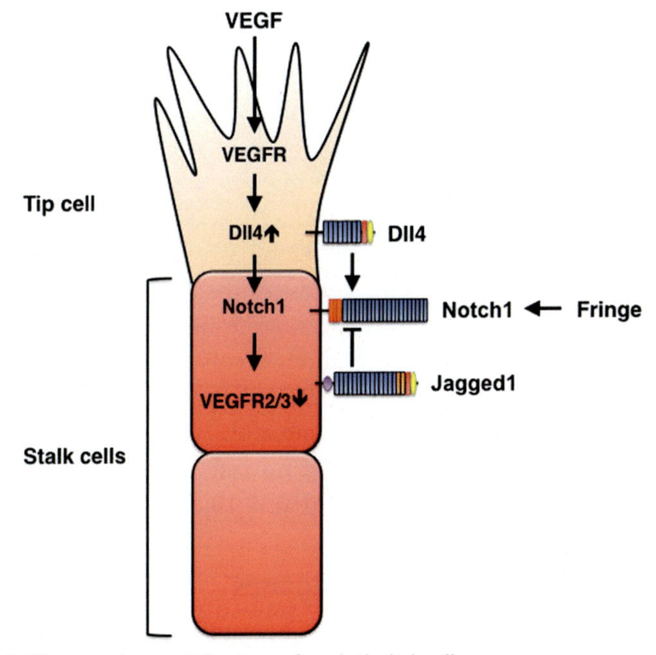

Figure 1.9 Phenotypic specialization of endothelial cells.

junctions with neighboring cells and synthesize basement membrane components. Only a standard balance between both processes establishes adequately shaped nascent sprouts [100]. During the transition from active sprouting to quiescence endothelial cells, the tip cell adopts a "phalanx" phenotype, resembling a phalanx formation of Ancient Greek soldiers, that is, lumenized, nonproliferating, and immobile cells, which promotes vessel integrity and stabilizes the vasculature through increased cell adhesion and a dampened response to VEGF [68].

1.4.2 Roles of vascular endothelial growth factor and vascular endothelial growth factor receptors

The VEGF signaling pathway has been established as the master regulator of angiogenesis. Endothelial cells express VEGFR-2, a tyrosine kinase receptor that positively drives the mitogenic and chemotactic responses of endothelial cells to VEGF. Tip cell migration depends on a gradient of VEGF, whereas stalk cell proliferation is regulated by VEGF concentration [81]. The leading tip cell responds to a VEGF gradient by migrating outward from the parent vessel up the gradient. VEGF induces the formation and extension of filopodia as well as the expression of DLL4 protein in the tip cells, and filopodia engages with those of a nearby tip cell to form a "bridge" and the formation of a new vessel. As the vessel elongates, the stalk cell proliferates in response to VEGF, creates a lumen, synthesizes a basement membrane, and associates with pericytes and increases the mass and surface of the growing vessel. VEGF stimulates tip cell induction and filopodia formation via VEGFR-2, which is abundant on filopodia, whereas VEGFR-2 blockade is associated with sprouting defects. Moreover, activation of Cdc 42 by VEGF triggers filopodia formation [101]. The Notch signaling induces the VEGFR-1 expression to minimize VEGF ligand availability, preventing tip cell outward migration. VEGFR-1 is expressed in stalk cells and is involved in guidance and limiting tip cell formation, and loss of VEGFR-1 increases sprouting and vascularization. During both mouse and zebrafish angiogenesis, VEGFR-3 is most strongly expressed in the leading tip cell and is downregulated by Notch signaling in the stalk cell [102] (Fig. 1.10).

1.4.3 Role of Notch and Notch ligands

Notch receptors are large transmembrane proteins that, when activated by their ligands expressed on adjacent cells, regulate cell fate in multiple

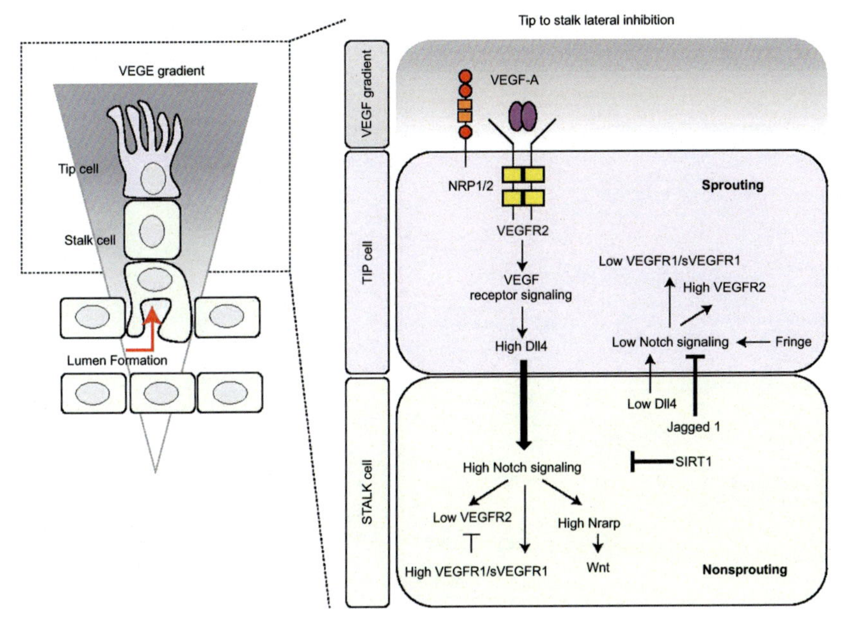

Figure 1.10 VEGF and Notch in tip and stalk cell selection.

lineages. The Notch signaling pathway is substantial for vascular development. The necessity of DLL-4/Notch-1 signaling in the endothelium has been well established, as loss of a single copy of DLL-4 or deletion of Notch-1 causes vascular defects and embryonic lethality. Notch-1 and Notch-4 and three Notch ligands, JAG-1, DLL-1, and DLL-4, are expressed in endothelial cells for the induction of arterial cell fate and for the selection of endothelial tip and stalk cell during sprouting angiogenesis. A loss of Notch signaling introduced during its activation reduces sprouting. Notch-1-deficient endothelial cells adopt tip cell characteristics, while Notch signaling activity is greater in stalk cells, in which activation of Notch by Dll-4 leads to a downregulation of VEGFR-2 and VEGFR-3 in these cells [103].

Blockade of Notch leads to widespread Flt4 expression, increases filopodia and sprouting, and promotes tip cell activity. Tip cells with low Notch activity have high VEGFR-2 and low VEGFR-1 expression, which results in higher levels of DLL4 expression and, hence through a higher production of DLL-4 than in neighboring cells, an increased ability to suppress its neighboring cells from becoming tip cells. Endothelial cells with higher levels of VEGF increase DLL-4 expression, which further

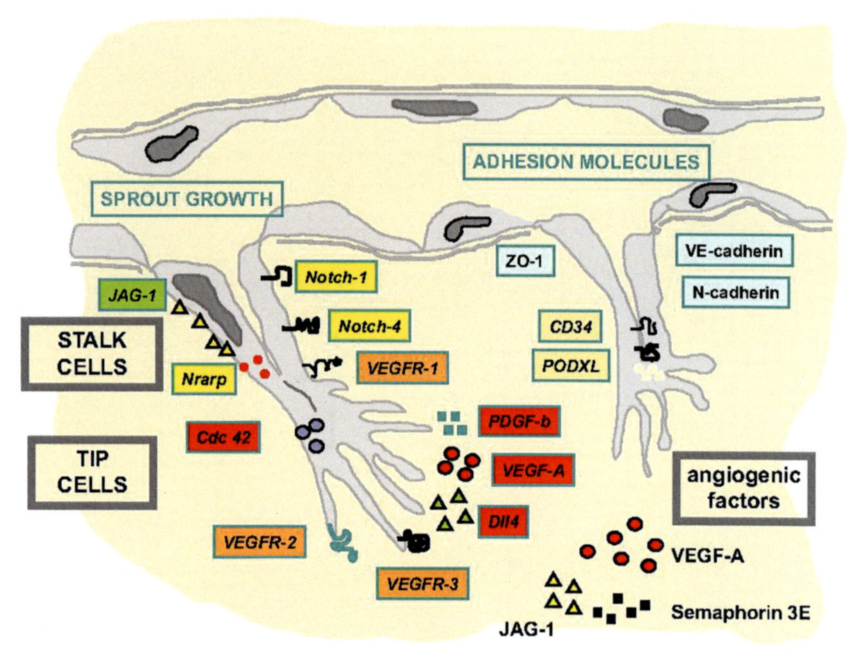

Figure 1.11 Role of Notch and Notch ligands.

increases the cell sensitivity to VEGF, and this cell becomes the tip cell selected for outward migration for the parent vessel. Stalk cells have high levels of Notch signaling activity and elevated expression of JAG-1 [104]. Stalk cell JAG-1 antagonizes DLL-4 activity, reducing the induction of Notch signaling in the adjacent tip cell, which therefore maintains (Fig. 1.11) its responsiveness to VEGF stimulation and migrates outward to establish a new branch [105]. Notch-regulated ankyrin-repeat protein (Nrarp) is a downstream of Notch that counteracts Notch signaling and is expressed in stalk cells at branch points. Silencing of Nrarp is responsible for a reduction in vessel density, due to lowered endothelial cell proliferation as a consequence of an upregulation of Notch and VEGFR-1, of poorly luminized vessels, and remodeling of endothelial junctions and vessel regression. By reducing Wnt signal in stalk cells, Nrarp is responsible for vascular stabilization and lowering Wnt signal induces vessel regression [106].

1.4.4 Lumen formation and perfusion

Many organs, including blood vessels, are composed of epithelial tubes with an epithelial surface lining the lumen that transports vital fluids. The

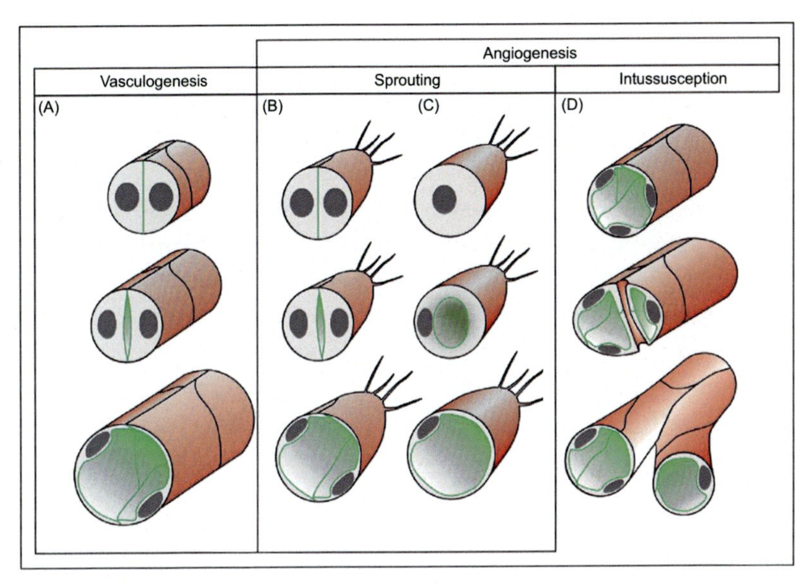

Figure 1.12 Lumen formation of blood vessels.

lumen of a blood vessel is vital for supplying blood to any given tissues. In the vascular system, lumen formation involves a complex molecular mechanism composed of endothelial cell repulsion at the cell–cell contacts within the endothelial cell cords, junctional rearrangement, and endothelial cell shape change. At the beginning of 18th century, scientists demonstrated that the capillaries send out sprouts, which later extend until they meet and anastomose with other sprouts or capillaries and into which a lumen advances [107]. Two different ways of lumen formation have been discussed: cord hollowing and cell hollowing. In the former, slit-like lumen formation takes place between facing endothelial cells and, in the latter, the ensuing lumen is the result of coalescence of intracytoplasmic vesicles. Clark and Clark also contributed other details to our knowledge of angiogenesis including the development of capillary lumina and fusion of sprouts [68] (Fig. 1.12).

During the final steps of capillary development, endothelial cell migration comes to a halt and endothelial cells form a lumen and reestablish functional adherens junctions. The formation of adherens junction was shown to be associated with the suppression of endothelial cell migration in monolayers. This process was shown to be mediated by VE-cadherin, a cell adhesion molecule belonging to the cell cadherin protein family, exclusively expressed in endothelial cells and facilitating their homotypic interaction. VE-cadherin is strictly required for the polarization of endothelial cells in vitro and in vivo, and VE-cadherin-based junctions are subjected to continuous reorganization, which renders them highly dynamic and sensitive to extracellular stimuli [108]. The lumen of the mouse dorsal aorta forms by organized changes in endothelial cell shape, while an alternative mode has been proposed in which intercellular and intracellular vacuoles form and fuse along connected endothelial cells [52]. Intracellular vacuolization is a rapid way to create endothelial cell luminal spaces. During this process collagen represents a promorphogenic factor, whereas basement membrane proteins are inhibitory for vascular morphogenesis. During lumen formation, two functionally distinct phenotypes of endothelial cells are recognized. The initial phenotype, represented by endothelial cells in mature blood vessels, is determined by an apico-basal polarity and junction-mediated contact inhibition. The second phenotype is found in activated tissues and is characterized by the loss of apico-basal polarity and adherens junctions, a spindle-shaped morphology, and the ability for guided migration. Endothelial cell polarization often begins with the delivery via exocytosis of de-adhesive apical glycoproteins, including CD34-sialomucins, such as CD34 and podocalyxin (PODXL), to the cell–cell contact [109]. The formation of apical cell surfaces and electrostatic repulsion of negatively charged apical glycoproteins are sufficient for the initial de-adhesion of adjacent endothelial cells and for slit formation, but are insufficient for the development of a patent vascular lumen. VEGF-A therefore induces cell shape changes that further separate the apical cell surfaces from each other. After lumen formation, the lumen diameter of vessel sprouting increases, and hemodynamic stimuli caused by shear stress significantly contribute to increased diameter [110]. Vascular lumen expansion is force-dependant and involves F-actin cytoskeleton and/or blood flow. It has been recognized that endothelial cells use other cells as migration tracks during capillary network remodeling. In addition to sprouting, a new type of endothelial cell migration has been defined, namely guided migration of endothelial cells along performed capillary-like structures. Endothelial cell-derived FGF2 is involved in the regulation of guided migration [68].

1.4.5 Network formation

Following the assembly of primitive vessels in the early embryo, remodeling transforms the plexus into an organized network of arteries, capillaries, and veins. The network formation represents a critical step in the process of angiogenesis and provides the growing tissue with a newly constituted apparatus of immature and rudimentary vascular channels. The development of a complex, interconnected meshwork of crude capillary tubules is the structural substrate upon which the fine-tuning process of vascular remodeling acts. Network formation originates from coalescence of sprouts or by IMG. IMG is an alternative or supplementary mechanism of capillary growth whereby the vascular network expands by insertion of newly formed columns of interstitial tissue (interstitial tissue structures) into the vascular lumen called tissue pillars or posts. They initially have small dimensions, with diameters between 0.5 and 2.5 mm, and then consecutively grow to a large size [111]. Interestingly, transluminal pillar formation is one particular way of expanding vessels, and thus it represents a crucial mechanism for vascular network formation. According to Burri's group, this mechanism proceeds through four steps: (1) protrusion of opposing capillary walls into the lumen and the creation of a contact zone between facing endothelial cells; (2) reorganization of their intercellular junctions and central perforation of the endothelial bilayer; (3) formation of an interstitial pillar core by invading supporting cells (myofibroblasts, pericytes) and deposition of matrix, with such pillars ranging in diameter from 1 to 2.5 nm; and (4) enlargement of the thickness of the pillars without additional qualitative alterations. The pillar formation sequence was observed during the initial remodeling of the capillary plexus into immediate pre- and postcapillary feeding vessels. This process is called IAR and provides a mechanism by which well-perfused capillary segments are transformed into terminal arterioles and veins. As a result of this process, a complex arterial and venous vascular tree arises from the primitive capillary plexus, forming a second layer. Other mechanisms involving sprout fusion, tip cell filopodia, and macrophages, participate in vascular network formation. Tip cell filopodia interact to initiate junction formation and reinforce sprout formation. Macrophages may also facilitate sprout fusion by bridging sprouts to their potential target [67].

1.4.6 Remodeling

The primary endothelial plexus generated by vascular sprouting consists of a homogeneous web of endothelial cell tubes and sacs. This plexus is created in excess and the final adjustment of vascular density involves the regression of unnecessary vessels through a process of vascular remodeling and pruning, which creates a more differentiated vascular network. Remodeling includes the growth of new vessels and the regression of capillaries in prechondrogenic regions to allow differentiation of cartilage, and the regression of the hyaloids vasculature to allow the development of the vitreous body in the eye [112]. Remodeling predetermines the formation of large and small vessels, the establishment of directional flow, the association with mural cells (pericytes and SMCs), and adjustment of the surrounding tissue. An interesting event which precedes vascular remodeling is the phenomenon of capillary retraction, described for the first time by Clark. Remodeling is a complex phenomenon which requires a wide range of molecular signaling. One of the earliest identified events involved in vascular remodeling was the interaction between the receptor tyrosine kinase tie-2 [68]. The formation and maturation of vitreal vasculature includes extensive initial vasculogenesis followed by combinations of vasculogenesis and angiogenesis events which are accompanied by pruning and remodeling of the maturing vitreal vessel network (Fig. 1.13).

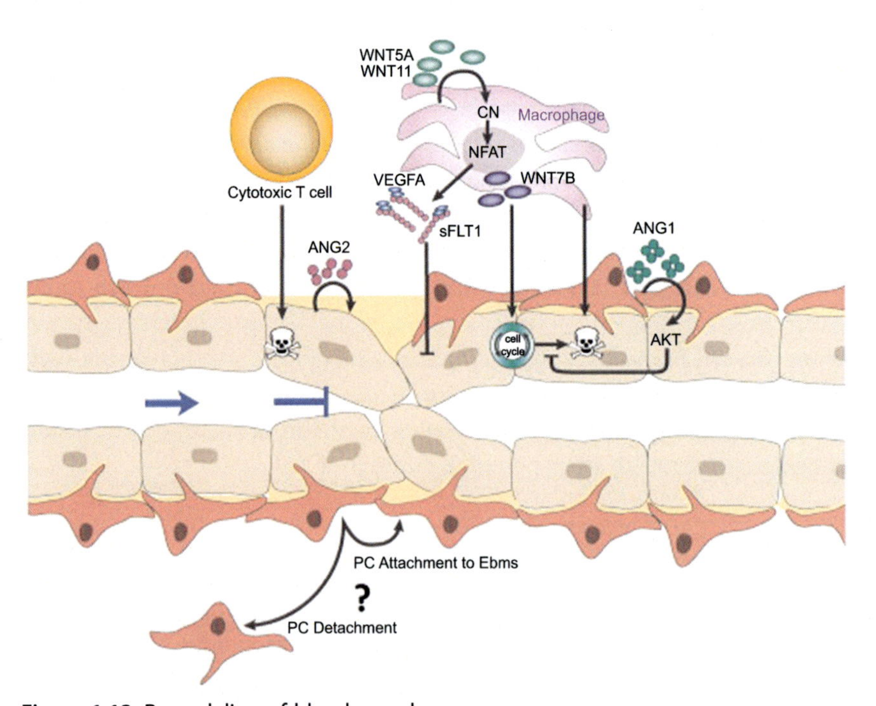

Figure 1.13 Remodeling of blood vessels.

1.4.7 Pruning

Pruning was first described in the embryonic retina and involves the removal of excess endothelial cells which form redundant channels. Simulations suggest how flow might affect vessel pruning, which, due to low wall shear stress, is highly sensitive to the pressure drop across a vascular network. In fact, the degree of pruning increases as the pressure drop increases. Pruning and remodeling of the vascular network may be stimulated by tissue-derived signaling molecules and blood flow conditions. Although the morphological hallmarks of IMG and IAR, critical mechanisms for generating and expanding a capillary network, were frequently observed in capillaries and terminal microvessels with diameters less than 25 nm, the establishment of transluminal pillars and folds in arteries and veins with diameters of upto 110 nm was a surprise. Such pillars were closely associated with bifurcations and were shown to be involved in the remodeling and pruning of larger microvessels [86]. This process hence was termed IBR. A fascinating model of vascular pruning is provided by modification of pulmonary vascularization under certain conditions. Chronic hypoxia caused by migration of native sea-level dwellers to high altitude or chronic lung disease leads to the development of increased pulmonary vascular resistance and pulmonary hypertension. The structural changes that are thought to underlie the increased vascular resistance can be broadly classified into two processes: first, remodeling of the walls of the pulmonary resistance vessels and, second, pruning of the total number of blood vessels in the lung. The structural changes include muscularization of nonmuscular arterioles, increased medial thickness of muscular arterioles, adventitial hypertrophy, and deposition of additional matrix components, including collagen and elastin, in the vascular walls [113]. The second major structural alterations caused by chronic hypoxia are loss of small blood vessels, which is said to increase vascular resistance by reducing the extent of parallel vascular pathways.

1.4.8 Maturation and stabilization (pericyte recruitment)

The stabilization of the newly formed vessel and the maintenance of the existing vasculature are late events in the angiogenic process. Blood flow is critically important for determining the fate of vessels with high flow which widen, while vessels with low flow regress, although recent advances in vascular biology strongly argue for the autonomous fate control achieved by blood vessels. More recently, Chenet et al. have demonstrated that lymphatic endothelial and vessel phenotypes are negatively suppressed by the flow of blood and that changes in hemodynamic forces can reprogram lymphatic vessels to blood vessels in postnatal life [45]. Several cellular and noncellular components in blood vessels, including endothelial cells, pericytes, SMCs, fibroblasts, glial cells, inflammatory cells, and the extracellular matrix, coordinate to control the maintenance of vessel integrity at varying degrees in different vascular beds. Pericyte adhesion to native capillaries and endothelial cells wrapped by surrounding pericytes are basic events in blood vessel stabilization and maturation. It has been hypothesized that concomitant with sprouting, endothelial cells direct the differentiation of mural cell precursors from the adjacent tissue by the secretion of soluble factors. A different view is that mural cells become associated with endothelial cells by migrating along newly made vascular sprouts [57] (Fig. 1.14).

A role for mural cells in maintaining vascular integrity was suggested by a number of gene knockout studies. This includes disruption of the genes encoding the endothelial specific receptors tie-1 and tie-2, the tie-2 ligand, (Ang-1), tissue factor system, and the PDGF-b/PDGF-b receptor system [114].

In the ocular model, changes in the state of ocular pericytes are known to be associated with vasculopathy in eye disease in humans. Pericyte deaths, thickening of the basement membrane, and changes in extracellular matrix expression pattern in proliferative diabetic retinopathy have been described. By studying postnatal remodeling of the retina

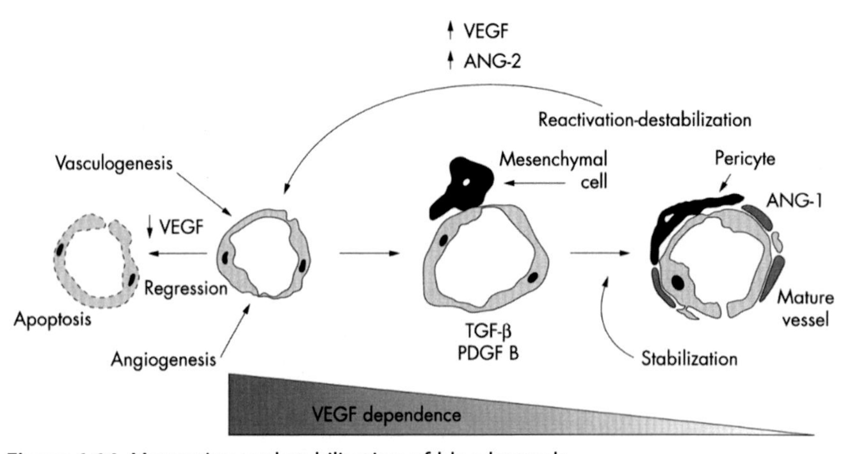

Figure 1.14 Maturation and stabilization of blood vessels.

vasculature, it has been shown that pericyte recruitment represents a crucial step in vascular maturation [115]. The transient existence of a pericyte-free endothelial plexus coincided temporally and spatially with the process of hyperoxia-induced vascular pruning, which is a mechanism for fine-tuning of vascular density according to the level of available oxygen [116].

CHAPTER 2
Lymphangiogenesis

2.1 Lymph vessel formation and physiology

The lymphatic system is comprised of a network of vessels interrelated with lymphatic tissue, which has the comprehensive mechanism to sustain the local physiologic environment for every cell in the body. The lymphatic system preserves extracellular fluid homeostasis that is favorable for superlative tissue function, removing substances that arise due to metabolism or cell death, and optimizing immunity against bacteria, viruses, parasites, and other antigens. A functional lymphatic vasculature is a requirement for the closed, high-pressure system for blood circulation in vertebrates, which leaks plasma components from capillaries and postcapillary venules. Lymphatic vessels retain normal tissue fluid volumes by returning the capillary ultrafiltrate and extravasated plasma proteins to the central circulation. Lymphatics also have a key role in the transport of lipids absorbed in the digestive tract. In addition, the widespread distribution of lymphatics throughout the body allows the rapid identification of antigens and immunological responses.

Interest in the contribution of the lymphatic system to health and disease has grown rapidly in recent years, due in large part to (1) the identification of specific genes and molecular markers to study lymphatic vessels; (2) evidence that lymphatic dysfunction contributes to cardiovascular diseases; and (3) increased awareness that lymphatic dysfunction in patients is a fairly common occurrence. For example, lymphedema is the most prevalent secondary disease found in patients who have undergone treatment for cancer [117]. In addition, lymphedema filariasis is one of the most common infectious diseases in the world, affecting millions of people worldwide. Many of the basic aspects of lymphangiogenesis and lymphatic transport have only recently begun to be clarified. Several major gaps in comprehension of the cellular and molecular mechanisms underlying lymphatic physiology persist, and the role of lymphatics in disease remains poorly understood [118].

The earliest known accounts of lymphatic vessels are contained in writings from 4 BCE by Hippocrates and Aristotle. Several centuries later,

Genesis, Pathophysiology And Management of Venous and Lymphatic Disorders
DOI: https://doi.org/10.1016/B978-0-323-88433-4.00011-5

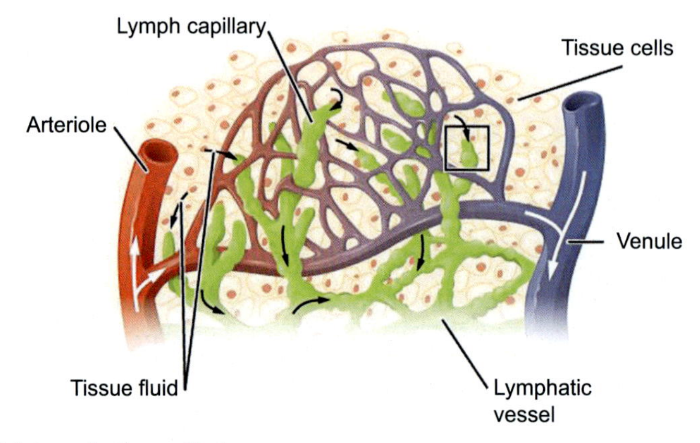

Figure 2.1 Lymphatic capillaries.

the Greek physician Claudius Galen described mesenteric lymph nodes filled with chyle, or lymph consisting of emulsified fats that originates from the small intestine. Contemporaries of Pecquet, Olaus Rudbeck and Thomas Bartholin, independently demonstrated that lymph flows from the liver into the thoracic duct, and that the lymphatic vessels are widespread throughout the body. Thomas Bartholin's naming of these vessels as "vasae lymphaticae" led to our current term "lymphatic vessels" [119]. In the 18th century William Hewson observed the rhythmic contractions of collecting lymphatics and noted numerous intraluminal valves to prevent backflow, while his colleague William Hunter identified that lymphatics are an "absorbent system." Carl Ludwig postulated that lymph was a filtrate of the blood. Later, Ernest Starling established that the balance between the hydrostatic and osmotic forces of the plasma and interstitium favored lymph formation. Subsequent advances combining cinematography and intravital microscopy, and later development of isolated lymphatic techniques in the second half of the 20th century allowed major advances in the physiology of lymph formation and propulsion. Likewise, the revolutionary discoveries in molecular biology and developmental biology laid the foundation for our understanding of lymphatic physiology [120] (Fig. 2.1).

2.2 Organization and anatomy of the lymphatic system

Like the blood circulation, the lymphatic system is a network of specialized vessels that perform exchange and transport mechanisms. Unlike the blood circulation, which has the heart as a central pump, the propulsion of lymph through the lymphatic vessel network is mediated by the forces driving the initial formation of lymph in the tissues, the intrinsic pump mechanisms that propel lymph forward through the system, and tissue pressures extrinsic to the system that favor forward flow. In addition, unlike the blood, lymph does not circulate per se, but rather is a filtrate formed in the tissues that the network delivers to the central circulation. The venous system enables the spontaneous return of the deoxygenated blood to the right heart for reoxygenation in the lungs to support metabolism throughout the body at a rate equal to the cardiac output, approximately 5 L of blood per minute in an average-sized human. In contrast, the lymph that enters the great veins accounts for less than 0.05% of venous return [121]. This much slower transport of lymph, approximately 8–12 L per day, half of which reenters the systemic circulation at lymph nodes, and the other half returning via the great veins, makes the system act as an additional reservoir for fluids generated by filtering of plasma through the microvascular walls and interstitial space. This reservoir is essentially a sample of filtered plasma available for the immune system surveillance for potential threats in local draining lymph nodes distributed throughout the system. In almost every tissue, lymphatic vessels have been found. Exceptions have traditionally included bone marrow, cartilage, the cornea, and the central nervous system. However, even within these tissues there has been evidence of lymphatic drainage under normal conditions, or lymphangiogenesis under pathologic conditions. Tracers injected into long bones have been found to drain into local lymph nodes with both radiological and histological approaches [122]. Sprouting of lymphatics into cartilage of the trachea has been reported in a mouse model of *Mycoplasma pulmonis* infection (Fig. 2.2).

Likewise, growth of lymphatic vessels into the cornea can be elicited by injury. Recent evaluation of the brain with advanced imaging and use of molecular markers confirmed the presence of lymphatics in the dura mater of the brain and the existence of prelymphatic channels that had previously not been acknowledged by certain research communities. Like many biological systems in nature that absorb and transport fluids to a central location, lymphatic networks have a largely fractal geometric organization [118]. This mode of distribution permits the smallest, most distal, blind-ended vessels to cover a large surface area within tissues to absorb fluids, serving as the site of lymph formation. By various conventions, these vessels are referred to as initial lymphatics because they are where

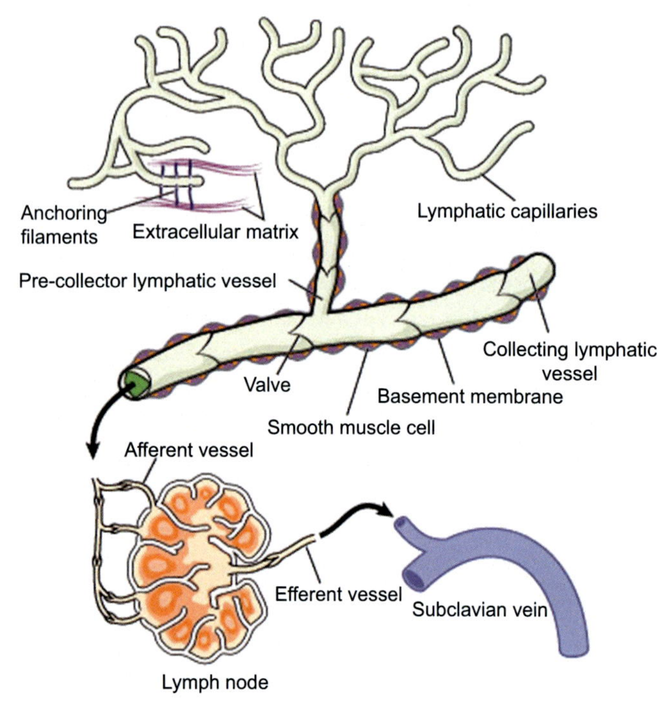

Figure 2.2 Structural biology of lymphatic system.

lymph initially forms, terminal lymphatics due to their blind-ended nature, or lymphatic capillaries' because like blood capillaries, they serve as a site of fluid exchange. In current literature, the terms "lymphatic capillary" and "initial lymphatic" are more frequently used. The initial lymphatics are located in close proximity to the microcirculation and consist of a single endothelial layer with a poorly defined basement membrane. These vessels may be like saccules, blind-ended, or they may form an interconnected network or plexus. The initial lymphatics drain into collecting lymphatics, which are distinguishable by the presence of a smooth muscle layer and one-way bicuspid valves to prevent retrograde fluid flow. In some cases, an intermediary lymphatic vessel type known as a precollector, lacking smooth muscle but having the one-way valves, is present between the initial and collecting lymphatics, establishes vessel tone and, unlike vascular smooth muscle, it also contracts physically. The prenodal collecting lymphatics, also called afferent lymphatics, transport lymph to the lymph nodes, where it comes into contact with a collective of antigen-presenting cells, T cells, and B cells. The lymph exists in the lymph nodes through postnodal collecting lymphatics, also called efferent lymphatics, although it is worth noting that this definition is relative to a particular node, as in some parts of the system the lymph passes through multiple lymph nodes in series. Eventually the collecting lymphatics throughout the body coalesce into the larger lymph trunks, of which the largest, the thoracic duct and right lymph duct, empty directly into the subclavian veins [123].

2.3 General initial lymphatic structure

The initial lymphatics are the site of lymph formation. Lymph within an initial lymphatic network is free to flow in the directions imposed by local hydrostatic forces, and solutes may diffuse freely within the vessels. The exit point from an initial lymphatic vessel or network is an intraluminal valve composed of endothelial cells and connective tissue that defines the border between the initial lymphatic vessel or network and downstream precollectors or collecting lymphatic vessels. In human skin, initial lymphatics are typically 35–70 nm in diameter and form interconnected networks. In contrast, the blind-ended lacteals in rat intestinal villi are only 15–30 nm in diameter. The initial lymphatics are composed of a single layer of endothelial cells, with a discontinuous and often indistinct basal lamina. In tissue sections, the endothelial cells of initial lymphatics largely resemble those of capillaries, and in classic histology lymphatic capillaries are distinguished from blood capillaries based on their different luminal dimensions, and also the contours of the cells often having an undulating course and cytoplasmic processes projecting both luminally and abluminally [124]. Reports on the morphology of initial lymphatics in the cat tongue, mouse trachea, rat mesentery, rat small intestine, guinea pig and rat uterus, and human tonsils have revealed specialized endothelial cells having a relatively flat, "oak leaf" shape. These specialized endothelial cells have apparent functional significance for lymph formation, as adjacent overlapping cells form structures that appear to be flaps. These flaps are thought to act as the microscopic one-way valves hypothesized to allow the formation of lymph from interstitial fluid. Schmid-Schonbein and colleagues named these "primary lymphatic valves" to differentiate them from the luminal valves observed in collecting lymphatics, which they termed "secondary lymphatic valves" [119] (Fig. 2.3).

Observations from studies of the localization of junctional proteins at these flaps support that they are indeed primary valves that favor lymph

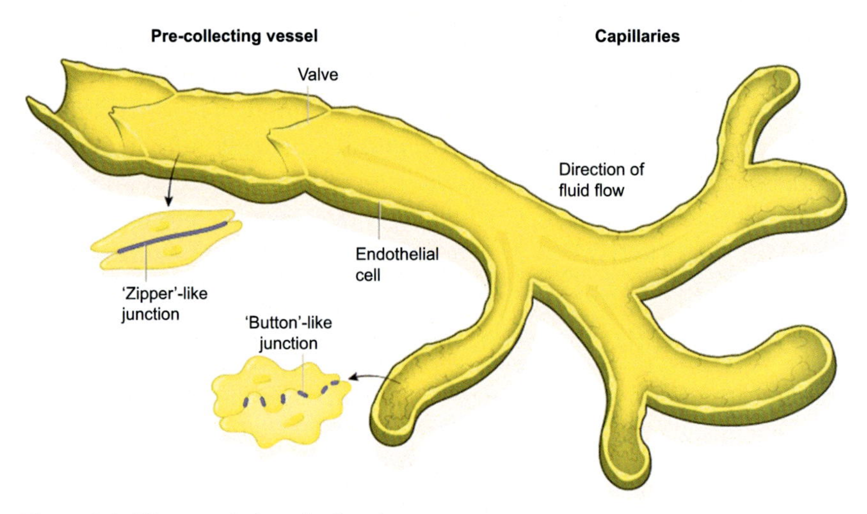

Figure 2.3 Microscopic lymphatic valves.

formation. vascular endothelial (VE)-cadherin, an adhesive protein, is required for establishing normal endothelial barrier integrity, as are several tight junction proteins (occluding, claudin-5, ZO-1 (zonula occludens-1), ESAM (endothelial cell-selective adhesion molecule), and JAM-A (junctional adhesion molecule-A protein)). platelet endothelial cell adhesion molecule 1 (PECAM-1) and Lyve1 also localize intermittently, but in the flaps, where VE-cadherin and the tight junction proteins are absent. Because PECAM-1 has a known role in leukocyte diapedesis, these PECAM-1-rich areas between the VE-cadherin "buttons" are also thought to be sites where lymphocytes may enter initial lymphatics with little resistance. The initial lymphatics also feature anchoring filaments that protrude into the surrounding interstitial spaces. The anchoring filaments of the initial lymphatics of human skin were reported to be composed primarily of fibrillin, and connect from the extracellular matrix (ECM) to the cytoskeleton via focal adhesions containing focal adhesion kinase and alpha1−beta1 integrin [125].

2.4 Precollector structure

Precollectors are defined as lymphatic vessels composed of a single endothelial layer but also having secondary valves to prevent backflow into initial lymphatics. In the precollectors, oak leaf-shaped endothelial cells may still be present in the distal regions near initial lymphatics, but in more rhombic shape, similar to venous endothelial cells. Accordingly, the junctions between endothelial cells begin to have continuous expression of PECAM-1 and VE-cadherin, suggesting that these vessels may act more as conduits than sites of lymph formation. Because there is no smooth muscle layer, the movement of lymph within precollectors depends highly on the inflow and outflow pressures of individual segments [124].

2.5 General collecting lymphatic structure

The collecting lymphatic vessel wall has an inner endothelial surrounded by a medial layer of circular smooth muscle cells (SMCs). An interesting example was reported in the diaphragms of 6-week-old rats, in which the collecting lymphatics had circular muscle where intraluminal valves are located, but primarily longitudinal smooth muscle between valves. Other cell types can also be identified on the adventitia of the collecting lymphatic vessel wall, such as dendritic cells, macrophages, and neurons. In addition, collecting lymphatics may have supporting microcirculation in close proximity to the vessel exterior, or vasa vasorum within the adventitia or smooth muscle layer, that provides oxygen and nutrients [126]. Also prominent in collecting lymphatics are the periodic secondary valves that organize the vessels into a chain of chambers that prevent backflow of lymph. Each valve leaflet is composed of a folded bilayer of endothelial cells with their apical sides facing away from each other. The basal sides of the intraluminal valve endothelial cells are separated by an inner supporting ECM containing elastin fibers. The single leaflet valves described in some cases may represent developing or regressing valves. In early studies of sectioned pulmonary lymphatics, a funnel-like structure was proposed [126]. In early studies of sectioned dermal lymphatics, unicellular valves that may have been newly developing valves were described and later confirmed. The secondary valves are spaced along the length of a collecting lymphatic vessel at semiregular intervals, forming chambers. Each chamber between two consecutive valves forms a functional contractile unit, called a lymphangion, meaning "lymph heart." Each lymphangion is capable of contracting either independently or in conjunction with its upstream and downstream lymphangions. Unlike initial lymphatics and precollectors, collecting lymphatics vessel networks are more consistently organized as a binary tree. At each bifurcation of the network, one or two secondary valves are often present. The networks of prenodal (afferent) collecting lymphatics lead to lymph nodes. Multiple prenodal collecting lymphatics may drain into lymph nodes. Typically, but not in all cases, one postnodal (efferent) collecting lymphatic exits the node [127] (Fig. 2.4).

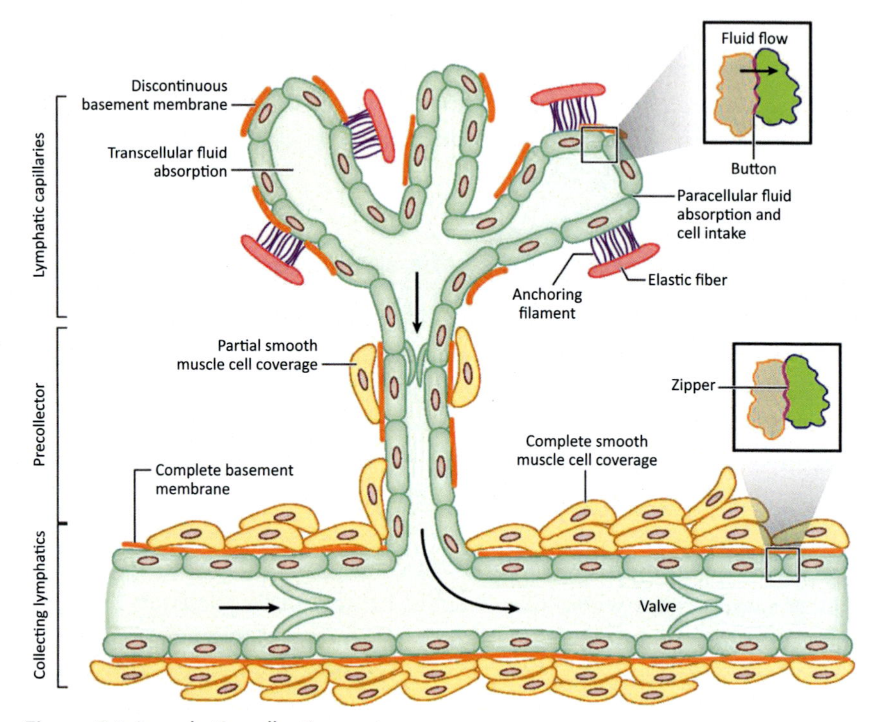

Figure 2.4 Lymphatic collecting system.

The larger lymphatic trunks also have the same generalized collecting lymphatic structure, but have functional differences with their smaller counterparts. The intestinal trunk and the lower lumbar lymphatic trunk drain into the cisterna chyli, as the base of the thoracic duct. Other peripheral lymphatic trunks from most parts of the body also drain into the thoracic duct. Lymphatics arising from the upper right thorax, right arm, the right side of the head and neck, and in some cases the lower lobe of the left lung drain into the right lymphatic duct, although in some cases there are tributaries from intrathoracic organs into the thoracic duct. The thoracic duct drains into the left subclavian vein, while the right lymphatic duct drains into the right subclavian vein [123].

2.6 Immunological markers of lymphatic endothelium

The discovery of several proteins that are generally found on lymphatic endothelial cells (LECs) but not blood endothelial cells has aided in: (1) isolation of these two broad types of cells for study in culture and (2) easier identification of lymphatic vessels by immunofluorescence labeling. These markers include a homolog of the *Drosophila melanogaster* homeobox gene prospero, known as prospero homeobox protein 1 (PROX1 or Prox1; abbreviations for human and rodent proteins, respectively), lymphatic vessel endothelial hyaluronan receptor 1 (LYVE1 or Lyve1), vascular endothelial growth factor receptor-3 (VEGFR3 or Vegfr3), and podoplanin. The Prox1 gene encodes a nuclear transcription factor described as the master control gene, allowing expression of lymphatic endothelial markers. Prox1 is highly expressed in initial lymphatics and precollectors. In collecting lymphatics, where the endothelial cell layer has a greater degree of basement membrane and is surrounded by an outer layer of SMCs, Prox1 expression is much lower [126] (Fig. 2.5).

However, Prox1 expression is very high in endothelial cells of the secondary valves in collecting lymphatics, where expression of forkhead box C2 (FOXC2), an important regulator of valve cell identity, is also very high. Although a reliable marker of LECs, prox1 can be detected in other nonendothelial cell types. These include heart valve endothelium, pancreatic epithelium, hepatocytes, bile duct cells, adrenal medullary neuroendocrine cells, megakaryocytes, cardiomyocytes, skeletal myocytes/satellite

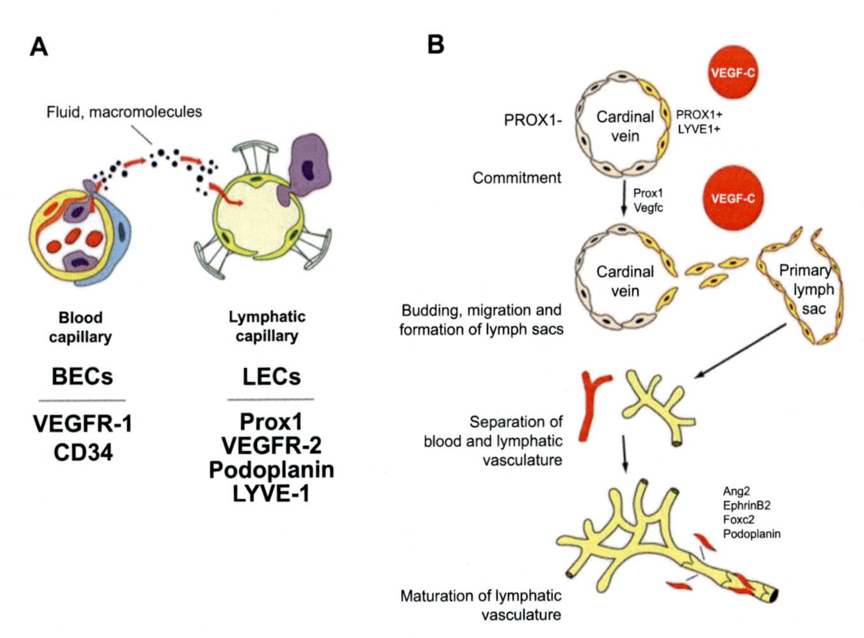

Figure 2.5 Developmental stages of lymphatic system.

cells, and platelets. LYVE1 is an integral membrane glycoprotein important for cell migration. Expression of LYVE1 is highest in initial lymphatics, while it is low or even undetectable in collecting lymphatics. Recently, LYVE1 was successfully used as a marker for sorting and culturing rat dermal LECs. While generally a good marker for lymphatic endothelium, LYVE1 has also been detected on macrophages and in liver and spleen sinusoid endothelium [128].

VEGFR3, also known as FLT4, was the first antigen marker used to identify lymphatic endothelium. VEGFR3 is a tyrosine kinase receptor for both VEGF-C and VEGF-D that mediates lymphangiogenesis. Podoplanin was originally described on rat kidney podocytes as a mucin-type transmembrane glycoprotein, and was later found to reliably identify lymphatic capillaries. Podoplanin has proved to be a useful target for sorting LEC populations [560] and is currently used commercially for producing cultures of LECs. Podoplanin expression persists throughout the lymphatic network. However, one report suggests that there are endothelial cells with low podoplanin expression in precollectors that contribute to the migration of CCR10 + T cells [129].

2.7 Lymphatic networks in different organ systems

2.7.1 Small intestine

The lymphatics of the small intestine have roles in dietary absorption, tolerance of symbiotic microflora, and immunity against pathogens in the gut. Lymph flow from the small intestine dramatically increases during absorption of nutrients, compared to nonabsorbent phases between meals. The lumen of the small intestine contains millions of villi that extend up to 1 mm from the surface of the mucosa. These villi, in combination with microvilli that compose the brush border of individual enterocytes, dramatically increase the surface area for absorption in the small intestine. In each villus, just beneath the absorptive surface of enterocytes, is a dense network of villus capillaries surrounding a lymph lacteal. Each lacteal is a blind-ended initial lymphatic in most mammals, although an exception is the rat, which has 1–10 lacteals per villus. The lacteals play an important role in the transport and distribution of absorbed dietary lipids. Dietary triacylglycerol is digested in the gut lumen and absorbed by enterocytes in the form of free fatty acids and 2-monoacylglyceraol. Medium-chain fatty acids are absorbed by enterocytes and are transported away by the hepatic portal vein [130].

Short-chain fatty acids are absorbed by colonocytes and are metabolized. In contrast, long-chain fatty acids absorbed from the diet are transported away from the intestinal mucosa in the form of triacylglycerol by the lymphatic system. Inside enterocytes, absorbed long-chain free fatty acids and 2-monoacylglycerol are reesterified and packaged into chylomicrons in the endoplasmic reticulum. After further processing in the Golgi complex, mature chylomicron particles are released by exocytosis into the intercellular space of the lamina propria. Apolipoprotein (Apo) B48, which is packaged into chylomicrons, is selectively expressed in the small intestine, and appears to be important for the overall efficient delivery of lipids into the lacteals. From there, a rate-limiting step for chylomicron transport is crossing the basement membrane, prior to entry into the lacteals. While the mechanism is unclear, junctional disruptions occur between the enterocytes at the tips of jejunal villi during lipid absorption, and poor tight junction integrity has also been reported due to a chronic high-fat diet [131] (Fig. 2.6).

The mechanism for entry of chylomicrons into lymph lacteals appears to be a transcellular route through LECs. Reports from multiple histological studies identify chylomicrons inside LECs, and data from a recent functional investigation with cultured LECs support an active transport mechanism. This mechanism occurs during the enterocyte nutrient absorption-induced fictional hyperemia that helps establish favorable interstitial fluid pressures to drive the increase in lymph formation [132].

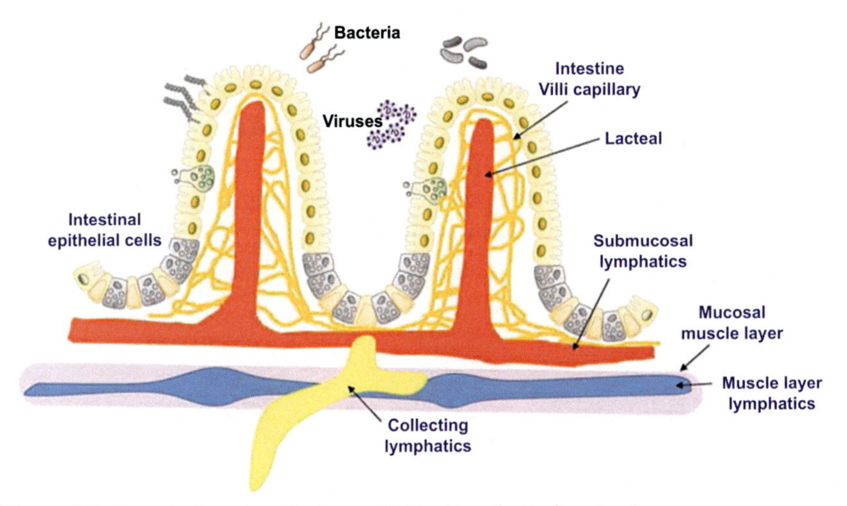

Figure 2.6 Lymphatic networks in small intestine (lacteal system).

SMCs parallel to and coming into contact with lacteals have also been shown. Their precise contribution is also undefined, but their function might be to continue to the piston-like contractions that have been described to drive flow of lymph out of the villi and into the submucosal lymphatics [133].

Several studies using genetically modified mice have generated results that highlight the importance of the normal development and function of the intestinal lymphatics to overall health. In Prox1 +/− mice, the lymphatic vessels become leaky, allowing chyle to accumulate and promote adipose growth. Likewise, in mice with an inducible deletion of T-Syn, important for synthesis of core-1-derived *O*-glycans in endothelial cells, misconnections between blood and lymphatic vessels develop, allowing chylomicrons to enter directly into the portal circulation, causing the development of fatty liver. From these studies, it is clear that disrupting the normal partitioning of dietary lipids out of the central circulation by intestinal lymphatics has an important impact on overall metabolism [134].

Another major function of the lymphatics and lymphoid tissues in the gut mucosa is immunity. The gut wall encounters a complex mixture of water and nutrients combined with both helpful and potentially harmful microbes. Moreover, the gut wall is a highly proliferative, constantly renewing layer of epithelium, and must be continuously monitored for transformed cells. Specialized areas named gut-associated lymphoid tissues perform these functions. In particular, Peyers patches in the small intestine sample contents from the gut lumen to discriminate between antigens and nonantigenic material. Specific enterocytes, M cells, and sample antigens, which are then processed in dendritic cells are presented to lymphocytes.

Data obtained from rat tissue show that lacteals are present in the villi projecting parafollicular regions. These lacteals drain into a submucosal plexus that is interconnected and form networks with many blind-ended lymphatics in the parafollicular areas, with many high endothelial venules in close proximity. These networks form the sinuses around each follicle. The perifollicular lymphatic sinuses then drain into submucosal collecting vessels [135].

2.7.2 Mammary glands

The lymphatic networks draining the mammary glands are extensive, complex, and intertwined with the outer dermal lymphatic networks. The mammary glands originate from the ectoderm and are situated between the outer dermis of the skin and the underlying fascia. The initial lymphatic networks of the mammary glands are found in the interlobular connective tissue surrounding the individual secretory alveoli. Unlike the blood capillary networks, the lymphatics do not penetrate into the alveolar lobules [136]. The distribution and ultrastructure of lymphatics in the rat mammary gland were compared between the virgin, pregnant, lactating, and postweaning periods. Lymphatic vessels were most abundant in the interlobular connective tissue during lactation, with few being found during other stages. Differences in ultrastructure included an increase in junctional gaps between endothelial cells during lactation, versus tighter junctions and a greater number and size of vehicles present in the endothelium during the virgin period [137] (Fig. 2.7).

In humans, the larger collecting lymphatics arising from the mammary glands follow the same paths as the arterial branches to the breast, namely the axillary and internal thoracic arteries, and to a much lesser extent the perforating branches of the intercostal arteries.

The mammary lymphatics participate in both physiological and pathological events in the mammary glands. For example, there is an increase in blood and lymphatic vessel abundance during lactation. The increased blood flow accommodates the increased exocrine function of the mammary gland, and also produces increased lymph flow. On the other hand, pathologic lymphangiogenesis is known to accompany breast tumors and

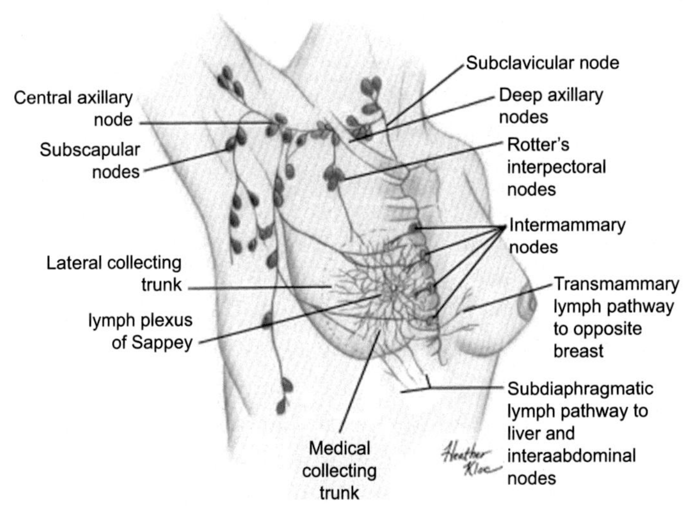

Figure 2.7 Lymphatic drainage channels of mammary glands.

facilitates metastasis. Understanding the molecular and physiological processes that promote lymphatic regression as the mammary gland transforms from lactation to a more dormant phase may help provide future therapeutic targets to combat breast cancers [137].

2.7.3 The skin

The skin protects the body's internal milieu from the outside environment. Dermal lymphatic networks play a critical role in this function. Findings from early investigations utilizing locally injected dyes in the skin by Tiechmann and Neumann, and later in 1908 by Unna, described the reticular networks of dermal lymphatics. Results from these early studies identified a polygonal initial lymphatic network just below the most superficial capillaries in the dermis, and a deeper plexus of lymphatics that included intraluminal valves (Fig. 2.8).

A two-dimensional polygonal network of initial lymphatics is apparent in the skin of humans and mice. In the dorsal skin of the human foot, this network was found between the papillary and reticular layers of the dermis. The vessels found in this network lack valves, are individually 10–30 nm in diameter but can reach 70-nm diameters at connecting nodes, and the width of the polygons of the network are 400–600 nm. In a recent study using human breast skin, the tips of initial lymphatics were found to appear starting at 25 nm beneath the surface of the dermoepidermal junction, connecting to the networks just beneath the superficial capillary networks of the papillary layer of the dermis [138]. Precollectors arise from the initial lymphatic network and travel more deeply into the reticular layer of the dermis, forming a three-dimensional network. These vessels have one-way valves to prevent retrograde flow and connect to collecting lymphatics that either appear in the dermis beneath the dermoepidermal junction or in the subcutaneous layer. The deeper into the dermis, the lymphatic vessels in these networks become larger in diameter, and the density of the networks decreases. In addition, there is a distinct loss of Lyve1 labeling on LECs, and the appearance of a smooth muscle layer at the point where collecting lymphatics appear in the deep dermis. While the initial lymphatics of the skin do not appear to be associated with arterioles, the larger collecting lymphatics that drain these networks were observed to follow alongside arteries in the rabbit ear [139].

The skin is rich in immune cells that serve to protect against potentially harmful antigens or pathogens that may infect the skin. Upon activation, these immune cells may enter lymphatics and migrate to lymph

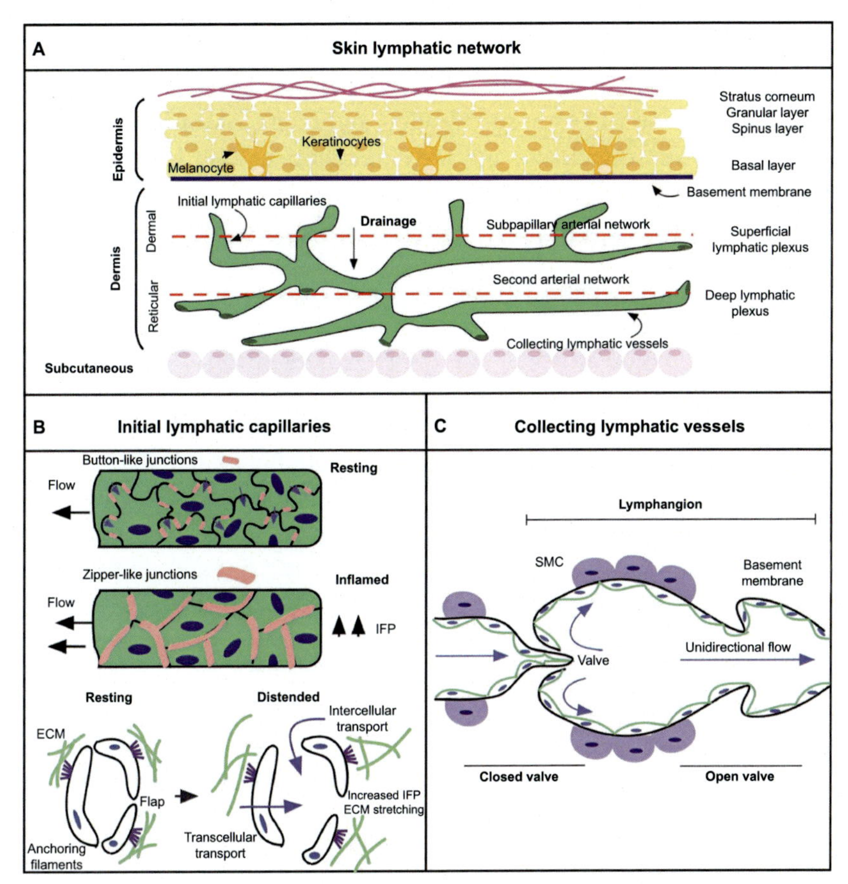

Figure 2.8 Lymphatic networks of skin.

nodes. A recent study utilizing whole-mount dermal sheet microscopy assessed the architecture of immune cell integration with lymphatic vessel networks [140]. From these results, it was estimated that in the first 30 nm below the dermoepidermal junction, for an average human with 1.8 m^2 of skin surface area, there are approximately 7.62×10^8 dendritic cells and 2.08×10^9 T cells. Throughout the dermis, dendritic cells, T cells, and macrophages were either located close to blood vessels or in the interstitial spaces (defined in this study as >15 nm from a vessel), but did not have a particular pattern in relation to lymphatic vessels. Despite the lack of an appearance of a pattern, dendritic cells functionally migrate toward, and enter, initial lymphatics, while marcrophages remain largely resident in the skin [141].

2.8 Development of lymphatic vessel networks

The formation of the lymphatic vasculature includes four important aspects: origination of LECs, specification and migration of LECs, maturation of lymphatic vasculature and SMC recruitment, and lymphatic valve formation [142].

2.9 Origination of lymphatic endothelial cell

Unlike the blood vasculature, the lymphatic vasculature is a translucent vasculature in the body, which until recently made it relatively difficult to study. Research on finding the origin of the developing lymphatic vasculature started in the 17th century. At the beginning of the 20th century, Florence Sabin first proposed that the lymphatic vasculature arises from the cardinal veins (CVs) from her research on pig embryos. Much later, with the transcription vasculature development, lineage tracing using Prox1 reporter mice revealed that the venous endothelial cells in the CVs did indeed rise to LECs. Similar results were seen in zebrafish embryos using live imaging techniques [143].

The formation of the mouse lymphatic vasculature occurs very quickly during development. From embryonic day (E) 9.5, when the specification of Prox1 + LEC progenitors first occurs in the CV, to E14.5 when the entire skin of the mouse embryos is covered by the lymphatic vasculature, the specification, proliferation, and migration of millions of LECs need to take place in a short period of time. In the past 5 years, this question has been investigated in many different ways. First, the intersomitic veins (ISVs) and the superficial venous plexus were identified to be an additional source of LEC progenitors in mouse embryos [144]. Prox1 + LEC progenitors are specified not only in the CV but also in the ISVs and the superficial venous plexus. These findings are consistent with the previous observations that LEC progenitors are derived from the veins. More significantly, other studies have discovered a nonvenous origin for some parts of the lymphatic vasculature [145].

For example, specialized angioblasts can differentiate into LECs within the CV in zebrafish embryos. Although George Huntington and Charles McClure suggested a nonvenous origin for at least some LECs in 1910, it was the first time that mesenchymal cells were shown to contribute to the formation of the lymphatic vasculature during embryonic development. Meanwhile, in the mouse dermal lymphatic vasculature, while the lymphatic vasculature of the cervical and thoracic skin is still derived from the veins, the lymphatic vessels of the lumbar and dorsal midline skin were found to form from nonvenous-derived progenitors [146].

Over the years, the diversity of the lymphatic vasculature in different tissues has drawn significant attention. In addition to the tissues that are well known to have a lymphatic vasculature, such as the skin and the mesentery, lymphatic vessels have been discovered in other tissues previously thought to be immune privileged, including the brain and eyes. For instance, while the main source of cardiac LEC is from extracardiac venous endothelium, the yolk sac hemogenic endothelium also contributes to a part of the cardiac lymphatic vasculature [133].

2.10 Specification and migration of lymphatic endothelial cells

The stepwise model of how the lymphatic vasculature forms with regard to the main genes that regulate LEC specification, budding, and migration during mouse embryonic development is well established and discussed in multiple reviews. Briefly, Prox1 + LEC progenitors are specified in the CVs and ISVs at E9.5–E9.75 and the majority of these cells bud off from the veins right after this specification. Immediately after LEC progenitors bud off the veins, they differentiate into mature LECs expressing an additional LEC protein podoplanin (Pdpn) [147]. These Prox1 + Pdpn + LECs proliferate, migrate, and assemble into a sac-like structure along the anterior and posterior axes of the embryo on the dorsal half of the CVs, called a lymph sac. Several reviews have summarized the function of the key factors that are involved in LEC specification including extracellular signal-regulated kinase (ERK), Sox18, CoupTFII, Prox1, Flt4, and Notch. ERK signaling activates the expression of SRY-box transcription factor 18 (Sox18) in the CVs, and then Sox18 and chicken ovalbumin upstream promoter transcription factor II (CoupTFII) both bind to the promoter region of Prox1 to activate its expression. The expression of Prox1 defines the specification of LECs and Prox1 activity is required in maintaining LEC identity throughout life. In addition to Sox18, CoupTFII, and Prox1, another main player in maintaining LEC identity is Flt4, which encodes Vegfr3 [148] (Fig. 2.9).

Flt4 and Prox1 interact with each other in a feedback loop, wherein Prox1 interact with each other in a feedback loop wherein Prox1 binds to the Flt4 promoter to maintain its expression and Flt4 is also necessary for

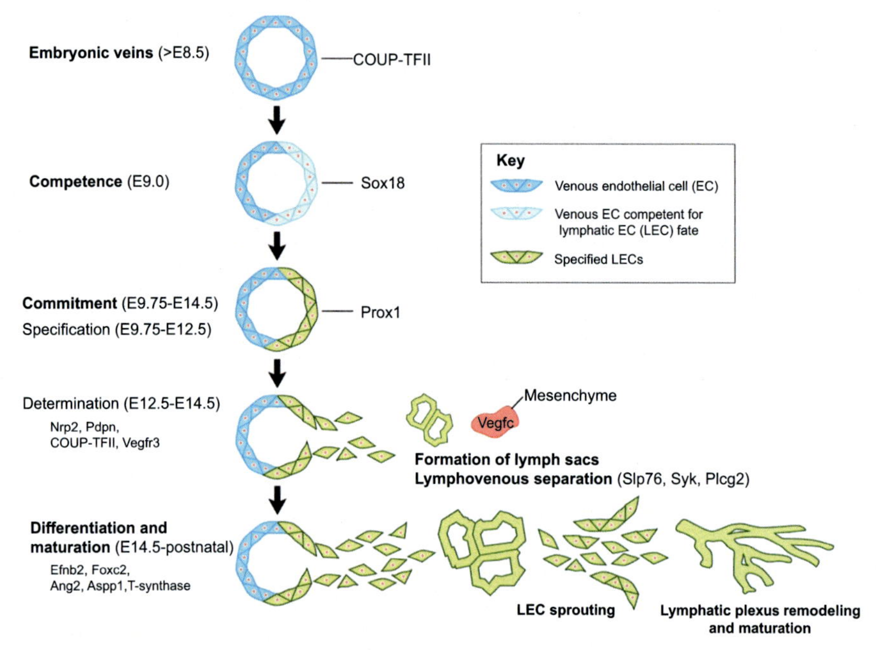

Figure 2.9 Lymphatic plexus remodelling and maturation.

maintaining the expression of Prox1, thus helping to maintain LEC identity. Contrary to Flt4, Notch1 functions as a negative regulator of LEC specification. Decreased Notch activity results in an increased number of LEC progenitors in the veins, which leads to malformation of the lymphatic vasculature. Moreover, the expression of Prox1 is modified by microRNAs. MiR-181a and miR-31 negatively regulate Prox1 expression in LECs [149]. Intriguingly, Wnt5b was recently found to induce LEC specification through β-catenin signaling in the zebrafish. On the other hand, β-catenin signaling is not critical for the specification of LEC progenitors during mouse embryonic development, but is essential for lymphatic vascular morphogenesis [149].

It is critical for LECs to bud off the veins and migrate to form the lymph sacs and this process requires Vegfc/Vegfr3 signaling. This signaling pathway has also been discovered to be crucial for mesenteric lymphatic vessel formation, intestinal lymphatic vessel (lacteal) formation, cardiac lymphatic vessel formation, central nervous system lymphatic vessel formation, and Schlemm's canal formation in the eyes [150]. Besides Vegfc/Vegfr3 signaling, many other signaling pathways affect LEC migration, such as the adrenomedullin signaling pathway, angiopoietin 2 (Angpt2)—Tyrosine kinase with immunoglobulin-like and EGF-like domains 1 (Tie1) (Angpt2 receptor) signaling pathway, PU.1 in the macrophages, Nfatc1 signaling and transcription factor Gata2-modified signaling [151]. Loss of adrenomedullin signaling results in hypoplastic lymph sacs, which indicates a defect of LEC migration. Lately, it has been found that the dosage and signaling of adrenomedullin signaling in the LECs leads to enlarged blood-filled lymphatic sacs and dilated dermal lymphatic vessels [150].

The metabolism of blood endothelial cells has been studied for many years in health and cancers. A recent study focused on how the metabolism of LECs affects the formation of this vasculature. Deletion of CPT1A, a rate-controlling enzyme in fatty acid β-oxidation in the LECs, leads to impaired lymphatic morphogenesis. Mechanistically, Prox1 directly upregulates CPT1A expression, which increases acetyl coenzyme A production. Acetyl coenzyme A is used by the histone acetyltransferase p300 to acetylate histones at lymphangiogenic genes, such as Vegfr3 to regulate LEC migration and identity maintenance [152] (Fig. 2.10).

As mentioned previously, Notch signaling is vital for the specification of LECs in the veins. It also regulates lymphatic vessel morphogenesis. LEC-specific deletion of Notch1 resulted in substantial lymphatic overgrowth with dilated lymphatic vessels. Increased cell proliferation and filopodia formation, and decreased cell death of LECs were seen in the Notch1 mutant mice. Planar cell polarity (PCP) signaling plays a significant role in many developmental processes [153]. The components of PCP signaling pathway are involved in the development of lymphatics also. Polycystin 1 (pkd1) is a newly identified gene that regulates lymphangiogenesis in two back-to-back studies. During lymphatic vessel formation, pkd1 is not involved in the budding of LECs off the veins, but controls LEC migration during vessel formation by regulating cell polarity. Loss of pkd1 leads to random migration of LECs and reduced lymphatic vessel identity [154].

2.11 Maturation of lymphatic vasculature and smooth muscle cell recruitment

As previously mentioned, the two general types of lymphatic vessels are the collecting lymphatic vessels and lymphatic capillaries, which have distinct morphology and function. The collecting lymphatic vessels are lined by SMCs and these SMCs are responsible for the contraction of the collecting lymphatic vessels to push the lymph forward. The collecting lymphatic vessels also have intraluminal valves to prevent lymph backflow.

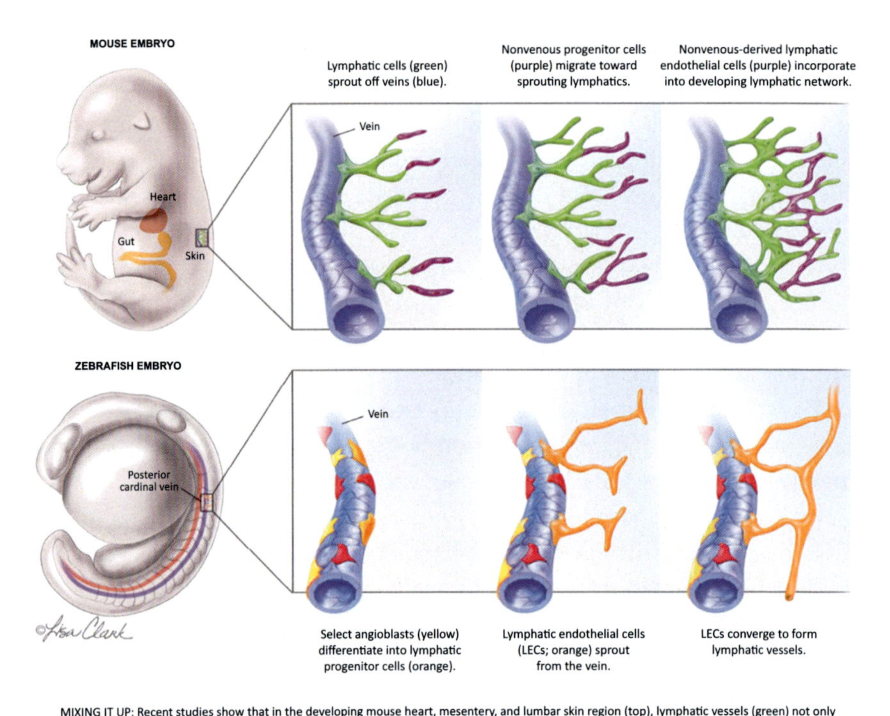

Figure 2.10 Lymphangiogenesis in embryo.

The collecting lymphatic vessels and lymphatic capillaries possess different types of cell–cell junctions [121]. The collecting lymphatic vessels have continuous zipper-like junctions similar to blood vessel endothelium, whereas the capillaries have discontinuous button-like junctions visualized by staining for the junction molecule VE-cadherin (Cdh5). Buttons on the lymphatic capillaries are considered to act as primary valves for fluid and cell entry into the lymphatic vessels. Angiopoietin 2 (Angpt2) is indispensable for the maturation of cell–cell junctions in the lymphatic capillaries since genetic deletion of Angpt2 suppresses the zipper-to-button junctional transformation [155] (Fig. 2.11).

More recent studies have been conducted on the maturation of collecting lymphatic vessels. Downregulation of the capillary genes in the collecting lymphatic vessels, SMC recruitment, and valve formation are important steps in the maturation of collecting lymphatic vessels. Foxc2/calcineurin/Nfatc1 signaling is a well-studied signaling pathway that is essential for the maturation of collecting lymphatic vessels. Foxc2 and

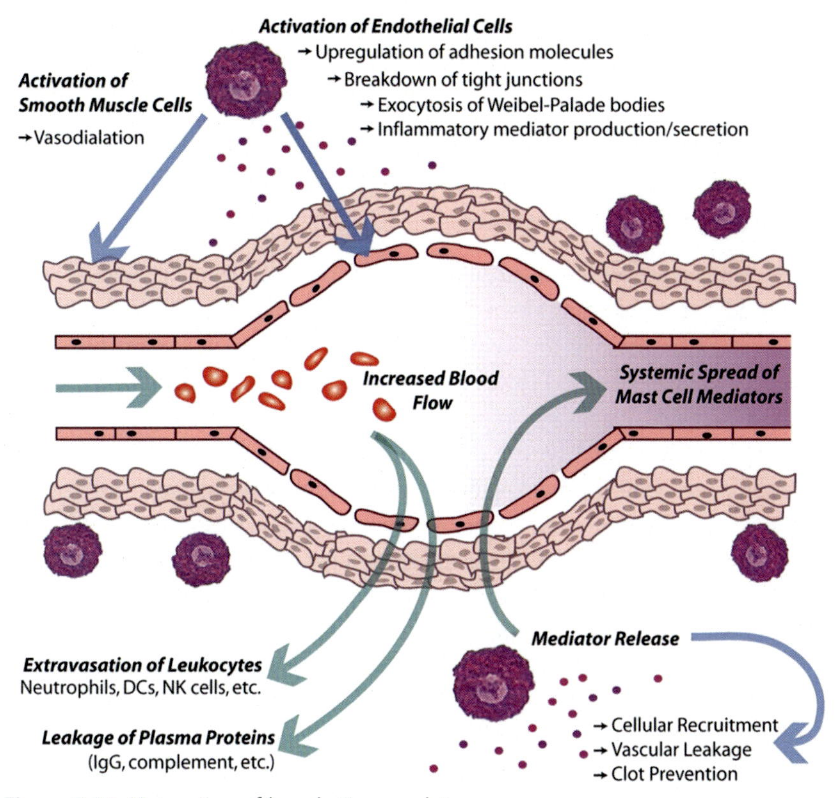

Figure 2.11 Maturation of lymphatic vasculature.

Nfatc1 cooperatively bind to the regulatory elements of lymphatic genes to control their expression [156]. In the collecting lymphatic vessel-rich mesentery, loss of Foxc2 leads to the retention of high expression levels of genes prox1, Flt4, and Lyve1 in the primitive lymphatic vessel plexus and these lymphatic vessels have excessive SMC coverage. Platelet-derived growth factor receptor-β (pdgfrb) is essential for the recruitment of pericytes by developing blood capillaries. Foxc2 knockout collecting lymphatic vessels exhibit unregulated expression of Pdgfrb, and these vessels are not able to differentiate into functional collecting lymphatic vessels and lack valves, indicated by the backflow of lymph [142].

2.12 Lymphatic valve formation

Lymphatic valves are an indispensable part of the lymphatic vasculature. They are critical for maintaining lymph transport in one direction. They are comprised of two valve leaflets, each with an ECM core that is lined by a layer of LECs on each side of the leaflets. Several signaling processes are important for the formation of the lymphatic valves, including transcription factors, junction proteins, ECM, and axonal guidance genes [157].

2.13 Lymphovenous valves

Lymphovenous valves are specialized valves where the thoracic duct empties into the blood vasculature at the subclavian veins. During lymph sac formation, most LECs bud off the veins and differentiate into more mature LECs to form lymphatic vessels. However, a scanty amount of LECs remain in the CVs and become the leaflets of the lymphovenous valves that make a junction between the jugular lymph sac and the jugular and subclavian veins during embryonic development [158]. One lymphovenous valve is located between the subclavian and external jugular veins, and another lymphovenous valve is located between the external jugular vein and internal jugular vein. The return of the lymph fluid to the blood circulation and also the restriction of blood flow back to the lymphatic vasculature are maintained by lymphovenous valves. When the lymphovenous valves are missing or malformed, blood-filled lymphatic vessels result. It has been reported that platelets are required for the separation of the lymph sac from the veins at their connecting points—the lymphovenous valves [159].

The binding between lymphovenous valve LECs expressing podoplanin and the platelet receptor Clec2 is crucial for the initiation of platelet aggregation. In Clec2-deficient mice, the missing fibrin-containing platelet thrombi at the lymphovenous valves leads to blood-filled lymphatic vessels. Furthermore, platelet thrombi are much larger in animals that lack lymphovenous valves and lymphatic valves, indicating that platelet thrombi can compensate for the loss of lymphovenous and lymphatic valves to maintain blood–lymphatic separation. Similar to the formation of collecting lymphatic valves, the formation of lymphovenous valves contains three main transformation steps of the LECs: delamination (E12.0), aggregation (E12.5), and maturation (E14.5–E16.5) [160]. Genetically, the main genes that are important for the formation of collecting lymphatic valves are also critical for lymphovenous valve formation (Fig. 2.12).

The transcription factors Prox1, Foxc2, and GATA binding factor 2 (GATA2) are highly expressed in the lymphovenous valves in addition to the lymphatic valves. Prox1 heterozygous animals develop one lymphovenous valve. Foxc2 heterozygous animals have a variable phenotype. Half of the

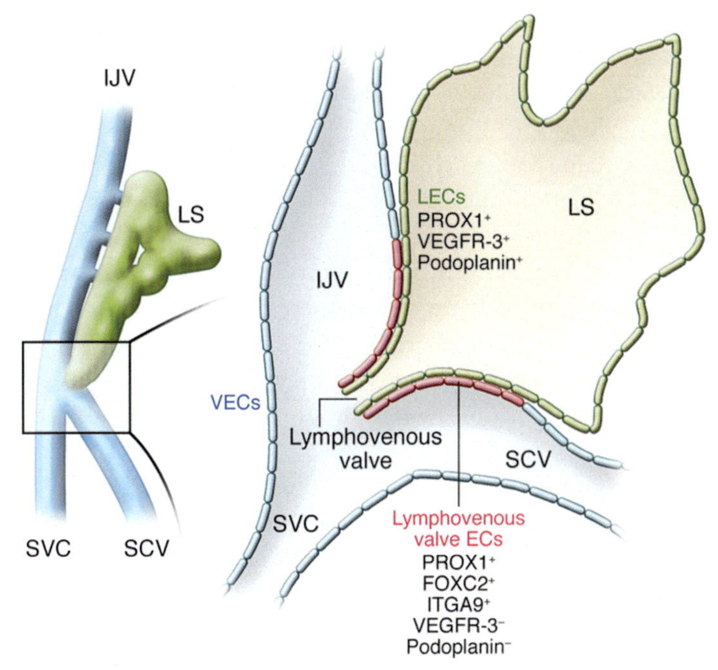

Figure 2.12 Lymphovenous valves.

animals develop one lymphovenous valve instead of two valves [161]. In addition, in Gata2 knockout embryos, properly differentiated lymphovenous valves are absent. Inactivation of Ephb4 in LECs leads to defective lymphovenous valve formation and consequent subcutaneous edema. As a main component of the valve, ECM also plays a vital role in lymphovenous valve formation. It was mentioned previously that lytic transglycosylases 9 (LTGA-9)-deficient mice develop defective lymphatic valves. However, the lymphovenous valves have been reported as normal in these mice. Instant Ltga5 mutants display defects in the formation of the lymphovenous valves. These mice develop dilated, blood-filled lymphatic vessels and the lymphatic capillaries are ectopically covered with SMCs [162].

2.14 Lymphatic development and diseases

Primary lymphedema is a major lymphatic disease in humans. Mutations in the LEC specification genes SOX18 have been identified as the cause of hypotrichosis–lymphedema–telangiectasia in humans. In the LEC specification-signaling pathway, mutations in RAF1 constantly activate MAPK/ERK signaling and are associated with Noonan syndrome, a disease that includes lymphangiectasia. Heterozygous mutations in human FLT4 that affect the VEGFR3 tyrosine kinase domains lead to lymphatic vascular defects and primary lymphedema (Milroy disease). VEGF-C/ VEGFR3 signaling acts through the Akt (protein kinase B) pathway [163]. Akt is required for lymphatic vasculature formation and valve development. Mutations in class 1 phosphoinositide 3-kinases (Pl3k) that lead to AKT hyperphosphorylation are associated with lymphatic malformations (LM) in humans. LEC migratory factor CCBE1 mutation in humans causes a type of lymphatic dysplasia known as Hennekam syndrome. Point mutations in human FOXC2 are associated with lymphedema—distichiasis (LD) syndrome, in which lymphatic valves are defective. Mutations in valve formation transcription factor GATA2 cause Emberger syndrome, myelodysplastic syndrome, acute myeloid leukemia, and "MonoMAC" syndrome with primary lymphedema [164].

Gap junction protein connexion-47 (Cx47) is highly expressed in the lymphatic valves. Mutations in CX47 are linked to primary lymphedema in humans. An inactivating mutation in the PCP pathway gene CELSR1 was recently found to be associated with hereditary lymphedema. Mutations that cause the loss of tyrosine kinase activity of EPHB4 (crucial for valve formation) are associated with an autosomal-dominant inherited form of lymphatic-related (nonimmune) hydrops fetalis (LRHF) [165]. Similarly, congenital chylothorax in human fetuses is caused by mutations in the valve formation gene ITGA9. A nonreceptor tyrosine phosphatase (Ptpn 14) can interact with Vegfr3 and affect lymphangiogenesis. Mutations in PTPN14 are linked to an autosomal-recessive lymphedema phenotype in humans. In humans, RAS P21 protein activator 1 (RASA1) mutations cause capillary malformation—arteriovenous malformation. More recently, homozygous and compound heterozygous mutations in Piezo type mechanosensitive ion channel component 1 (PIEZO1) (encodes a mechanically activated ion channel) resulted in an autosomal-recessive form of generalized lymphatic dysplasia with a high incidence of nonimmune hydrops fetalis and childhood onset of facial and four-limb lymphedema, which indicates the potential function of PIEZO1 in lymphatic development [166].

2.15 Lymphangiogenesis and relationships with angiogenesis during microvascular network remodeling

Lymphatic network growth and dysfunction associated with lymphedema, inflammation, cancer metastasis, and other pathological conditions highlight the need to better understand the dynamic nature of lymphatic network structure at the vessel, cellular, and molecular levels. In recent years, whole-mount tissue models have provided an insight into the organization of lymphatic networks and the mechanisms involved in lymphatic vessel growth, that is, lymphangiogenesis. Highlighting the functional importance of lymphatic networks on tissue homeostasis and immune cell trafficking, the design of molecular-based therapies aimed at manipulating the microcirculation requires understanding the relationships between lymphangiogenesis and angiogenesis [167].

2.16 Lymphangiogenesis

Lymphangiogenesis is the process of lymphatic growth during which new initial lymphatic vessels are formed from preexisting vessels. While modes of lymphangiogenesis might include intussusception (i.e., vessel splitting) or other less characterized dynamics, such as endothelial cell migration and reconnection, the relative contributions of these modes to network growth are unclear. Currently, LEC sprouting is the best described mode of growth. Lymphatic capillaries first form fine filopodia which proceed to sprout formation, similar to those in the early process of angiogenesis from existing blood vessels [151]. These sprouts proliferate and extend to form new mature, blind-ended lymphatic vessels. Current evidence suggests that lymphatic sprouting is primarily mediated through VEGFR3 expressed by LECs. VEGFR3 is a tyrosine kinase receptor activated by VEGF-C and VEGF-D. VEGF-C and D also bind to neuropilin-2 (Nrp2), a transmembrane receptor that plays a role in the developmental axon guidance. Nrp2 is thought to act as a coreceptor for VEGFR3 and a mediator of lymphatic sprouting. VEGF-C has also been shown to stimulate angiogenesis by activating VEGFR2 [151] (Fig. 2.13).

2.17 Relationships between lymphangiogenesis and angiogenesis

Analysis of microvascular networks using intravital microscopy and contrast media have historically suggested that lymphatic networks remain distinct from blood microvascular networks in adult tissues. The concept of lymphatic/blood vessel mispatterning is further supported by the phenotypic similarities between lymphatic and blood endothelial cells during quiescent and certain pathological conditions. More recent observations

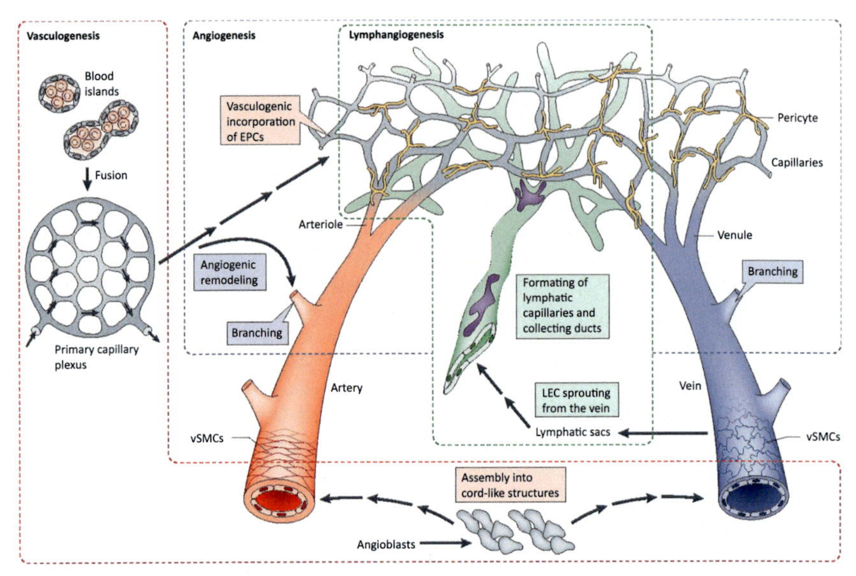

Figure 2.13 Lymphangiogenesis.

suggest that endothelial cells are in fact capable of forming connections, defined by the submicron continuous junctional labeling of endothelial cell adhesion molecules, between lymphatic and blood vessels at the capillary level in adult rat microvascular networks [168].

The attached coordination between blood and lymphatic vessels starts during development. As the blood vascular system is developing, a subpopulation of venous endothelial cells begins expressing the lymphatic-specific transcription factor Prox1 and buds off from the CV, forming the first lymph sacs that later mature into the lymphatic vascular system. Evidence for this was given by Benest et al., who demonstrated that the presence of lymphatic vessels influenced VEGF-C-induced angiogenesis. Additionally, Nakao et al. suggested that VEGFR-2 expressed by angiogenic blood vessels impeded lymphangiogenesis by sequestering VEGF-C. Indeed, angiogenic molecular signals such as VEGFs and the angiopoietins play similar roles in lymphangiogenesis. The intracellular domain of Ephrin B2 was found to be critical for mouse lymphatic vessel patterning, suggesting that ephrins also represent a patterning cue shared by both the lymphatic and microvascular systems [145]. Ephrin B2 and its ligand EphB4 are important regulators of arterial and venous cell fate during embryonic development. In adults, bidirectional signaling has been linked to vascular cell guidance and recruitment of perivascular support cells, yet the roles of these molecules in vessel identity remain unclear. Nitric oxide has been shown to permit hypoxia-induced lymphatic blood perfusion in zebrafish at sites of arterial lymphatic anastomoses. The findings of this particular study implicate a molecular cue that is sufficient to induce conductance between the lymphatic and blood microvasculature. Increased nitric oxide production has also been directly linked to lymphatic metastasis and implicated nitric oxide as a mediator of cell flux into lymphatic vessels [169]. Lymphatic—venous connections at the large vessel level outside the entry points to the subclavian and thoracic ducts have been reported, particularly in situations of increased pressure due to vessel occlusion. Lymphatic—venous anastomoses at the large vessel level were also reported in the heart 7 and 14 days after lymphatic occlusion. Endothelial cell phenotypic identity also provides an example of the interrelationships between blood and lymphatic systems. Immunolabeling of VEGFR3, LYVE1, and PECAM offers evidence for both the similarity and delineation of lymphatic and blood endothelial cell types [170] (Fig. 2.14).

During development, LECs originate and bud off the venous system. Macrophages have recently been implicated as paracrine regulators of both angiogenesis and lymphangiogenesis. Their phenotypic overlap also highlights the emerging area of research focused on LEC lineage during lymphangiogenesis in adult tissues and the potential for macrophages or circulating progenitor cells to differentiate into both blood and LECs. Undoubtedly, the advances in our understanding of macrophages as a cell regulator and source for angiogenesis and lymphangiogenesis implicates the potential for new discoveries related to endothelial cell plasticity and the potential for blood to LEC phenotype plasticity, a concept that has already been proven possible in vitro [171].

2.18 Lymphangions and the lymphatic pump mechanism

In a standing human being, for collecting lymphatics originating in the feet to deliver the lymph to the great veins in the upper thorax, they must overcome a significant hydrostatic pressure gradient. This is overcome by serial organization of individual lymphangiogenesis, the key collecting lymphatic pumping units, dividing the overall pressure gradient into a series of many smaller steps (Fig. 2.15).

This was initially shown with measurements of pressures within lymphatic networks of exteriorized mesentery, which revealed that with each

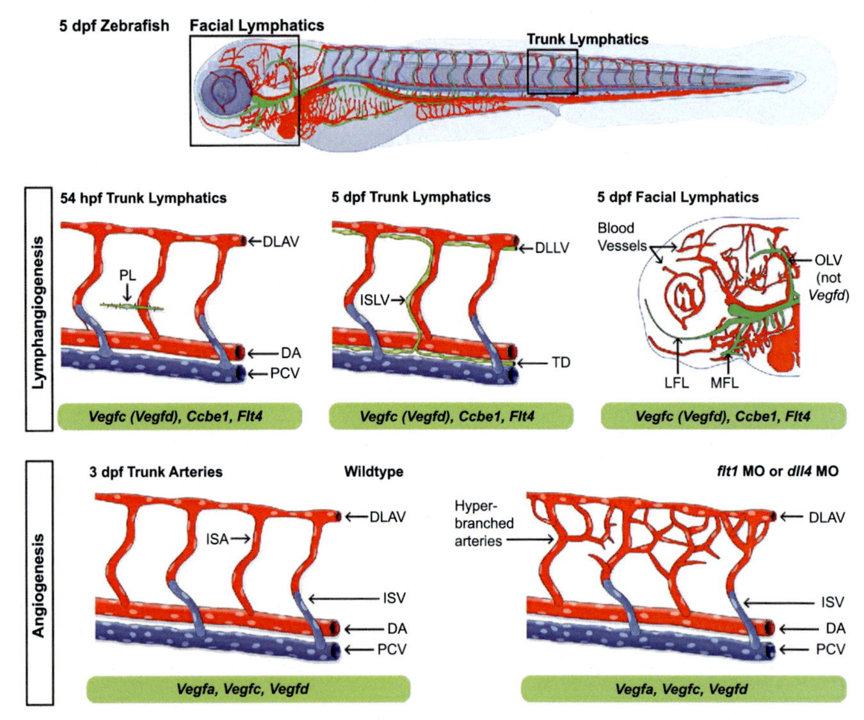

Figure 2.14 Lymphangiogenesis and angiogenesis.

passage across a secondary valve to a downstream segment, intraluminal pressure increases gradually. This concept was later confirmed in the rat tail with noninvasive near-infrared imaging and computational modeling [894]. The secondary valves between lymphangions prevent backflow of lymph. A constriction of a lymphangion sufficient to open the downstream valve will move lymph forward to the downstream lymphangion. The pumping process controlled by lymphangions represents the primary mechanism to propel lymph back to the central circulation [139].

2.19 Lymphatic smooth muscle

The smooth muscle layer of lymphatic vessels, often simply referred to as lymphatic muscle due to its unique properties, generates the lymphatic contractile cycle required for normal lymph flow. This cycle can be characterized in a similar fashion as the cardiac cycle, with systolic and diastolic phases. Lymphangion systole is produced by periodic phasic contractions of the lymphatic muscle. During diastole, the lymphatic muscle does not

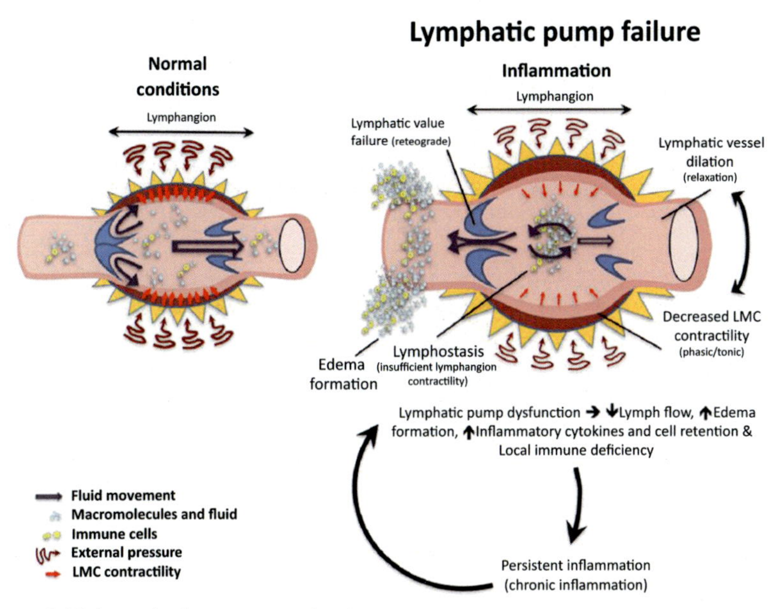

Figure 2.15 Lymphatic pump mechanism.

completely relax, maintaining a certain degree of tone. Lymphatic muscle function has been characterized by measurements of contractile force to determine length–tension relationships, or vessel diameter to study pump action. Because lymphatic muscle has functional properties resembling both cardiac and vascular smooth muscle, the parameters typically used to describe both cardiac and vascular smooth muscle function in arteries and arterioles have been adopted to describe lymphangion function. The parameters analogous to cardiac parameters include the phasic contraction frequency, end-diastolic diameter, end-systolic diameter, and others derived from these measures [121].

2.20 Filtration at lymph nodes

With few exceptions, lymph is delivered by collecting lymphatics to at least one lymph node prior to entry into the thoracic duct. Here, immune cells can sample the lymph contents for antigens. This biological filtration through phagocytosis and retention of certain cells in the nodes causes postnodal lymph to be dramatically different from prenodal lymph in terms of cell types present. In prenodal lymph, monocytes, macrophages, and dendritic cells are often present, but these are largely absent in

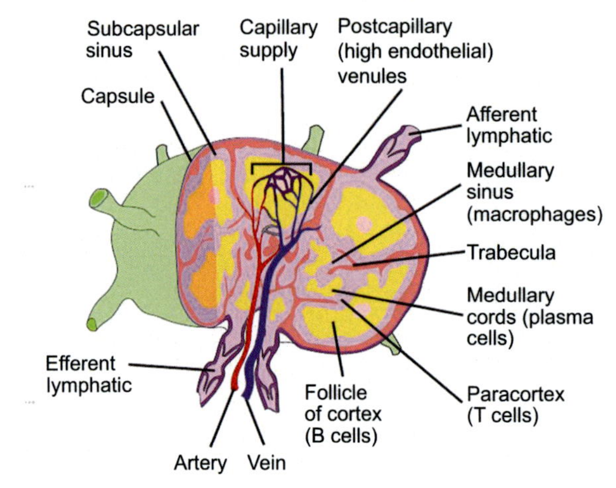

Figure 2.16 Filtration at lymph nodes.

postnodal lymph. In contrast, lymphocyte counts can be higher in postnodal lymph due to exiting from the lymph nodes into the postnodal lymphatic vessels (Fig. 2.16).

In addition, lymph nodes also mechanically filter the lymph. The blood–lymph barrier within the nodes allows passage of protein-free fluid according to the Starling forces. The result is that fluid is lost from the prenodal lymph into the blood, which increases the protein concentration to be higher in postnodal lymph than in prenodal lymph [127].

2.21 Lymphedema

The lymphatic vessels are essential for regulating tissue fluid balance and, as a result, failure of lymphatic networks to sufficiently perform this function manifests in lymphedema. Lymphedema is categorized as primary if it results from inherited genetic mutations and secondary if it results from trauma, such as surgery, radiation, or obstruction (e.g., filarial parasites). While lymphedema is not typically life threatening, it is a lifelong disease that, depending upon its severity, can feature a range of mobility problems, pain, and susceptibility to skin infections. Treatment is through compression garments and therapy. Because the disease is lifelong, management typically needs a close physician–patient relationship [172]. Classifications of primary lymphedema are traditionally based upon clinical features, stage of life at which it presents (birth, puberty, or adulthood), and lymphoscintigraphy findings. As mentioned earlier in the section on

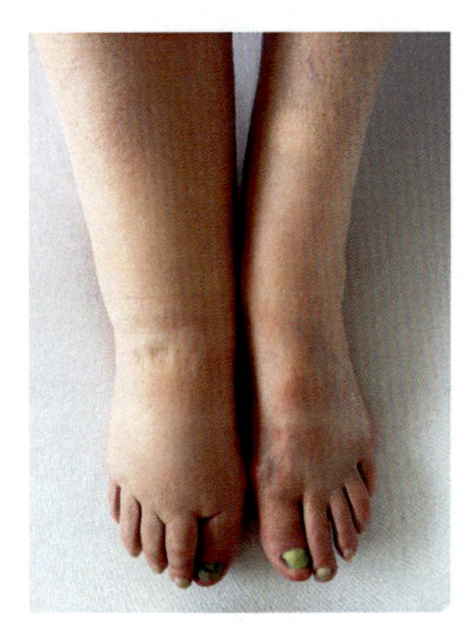

Figure 2.17 Lymphedema.

lymphatic development, many of the gene mutations involved in lymphedema contribute to the development of valves or lymphatic endothelial integrity. In the case of Milroy disease, caused by mutations of either VEGFR3 or its agonist VEGF-C, impaired fluid absorption in the initial lymphatics was proposed as a disease mechanism. However, more recent findings that VEGFR3 contributes to lymphatic valve formation and function suggest that valve dysfunction may also be a critical factor [173] (Fig. 2.17).

In LD syndrome, caused by mutation of FOXC2, patients present with aberrant eyelashes at birth, but the onset of lymphedema typically occurs at puberty or later in life, sometimes as late as over 50 years of age. The mutation affects both lymphatic and venous valves and can also cause abnormal appearance of smooth muscle around initial lymphatics. Mutations of these genes and the several others associated with other types primary lymphedema point to the critical physiological function of lymphatic valves for maintaining a normal "safety factor" against the development of tissue edema. Secondary lymphedema is much more frequent than primary lymphedema, and is commonly observed after removal of axillary lymph nodes in association with breast cancer surgery. The result is severe arm swelling that develops months to years after the operation. The onset and degree of secondary lymphedema in breast cancer patients who have undergone surgery is highly variable, and the pathophysiology is an intense area of investigation [174]. One major risk factor for the development of secondary lymphedema is obesity. Genetic risk factors are only beginning to be investigated. The same genes involved in primary lymphedema, that is, those involved in lymphatic valve development and maintenance, may contribute to susceptibility. To date, gene variants of FOXC2 and FLT4 (encodes VEGFR3) have been identified in lymphatic filariasis, the most common form of secondary lymphedema worldwide, and caused by infection with filarial worms, primarily *Wuchereria bancrofti*. In addition, SNPs in the genes that encode VEGFR3, MMP-2, and carcinoembryonic antigen-related cell adhesion molecule-1 (CEACAM-1), all of which have roles in lymphangiogenesis, have also been identified as having significance in lymphatic filariasis [175].

Several animal models of secondary lymphedema have been developed to better understand the pathophysiological changes in lymphadenomatous tissues. In earlier studies, secondary lymphedema was produced surgically in the dog and rat hindlimb or rabbit ear. These models involve excision of skin and subcutaneous tissue to disrupt lymph flow, and can cause significant tissue volume increases. While the larger animal models may have more relevance to human anatomy, mouse models offer the ability to utilize genetic approaches not readily available in other species. Thus, the skin excision approach has also been widely used with the proximal mouse tail, producing significant edema, lymphatic hyperplasia, and elevated expression of VEGF-C. This model has remained quite popular and has recently been used in efforts to ameliorate lymphedema through antiinflammatory approaches [176].

2.22 Obesity, lymphatics, and lymphedema

Obesity is an epidemic of the 21st century, with worldwide prevalence vastly increasing and most of the adult population being either overweight or obese. There is growing evidence of a reciprocal connection between lymphatic dysfunction and obesity. Obesity has long been known to be a risk factor for secondary lymphedema, but more recently, it has also been shown that extreme obesity can cause lymphedema in the absence of other risk factors. There is also increasing clinical evidence to suggest that obese patients are predisposed to secondary lymphedema and primary lymphedema can induce adult-onset obesity [177]. Animal models also show a connection between obesity and lymphatic dysfunction. Obese mice were shown to have impaired lymphatic transport, and decreased dendritic cell migration to lymph nodes. Additionally, in a mouse model of lymphedema obese mice were shown to have impaired lymphatic function associated with increased inflammation, fibrosis, and adipose deposition. Obesity-induced inflammation likely begins in the fat cells themselves [178].

Lymph is a fluid rich in emulsified lipids and fat deposition develops first along the lymphatic structures, and all lymph nodes are embedded in adipose tissue. However, the casual relationship between fat deposition and lymph remains unclear. A disruption or malfunction in lymphatic drainage, that leads to lymphedema and increased interstitial fluid that is chronic, results in a predisposition to increased fibroblasts, adipocytes, and keratinocytes in the tissue. Accumulation of this fluid leads to fibrosis and decreased oxygen tension and macrophage function. With lymphedema, the lymphatic stasis initiates a cycle of inflammation—progressive tissue fibrosis—worsening lymphatic function, which over time leads to end organ failure of the lymphatic system [134].

2.23 Gastrointestinal lymphatics and inflammation

Inflammatory bowel diseases include ulcerative colitis, an inflammatory disorder along the length of the colonocyte layer of the colon, and Crohn disease, which can affect any part of the gastrointestinal (GI) tract and involves inflammation of multiple tissue layers. Increased lymphatic vessel numbers, along with lymphangiectasia, are observed in both ulcerative colitis and in ileal and colonic Crohn disease. Lymphocytic and granulomatous lymphangitis of the gut lymphatics was one of the earliest noted pathologies in Crohn disease. The leakage of lipids from damaged lymphatics was proposed to act as a chronic irritant that perpetuated the inflammation. Such leakage may also be connected to the expansion of normal mesenteric fat beyond the mesentery and onto the intestinal wall, known as creeping fat and a hallmark of Crohn disease-affected areas [179] (Fig. 2.18).

Another important role of GI-derived lymph is its role in the development of systemic inflammation following traumatic injury. It has long been known that cannulating the thoracic duct and diverting its lymph out of the body can cause immunosuppression. This finding was later used to demonstrate that lymph derived from the gut following shock could promote systemic inflammation, lung injury, and multiple organ failure. Mesenteric lymph collected after experimental shock has been shown to cause increased inflammatory adhesion molecule surface expression on endothelial cells, endothelial apoptosis or injury, priming of

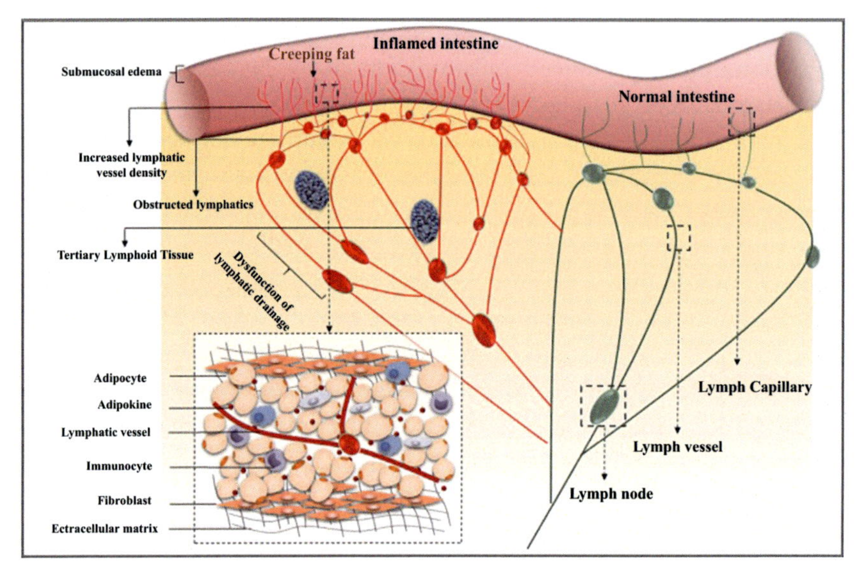

Figure 2.18 Gastrointestinal lymphatics and inflammation.

neutrophils, and suppression of hematopoiesis. Other types of potential signals such as microrNAs, bioactive lipids, or metabolities have also been areas of investigation in gut injury or pancreatitis [180].

2.24 New directions: microparticles, nanoparticles, and immunotherapy

Cells are known to release membrane vesicles called exosomes or microparticles into the extracellular environment. These extracellular vesicles serve as a mode of intracellular communication through the transfer of proteins, lipids, and RNA between cells. Exosomes from different cell types contain distinct endosome-associated proteins and mRNA and noncoding RNA. Extracellular vesicles are involved in numerous physiological processes and have an important role in immune regulation. Notably, they can mediate immune stimulation or suppression and drive inflammatory, autoimmune, and infectious disease pathology [181].

This property and the fact that exosomes can carry small pieces of genetic information make them potential therapeutic devices, possibly for future gene therapy approaches. Exosomes have been shown to preferentially enter lymphatic vessels and can be transported to lymph nodes within minutes. The functional role may be to prime cells in the lymph node ahead of the arrival of antigen-presenting cells or other immune cells. Exosomes isolated from the conditioned media of Hairy/enhancer-of-split related with YRPW motif (HEY) cells treated with the toll-like receptor (TLR3) agonist polyinosinic:polycytidylic acid (poly-IC) were found to be retained in lymph nodes when compared to control exosomes, and enhanced lymph flow and neutrophil recruitment to the lymph nosed [182].

Due to the importance of lymphatic vessels in immunity, they present an important transport route in drug delivery. The optimum range for lymphatic uptake of subcutaneously injected particles is 10–80 nm in diameter, which is very similar to the diameter range of extracellular vesicles, which is 30–100 nm.

CHAPTER 3

Lymphatic system of the lower limb

The lymphatic system has a significant role in the progress of long-term complications of varicose veins. Hence, the lymphatic system requires special attention. Progressive venous hypertension triggers massive production of tissue fluid, and at a certain point the lymphatic channels fail to drain the tissue fluid from tissue spaces. Homeostasis, which is normally maintained, becomes disturbed and swelling starts to appear on the foot with prolonged standing. After the leg is elevated for some time the edema subsides. Continuous stasis of lymph at tissue spaces can cause infection and predisposes the individual to extensive cellulitis. Chronic stasis will produce progressive cellular changes and damage, finally resulting in chronic nonhealing ulcers.

3.1 Importance of the lymphatic system

Though a very deep knowledge of the lymphatic system is important for clinicians, it is generally overlooked. For the surgeon it has a very vital role in diagnosing and staging of many diseases; this applies also to pathologists and radiotherapists (Fig. 3.1).

The lymphatic system is a specialized component of the circulatory system. It differs from the circulatory system in that the lymphatics do not form a closed ring or circuit, instead they begin blindly in the intercellular spaces of tissues of the body. It consists of lymph, which is derived from the blood, tissue fluid, and a group of channels called the lymphatic vessels, which return lymph to the blood. Lymphatic vessels run parallel to the veins.

The lymphatic system includes:

1. The lymphatic capillaries which originate as microscopic blind-end vessels.
2. The lymph vessels running parallel to the veins.
3. Lymph nodes located along the paths of collecting vessels.
4. Specialized lymphatic organs such as the tonsils, thymus, and spleen.

Genesis, Pathophysiology and Management of Venous and Lymphatic Disorders
DOI: https://doi.org/10.1016/B978-0-323-88433-4.00013-9

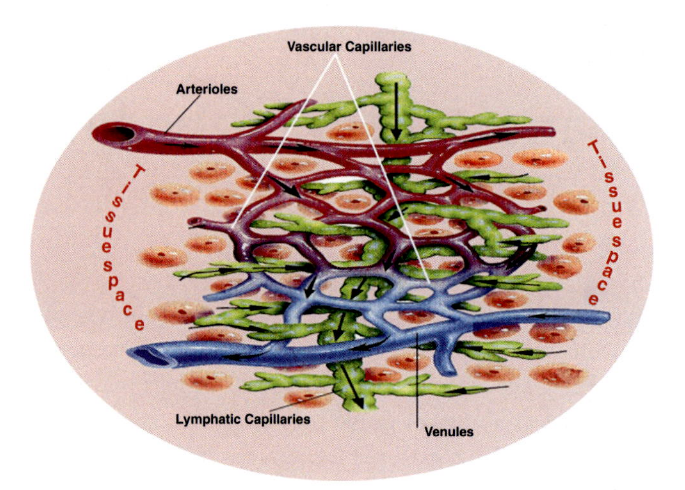

Figure 3.1 Micro lympho-vascular system.

5. Lymph and interstitial fluid. Lymph is a clear, watery fluid found in the lymphatic vessels. Interstitial fluid fills the spaces around cells.
   a. In some cases, it is part of the semifluid ground substance.
   b. In others, it is bound water in a gelatinous ground substance.

3.2 Structure of lymphatic vessels

Lymphatic vessels resemble veins in structure with the following exceptions:

1. Lymphatics have thinner walls.
2. Lymphatics contain more valves.
3. Lymphatics contain lymph nodes located at certain intervals along their course.

3.2.1 The lymphatic pump

Although there is no muscular pumping organ connected with the lymphatic vessels to force lymph onward as the heart forces blood, lymph moves slowly and steadily along its vessels, with the skeletal muscles helping to some extent.

3.2.2 Functions of lymphatic vessels

Lymphatics play a critical role in maintaining homeostasis in fluid and electrolyte balance. The high degree of capillary permeability permits high-molecular-weight substances which cannot be absorbed by the blood capillary to be removed from the interstitial spaces. Proteins from the interstitial spaces can return to the blood only via lymphatics. This is of great clinical importance, as any block in lymphatic return will cause blood protein and blood osmotic pressure to fall below normal, with fluid imbalance and death ensuing. Lacteals (lymphatics in the villi of the small intestine) serve in the absorption of fats and other nutrients. The lymph found in lacteals after digestion contains 1%–2% fat, is milky in appearance, and is called *chyle*.

3.2.3 Important functions

1. Maintenance of fluid balance and electrolyte
2. Immunity

3.3 Importance of the lymphatic system in maintaining fluid balance

Tissue fluid (Lymph) secretion is controlled by a regulatory mechanism whic is effected by the arteriolar sphincter and the microscopic venous valves. The capillary is bridging the gap between the above mentioned two structures. The inflow of blood into the capillary is regulated by the arteriolar sphincter and the microscopic venous valves at the other end of the capillary is control the back flow of blood from the vein to the capillary. The capillary filteration pressure is thus controlled just to suit to filter sufficient quantity of tissue fluid. Any imbalance will upset the microcirculatory system. The lymphatic system starts at the interstitial spaces, where the tissue fluid is stored for tissue utilization. The excess tissue fluid and waste product are drained through the lymphatic channels by the lymph sacs.

1. Plasma filters into the interstitial spaces from blood flowing through the capillaries.
2. Much of this interstitial fluid is absorbed by tissue cells or reabsorbed by the blood before it flows out of the tissue. A small amount of interstitial fluid is left behind. If this were to continue for even a brief period of time, the increased interstitial fluid would cause massive edema.
3. This edema would cause tissue destruction or death.

This problem is avoided by the presence of lymphatic vessels that act as "drains" to collect the excess fluid and return it to the venous blood just before it reaches the heart. Lymph flows through the thoracic duct and reenters the general circulation at the rate of 125 mL/hour. Though most of this flow is against gravity, or "uphill," it moves through the system in the right direction because of the large number of valves that permit fluid flow only in one direction. The movement is helped by breathing and skeletal muscle contractions. Activities that facilitate the central movement or flow are called *lymphokinetic actions*. Imaging techniques show that lymph pours into the central veins most rapidly at the peak of inspiration. The mechanism of inspiration, resulting from the descent of the diaphragm, causes intraabdominal pressure to rise as intrathoracic pressure fall, simultaneously causing pressure to increase in the abdominal portion of the thoracic duct and to decrease in the thoracic portion. The thoracic duct is literally "pumping" lymph into the venous system during each inspiration. The rate of flow of lymph into the venous circulation is proportional to the depth of inspiration. The total volume of lymph that enters the central veins during a given time period depends on both the depth of inspiration and the overall breathing rate. Contractions of skeletal muscles also exert pressure on the lymphatics to push the lymph forward. During exercise, lymph flow may increase by as much as 10–15 times. In addition, segmental contraction of the walls of the lymphatics themselves results in lymph being pumped from one valve segment to the next.

Other pressure-generating factors include:

1. Arterial pulsations,
2. Postural changes,
3. Passive massaging of the body/soft tissues.

3.4 General principles of tissue fluid formation

The main function of the circulatory system is to deliver efficiently food and fuel to the tissues for their proper functioning and to take back the unused food and waste products to the heart for recycling, purification, or excretion. This process at the tissue level is usually referred to as microcirculation. The capillaries, cells, and tissue spaces form the field of action. The arteries pump in blood to the arteriolar end of the capillaries. With the help of the sphincters in the arteriolar ends, it regulates the blood flow to the capillaries for efficient filtration to conserve adequate tissue fluid in the intercellular (interstitial) spaces for proper cellular nutrition and functioning. The fluid is filtered out through the micropores in the capillaries and the excess fluid is partly absorbed at the venular end of the capillaries for recycling. The lymphatic channels are vital and of great importance in the transfer of the excess fluid from tissue spaces and help the vein in the collection of this fluid for delivery to the heart. The lymphatic channels follow the same course in parallel to the venous channels, just like twins. The important role of the lymphatics is usually underestimated or even ignored by clinicians and hence it role is emphasized here.

The lymphatic channels begin their course at the tissue spaces as tiny bulbar structures, and closely follow the course of the venous channels. The anatomic and histological structures of lymphatic vessels are exactly the same as in the venous channels. They are not easily identifiable to the naked eye as they are colorless and only distinguishable when they drain from the intestine due to the emulsified fat, known as *chyle*, which appears milky. From a surgical point of view this is very important, especially in thoracic duct injuries. Another important point, which has been rarely studied and is poorly understood, is injury to the lymphatic trunks accompanying the saphenous veins and their tributaries during Trendelenburg's operation, the sapheno-femoral or sapheno-popliteal or perforator ligations, or multiple excision and ligations, or even stripping of varicose veins in the leg. Thus severed lymphatic channels predispose to lymphedema in the leg.

Edema means collection of fluid in the tissue spaces. The fluid in the body remains in two compartments, namely the *intravascular* and *extravascular* compartments. Fluid, electrolytes, and gases exchange between these two compartments based on physical laws, including *hydrostatic pressure*, *osmosis*, and *diffusion* and help the body to maintain fluid and electrolyte balance. The movement of fluid and electrolytes from the intravascular compartment to the extravascular compartment, namely the tissue spaces (interstitial spaces), through the capillaries is called *filtration* and the same action in reverse is called *reabsorption*. Reabsorption is shared by the capillaries and the lymphatics. When there is an alteration to these physical forces or barriers, fluid and electrolytes imbalance sets in, in either of the two compartments. In certain conditions total fluid volume may be increased, as in renal shutdown or excessive intravenous fluid infusion. There can also be fluid and electoral imbalance as in severe vomiting and diarrhea, and blood loss. When fluid accumulates in the skin and subcutaneous tissues, it can manifest in two ways as *pitting* or *nonpitting* edema. Edema is usually classified as (1) *generalized*, (2) *regional*, or (3) *localized*.

3.5 Generalized edema

Causes:

1. Hypoproteinemia,
2. Right ventricular failure,
3. Renal failure,
4. Overhydration by excessive fluid infusion,
5. Drugs, especially corticosteroids.

Hypoproteinemia is one of the most common causes of generalized edema and is called anasarca. The second common cause is right ventricular failure, where there is a failure of the central pumping mechanism resulting in stagnation in the venous system. This in turn increases venous pressure and leads to increased pressure at venules and tissue fluid production is enhanced. All the fluid thus formed cannot be drained by the lymphatic channels with the resultant fluid accumulation at tissue spaces.

3.5.1 Regional edema

When fluid accumulation is confined to a specific part of the body, such as a lower limb, upper limb, brain, intestines, or lungs, it is called regional edema. A venous block, lymphatic block, or both can cause this type of edema, such as in chronic venous disease, intestinal strangulation, and pulmonary edema.

3.5.2 Localized edema

Local trauma, toxins (insect bites), or infections are the most common causes of localized edema. The release of histamine and other humoral factors helps to increase capillary permeability and more tissue fluid is secreted locally. Infection produces inflammation and as resultant, edema occurs.

3.6 Precipitating factors for edema

Several factors contribute to precipitating edema. They may act singly or in combinations at different levels: at the capillary, venous, or lymphatic level. They are also influenced by physical or mechanical factors, the main ones of which are:

1. Increased capillary permeability,
2. Increased venous pressure,
3. Poor lymphatic drainage,

4. Decreased osmotic pressure due to derangement of crystalloids and colloids in plasma.

3.7 Lymphedema

Tissue fluid is also called lymph. Fluid accumulation in the tissue spaces (interstitial spaces) due to impaired lymphatic drainage is described as *lymphedema*. This is usually seen in the *hanging parts of the body, for example, the upper limbs, lower limbs, breasts, and scrotum.*

Lymphedema is classified into two types: *primary* and *secondary*.

1. *Primary*: This is due to maldevelopment of lymphatic channels and is congenital, and is manifested at different stages of life and accordingly it is subclassified into three subtypes.
   a. *Congenital lymphedema*: This is present at birth and is usually seen in females, contributing to around 10%–25% of primary lymphedema and has a genetic predisposition (familial sex-linked pattern). This is usually described as Milroy's edema.
   b. *Lymphedema precox*: This is the most common variety (60%–80%). It may make its appearance at any time after birth and before 35 years of age, with the maximum incidence at puberty. It is also seen commonly in females.
   c. *Lymphedema tarda*: This appears after the age of 35, is also called Meige disease, and is less common than the two conditions described above (10%).
2. *Secondary lymphedema*: Secondary lymphedema develops with a primary cause. These patients had an evidently normally developed and functioning lymphatic system. Later, a definite cause of interference in the normal mechanism of lymphatic drainage develops. This could be due to acute or chronic infections, primary malignant change in lymph nodes, malignant infiltration, or injuries following extensive trauma or surgical interventions with either accidental or block dissections, or irradiation. Increased venous hypertension, as happens in varicose veins, predisposes to increased capillary filtration and progressive interstitial fluid retention with resultant infections (cellulitis) and fibrosis. All the above-mentioned factors ultimately produce blockage of the free flow of lymph.

Filarial lymphedema produces unilateral or bilateral lymphedema of the legs due to the parasitic infestation of lymph nodes by *Wuchereria bancrofti* and associated bacterial infections and lymphangitis, and is found globally.

Trendelenburg's operation and multiple vein ligation or excision for treating varicose veins, or stripping of long saphenous veins for any purpose, ultimately produces venous hypertension and lymphatic block.

Lymphedema precox is most often mistaken for filarial lymphedema. Pretibial myxedema of hypothyroidism is also confused with the previous two conditions. In all these conditions, because of the chronic tissue changes, there is fibrous tissue proliferation, and the edema is nonpitting or stripping.

CHAPTER 4

Diseases of the venous system

The development of the human body is initiated by the union of a sperm with an ovum to form a cell, which inherits all the genetic qualities of the parents. That cell, by repeated multiplication and remodeling, transforms into different types of tissues becoming a very complex multisystem structure providing different physiological functions. The viability and vitality of the cell sustains life and for that proper food and fuel are required and the circulatory system is hence a vital system to perform that function. The circulatory system is constituted of four parts; the *heart* as the pump, the *arteries* as the outlet pipes, the *veins* and *lymphatics* as the inlet pipes, and the *capillaries* as the target place of microcirculation, where transfer of nutrients and gases takes place. All four components are equally important. Physiological abnormalities can happen anywhere and these pathologies can be fatal or result in lifelong complications (Fig. 4.1).

The pathological conditions of the heart and its corrective procedures have been always overemphasized by clinicians. Arteries were given second priority, and the veins and lymphatics were always undervalued.

Of the different types of venous diseases, the congenital malformations are most uncommon and produce cosmetic and functional problems. On the other hand, chronic venous disease (CVD) is not only a troubling problem resulting in prolonged difficulties, but also one which produces irreparable impacts on psychological, social, and family set up. Recent research into venous diseases in the field of venous structural pathologies in relation to the venous wall and valves and also genomic studies to identify the genetic relationship, the changes in microcirculation, and changes in venous ulcer and healing, has been a great boon to future developments in CVD management. On the basis of a study of 34,000 patients and follow up, the following classification is proposed:

Classification of venous diseases

1. *Primary varicose veins*
    a. Valve/vessel pathology of the superficial system of lower limbs (LLs)
    b. Valvular pathology of the deep venous system (deep venous reflux syndrome)

(*Continued*)

Genesis, Pathophysiology and Management of Venous and Lymphatic Disorders
DOI: https://doi.org/10.1016/B978-0-323-88433-4.00002-4

Classification of venous diseases

2. *Secondary varicose veins* (primary pathology is blockage of the deep veins of a LL reflecting on the superficial system)
3. *Congenital venous malformations*

---

1. Primary venous diseases (congenital)
2. Secondary venous diseases (acquired)

The primary diseases include all the congenital disorders so far described and the primary varicose veins; and the secondary diseases

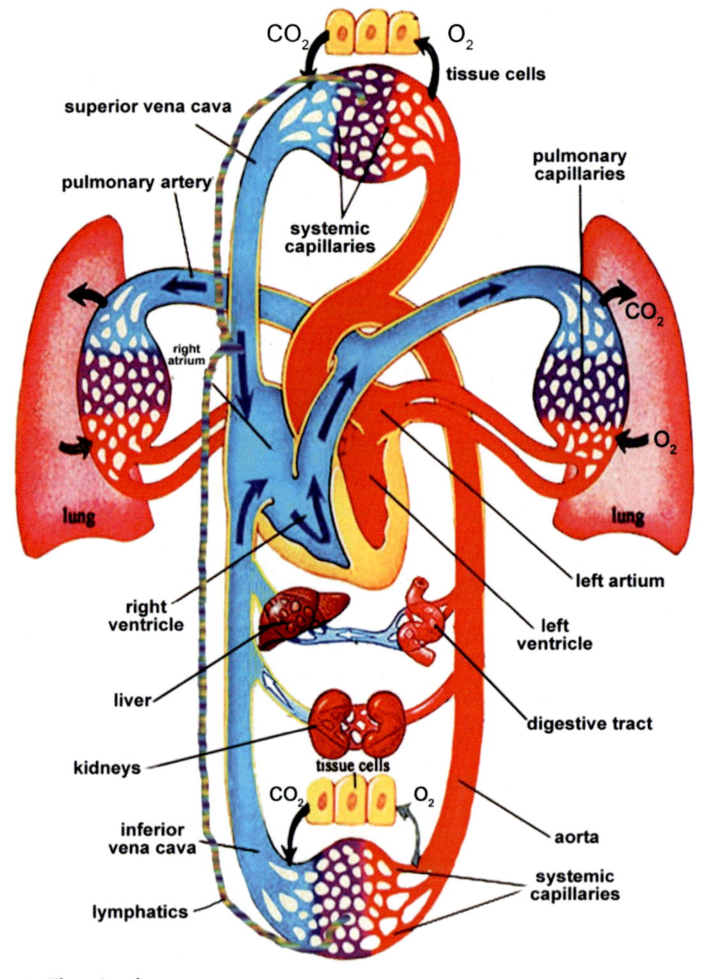

Figure 4.1 The circulatory system.

include varicose veins due to extrinsic and intrinsic factors producing obstruction in the deep vein and pelvic veins. Statistical analyses have shown that more than 60% patients have a clear family association, 30% ignorance in identifying varicose veins, and 10% not viewable from outside *but* remaining symptomatic.

The congenital conditions include:

1. Congenital malformations
2. Hemangiomas
3. Lymphangiomas
4. Lymphatic cysts
5. Arteriovenous fistula/congenital arteriovenous malformations
6. Klippel–Trenaunay syndrome
7. Primary varicose veins

The secondary conditions include:

1. Deep vein thrombosis
2. Tumors in the pelvic cavity
3. Malignant infiltration to the iliac veins

Treatment of *primary venous diseases* is almost the same as for primary varicose veins and is preferably by microfoam sclerotherapy followed by excision for cosmetic appearance.

4.1 Venous diseases of the lower limb

In an average individual the heart is located about 1.5 m above the ground level, and the veins in the LL have to pump up the venous blood and the lymph to this great height. The negative pressure achieved in the leg vein due to diastole is insufficient to pull up the blood from the LLs. To aid in this process, the LL veins are equipped with nonreturn valves. While the superficial venous system is provided with only nonreturn valves, the deep venous system has been provided with nonreturn valves and a *peripheral pump*, the calf muscle. The nonreturn valves open up during diastole and allow the blood to pass toward the heart, while during systole the valves remain closed and thus prevent the blood from descending.

Primary varicosity affects only the superficial system of veins. The incompetence of the valves produces backflow and results in venous stagnation and increased pressure effects on the venous walls, resulting in dilatation and tortuosity. The venous hypertension thus resulting leads to increased capillary pressure and more tissue fluid (lymph) formation. The increasing hydrostatic pressure progressively expands the venous walls as it has less collagen and

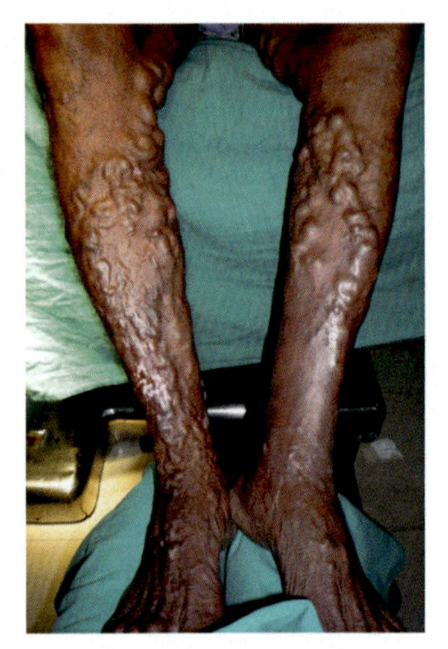

Figure 4.2 Classical varicose veins of the lower limb.

elastin content and this in turn creates more gaps at the valvular level. Thus there is successive deterioration in valvular function (Fig. 4.2).

The progressive valvular dysfunction slowly increases venous hypertension and when it is reflected at the tissue level pathological changes start with tissue fluid accumulation and progressive tissue damage.

4.2 Anatomy

A precise knowledge of the surgical anatomy of the venous system of the LL is very important in deciding and instituting the right surgical corrective procedures for venous diseases.

The venous return from the LL is facilitated by two parallel columns of blood vessels which are interconnected. One is the main channel, the deep venous system (DVS), and the other is the subchannel, the superficial venous system (SVS). The flow of blood from the SVS to the DVS is regulated by nonreturn valves. The intercommunicating channels are called the perforators.

The drainage of blood from the feet to the heart is a very difficult process as the venous blood has to ascend to a significant height against gravity. To prevent the blood from descending during systole, the veins are provided with many nonreturn valves. The two erect columns have maximum hydrostatic pressure at the lowermost part and hence the nonreturn valves are arranged very close to the lowermost part of the venous system. As it ascends, the gap between the valves increases and at the beginning of the inferior vena cava the valves are absent. In addition to the valves, the DVS is provided with an extra muscle pump at the lower end, mainly the calf muscle or *peripheral heart*, the bellowing action of this pump pushes the blood in the DVS very forcibly, while the heart tries to pull it up. Unfortunately, the SVS has no such support, except the valves, and hence varicose veins develop only in the SVS.

4.3 The superficial venous system

The SVS is formed as a network that connects the superficial dorsal veins of the foot and the deep plantar veins. The dorsal metatarsal veins drain into the dorsal venous arch, which is continuous with the long saphenous veins and the short saphenous vein laterally. The long saphenous vein ascends anterior to the medial malleolus, passes medial to the knee, and ends at the common femoral vein. Before its termination, it also receives a few tributaries—the superficial circumflex iliac, the superficial external pudental, and the superficial inferior epigastric veins. The posterior arch vein drains the area around the medial malleolus, ascends up the posterior medial aspect of the calf, receives the medial perforators (Cockett's), and joins the long saphenous vein at or below the knee. The short saphenous vein arises from the dorsal venous arch at the lateral aspect of the foot and ascends posterior to the lateral malleolus and drains into the popliteal vein (Figs. 4.3–4.6).

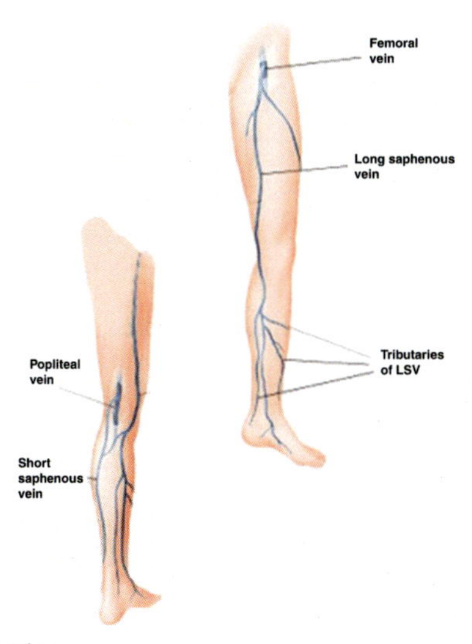

Figure 4.3 Superficial veins.

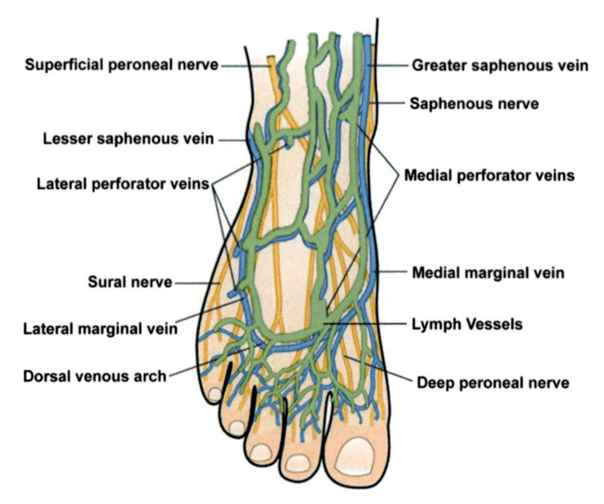

Figure 4.4 Veins, lymph vessels, and nerves.

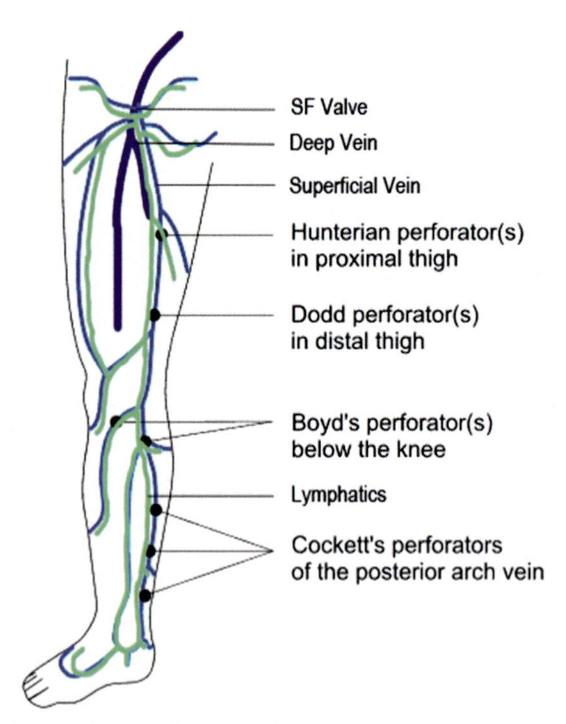

Figure 4.5 Perforators long saphenous vein.

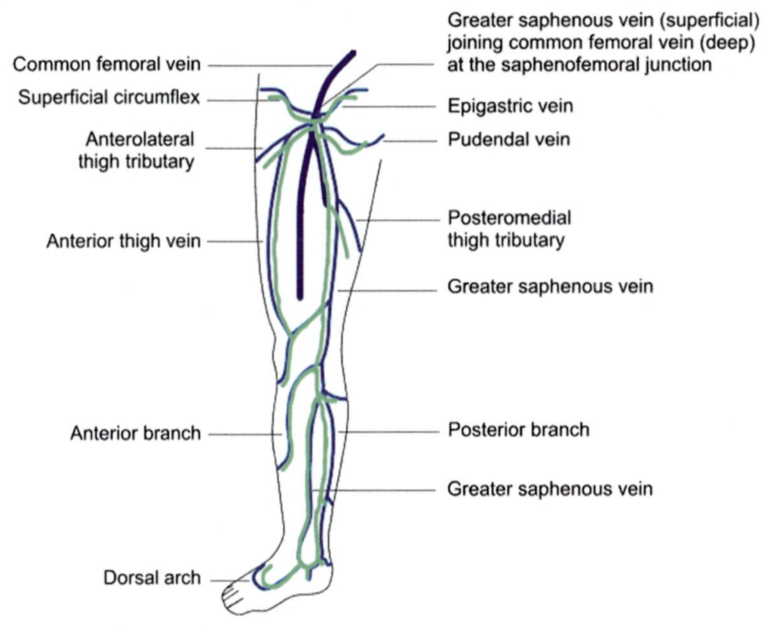

Figure 4.6 Perforators short saphenous vein.

4.4 The deep venous system

The plantar digital veins drain into a network of metatarsal veins which form the deep plantar venous arch. This continues into the medial and lateral plantar veins which join the posterior tibial veins. The dorsalis pedis vein joins the anterior tibial veins at the ankle. The posterior tibial vein, as it ascends, receives drainage from the *soleal sinuses*, peroneal veins, and gastrocnemius vein, and ends in the popliteal vein. As it enters the adductor canal it is called the femoral vein (superficial femoral vein).This is joined by the profunda femoris vein and becomes the common femoral vein. As it crosses the inguinal ligament it becomes the external iliac vein (Fig. 4.7).

The following points are critical:

1. The DVS is within the muscular compartment, whereas the SVS remains outside the muscular compartment and below the skin.
2. In addition to the four perforators in the leg and the two perforators in the thigh, there are several intercommunicating veins between these two systems of veins in and around the foot and also many intercommunicating venous plexuses.
3. At the tissue level, the smaller veins remain interconnected with the two systems.

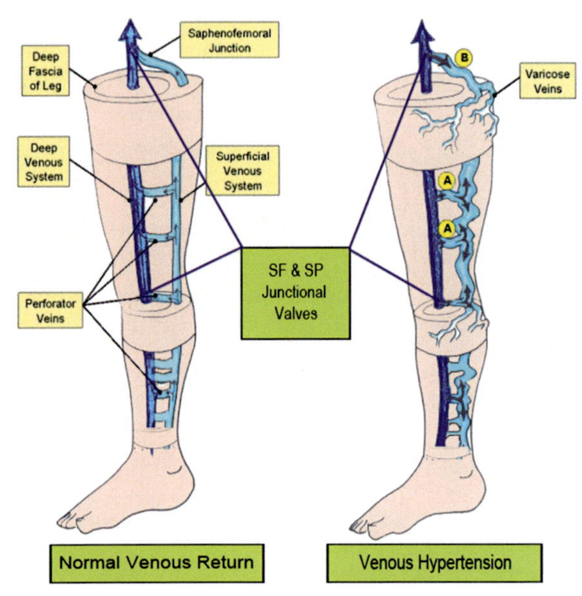

Figure 4.7 Saphenofemoral–saphenopopliteal valves and perforators.

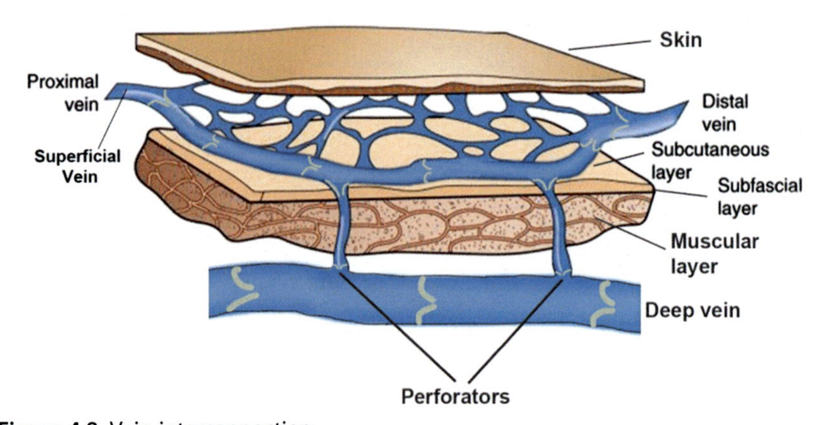

Figure 4.8 Vein interconnection.

4. The input of blood to the tissues is carried out by the arterial system, but the drainage is not only carried out by the venous system but also by the lymphatic system, and is very important in the pathogenesis. The progressive venous hypertension leads to lymphatic system failure and contributes to morbidity. *The lymphatic system is the sister to the venous system.* Carbon dioxide and some tissue fluids are carried by the venous system and the major part of the tissue fluid is carried by the lymphatic channels. The veins and lymph vessels are similar to uniovular twins (brother and sister) in many respects (Figs. 4.8 and 4.9).

Figure 4.9 The veins on the leaf (leaf skeleton) simulate the venous networking; the larger one corresponds to the trunkal veins, whereas the smaller one corresponds with that of the interconnecting venous network.

4.5 Structure of microcirculation

Four structures contribute to the microcirculation: arterioles, capillaries, venules, and lymph channels (Figs. 4.10 and 4.11).

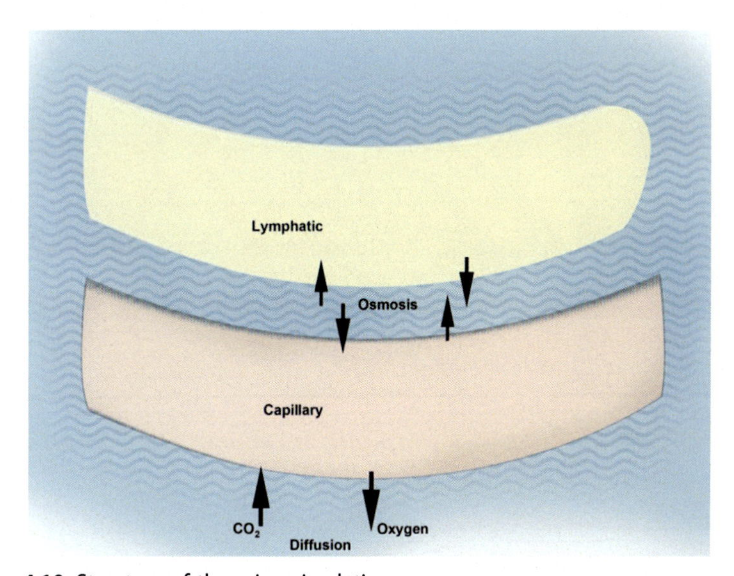

Figure 4.10 Structure of the microcirculation.

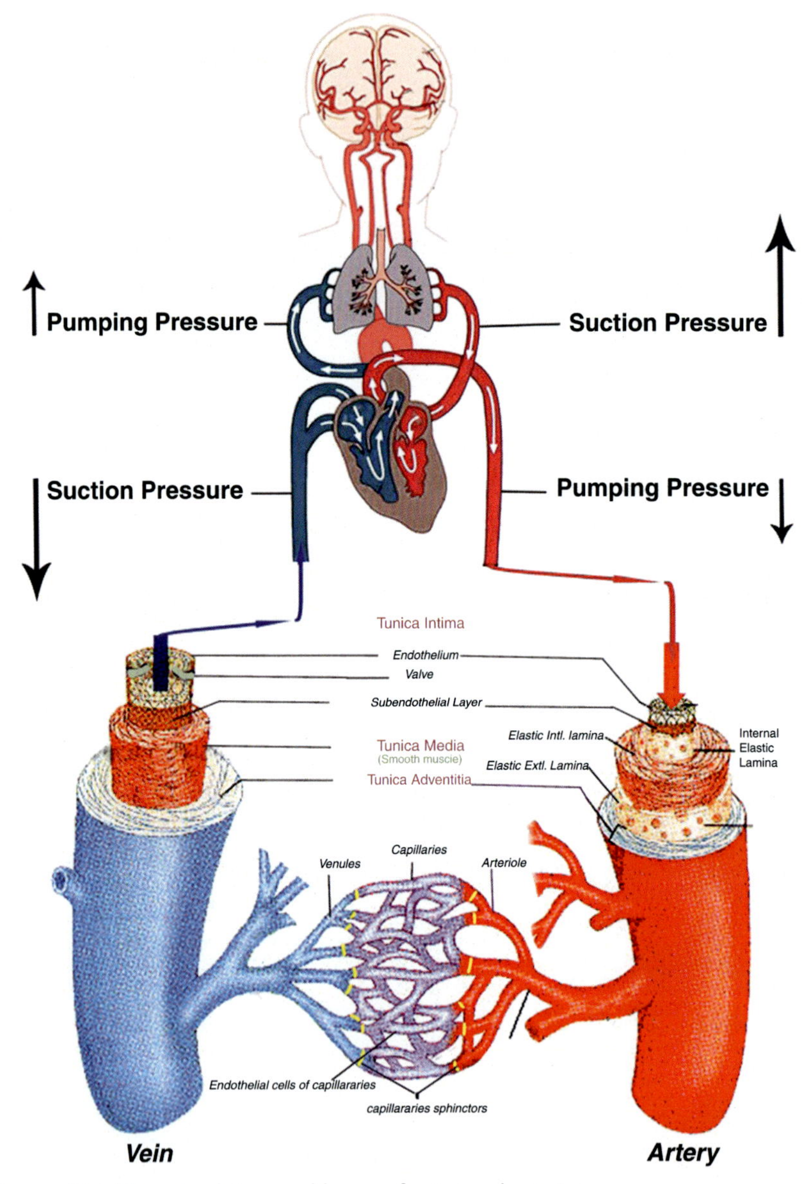

Figure 4.11 Microcirculation and layers of veins and arteries.

4.6 Arterioles

Arterioles are about 0.1–0.5 mm in diameter, and regulate the flow of arterial blood to the capillaries. The inner layer is composed of endothelium surrounded by muscles and elastic fibers, with rich innervation and chemotactic receptors. They also regulate systemic vascular resistance and capillary hydrostatic pressure, and thus influence the exchange of fluid at the capillary level. Like the arteries they have three layers: the inner tunica intima, composed of an endothelial lining and elastic tissue, the middle tunica media, composed of muscles and elastic fibers, and the outer tunica adventitia, composed of elastic fibers and connective tissue.

4.7 Capillaries

These are the smallest blood vessels in the body, having a diameter of about 0.01–0.05 mm. They begin at the arteriolar end and end at the start of the venules. On the inner side is the endothelial lining on a basement membrane. There are pores in the endothelium which allow the transfer of fluids, electrolytes, macrophages, and macromolecules to the interstitial spaces and their return. Exchange of gases also takes place here, and they have a precapillary sphincter to regulate the blood flow.

4.8 Venules

There are two types of venules: small and large. The small venules have an endothelial lining on a basement membrane and the larger venules have, in addition to the endothelial wall, a muscular wall. Fluid and macromolecular exchange take place mostly at the beginning of the venules. The sympathetic innervation of the muscles controls the tone of the larger venules and facilitates control of hydrostatic pressure. In the general structure of the venules it is very similar to the arterioles with the exception that the muscle and elastin are reduced and hence thinner. *As the size of the vein increases the muscular layer gets thinner and the collagen is increased.* The intimal folds are the circular valves of the veins and these are very closely placed toward the venules with the distance increasing closer to larger veins. At the lowermost part of the LL the valves are closely placed and toward the iliac veins they are reduced, finally being absent near the inferior vena cava.

4.9 Lymphatic vessels

The arteries travel alone, but the veins have a companion, the lymph vessels. The lymphatic capillaries begin at the interstitial space with blind ends. They travel almost entirely along the veins, carrying excess tissue fluid and waste materials, and ascend with the help of valves, as in the veins, finally becoming the lymphatic trunk and ending as the thoracic duct empties the contents into a great vein near the junction of the left subclavian vein and the superior vena cava. The lymphatics drainage of the LLs plays a great role in the formation of chronic venous ulcers of the legs and feet. Failure of the lymphatic system produces impedance of the microcirculation.

4.10 Valves in veins

Valves are unique to the veins of the LLs. In order to assist the smooth pumping of venous blood from the feet to the heart against gravity, they prevent backflow during contraction of the heart. The number of valves varies depending on the site, size, and level of the vein. At the level of the feet, where the venous pressure is at its maximum the valves are arranged very closely, and toward the thighs the distance is increased, so that at pelvic region they are very scarce, and they are totally absent above the level of the umbilicus. The valves are unidirectional and allow blood to pass only toward the heart. They are circular in appearance and are developed by modification of the inner layers of the venous wall to maintain the same histology.

4.11 Comments

The treatment of varicose veins of the LLs remains an unsolved enigma. To obtain a clear insight into the problem one has to have a thorough knowledge and understanding of the detailed anatomy of the venous system of the LLs, and the pathophysiology and biophysics behind the causation of this illness. A treatment based on the biomechanics of the problem alone can cure the disease.

However, medicine is an ever-changing science, and as new research and clinical experience broaden our knowledge, periodical changes in treatment and drug therapy occur.

CHAPTER 5

The pathophysiology of varicose veins of the lower limb

The blood pressure of a human at different levels of the blood vessels in the leg is as follows:

Normal arterial pressure: ↓100–140 mm Hg ↑70–100 mm Hg

Mean arterial pressure: 40 mm Hg

Capillary pressure: Arterial end: 30 mm Hg Venous end: 10 mm Hg

Mean capillary pressure: 20 mm Hg

Venous pressure: Venules: 3–7 mm Hg Great veins: <1 mm Hg

Against the normal pressure gradients mentioned above, in varicosity of the superficial veins of lower limb (LL), there are abnormal variations in pressure in the superficial system of veins in the LL. Due to the malfunctioning of the valves in the veins there is valvular incompetence, which prevents the smooth upward flow of blood through the superficial veins. While the venous pressure in the deep vein remains at about <6 mm Hg during systole and at <1 mm Hg during diastole, the venous pressure in the superficial venous system (SVS) rises almost to the level of arterial pressure. As a result of this abnormally increased venous pressure, there is stretching of the walls of the veins, which subsequently results in abnormal distension, thinning, and tortuosity of the venous channels. There is a lateral circumferential thrust on the vein wall and a downward thrust due to the unsupported erect blood column.

A study of the biophysics of the venous circulation is very important in understanding the proper maintenance of venous return in varicose veins (VV) of the LL. This will help in understanding the cause of failure of the currently practiced surgical management of this venous disease and also to institute the correct line of surgical management of this tricky disease.

5.1 Biophysics

There are two erect fluid columns placed in parallel to each other with multiple intercommunications: the deep vein and saphenous vein systems. The pressure gradients in venous disease are as follows:

Genesis, Pathophysiology and Management of Venous and Lymphatic Disorders
DOI: https://doi.org/10.1016/B978-0-323-88433-4.00012-7

The average venous pressure in the deep vein is around 3 mm Hg (between 6 and <1 mm Hg) at the ankle region and toward the upper thigh about -4 mm Hg. The venous pressure in SVS is around 60 mm Hg at the ankle level, 45 mm Hg at knee level, and 30 mm Hg at the upper thigh level, which is very high.

Anatomically, the involvement of the part of the venous segment is very important. The venous system has two functional parts:

1. The *collecting* part (tributaries);
2. The *conducting* part (trunk).

The collecting part forms the *tributaries* (venules and small veins) which range in size between 0.02 and 3.00 mm and the conducting part forms the *truncal veins*. There can be valvular dysfunction at the sapheno-femoral (SF) junction or sapheno-popliteal (SP) junction, in the perforators, in the truncal part or tributaries, or in multiple combinations.

In venous hypertension, during diastole, the venous pressure in the deep veins falls to <1 mm Hg and the venous return from the superficial vein, which has a pressure of >60 mm Hg, is facilitated by an abnormal gush of blood into the deep venous system through the SF valves, SP valves, and the perforators in the leg and foot. In systole, though there is reflex at major valve levels due to *inactive* valves, the great pressure in the superficial veins prevents the entry of blood. It has to be especially borne in mind that the input to the capillaries at the tissue level, whether draining to the superficial system or to the deep system, remains constant (Figs. 5.1 and 5.2).

The incompetence of the valves in the superficial system produces back pressure in the venules and on the vessel wall, and as a result there is an increase in filtration pressure at a capillary level which results in a large quantity of interstitial tissue fluid formation. As it progresses, the lymphatic channels fail to drain the large quantity of fluid leading to pathological changes at a cellular level. In addition to tissue fluid secretion, the red blood cells (RBCs) will also seep into the tissue spaces and the damaged RBCs will contribute to further cellular changes.

One factor needs special mention. The valvular involvement in the venous pathology is very interesting. The cellular pathology is produced only in a particular way. One group of VVs appears only in truncal veins. The second group involves only the smaller veins (the tributaries). The third group has involvement of both the truncal veins and tributaries in varying proportions. It is surprising that tissue damage is noted only in cases where the tributaries are involved and never in truncal involvement

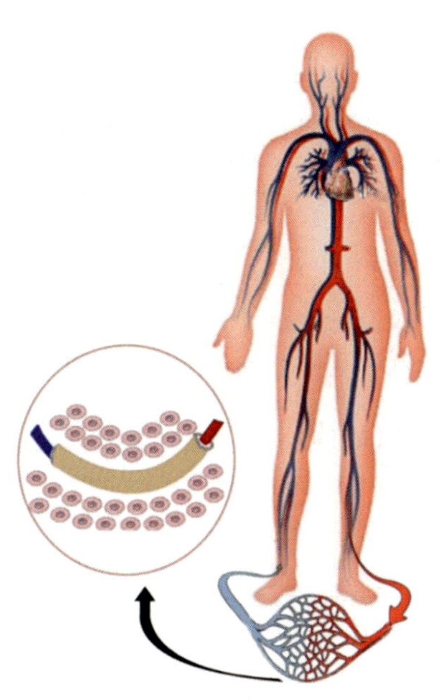

Figure 5.1 The circulatory system.

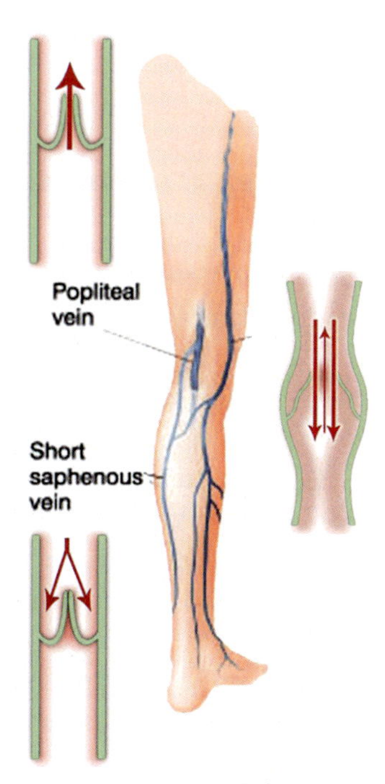

Figure 5.2 Functioning of the valves of veins: (left) normal; (right) diseased.

alone. The reason for this is that when the valves of the tributaries are normally functioning, the increase in the venous pressure is not transmitted to the tissue level and blood is diverted via interconnecting channels to the deep venous network system.

5.2 Cellular changes

In venous hypertension, the *pathophysiology of microcirculation changes* is as follows: The increased venous pressure produces dilation of venules and capillaries, increased capillary permeability, increase in tissue fluid formation and decrease in fluid absorption, progressive lymphatic failure; an increase in the size of the capillary pores allows blood cells to enter the tissue space and macromolecules to enter the circulation. There is pericapillary fibrin cuff formation on one side, and on the other side blood cell trapping, adhesion and activation of white blood cells produce impedance of microcirculation, release of free radicals, proteolytic enzymes, chemotactic agents, and cytokinins. RBCs disseminate the hemosiderin pigments which produce discoloration and irritation. All these contribute to impairment of tissue perfusion and oxygenation with cellular vitality. Any minor trauma can trigger ulcer formation with the final *eroding chronic venous ulceration.*

5.3 Chronic venous disease: from symptoms to microcirculation

The recently published European Venous Forum (EVF) guidelines 2018 update on the management of chronic venous disorders (CVeDs) of the LLs focused on several new aspects: a new place for early symptoms, new data on microcirculation alterations, and a reevaluation of veno-active drugs (VADs). The chronic venous disease (CVD) symposium held at the annual meeting of the EVF on June 28, 2018 in Athens, Greece, highlighted this perspective by answering three questions: What do symptoms mean and how do they influence our choice of investigations? Is there a link between symptoms and microcirculation alterations? How does one choose the right VAD for the right patient based on the updated EVF guidelines? The answers given led the speakers to three conclusions: early symptoms reveal the initial stage of CVD and patients with C_{0s} disease should be properly diagnosed, investigated, and treated; damage to the microcirculation is likely to be the first evidence of the onset of venous disease; *Ruscus* + HMC (Hesperidin Methyl Chalcone) + VitC has proven efficacy in randomized controlled trials, and has been given a strong recommendation (grade 1A) by the 2018 EVF guidelines for treatment of pain, heaviness, feeling of swelling, paresthesia, and edema, and should be considered as one of the preferred treatments to relieve these symptoms in CVD patients.

Recent data from the literature highlight the role of the microcirculation in CVD and the importance of investigating early symptoms when signs are not yet present. This clinical situation is described as C_{0s} in the clinical, etiological, anatomical, and pathophysiological (CEAP) classification of CVeDs, which defines the following classes: C_0: no visible or palpable signs of venous diseases; C_1: telangiectasia or reticular veins; C_2: VVs; C_3: edema; C_{4a}: pigmentation and/or eczema; C_{4b}: lipodermatosclerosis and/or atrophie blanche; C_5: healed venous ulcer; and C_6; active venous ulcer. Each clinical class is further characterized by a subscript for the presence of symptoms (S—symptomatic) or absence of symptoms (A—asymptomatic). The term CVD includes any morphological and functional abnormality of long duration affecting the venous system, manifested by symptoms and/or signs indicating a need for investigation and/or care. What do symptoms mean and how do they influence our choice of investigations? Is there a link between symptoms and microcirculation alterations? How does one choose the right VAD for the right patient, based on the updated EVF guidelines? This chapter reports the answers given to these interesting questions.

5.4 Significance of symptoms and how they influence the choice of investigation

Although earlier, the difference between the symptoms and signs encountered in patients with CVD were sometimes neglected, these aspects are now considered as separate and valuable information. Symptoms reveal what the patient feels, and usually correspond to an unpleasant sensation. Symptoms include pain or aching, throbbing, tightness, heaviness, feeling of swelling, muscle tiredness, itching, cramps, burning sensations, restless legs, tingling, and venous claudication. Signs correspond to what is clinically visible and can be observed by the physician, and the major signs include telangiectasias, reticular veins, VVs, edema, and skin changes such as pigmentation, corona phlebectatica, eczema, lipodermatosclerosis, atrophie blanche, and ulceration. Venous symptoms were hardly mentioned in textbooks, poorly studied, and infrequently reported until recently. Therefore, a working group of experts, the SYM (Vein Consensus statement (consensus statement on venous symptoms— SyM Vein)) vein group, decided to fill this gap and produced a valuable document, the "SYM vein consensus statement," which clearly defined all of these numerous venous symptoms and explained, at least in part, their origin, their attribution to venous disorders, and how the physician could deal with them.

Analyzing the pathophysiology of venous symptoms helps us to understand the deleterious processes that generate these symptoms. Briefly, venous hypertension, often induced by reflux, is a key factor with mechanical and biological consequences. Mechanical consequences on large and smaller veins include wall remodeling, resulting in capillary dilation and leakage, as well as liquid overload in lymphatics. In patients with advanced CVD, tissue fluid accumulates in the interstitium and causes edema. Venous hypertension is also responsible for biological changes. It slows blood flow in the capillaries, prompting leukocyte adhesion [183]. Due to altered shear stress, leukocytes begin rolling and sticking to the vein wall and then migrate out of the capillary, where they release inflammatory mediators, triggering inflammation, and participating in the remodeling of the venous valve and wall, which aggravates venous hypertension. Leukocytes are involved at the onset of CVD and inflammation is a key point to understand venous symptoms and signs, and together with the excess of fluid, explain most of the venous symptoms, especially in the early stages. VADs have been shown to significantly reduce heaviness and swelling without changes in the venous filling index, ejection fraction, and residual volume fraction. This suggests that before the larger vessels become involved, the smaller vessels and microcirculation may play a role in venous symptoms. Studying the effect of VADs on the microcirculation in in vivo animal studies showed that VADs decreased microvascular permeability, leukocyte rolling, and adhesion to the endothelium. Apart from studying the microcirculation, investigating lymphatics and small superficial veins may open new perspectives to understand venous symptoms. Fluorescence microlymphography visualizes the superficial lymphatic capillaries and shows alterations in the initial lymphatics of microlymphangiopathy. Small superficial veins of the human LL were shown to contain typically bicuspid venous valves, present in vessels as small as 18 μm. Recently, Vincent et al. performed retrograde resin venography in amputated legs affected by arterial or mixed arterial/venous disease and proved the existence of small microvenous valves which were incompetent, while the large superficial trunks were competent.

Symptoms are not a stand-alone entity; other factors will have to be considered. In C_{0s} and C_{1s} disease, symptoms are predominant, and because of the absence of visible or palpable VVs, duplex ultrasound (DUS) is the gold standard to detect or exclude the presence and delineate the anatomic extent of reflux in superficial veins and tributaries, as reflux often precedes the clinical manifestation of varices. In certain cases (C_{0s}), ultrasound scanning may have to be repeated at the end of the day, which may reveal reflux on this occasion. Other investigations, such as laser Doppler, capillaroscopy with cytoscan, or superb microvascular imaging (SMI), are used for research. Initially used for tumor evaluation in oncologic settings, SMI has been applied to small vessels in a recently published study comparing SMI images to color DUS and allowed better visualization of the smaller vessels and potential reflux. Compared with color Doppler imaging, SMI detected a significantly greater number of refluxing vessels and smaller diameter vessels, predominantly located in the superficial layer (<3 mm from the skin surface). In C_{2S} class, corresponding to symptomatic VVs, symptoms may vary considerably, and do not determine the extent of the investigations required. Extensive DUS of the superficial and deep venous systems may be used in view of the treatment strategy, while additional investigations such as venography, computed tomographic venography, or magnetic resonance venography may be proposed in cases of a major discrepancy between the DUS findings and symptoms. In patients suffering from C_{3S} disease with edema and symptoms, extensive DUS of the superficial and deep venous systems is mandatory, and potentially accompanied by plethysmography. This allows the measurement of peak reflux velocity, refluxing volume flow in mL/s, and venous filling index. If there is neither reflux nor obstruction, lymphoscintigraphy and/or fluorescent microlymphography must be considered to exclude lymph edema. In C_4 to C_6 classes clinical signs are preponderant, even though symptoms are also present in the majority of cases. The proposed investigations are extensive DUS of the superficial and deep venous systems, and possibly also plethysmography. In conclusion, venous symptoms remain poorly understood, although the microcirculation seems to play a key role in their occurrence. In the case of venous symptoms, investigations have previously focused on the larger vessels; however, recent research on C_{0s} and C_{1s} limbs has drawn attention to the smaller veins, which are now considered as the probable site of alterations, especially in the early stages of CVD, and therefore, the microvessels and microcirculation should receive full attention.

5.5 Different microcirculatory patterns in symptomatic versus asymptomatic patients

The vein wall is relatively thin compared to that in the arteries, and is composed of three histological layers: tunica intima (mainly endothelial cells), tunica media (internal elastic lamina and vascular smooth muscle cells—VSMCs), and tunica adventitia (fibroblasts embedded in an extracellular matrix (ECM) containing collagen, elastin, and other proteins). The vein wall structure and function are regulated by signaling ions, molecules, and enzymes, among which the matrix metalloproteinases (MMPs), endopeptidases, Zn^{+2} dependent, secreted by fibroblasts, vascular smooth muscle cells, and leukocytes, can be considered capable of degrading proteins of the ECM and regulating G-protein-coupled receptors and cell signaling. MMPs could also be regulated by endogenous tissue inhibitors of metalloproteinases (TIMPs), and the imbalance in the relationship MMP/TIMP could play a role in the regulation of venous structure and function. Chronic venous hypertension increases hypoxia inducible factors leading to increased MMP expression/activity, degradation of ECM proteins, and consequently venous dilation and valve degradation. On the other hand, *Ruscus aculeatus* (butcher's broom) extract is used in association with the flavonoid hesperidin methychalcone to increase peripheral venous tone in CVD. In isolated cutaneous veins, *Ruscus* extract causes contractions owing to activation of both postjunctional α1-α2-adrenoceptors and through direct action on venous smooth muscle cells. Since 2006, the BioVasc laboratory in the State University of Rio de Janeiro has been studying the microcirculation in vivo in animal models and in patients with CVD. It has shown different degrees of microcirculatory dysfunction according to the CEAP clinical classes: the cutaneous capillaries become progressively enlarged and tortuous, losing their hairpin/paperclip shape and forming the bulks or skeins described in the literature as "glomerulus-like" capillaries.

Recently, further investigations have been conducted in patients with C_{0s} disease to evidence potential differences between symptomatic and asymptomatic C_0 limbs (C_{0s} and C_{0a}). In comparison with a typical cytocam image of C_{0A} limbs showing normal capillaries, images from C_{0s}

limbs showed several modifications: the capillaries appeared morphologically different, as if the arrangement of the capillaries was disturbed, with less capillaries per field, and the dermal papillae appeared larger. "Dermal papillae dilation" could be the first clue to a potential response to an injury, such as venous hypertension. The message for the future could be that symptoms and signs of CVD are not only a consequence of wall abnormalities and valvular dysfunction in the larger superficial veins, but are also associated with changes in the microvenous circulation. Alterations to the subdermal valvular plexus could explain the incongruities between symptoms and clinical class. It is a common finding that some patients without VVs have many symptoms, while others with VVs have no symptoms at all. The initial microvascular injuries observed in patients with C_{0s} limbs could be treated as soon as symptoms appear, even in the C_{0s} stage, in order to delay the evolution of disease. In conclusion, three points can be highlighted: capillary impairment is linked to the clinical class of the CEAP classification; the microcirculatory changes observed in the early stages of CVD show atrophic and inflammatory patterns that may be responsible, at least in part, for the observed symptoms; and acting on the microcirculation and capillaries at any stage of CVD may become essential in the future.

5.6 Choosing the right veno-active drug for the right patient

In the past, the choice of VAD to treat patients with CVD was related to the physician's experience, or sometimes on the basis of the relative effectiveness of a drug on a single symptom; for example, selecting an antihistamine for treating itching. Venous symptoms are present in various combinations, making each patient unique. In previous guidelines, the symptoms to be considered were not exhaustively detailed. The 2015 European Society for Vascular Surgery guidelines on CVD stated that VADs should be considered as a treatment option for swelling and pain, without mentioning any other symptoms. The 2014 International Union of Angiology (IUA) guidelines on CVD stated that "VADs may be used to relieve CVD-related symptoms and edema in patients at any stage of disease," and considered only four specific symptoms. Therefore, when preparing for the 2018 EVF/IUA guidelines on CVD, the faculty decided to scrutinize both old and new meta-analyses providing data to determine the level of evidence on the magnitude of the effect that each VAD has on each well-defined symptom. The grading of recommendations, assessment, development, and evaluation system was used to assess the effectiveness of each VAD. Level A evidence implies that further research is very unlikely to change our confidence in the estimate of effect and derives from two or more scientifically sound RCTs (Randomized controlled trials) or systematic reviews and meta-analyses in which the results are clear-cut and are directly applicable to the target population. A strong recommendation is made if the benefits outweigh the risks, and a weak recommendation is made if the benefits and risks are closely balanced or if there is uncertainty about the magnitude of the benefits and risk. A strong recommendation can apply to most patients in most circumstances without reservation. Another important measure is the number needed to treat (NNT), which corresponds to the number of patients who needed to be treated to prevent one additional bad outcome. It is also defined as the inverse of the absolute risk reduction. The rigorous methodology of the Cochrane library meta-analytic method was used for recent meta-analyses; notably, the meta-analysis of randomized control trials of *Ruscus* + HMC + VitC published in 2007 by Kakkos et al. With the use of this strict methodology for rewarding a grade to the most frequent symptoms, *Ruscus* + HMC + VitC received a level A for pain as the reported NNT was 5 and the standard mean difference (SMD) versus placebo was -0.83^{-33}, a level A for heaviness, with an NNT of 2.4 and an SMD of −1.23; and a level A for feeling of swelling, with an NNT of 4 and an SMD of −2.27. Further, a level A was also attributed for paresthesia, with an NNT of 1.8. In comparison, micronized purified flavonoid fraction (MPFF) was also granted a level A for pain, heaviness, and sensation of swelling, with NNT values of 4.2, 2.9, and 3.1, respectively. Oxerutins obtained a level B grade for pain and heaviness; horse chestnut seed extract, a level A grade for pain, and calcium dobesilate, a level B for pain and a level A for heaviness. With *Ruscus* + HMC + VitC, the NNT for the main symptoms is very low and the SMD is good. The most common symptoms reported by patients with CVD are those for which *Ruscus* + HMC + VitaC has been proven to be effective. Global symptoms were significantly reduced with the use of *Ruscus* + HMC + VitC compared to placebo when assessed as a categorical variable (RR (Risk Ratio) 0.54). This VAD could also be chosen for its effectiveness for leg edema: *Ruscus* + HMC + VitC showed strong evidence of statistically significant efficacy on foot/leg volume (SMD = 0.61) and ankle circumference (SMD = −0.74) and was awarded a level A grade of evidence for these clinical signs [184].

In conclusion, *Ruscus* + HMC + VitC is a splendid choice for most patients with CVD based on its effectiveness across a wide range of symptoms and has been awarded a grade 1A recommendation in the international guidelines on CVeDs. The NNT to significantly reduce or abolish venous symptoms with *Ruscus* + HMC + VitC is small, indicating that a large proportion of patients treated with this combination will benefit from its effectiveness.

The conducted symposium provided the opportunity to highlight three messages: early symptoms reveal the initial stage of CVD and thus patients with C_{0s} disease should be properly diagnosed, investigated, and treated; microcirculation damage is likely to be the first evidence of the onset of venous disease; having proven its efficacy in RCTs, *Ruscus* + HMC + VitC has been given a strong recommendation (grade 1A) for pain, heaviness, feeling of swelling, paresthesia, and edema by the 2018 EVF guidelines and should be considered as one of the preferred treatments for symptomatic relief in CVD patients.

5.7 From the first symptoms to the advanced stages: chronic venous disease progression

Risk factors for the development of CVD and VVs progression are widespread and include advanced age, excess body weight, and family history. Physicians and patients should be aware of the risk factors for CVD, the treatments and measures available to slow disease progression, and the serious consequences of allowing the disease to progress unchecked.

CVD progression is defined as a change in clinical signs that indicate an increase in CEAP class, a change in symptoms, or changes in both. Progression of CVD is highly variable and proceeds along different pathways in different patients, though some patterns are evident. Chronic venous hypertension and dilation lead to a number of pathophysiological changes in the vein wall and surrounding tissues. Altered shear stress and an initial inflammatory process at the vein wall lead to endothelial cell (EC) activation, which increases the permeability of the endothelium. In addition, abnormal hemodynamics allow multiple other factors to come into play, such as hypoxia, dysregulated apoptosis, and ECM changes, all of which can influence VV development at the cellular level. Following RBC extravasation through the damaged endothelium and vein wall and RBC breakdown, fibrin and hemosiderin concentration increases in the tissues and activate mast cells and macrophages, which contribute to further EC activation and inflammation [183]. Activated endothelium also triggers the adhesion and migration of leukocytes through the vein wall, where they release Transforming growth factor-beta 1 (TGF-β1) and proinflammatory cytokines, all of which stimulate collagen synthesis by fibroblasts, increase vascular wall thickening and remodeling, and potentiate inflammatory cascades. One of the critical events that begins this vicious cycle of inflammation and tissue destruction is the infiltration of leukocytes through the venous endothelium into the vein wall and surrounding tissues. This occurs at EC junctions through a process known as diapedesis in which it appears that leukocyte signaling directs the rearrangement of cell–cell junctions such as adherens junctions formed by vascular endothelial cadherins and tight junctions formed by claudins. Chronic inflammation alone can have deleterious consequences on venous flow. Arterial and venous flow patterns in legs of patients with C4–C6 CVD were found to be similar to those in patients with leg inflammation caused by cellulitis [185].

The microcirculation also becomes increasingly impaired in CVD. Vasomotor activity in the skin microvasculature of CVD patients tends to decrease with increasing CVD class, and the amplitude of vasomotor cycles is significantly higher in C4–C6 patients than in C1–C3 patients. As CVD progresses in severity, the microvasculature network in ulcerated skin becomes very dense and highly disorganized. In CVD progression, the risk factors that lead to chronic venous hypertension and dilation, and the consequent inflammation, also create a vicious cycle of pathologies that worsen abnormal venous flow, further increase local venous pressure, damage venous valves, and potentiate inflammation [186]. Local inflammation brought on by leukocyte adhesion and infiltration leads to structural alterations in the walls and valves of veins, which increase venous reflux, capillary hypertension, leakage, and edema, all of which contribute to inflammatory responses in the surrounding tissues and skin, and can eventually lead to ulceration [187].

However, some case histories illustrate that severe GSV reflux alone, without VVs or obstruction, can lead to symptoms. Epidemiologic evidence from large cohorts shows that CVD progresses at a substantial rate, portending marked increases inpatient burdens in coming decades as the population ages and life expectancies become longer. The lifestyle and clinical factors most strongly related to progression were advanced age, family history of VVs, and history of deep vein thrombosis, high BMI (Body Mass Index), and venous reflux. In particular, reflux in the small saphenous vein coupled with deep reflux was likely to be associated with a higher risk of progression. The venous reflux, particularly superficial combined with deep venous reflux, was associated with greater progression and suggests that duplex scanning could be a useful prognostic tool to identify patients at risk for rapid progression. Though prevention is not currently possible, patients with risk factors should be monitored for signs of early-stage disease and treated appropriately to manage disease progression. In all applicable cases, lifestyle changes are warranted to increase physical activity, stop smoking, and reduce excess body weight. In patients with subclinical segmental disease, compression stockings with or without MPFF are recommended for symptomatic relief. Therapy with veno-active drugs may also be recommended, many of which have demonstrated efficacy for the signs and symptoms of CVD and can slow progression. In patients with advanced CVD, such as those presenting with edema, skin changes, or ulceration, aggressive interventions are necessary and physicians should consider procedures to treat reflux and obstruction, while compression therapy and veno-active drugs have been shown to promote ulcer healing [188].

There is no prophylaxis for VVs, though some risk factors are modifiable. Without appropriate treatment, CVD will progress in most cases because of a cycle of chronic inflammation that leads to further deterioration in venous flow, venous hypertension, and edema. Preventive programs may help to inform patients about the risk factors for CVD, how they can reduce risk, the early symptoms, and the serious consequences of progression.

5.8 Role of TGF-β1 in vascular wall pathology

Various theories have been suggested, yet the molecular sequences which lead to the formation of venous incompetency are poorly understood. TGF-β1 is a highly complex polypeptide with multifunctional properties that has an active role during embryonic development, in adult organ physiology, and in the pathophysiology of major diseases, including tumor and various autoimmune, fibrotic, and cardiovascular diseases. Therefore focusing to understand its signaling pathways (and possible disruptions) will be an essential requirement for a better comprehension and management of specific diseases. This part of the discussion aims at shedding more light on venous pathophysiology by describing the TGF-β1 structure, function, activation, and signaling, and stipulating an overview of how this growth factor and disturbances in its signaling pathway may contribute to specific pathological processes concerning the vessel wall which, in turn, may have a role in chronic venous insufficiency (CVI) [189].

The term CVeD includes the full spectrum of morphological and functional abnormalities of the venous system, from telangiectasia to venous ulcers. Functional abnormalities of the veins of the lower extremities providing edema, skin changes, or venous ulcers are clinically known as CVI—a term reserved for advanced CVeD. Although it is extremely common, the exact prevalence of CVeD remains unknown. Reports of prevalence of CVI vary from <1% to 40% in females and from <1% to 17% in males. CVI not only affects a significant proportion of the population, but also causes considerable morbidity and adversely impacts the quality of life (QoF) of those affected. The recurrent nature of the disease, the socioeconomic burden, and the ineffectiveness of treatment modalities demands the need for more CVI-related research. Several theories about venous pathophysiology and varicosity genesis in the LLs (e.g., venous stasis theory, arteriovenous fistula theory, diffusion block theory) are outdated or have been refuted. A more recent hypothesis proposed that a dysfunctional venous system follows venous wall and valvular damage, which are triggered by venous hypertension and are the result of sterile inflammatory reactions. However, the molecular sequence of events that lead to venous wall remodeling and structural weakness remains poorly understood. It has been suggested that venous hypertension and/or wall hypoxia originates endothelial activation and expression of growth factors, adhesion, and signaling molecules, which lead to leukocyte activation and migration. This inflammatory process is responsible for the secretion of mediators that may trigger a local dysregulation of MMP/TIMP ratio that prompts abnormalities in ECM structure, leading to decreased elasticity and increased distensibility of the venous wall. Tissue remodeling is a complex process that is controlled by a great variety of factors, including TGF-β1 [190].

TGF-β1 is a highly intricate polypeptide that belongs to the superfamily TGF-β1, which contains more than 30 structurally related polypeptide growth factors in mammals. In general, the family members are subdivided into two functional groups: (1) the TGF-β-like group that includes TGF-β (1−3), activins, inhibins, and some growth differentiation factors (GDF); and (2) the bone morphogenetic protein (BMP)-like group comprising BMPs, most GDFs, and anti-Mullerian hormone [191]. The TGF-β superfamily members share a conserved cysteine knot structure, are ubiquitously expressed in diverse tissues, and function during the earliest stages of development and throughout the life-time of humans. Disturbances in TGF-β superfamily pathways, including either germline or somatic mutations or alterations in the expression of members of these signaling pathways, often result in several pathological conditions [189].

TGF-β1 is the most important isoform of the family in the cardiovascular system and is present in ECs, VSMCs, myofibroblasts, macrophages, and other hematopoietic cells. Knockout studies of TGF-β1 signaling components in mice offered the first indication of their crucial role in vascular development and function. Both EC and their supporting cells (VSMCs and pericytes) are needed to form a mature vascular network and TGF-β1 has been proposed not only to affect ECs and VSMCs (e.g., proliferation, differentiation, migration), but also to regulate the interaction between them. TGF-β1 is able to act as a promoting and inhibitory factor of angiogenesis [192].

In addition to the angiogenic effect, TGF-β1 can induce a process called endothelial-to-mesenchymal transition (EndMT), by which ECs lose apical—basal polarity and acquire a mesenchymal migratory phenotype. EndMT is essential during embryonic development and tissue regeneration/wound-healing, playing a role in pathological conditions like fibrosis or contributing to the generation of cancer-associated fibroblasts that are known to influence the tumor-microenvironment to make it favorable for tumor cells. TGF-β1 has also been shown to be a key regulator of ECM synthesis and remodeling. Specifically, it has the ability to induce the expression and deposition of ECM proteins, as well as to stimulate the production of protease inhibitors that prevent their enzymatic breakdown. Abnormalities found on the structural matrix components (e.g., collagen, elastin) and the resultant increasing of ECM stiffness/loss of elasticity are common observations in cardiovascular and venous diseases. Therefore the participation of TGF-β1 in the pathogenesis of vascular pathologies associated with matrix remodeling and fibrosis is not surprising. The widespread expression profile of TGF-β1 receptors in all immune cell types suggests TGF-β1 participation in broad activities of the immune system [193].

According to Goumans et al., it delicately regulates the tolerogenic versus immunogenic arms of the immune system to balance adequate host defense while limiting collateral inflammatory tissue damage. It is worth mentioning that this multifunctional growth factor is also known by its dual action [193]. Indeed, the TGF-β1-elicited response is highly context-dependant throughout development and across different tissues. For instance, some authors reported that TGF-β1 protects bovine aortic ECs from apoptosis, in contrast, others showed the opposite effect on porcine microvascular ECs. While at early stages of tissue repair TGF-β1, as a major orchestrator of the fibroproliferative response, stimulates the chemotaxis of repair cells, modulates immunity and inflammation, and induces matrix production; at early stages, it negatively regulates fibrosis through its strong antiproliferative and apoptotic effects on fibrotic cells [194]. Likewise, a dual role of TGF-β1 in the tumor-microenvironment was described; it seems to prevent tumor growth and angiogenesis at early phases of tumor progression. Moreover, numerous in vitro studies also showed a dose-dependent (and timing-dependent) action of TGF-β1; for example, EC invasion and capillary lumen formation are inhibited by high concentrations of TGF-β1, whereas lower concentrations potentiate the effects of basic fibroblast growth factor (FGF) and vascular endothelial growth factor-induced invasion [195].

5.8.1 TGF-β1 signaling pathways: major components and regulation

TGF-β1 is synthesized as an inactive protein precursor (i.e., preproprotein) consisting of a signal peptide, an N-terminal prodomain, and a C-terminal biologically active peptide. Early on, it was realized that TGF-β1 is secreted from cells in a latent complex with its prodomain (latency-associated peptide or LAP), indicating that during synthesis noncovalent interactions are formed between the prodomain and the mature domain, and a small latent complex comes into existence [193]. During the secretory process, the prodomain of TGF-β1 interacts covalently with a latent TGF-β-binding protein (LTBP) to form a large latent complex, which is then secreted into the ECM. LTBP associates with the TGF-β1 prodomain via the signature 8-Cys region, which is unique to those proteins. LTBP is required for secretion and correct folding of TGF-β1. The association with LTBP results in the storage of latent TGF-β1 in ECM structures rapidly after secretion. It remains inactive in these structures (a disulfide bond prevents it from binding to its receptors) until there is a proteolytic cleavage of LAP or a conformational change in LAP (induced by contractile forces)-critical events for protein activation [196].

5.8.2 TGF-β receptors and Smads

TGF-β1, as any other TGF-β family member, elicits its cellular effects by binding to different receptors. These are heteromeric complexes comprised of type 1 (activin-like kinase or ALK: ALK1, ALK2 and ALK5; ALK5 is also termed TGF-βR1) and type 2 (TGF-βR2) transmembrane serine/threonine kinase receptors. Type 2 receptors are constitutively active kinases capable of binding TGF-β1 and regulate this growth factor binding to its corresponding receptors, though they do not signal directly [49]. The type 3 receptor binds differing profiles of TGF-β family members. For instance, TGF-βR3, but not endoglin, can bind TGF-β2, an important distinction as TGF-β2 cannot otherwise bind to type 2 receptors. Thus, cells lacking TGF-βR3 are insensitive to TGF-β2 [197].

Through intracellular mediators, known as Smads, the TGF-β1 pathway can directly transduce extracellular cues from the cell-surface transmembrane receptor to the nucleus. This well conserved family (eight in mammals) can be divided into three functional classes: receptor-regulated Smads (R-Smads: Smad1, Smad2, Smad3, Smad5, and Smad9, which is mostly known as Smad8), common-mediator Smads (Co-Smad: Smad4) and inhibitory Smads (I-Smad: Smad6 and Smad7) [198]. All R-Smads have a C-terminal SSXS motif, within which the last two serines are directly phosphorylated by the type 1 receptor, while I-Smads lack the C-terminal SXS phosphorylation motif and thus act as inducible inhibitors (negative regulators) of the pathway [199].

The TGF-β1 initiates its signaling by binding to high-affinity cell-surface receptors, type 1 and type 2 receptors. TGF-β1 binds to TGF-βR2, resulting in conformational changes that induce recruitment and complex formation with an appropriate type 1 receptor. Within the heterotetra-complex just formed, two type 2 receptors transphosphorylate two type 1 receptors in the glycine serine-rich domain, activating their serine/threonine kinase activity. In turn, the activated type 1 receptors mediate cellular effects through interaction and phosphorylation of R-Smads (ALK5 mediates the phosphorylation of R-Smad2/3, while ALK-1/2 mediate the phosphorylation of R-Smad 1/5/8) [197]. Upon phosphorylation, two activated R-Smads form a complex with Co-Smad, and this complex moves into the nucleus, where it combines with transcriptional activators and repressors, modulating target gene expression in a cell type-dependent manner. The activation of R-Smads can be inhibited by I-Smads, which can compete for TGF-R1 interaction, recruiting specific ubiquitin ligases or phosphatases to the activated receptor complex [200].

5.8.3 Role of TGF-β1 signaling pathways in vessel wall pathological processes

5.8.3.1 TGF-β1 and vascular wall shear stress

Mechanical forces imposed by pulsatile flow of blood, which include frictional wall shear stress, circumferential distention, and blood pressure, play an important role in maintaining vessel structure and function. ECs lining the vasculature are continuously exposed to shear stress, leading to reorganization of their cytoskeleton, morphological alterations, and the production of a variety of substances that act on ECs themselves and on surrounding cells (e.g., VSMCs). Failure to adapt to shear stress results in endothelial damage, which may lead to the generation of atherosclerotic plaques or abnormal vessel repair [201].

Several studies were able to link TGF-β1 production and vascular remodeling induced by shear stress. It was demonstrated that human arterial and venous VSMCs exposed to chronic cyclical mechanical strain responded, in a "dose-dependent" fashion, with an increase of TGF-β1 mRNA expression and matrix accumulation. According to the author's proposal, this would most likely represent the main biological mechanism whereby hypertension promotes cardiovascular matrix accumulation [202]. Other authors could confirm that low shear stress (which occurs preferentially at vessel branch points, bifurcations, and regions of high curvature) was a pathological inducer for vascular remodeling by upregulated migration and proliferation of ECs and VSMCs [198]. An increased paracrine secretion of platelet-derived growth factor-BB (PDGF-BB) and TGF-β1 from ECs induced by low shear stress was found, as well as the activation of extracellular signal-regulated kinase (ERK)1/2 and affected expressions of LOX and lamina A-processes that were suggested as having a possible role in the effects of PDGF-BB and/or TGF-β1 on cellular migration and proliferation [203].

Furthermore, observations on vein graft remodeling have identified hemodynamic forces (wall shear and tensile stresses) as the primary stimuli that induce active reorganization of the graft wall, and a role for TGF-β1 in this event. Evidence supporting the concept that increased wall stress after vein graft implantation induces the recruitment of adventitial fibroblasts, mediated by a connective tissue growth factor (CTGF, also known as CCN2) and TGF-β1, and the conversion to a myofibroblast phenotype has been provided. This adventitial adaptation, despite being important in the maintenance of wall stability, limits the early outward remodeling of the vein conduit and may have a detrimental effect on maintaining the long-term vein graft patency [189].

Recent findings (such as suppression of TIMP1, enhanced expression of TGF-β1 and BMP-2 mRNA, or upregulation of microRNA-138/200B/200C) were consistent with the previous results and suggested a role of arterial-like wall strain in the activation of propathological pathways, resulting in adventitial vessel growth, activation of vasa vasorum cells, and upregulation of specific gene products associated to vascular remodeling and inflammation. The potential role of TGF-β signaling in mediating the protective effects of physiological shear stress on ECs has been also studied. Interestingly, the results revealed that shear stress induced TGF-β3 signaling and a subsequent activation of Kruppel-like factor 2 and nitric oxide (NO), indicating that TGF-β3 (but not TGF-β1) has a critical role in the maintenance of endothelial homeostasis in a hemodynamic environment [204].

5.8.3.2 TGF-β1 and vascular wall fibrosis

In physiological conditions, fibrosis is a process of normal wound-healing and repair, activated in response to injury, to maintain the original tissue architecture and its functional integrity. It involves a complex multistage process with recruitment of inflammatory cells, release of fibrogenic cytokines and growth factors (such as TGF-β1), and activation of collagen-producing cells. However, prolonged chronic stimuli lead to a long-term activation of myofibroblasts (a specialized type of fibroblasts that is normally activated during wound-healing, which in turn may result in excessive and abnormal deposition of ECM and fibrosis). If the build-up of ECM occurs in organs (e.g., lungs, liver, kidneys, and skin) it can interfere with their function and, if it continuous unabated, leads to organ failure [205]. As mentioned before, TGF-β is a key regulator of ECM, thus its excessive signaling has long been implicated in the pathogenesis of vascular fibrosis and other fibrosis-related diseases. Moreover, it acts as a mediator of vascular fibrosis induced by several agents involved in vascular diseases (e.g., mechanical stress, angiotensin II, high glucose, advanced glycation products). The heat shock protein 70, whose primary function is to repair denatured proteins through folding/unfolding steps and thus achieve correct functional configuration, is another example of an agent believed to stimulate TGF-β1-induced ECM accumulation and to contribute to the inflammation and fibrosis present in fibrosis-related diseases [206].

Several genes encoding ECM proteins that are known to be important in driving fibrosis are directly regulated by TGF-β signaling, through Smads [85–87] as well as with the involvement of MAPKs, Rho family members, and reactive oxygen species. TGF-β1, at low concentrations, increases the synthesis of ECM proteins, such as fibronectin, collagens, and plasminogen activator inhibitor one in VSMCs, ECs, and fibroblasts.

The synthesis of fibronectin by VSMCs, via the TGF-β1/Smad3 signaling pathway, leads to the deposition of ECM in the neointima. Other authors have shown that in a high-phosphate environment, the upregulation of fibronectin in cultured VSMC takes place via TGF-β1 production. The reduction of types 1 and 3 collagen secretion by VSMCs is induced through TGF-β1/Smad3 signaling pathway inhibition [207].

It has been suggested that TGF-β1 and CTGF synergize to promote chronic fibrosis. CTGF is expressed predominantly in embryonic and adult vasculature, regulates various biological apoptosis, ECM production, and angiogenesis, and was found to be upregulated in a variety of fibrotic disorders. The enhancement of TGF-β1 activity via CTGF may occur through the following mechanisms. CTGF not only increases the affinity between TGF-β1 and its receptors. Evidence from in vitro studies indicated that CTGF might act as a downstream target of TGF-β1, as treatment with exogenous TGF-β or exogenous CTGF significantly upregulated the TGF-β/CTGF pathway and increased the expression levels of ECM components. In addition, studies with animal vein bypass grafts showed that enhanced signaling via TGF-β/CTGF, coupled with the reduced MMP2 and MMP9 activity, promotes progressive ECM accumulation and neointimal fibrosis during late neointimal expansion in vein grafts [208].

5.8.3.3 TGF-β1 and venous wall abnormal morphology and functioning

Healthy veins of the lower extremities are equipped with efficient walls, contractile VSMCs, and competent valves in order to withstand the high hydrostatic venous pressure in the LLs and allow unidirectional movement of deoxygenated blood toward the heart. In contrast, VVs (a common clinical manifestation among patients with CVI) appear to be dilated, elongated, tortuous, and often show incompetent venous valves and a measurable venous reflux. Moreover, structural and histological evidence suggests that VVs have both hypertrophic (with abnormal VSMC shape and orientation and ECM accumulation) and atrophic (with ECM degradation and an increase in inflammatory cell infiltration) regions and no clear boundaries among vascular layers—irregular distribution of collagen bundles or thickened and fragmented elastic fibers may be found throughout the vein wall making it difficult to distinguish tunica intima, media, and adventitia [209]. The primary reason for this extensive ECM remodeling and structural weakness of the vein wall has still not been scientifically explained, but numerous factors, including TGF-β1, seem to be implicated not only in the pathogenesis of VVs, but also in numerous complications associated with VVs (e.g., thrombophlebitis, lipodermatosclerosis, venous ulcers).

Some studies have reported unchanged TGF-β1 levels in cell cultures from VVs and comparable amounts of TGF-β1 mRNA levels or TGF-β1 active form in normal and VVs. By contrast, others demonstrated increased TGF-β1 mRNA levels, protein expression, immunoreactivity, and total content in the walls of VVs, or decreased protein expression of TGF-β1 latent and active forms in VVs when age-related differences were controlled for. Furthermore, as tissue responsiveness to TGF-β1 depends on several factors (e.g., TGF-β signaling components availability) rather than on its activation/availability alone, a few studies evaluated mRNA expression and protein expression of TGF-β R2/3 and Smad2/3 in VV walls and achieved, once more, opposite results [210]. The conflicting results regarding TGF-β1 (and its signaling components) expression/activity may be partially explained by important methodological differences: for example, (no) use of effective control samples (i.e., control and VVs were harvested from patients with CVI), when some argue that CVeD is a generalized disorder in the venous system; (no) control of individual differences between control and experimental groups, when there is evidence that aging induces dysregulation of TGF-β1; (no) distinction between hypertrophic and atrophic varicose segments, when considerable heterogeneity regarding cellular and matrix components presence was already observed; (no) separation of specimens based on anatomic harvest site, when vein source and location seem to be a factor of variability [210].

As explained earlier, hemodynamic forces (wall shear and tensile stresses) and inflammation are among the potential factors that could modulate the expression/activity of TGF-β1 in the vascular wall. An increase in venous hydrostatic pressure in the lower extremities (caused by certain genetic, environmental, and behavioral risk factors) could lead to EC injury, increased permeability, activation of adhesion molecules, and leukocyte infiltration—collectively these factors could contribute to vein wall inflammation [210]. In response to inflammation or injury, the active form of TGF-β1 can be released by a variety of mechanisms, including enzymes such as proteases and glycosidases secreted by leukocytes and mast cells. Enhanced infiltration of mast cells was noted in VV walls particularly from elderly subjects, suggesting that degranulation of mast cells may release enzymes into the ECM (e.g., tryptase, hydrolases, oxidative enzymes, carboxypeptidases) and contribute to TGF-β1 maturation or activation, as well as to venous wall diminished structural integrity (e.g., mast cell tryptase catalyzes the degradation of different matrix components such as type IV collagen, elastin, fibronectin, and extracellular proteoglycans) [211].

Similarly, while studying the relationship between NO production (an important cellular-signaling molecule that regulates vascular tone and has diverse pathophysiologic functions such as inhibition of platelet adhesion/aggregation, mediation of the inflammatory cascade, among others) and TGF-β1 expression in VVs, it was found that TGF-β1 overexpression was correlated with overproduction of inducible NO synthase and with monocyte/macrophage infiltration in both tortuous and nontortuous VVs. It was then speculated that NO released by the upregulation of NO synthase enhances the production of TGF-β1 in VVs, which in turn may be related to dysregulated cell cycle (e.g., decreased apoptosis) and to progressive hypertrophy of the vein wall [212].

Due to TGF-β1's broad influence on ECM remodeling, vein wall imbalance of elastin and collagen content and the growth factor expression/activity (mostly in association with MMP/TIMP ratio changes) have also been objects of study. Depletion of vein elastic components is a common observation in the venous wall of elderly and CVI patients and was related to decreased elastin synthesis and increased elastase activity, which in turn could be a consequence of elastic tissue modulators' (such as LTBP and TGF-β1) attempts to stabilize elastin expression in regions of extensive injury [213]. Changes in TGF-β1 expression/activity have also been associated with CVeD progression and with numerous CVI-related complications. Several studies concerning advanced forms of dermal pathology in CVI patients (i.e., dermal fibrosis and venous ulcers) have identified CVI dermal fibroblasts as an important target for leukocyte-derived TGF-β1, and established a link between CVI progression and increased tissue levels of TGF-β1, MMP2 activity, and decreased TGF-β1-induced mitogenic responses of fibroblasts. Moreover, two other studies on CVI-induced dermal changes suggested a connection between TGF-β1 signaling and cellular senescence [214].

TGF-β1 plays a vital role in many biological and pathological processes, some of which have an impact on cardiovascular homeostasis. Considerable progress has been made over the past several years in the understanding of biomechanical and structural aspects of this growth factor, including its activation and intracellular signaling (canonical and noncanonical) pathways. CVeD is far from being completely understood for some major reasons: a multifactorial etiology and a great difficulty in identifying the primary stimulus and mapping the sequence of pathological events (by the time CVI symptoms present clinically, a vicious cycle of pathological events is already in motion) [193]. The vein wall changes and valvular dysfunction (appointed as primary events) are triggered by blood stasis and venous hypertension and are the result of sterile inflammatory reactions [193].

All the main mechanisms postulated to contribute to the disease onset and development (i.e., shift in hemodynamic forces, wall inflammation, ECM degradation/deposition, venous tone alteration) seem to affect or be affected by TGF-β1 expression and activity, which can be dysregulated. Given the great number of components and regulation factors involved in TGF-β1 signaling, future studies should focus on the expression/activity of molecules beyond TGF-β1 itself (e.g., LTBP isoforms, TGF-β receptors, Smads, MAPKs, integrins). Among CVeD treatment modalities, the outlook for pharmacotherapy targeting the TGF-β1 signaling pathway seems promising, yet challenging—current drugs targeting TGF-β1 and its signaling components are not selective for pathological signaling pathways and require careful monitoring of side effects. It is essential to develop further studies which could be able to dissect TGF-β1 mechanisms of action as well as to identify new cell type regulators (that could be safely used for anti-TGFβ signaling therapy), in order to improve the effectiveness of CVeD prophylaxis and treatment [215].

5.9 Role of mitogen-activated protein kinase (ERK1/2) in venous reflux in patients with chronic venous disorder

CVeD is a disorder in which there is an alteration in the conditions of blood return to the heart. This disorder may arise from incompetent valves and the resultant venous reflux (CVI). The mitogen-activated protein kinase (MAPK) enzymes mediate a wide array of pathophysiological processes in human tissues. In this family of proteins, ERK1/2 plays a direct role in the cell homeostasis that determines the viability of mammalian tissues. This study sought to examine whether ERK1/2 plays a role in venous reflux. A prospective study was conducted on 56 participants including 11 healthy controls. Of the CVeD patients, 23 had venous reflux with CVI (CVI-R) and 22 had no reflux (NR) [216]. Great saphenous vein specimens were subjected to gene (real-time polymerase chain reaction), and protein (immunohistochemistry, IHC) expression techniques to identify ERK1/2. The Mann–Whitney U test was performed to compare the data between groups. Patients with CVI showed significant gene activation of ERK1/2 protein, and, in those with venous reflux, the expression of this gene was significantly greater. The CVI-R group <50 years showed significantly greater ERK1/2 gene expression than their age-matched controls. Expression patterns were consistent with IHC findings. The studies suggested that ERK1/2 expression is involved in venous vascular disease. The intensity of ERK1/2 protein expression was significantly greater in the veins of patients with CVeD ($P = .0007$) [216].

The histopathology study revealed a lack of ERK1/2 protein expression in the controls <50 years, while controls >50 years showed mild expression levels in the vein tunica media which was visualized around the muscle fiber bundles. In the older CVeD patients without valve incompetence (NR > 50), ERK1/2 protein expression was detected in all three vein wall tunicae. These patients displayed a similar intensity of expression as their younger counterparts (NR < 50 years). ERK1/2 appeared in the ECM of the tunica media, around the muscle fiber bundles, and also in the adventitial layer. The endothelium was well preserved and showed strong reactivity for the ERK1/2 protein. In patients with venous reflux, ERK1/2 protein expression patterns differed according to age. CVI-R patients <50 years showed greater ERK1/2 labeling intensity in their samples in the intimal and medial layers, with areas of hypertrophy in the tunica media and hyperplasia in the tunica adventitia [217]. Expression was especially intense in the muscle fiber bundles of the tunica media where blood capillaries showed greater reactivity for the protein. The tunica intima showed reactivity throughout the endothelium, though most of this layer was not preserved, unlike the observations in the NR group. These samples from the younger patients (CVI-R <50 years) stood out because of their constant ERK1/2 expression throughout the whole of the tunica intima and tunica media.

Various mechanisms have been proposed to explain its etiology, but one of the most accepted is valve incompetence and elevated venous pressure. Classically, it was described that VVs arose because of increased venous pressure produced by a lack of valves in the great saphenous vein. However, today, the condition is considered multifactorial, whereby factors such as intrinsic structure, biochemical changes, changes in cell mechanisms and smooth muscle, as well as enzyme activities play an important role in its etiology [190].

Pappas et al. examined the role of ERK1/2 in venous disease. They were able to correlate its activation with disease progression in patients with CVI, and they proposed a key role for this factor in ECM contraction to promote regulation of the activity of smooth-muscle actin (α-SMA) in fibroblasts. The increased synthesis and secretion of ERK1/2 proteins induce tissue milieu modifications that may affect tissue structure and function. The results strongly suggest a link between the hypertrophy and hyperplasia observed in our patients, showing the greater gene and protein expression levels of this MAPK. Several studies have shown how increased ERK1/2 signaling may be implicated in a senescence response of fibroblasts, likely affecting vein ulcer healing, hypertrophy, and hyperplasia. Many lines of evidence point to a role of integrins as mechanosensors and mechanotransducers of extracellular signals, with consequences for overall vein physiology. Integrins are heterodimer cell-surface transmembrane receptors that join the ECM to the intracellular cytoskeleton of actin. Forces exerted on the extracellular membrane are thus converted into chemical signals via interactions with integrin receptors in vein remodeling [216]. A triggering event for such interactions could be the vein hypertension that exists in CVI. In this setting, inflammatory factors will promote matrix contraction, which is also linked to greater α-SMA and the development of a more contractile muscle fiber phenotype. Pascual et al. and Serralheiro et al. showed that the proinflammatory factor TGF-β is activated in patients with CVI, thus indicating a relationship between the environment generated by vein hypertension and changes produced in cell transduction mechanisms in vein insufficiency. Similarly, our data suggest that the venous reflux in CVI is related to ERK1/2-driven differentiation. Hence, the development of a contractile phenotype increases mechanical stress in the matrix as revealed by histological findings. According to the data, venous hypertension could affect other cell repair mechanisms like apoptosis. ERK1/2 activation has an impact on the cell in several ways. The described stimuli direct ERK1/2 toward specific action sites so that they can perform functions in the membranes, cytoskeleton, and/or nucleus. The important role that ERK1/2 possibly plays in long-term apoptic activity through its phosphorylation was recently described [216].

The study was designed to examine the possibility of the differential expression of ERK1/2 protein related to the cytoarchitectural variations observed in the different vein tunicae in patients with CVeD and/or valve incompetence. Our results show that patients with CVeD have higher gene and protein levels of ERK1/2, and that when there is venous reflux, younger individuals show greater ERK1/2 expression, although the levels (of both protein and gene expression) were similar to those detected in the older CVeD patients but with competent valves [216]. We propose that in young patients with valve incompetence, the aging process is altered and, as described by Serra et al., there are possibly many genes that will trigger CVeD. This significant increase in expression may be related to clinical signs, since we saw that it was increased with the process of incompetent valves in young patients with venous reflux. These factors in CVD should be further investigated as this was one of the limiting points of the study conducted. Future studies should include a better understanding of the cellular mechanisms of both the great and small saphenous veins. Comparison of different segments of the great saphenous vein (thigh vs leg in CVeD vs controls, respectively), also a limitation of our study, was conducted. Nevertheless, for the very first time, the involvement of ERK1/2 in CVI is evident [218].

Many ERK1/2 proteins are transcription factors. MAPK translocation to the nucleus is therefore important for mitogenic induction. This determines that the effect of ERK1/2 is exerted through its capacity to induce differentiation, angiogenesis, proliferation, survival, and protein synthesis, that is, it plays an essential, physiopathological role. Current preclinical trials are being directed towards Ras/Raf and MEK kinase inhibitors as blocking agents of ERK signaling as, so far, inhibitors of ERK1 and ERK2 mutation activation have not been detected yet, in future work, the use of these inhibitors to modulate cell actions in CVeD, and especially in venous reflux, needs to be explored. The mechanisms responsible for modulating the ERK1/2 in CVeD should also be investigated [219].

5.10 Human venous valve disease caused by mutations in FOXC2 and GJC2

Venous valves (VVs) are widely distributed throughout veins and venules and facilitate unidirectional blood flow back to the heart, which acts to reduce the peripheral venous blood pressure. VV failure is a central feature of the venous reflux that is seen in upto 40% of adults. Reflux leads to chronic venous hypertension (particularly in the LLs), which can cause pain, edema, hyperpigmentation, skin damage, and chronic intractable ulceration. Our understanding of the molecular mechanisms of VV development and subsequent maintenance is limited, and there are few therapeutics options to treat VV dysfunction. Elucidation of these mechanisms and understanding of how their dysfunction may lead to VV failure could facilitate the development of novel therapies to treat this condition. Clinical studies have suggested a link between venous reflux and primary lymphedema, but its cause (i.e., a direct VV defect or an indirect effect, such as vein dilation) has not been elucidated. We have previously shown how genes (Itga9, Efnb-2, Fn-EIIIA) regulating lymphangiogenesis also control VV formation and maintenance [220].

Several transcription factors (Prox1, Foxc2, Nfatc1) and gap-junction proteins (connexion [Cx]37, Cx43, and Cx47) have been implicated in the development of lymphatic valves (LVs), cardiac valves, or VVs. Mutations in the genes encoding FOXC2, (FOXC2) CX47 (GJC2), and CX43 (GJAI) cause primary lymphedema in humans. $Gjc2^{-/-}$ and $Gja4^{-/-}$ (Cx37) mice have VV defects during VV maturation, but the timing of onset of those abnormalities during VV formation and the developmental process underlying absent VVs have not, to our knowledge, been studied. Expression of Foxc2 and Nfatc1 is segregated to opposite leaflet surfaces during valve maturation, but their expression during VV initiation and potential cooperative signaling has not been studied. Similarly, during leaflet maturation Cx37 and Nfatc1 are not coexpressed in the same valve leaflet ECs, but expression patterns in VV initiation remain undetermined [165].

In this study, we quantified the VV defects in the limbs of patients with primary lymphedema caused by mutations in GJC2 or FOXC2 and examined the earliest underlying mechanisms of VV failure caused by loss of those genes in mice. Reduced numbers of VVs and shorter VV leaflets in patients with lymphedema were caused by mutations in FOXC2 or GFC2. The loss of connexion-encoding genes Gja1, Gja2, or Gja4 similarly resulted in a failure of VFC (Valve-forming endothelial cell) organization, and for Gjc2 and Gja4, that resulted in reduced VFC proliferation. Both Nfat and Foxc2 signaling and blood flow were required for VV maturation to postnatal day 6 (p6), unlike its role in LVs, we show that Foxc2 is not required in VV ECs for valve maintenance [221].

5.11 FOXC2, GJC2, and GFA1 mutations in human venous valve disease

Understanding the relationship between primary lymphedema and VV dysfunction has been hampered by our inability to noninvasively quantify the presence and morphology of VVs in patients. Although VVs have been seen by ultrasonography, they have not been quantified on a systemic basis. Conventional ultrasonography was used to visualize VVs in humans. Quantification was reproducible for the number of VVs/vein in repeated scanning by different operators ($n = 28$ veins, intraclass correlation coefficient = 0.90, 95% confidence interval [CI] 0.63–0.91, $P < .0005$) and for leaflet length in repeated measurement by different operators ($n = 15$VV, intraclass correlation coefficient = 0.94, 95% CI 0.83–0.98, $P < .0005$). This allowed quantification of the number of VVs and VV leaflet length in the upper and lower limbs. To limit scan duration, only the short saphenous, popliteal, basilic, and brachial veins were scanned [220].

The results were surprising because of the reduced numbers of VVs in patients with mutations in both FOXC2 and GJC2 compared with age- and sex-matched controls. Those VVs that were present in the patients with FOXC2 or GJC2 mutations had shorter leaflets. VV defects were present in the veins of both upper and lower limbs in patients for both FOXC2 and GJC2. There also appeared to be fewer VVs in the single patient identified with lymphedema caused by a mutation in GJA1 (CX43). Consistent with these findings and in agreement with their expression pattern in mouse VVs, FOXC2 and CX43 were localized on adult human VVs, suggesting that human VV phenotypes may result from a tissue-autonomous effect of the patient's mutations [220].

CX43 is predominantly located on the sinus surface. It was not able to obtain typical punctate CX47 staining using commercially available antibodies in human VVs. It is possible that the CX47 expressed in VVs during development and early life is subsequently downregulated in the adult human VV, as has been found in mice for VEGFR3. To investigate possible underlying mechanisms for these human phenotypes, we next analyzed VV development in mice and focused on the proximal femoral vein (FV) because of the critical role of proximal deep VVs in human disease and because of VVs most consistently develop at this position in the mouse, allowing reliable identification of the earliest stages in its development. The mechanisms by which mutations in FOXC2 and GJC2 lead to defective VVs are unknown [222].

5.11.1 Maturation of leaflets and commissures

The importance of maturation of VVs to P6, and influence Foxc2 or calcineurin-Nfat signaling for development of VV leaflets and commissures were analysed. In contrast to their cooperative roles in patterning the initial VV ring, calcineurin inhibition with CsA, loss of CnB1 ($Ppp3rl^{lx/lx}$ deletion) at P0, or deletion of FOxc2 alone were each sufficient to cause significant defects in leaflet/commissure development through to stages 3–4. There was no loss of the ring of rotated $Prox1^{hi}$ free-edge cells, which remained intact even at P6, demonstrating that neither Nfatc1 signaling nor Foxc2 are required for maintenance of this initial valve structure and the free-edge phenotype of these cells. Cx37 remained expressed in VV leaflets at P6. Unlike the phenotype seen with loss of Ppp3r1, FOXC2 and to some extent, Itga9 or Efnb-2, analysis of $Gja4^{-/-}$ VVs at P6 revealed complete absence of both leaflet structure and $Prox1^{hi}$ and $Foxc2^{hi}$ cells, demonstrating that the early development of VV seen at P0 in $Gja4^{-/-}$ (Cx37) mice is subsequently lost, and Cx37 is required to maintain the free-edge cell phenotype. This was associated with failure to form the gap in SMA-expressing cells normally seen around the valve. As expected, VV in $Gja4^{+/-}$ mice developed normally. These results are consistent with previously reported findings in LVs and with the absence of VVs at P4 in $Gja4^{-/-}$ mice [223].

Because several genes required for VV development, including FOXC2 and Gja4 (Cx37), are upregulated by fluid shear stress, it was hypothesized that blood flow may be required for normal VV leaflet growth. To alter blood flow across the developing VV, the FV was ligated and divided at P0, and the VV analyzed at P6. No thrombosis was seen, and pups operated on gained weight normally ($P > .05$ vs unoperated littermates). The FV was ligated initially at P0, which did not result in visible diversion of blood flow (likely because of the small collaterals seen at the site of the ligation; not depicted), and valve development was unaffected. When the FV is ligated twice and divided, it resulted in rerouting of blood via collaterals. VVs exposed to those altered flow conditions showed reduced progression to the later stages of leaflet development compared with VVs exposed to unaltered flow conditions on the unoperated, contralateral slide, and were smaller. Because patients with heterozygous FOXC2 mutations showed severe VV defects, and FOXC2 is required for LV maintenance, we next asked whether Foxc2 was also needed for VV maintenance as well as development [220]. Induction of deletion in $Foxc2^{lx/lx}$ mice at 4 weeks resulted in absent Foxc2 immunosignal 4 days later (not depicted), but interestingly, VV length at 6 weeks ($Foxc2^{lx/+}$ or $Foxc2^{lx/lx}$) or 28 weeks ($Foxc2^{lx/+}$) was similar to littermate controls, demonstrating that EC Foxc2 expression is not required for VV maintenance at the timepoints analyzed [220].

Profound structural VV defects were identified in patients carrying mutations in the genes encoding the transcription factor FOXC2 and the gap-junction protein CX47. VV development in mice to the earliest endothelial events were studied in detail and identified a temporospatially regulated pattern of transcription factor and gap-junction protein expression around $Prox1^{hi}$ VFCs. This is required for the organization of an initial ring of VFCs that is then critical for ongoing VV development [220]. Using a genetic loss-of = function approach combined with drug inhibitors, we have demonstrated requirements for Cx37, Cx43, and Cx47 and provided evidence for Foxc2–Nfatc1 cooperative signaling in the regulation of VFC organization at P0. At later stages, the requirements for Foxc2 and calcineurin–Nfat signaling and blood flow in regulating the maturation of valve leaflets were identified.

5.11.2 Human venous valve phenotypes

Valve defects were particularly severe in patients with FOXC2 mutations, with 75% fewer valves, and the remaining valves were almost half the length of valves in matched controls. This finding is consistent with the increased incidence of CVI, which was previously identified in the same group of patients. Patients with mutations in GJC2 had 62% fewer valves, and those present were substantially shorter than those seen in controls. Again, this is consistent with the studies of incompetent veins in those patients. The human GJC2 mutations studied here produce an aa substitution in the first extracellular loop. The mutation in GJA1 (Cx43) results in a substitution (K206R) in the highly conserved SRPTEK motif of the second extracellular loop and may exert dominant negative effects on connexion docking, gap-junction functions, and levels of WT (Wild Type) protein. The finding of fewer VVs in a single patient (albeit without statistical significance) is consistent with the VV developmental defects seen in mice after Gja1 deletion and requires exploration as more patients with GJA1 mutations are identified [224].

In lymphatic endothelium, Foxc2 and Nfatc1 are coexpressed, physically interact, and signal cooperatively to pattern vessel maturation. The data indicate differential requirements for maintenance of the free-edge cells during later development beyond P0. As expected, at P6 a severe phenotype was seen in both mutants with complete loss of VV leaflets but with some residual Prox1^{hi} VFCs remaining. By comparison, valve cells were completely absent at P2 with loss of Cx47, and at P6 with loss of Cx37. This phenotype was more severe than that identified with loss of integrinα9, ephrin-B2, or CnB1. This restricts the requirements for Nfatc1 to extension of the VV leaflets and not to maintenance of VFC identity. The results suggest that Cx37 and Cx47 are required for the early stability of the VFC ring phenotype, as has previously been suggested for Cx37 in LVs [225].

These results are also in accordance with a study of failed invagination of lymphovenous valves in Gja4$^{-/-}$ mice. Cx37 remains highly expressed by VVs and may additionally be required for VV maintenance. In Gjc2$^{GFP/+}$ mice, Cx47 could also have ongoing roles in maintenance. The role of Cx43 was examined by conditional deletion in ECs, and we found that Connexin was required for early but not later stages of valve development. Gja4$^{-/-}$ and GjC2$^{GFP/GFP}$ mice failed to develop an area of reduced SMA-expressing cells around the valve, most likely secondary to loss of the valve ECs [220].

Calcineurin–Nfat signaling was critical for leaflet elongation and commissure formation, consistent with its requirements for leaflet growth in the aortic valve and LVs. Conditional homozygous deletion of Foxc2 similarly produced a phenotype during leaflet maturation. In contrast to the requirement for Foxc2 in LV maintenance, we found that Foxc2 was not important for VV maintenance during the time frame investigated. The timing of the onset of VV failure in patients with FOXC2 mutations is unknown; further work should explore whether VVs in those patients develop normally and later regress or fail to develop at all. We speculate that other transcription factors (e.g., FOXC1) could regulate VVs and compensate for loss of FOXC2 [221].

5.11.3 The role of blood flow

The regulators of Cx37, Cx43, and Cx47 in VV formation remain unclear. In vitro and in vivo experiments indicate that Foxc2 regulates expression of Cx37 in lymphatic endothelium. Although both are expressed by VFCs, the lack of a VFC phenotype seen after Foxc2 deletion suggests that other factors can compensate for the loss of Foxc2 in these cells at P0. For example, Klf2 regulates Cx37 expression in blood endothelium in regions exposed to high laminar shear stress. Alternatively, induction of a GATA2/Foxc2/Prox1 pathway by oscillatory shear stress has been proposed as a mechanism for initiation of LV formation. GATA2 is expressed by venous VFCs and a similar process may regulate VV initiation. The early protrusion of VFCs into the lumen and the associated shear exposure may provide a further mechanism to regulate expression in those cells. The findings of smaller, less-developed VVs after reduction in blood flow through the FV (albeit during VV maturation, not initiation) is consistent with that concept [220].

Patients with mutations in Foxc2 and GJC2 have a globally reduced number of VVs and shorter VV leaflets. Foxc2 and calcineurin–Nfatc1 signaling cooperate to organize the initial ring of VV-forming cells. Cx37, Cx43, and Cx47 are critical for early organization of VFCs at P0, and failure of this process likely underlies abnormal VVs identified in patients with mutations in GJC2. Foxc2 expression in valve ECs is not required for VV maintenance [225].

5.12 Pathophysiological mechanisms of chronic venous disease and implications for veno-active drug therapy

Estimates for the prevalence of CVD in adults, typically diagnosed by the presence of VVs, have been reported to vary between 5% and 65%, depending on the population, and tend to be higher in Western countries than in developing countries. However, the vein consult program, an international, observational, prospective survey that included over 90,000 patients across different geographic regions, demonstrated that the prevalence of symptomatic CVD (C0s or higher) was roughly similar around the globe, with a prevalence of 78% in Western Europe, 87% in Eastern Europe, 88% in Latin America, 85% in the Middle East, and 87% in the Far East. VVs and CVD are generally reported to be more common in women than in men; a notion that has been reinforced by several cross-sectional epidemiological studies. CVD C1–C3 but not CVD/CVI (C4–C6) was more common in women in the vein consult program, while in the Bonn Vein study VVs, CVI, and leg symptoms were all more prevalent in women than in men [226].

Indeed, established risk factors for CVD are increasing age and previous pregnancy. Previous pregnancy was associated with an odds ratio of 1.98 for VVs and 1.20 for venous reflux, whereas the risk of CVI was found to increase with the number of pregnancies. Less consistent risk factors have been reported to be female sex alone (without previous pregnancy), family history of CVD, high body mass index, and occupations that involve prolonged standing or sitting. The symptoms of CVD can be a daily challenge for patients and can measurably decrease QoL. Symptoms also limit the ability of patients to participate in social, physical, or occupational activities, with financial consequences due to lost work time or disability. CVD also represents a substantial burden for healthcare systems. Cost accrues from inpatient and outpatient hospital care, treatments, and losses in productivity and work days due to disability, especially in cases with venous ulceration [183]. Total annual healthcare costs for venous and lymphatic disorders have been estimated to be €2.18 billion for Germany in 2006 and €2.24 billion for France in 1991. Da Silva et al. reported that, in Portugal, patients with venous ulcers required an average of 32 dressings and approximately 40 days of disability, with 12.5% taking early retirement. The total healthcare system costs of CVD have been estimated at 1%–2% of the healthcare budgets in Western European countries and the United States, with 22% of nurses' time spent on the treatment of leg ulcers.

5.13 Pathophysiology of chronic venous disease

Individuals with venous hypertension, which may occur consequent to predisposing factors such as advanced age, obesity, previous pregnancy, family history, and/or environmental or occupational factors, are at high risk of developing in CVD. In most cases, this hypertension is associated with poorly functioning venous valves and venous reflux, though it may also result from previous venous obstruction. In normal healthy individuals, blood is pumped out of the legs by calf muscle action. Ineffective calf muscle pumping due to leg immobility from prolonged sitting or standing, inactivity, or obesity contributes to venous blood pooling. The subsequent venous hypertension in the lower extremities initiates a circular loop of vascular and inflammatory phenomena that potentiate the hypertension [226]. There is increasing evidence to support the notion that among several possible trigger mechanisms, CVD is primarily a blood pressure-driven inflammatory disease, although the sequence of events is not fully understood and may differ depending on the risk factors involved. Elevated venous pressure and a shift in fluid shear stress generate an abnormal biomechanical environment in veins, vein walls, and valves.

These hemodynamic and biomechanical abnormalities may induce endothelium dysfunction, leading to early release and activation of enzymes that degrade the ECM and, in turn, set in motion a cascade of leukocyte infiltration and inflammation [227].

As a consequence of the EC activation, several growth factors are secreted, inducing hypertrophic effects in VSMCs, ECs, and ECM. Vascular endothelial growth factor (VEGF), platelet-derived growth factor, angiotensin 2, endothelin 1, and FGF-β, are some examples of the growth factors found to be elevated in VVs, all of which contribute to VSMC proliferation, which in turn disrupts the normal structure of the vein wall. Since VSMC must dedifferentiate to undergo proliferation, proliferating VSMCs have limited functionality and contractility, and are therefore poorly suited for maintaining vessel tone in the context of chronic high pressure. Imbalanced collagen synthesis by VSMCs has been implicated in the altered elasticity and increased rigidity of VVs, which weakens the vein wall and renders it incapable of maintaining shape and integrity against high pressure [228].

Elevated concentrations of TGF-β1 also contribute to the disruption of ECM homeostasis, as TGF-β1 stimulates the release of TIMPs responsible for degrading ECM proteins. Under conditions of normal blood flow and high shear stress, leukocytes remain unattached or roll along the lumen of the vein by adopting a rounded shape without pseudopods and by producing low levels of cell-surface adhesion molecules. Additionally, venous ECs respond to pulsatile laminar shear stress by producing antithrombotic and antiinflammatory factors, such as nitric oxide and prostacyclin, and by limiting the proinflammatory action of tumor necrosis factor alpha [229]. When venous reflux and altered hemodynamics with elevated pressure are present, reduced and abnormal fluid shear stresses are produced in the vein wall, disrupting normal venous homeostasis and activating the production of inflammatory factors by ECs, such as monocyte chemoattractant protein 1 (MCP-1) and bone morphogenic protein 4. ECs also increase production of cell adhesion molecules, such as intracellular adhesion molecule 1 (ICAM-1) and vascular cell adhesion molecule 1 (VCAM-1), which promote leukocyte adhesion and migration [230].

A protective layer of glycoprotein and ECM components, called the glycocalyx, overlies the EC layer and is a critical component of the vascular lumen important for mechanotransduction of shear stress and vein integrity. Damage to the glycocalyx may occur through chronic distention due to hypertension, degradation due to low shear stress, or enzymatic cleavage by MMPs may trigger prothrombotic processes, as well as increased permeability and leukocyte adhesion. All these features combine in CVD to produce a persistent proinflammatory and prothrombotic environment that leads to leukocyte activation, attachment, and transmigration, as well as secretion of other proinflammatory cytokines [231]. Many studies confirm the role of inflammation in CVD, reporting increased expression of inflammatory markers, in vitro, in preclinical studies, and in patients with CVD. Vascular EC cultures obtained from VVs were found to exhibit high levels of proinflammatory surface markers (CD31, CD146, and ICAM-1) and cytokines (osteoprotegerin and VEGF). Sola et al. reported distinct chemokine expression patterns in VVs due to significant upregulation of MCP-1 and interleukin (IL)-8/CXCL8 and elevated expression of interferon-γ-inducible protein-10/CXCL10, RANTES/CCL5, macrophage-inflammatory protein-1a/CCL and -1b CCL4 mRNA [50]. Poredos et al. demonstrated that the circulating inflammatory markers, high-sensitivity C-reactive protein and IL-6, as well as the fibrin degradation product, D-dimer, and von Willebrand factor were significantly higher in the blood samples taken from VVs, compared to those from systemic blood [232].

The idea that venous leukocyte activation and adhesion are increased in CVD is consistent with observations of enhanced leukocyte trapping in the lower extremities of CVD patients. Even in healthy persons, blood returning from the feet after an extended period of inactivity contains 15%–20% fewer leukocytes than the arterial blood entering the feet. However, this trapping effect is more pronounced in CVD patients, whose leukocyte counts in returning venous blood were 24% lower than in normal subjects. In addition, immunohistochemical analysis of skin biopsies from the affected tissues of CVD patients revealed elevated numbers of leukocytes, especially T lymphocytes and macrophages [183]. Persistent accumulation of leukocytes in the legs, many of which migrate through the activated endothelium and the distended walls of the small veins and postcapillary venules into the surrounding tissue, is believed to be the basis for the chronic inflammation and lipodermatosclerotic skin changes associated with advanced CVD. The altered hemodynamics and reduced venous return in the large veins of the legs result in the shunting of blood and fluid to smaller veins and capillaries of the skin, where pressure is also increased [233].

As a consequence, capillary beds and microvasculature become chronically dilated, dense, elongated, and tortuous. Damage to the glycocalyx and endothelium in these abnormal pressurized capillaries leads to increased permeability, edema, leakage of RBCs, and leukocyte infiltration and activation. Elevated levels of MMPs in the chronically inflamed dermal tissues contribute to excessive breakdown of ECM and collagen, which can lead to impaired healing and ulceration. Abnormal capillary remodeling and permeability have also been linked to the high plasma levels of vascular endothelial growth factor found in CVD patients, especially those with skin changes. Development of dermal tissue fibrosis may be due to the high levels of TGF-β1 present in the skin of the lower legs of CVD patients because TGF-β1, produced by activated leukocytes, stimulates the production of excess fibrinogen and collagen, which leads to fibrosis. Finally, the breakdown of extravasated red cells and the subsequent release of hemoglobin and ferric iron into the surrounding structures increase the oxidative state of the tissue, which increases MMP activity, exacerbates tissue damage, and further impairs wound-healing [234].

5.14 Genetic biases related to chronic venous ulceration

Genetic studies, together with insights into biochemical, physiological, and pathophysiological processes related to wound-healing on the molecular level have broadened our understanding of chronic venous ulcer etiology. History has taught us that new pathophysiological findings may lead to new and improved treatment methods. For example, findings in the treatment of chronic wound dressings lead to a significant reduction in the necessary treatment period, and therefore overall treatment costs. Nevertheless, to further increase the healing rate of such wounds without additional cost to the patients, hospitals, or insurance companies, a profound understanding of the chronic wound pathophysiology and extensive knowledge of available state-of-the-art wound dressing materials are required. An increasingly important factor in the management of such wounds is the genetic aspect, which is the focus of this review [235].

Many different theories have emerged about the etiology and mechanisms related to CVI. It is generally accepted that venous reflux and venous hypertension are responsible for CVI. Once venous hypertension is established, a disruption of the hemodynamic balance causes a failure of the venous valves via a blood flow-mediated inflammatory reaction. The changes in the blood flow in veins allow activation of the leukocytes, diapedesis into the venous parenchyma, release of enzymes, and remodeling of the vascular wall, ending in venous valve destruction and possible loss of function. This theory is based on different histological findings, which demonstrate that extensive ECM remodeling with decreased elastin content and an altered molecular differentiation of collagen and growth factor expression takes place in CVI [236]. Other studies have suggested that overexpression of the acidic FGF in VV walls via fibroblast growth factor receptors (FGF-r), and the MAPK pathways can influence the expression of enzymes involved in extensive ECM remodeling, and play an important role in the pathogenesis of CVI. Another finding in patients with CVI that could explain the cause of the disease lies in overexpression of inducible nitric oxide synthetase and TGF-β1, as well as an increase in the presence of CD68 + monocytes/macrophages. Regardless of the influence of the above-mentioned molecular mechanisms, genetic aberrations are suspected to be the underlying cause of CVI. The exact pathophysiological mechanism of CVI development is nevertheless still to be identified [237].

In the hope of revealing the genetic basis of the CVI pathophysiology, Markovic et al. analyzed gene expression of the entire human genome. There was significant association between patients with CVI and patients unaffected by superficial reflux disease. Furthermore, they were able to show that patients with CVI had a significantly higher expression of the HPGD (hydroxyprostaglandin dehydrogenase) gene. This gene encodes a member of the short-chain nonmetalloenzyme alcohol dehydrogenase protein family. The encoded enzyme is responsible for the metabolism of prostaglandins that function in a variety of physiologic and cellular processes, such as inflammation, blood clotting, and the sensation of pain. By catalyzing the conversion of the 15-hyrdroxyl group of prostaglandins (PG) into the ketone group, the HPGD enzyme significantly reduces the biological activity of PGs and reduces the inflammation activity in vivo [238]. A possible explanation for the upregulation of the HPGD gene in CVI patients is that the increase in inflammatory mediators induces the expression of HPGD. If this is the case, then the HPGD enzyme would oppose the inflammation that occurs as a result of the impaired hemodynamic activity. Studies should be conducted to determine if the expression of the HPGD gene in patients with severe CVI or CVU (Chronic venous ulceration) is too low to counter the inflammation in the surrounding tissue. Markovic et al. have also shown that the relative expressions of the collagen type 13α1 (COLI3A1) and collagen 27α1 (COL27A1) genes were downregulated in CVI patients [239].

This finding supports data from other studies that propose that altered molecular differentiation of collagen and ECM remodeling takes place in the wall of VVs. Gemati et al. were, on the other hand, investigating single-nucleotide polymorphisms (SNPs) in several candidate genes. They were able to show that polymorphism in high iron Fe gene coding for the human hemochromatosis protein, FPN1 (ferroportin-1), MMP12 (matrix metalloproteinase-12), and FX111 (Factor XIII) genes coincide with the severe form of CVI [240].

Although the exact pathogenesis of venous ulceration remains unknown, several hypotheses, describing different underlying mechanism, exist. There is a growing acceptance that chronic inflammation and tissue remodeling play an important role in the development of CVU. Moyses et al. and Thomas et al. have shown trapped leukocytes occurring in limbs with venous dysfunction. This observation is analogous with other evidence demonstrating the importance of the inflammatory process. Coleridge-Smith et al. studied leukocyte activation and its relationship to hypertension extensively. They have shown an increase of neutrophil elastase and lactoferrin, an important marker of neutrophil activation, in patients under transient conditions of venous hypertension and CVI [235]. Coleridge et al. proposed that venous hypertension leads to capillary distention and widening of endothelium pores that allow leakage of macromolecules, resulting in the formation of pericapillary fibrin cuffs. These cuffs then prevent the passage of oxygen and nutrients. Another hypothesis, the so-called growth factor trap, describes the leaking of macromolecules, specifically fibrinogen and α2-macroglobulin, into the surrounding dermis. These macromolecules form a so-called "TRAP (Thrombospondin-related anonymous protein)" for different growth factors (i.e., TGF-β1, FGF-r) that are necessary for tissue repair and integrity [241].

In patients with CVI, biopsies of the dermal tissue demonstrate an increased presence of leukocytes in both lipodermatosclerotic and healing, but ulcerated, skin. In the histologic evaluation of the different leukocyte cell subtypes, it has been shown that the predominantly occurring cells are T-lymphocytes and macrophages. It was additionally revealed that the endothelial expression of ICAM-1 that promotes leukocyte adherence to the endothelium has been elevated. It has been postulated that shear stress plays a critical role in leukocyte activation. Shear stress is described as a tangential force (per membrane area) produced by the moving blood while acting on the endothelial surface. It is a function of the velocity gradient of the blood flow near the endothelial surface. Cessation of blood flow and stasis with lower shear stress may activate leukocytes, as shown by the projection of pseudopodia on the surface of the leukocytes [242].

As a result, the expression of ICAM, VCAM-1, E-selectin, and L-selectin, takes place on the endothelial surface. All these adhesion molecules can act as intermediaries in the adhesion process, binding specific leukocyte receptors (i.e., very late activation antigen 4 and/or leukocyte-function-associated antigen 1). Another observed and reported phenomenon is transendothelial migration into the surrounding tissue. Red cells and platelet aggregates may fill the capillary lumen and can be found in the pericapillary space [243]. In the subepithelial layer, the forming of granulation tissue composed of lymphocytes, plasma cells, macrophages, histiocytes, and fibroblasts takes place. As a result of the forming granulation tissue, deposition of fibrin occurs. Through the mentioned processes, collagen fibers have been shown to lose their normal orientation in the cutaneous tissue.

Regardless of the many available theories on venous ulcer formation, none explains why some patients with CVI develop CVUs and some do not. There exists a subset of patients with refractory ulcers that, over time, become larger despite standard care using multilayer compression bandages. In fact, the size and duration of a chronic venous leg ulcer represent important risk factors for predicting venous ulcers.

5.15 Gene expression of the epidermal wound bed

Charles et al. studied different gene expressions in keratinocytes present at the wound edge of a nonhealing venous ulcer. They were able to show 15 different genes that were differently expressed, when comparing healing and nonhealing venous ulcers. The most upregulated gene in the nonhealing wound was the gene that codes for secreted frizzled-related protein 4 (SFRP4), a mediator of Wnt signaling. The Wnt signaling pathway is responsible for the transmission of the signals from outside the cell to inside the cell. It is involved in many different physiological pathways including embryogenesis, carcinogenesis, angiogenesis, and mobility of the epidermal cells [244]. SFRP4 has a strong association with apoptosis of cells and is linked with diseases where cell death is part of the disease pathology. Another upregulated gene is the branched-chain aminotransferase 1. This gene encodes the cystolic form of the enzyme branched-chain amino acid transaminase. By catalyzing the reversible transamination of branched-chain alpha-keto acids to branched-chain L-amino acids, it plays an important role in cell growth. These findings were consistent with another research study that showed a connection between the activated β-catenin and c-myc (regulator gene that is pivotal in growth control, differentiation, and apoptosis) in keratinocytes of the nonhealing edges of venous ulcer [245].

Stabilization of nuclear β-catenin can block the epidermal growth factor (EGF) response, induce c-myc, and repress the K6/K16 keratins (cytoskeletal components that are important for migration). By doing so, activation of the β-catenic/c-myc pathway contributes to impairment of healing by inhibiting keratinocyte migration and altering their differentiation. A novel finding from the work of Charles et al. revealed upregulation of cytochrome P450 and 17-β-hydroxisteroid dehydrogenase. The authors suggest that the upregulation of the steriodogenesis-associated genes, steroid synthesis, and metabolism, may participate in the pathogenesis of the nonhealing venous leg ulcer. All top 20 genes that were found to be downregulated in the nonhealing wound edges code for proteins that are considered crucial to epidermal structural integrity or are associated with epidermal injury, repair, hyperproliferation, differentiation, or a combination of some or all of these [246]. The same study also showed that the most downregulated gene group is the one that codes for keratins. In this group, the one gene that stood out was the gene that codes for keratin 16, a gene product that has been associated with epidermal injury and a timely epidermal repair. The authors of the study also pointed out that not only is the gene that codes for keratin 16 downregulated, but also the connected genes that code for K6A, K6B, K17, K1, and K14.

The conclusion, based on the downregulation findings, was that the biological processes, which are essential for the maintenance of the epidermis, are seriously affected. An additional and interesting finding from Charles et al. was that no genes in close connection with stimulation of wound repair are significantly downregulated. Genes that code for the platelet-derived growth factor or keratinocyte growth factor had a similar expression in the nonhealing and normally healing ulcer [95].

5.16 Gene expression of the dermal wound bed

Gene expression in the dermal layer differs from gene expression in the epidermis. Most of the upregulated genes in the nonhealing venous wound code for proteins that have been associated with tissue injury, ECM formation, and the wound-healing process. Propedin is a protein that plays an important role in tissue inflammation. The protein promotes the association of C3b with factor B (C3Bb), thereby providing a focal point for the assembly of C3Bb on the surface. The gene that codes for propedin is upregulated in nonhealing venous leg ulcers. As in the study of the epidermal gene expression, the group of Charles et al. was able to identify genes in the dermal wound bed that were significantly downregulated in the nonhealing venous ulcer.

Heparin-binding epidermal growth factor (HB EGF, an EGF-like growth factor) is required in cutaneous epithelialization. Its mitogenic and migratory effects on keratinocytes and fibroblasts promote dermal repair and angiogenesis [244]. The downregulation of the gene that codes for HB EGF in nonhealing leg ulcers can partly explain the phenotype of nonhealing wounds. Recent studies on complementary DNA-microarray have demonstrated diversity in the genetic expression of healing versus nonhealing wounds. Many different abnormalities have been found in different genes. The most important seem to code for structural factors, mediators, or inflammation and apoptotic pathways. It has been shown that there is a different genetic physiology in healing than for nonhealing wounds.

5.17 Gene polymorphism in chronic venous ulcer

As mentioned earlier, Gemmati et al. studied selected SNPs of candidate genes potentially involved in the pathophysiology of CVU. They were able to show that polymorphisms of the HFE (Human homeostatic iron regulator protein), FPN1, and MMP12 are connected with susceptibility for developing a venous ulcer. Polymorphisms of the gene FXIII were associated with healing time, ulcer size, and response to surgery. It is known that iron overload plays an important role in CVI and the development of CVU. A gene that plays an important role in the metabolism of iron is the HFE gene. Mutations of the HFE gene in patients with CVI facilitate the occurrence of skin lesions. The mutations lead to increased efflux of iron, modifying the stability of the ferritin deposition and efficiency of the hepcidin regulation system. Increased iron reflux could also positively affect bacterial growth, making the environment of a nonhealing wound even more likely to become infected [247].

FPN1 is a gene that codes for the protein ferroportin, the only protein responsible for the transport of iron from the inside of the cell to the outside. Expressed by macrophages, the process is regulated at several different levels. FPN1 mRNA contains, in the 5'-UTR (untranslated) region, the IRE (iron-responsive element) sequence, which interacts with the iron regulatory protein. The existence of an IRE region in the 5'-UTR of FPN1 results in increased expression of the protein proportional to the cellular iron loading, thereby modulating the iron export. The polymorphism−8CG in the FPN1 gene can be found near the region coding for IRE. Gemmati et al. suggested that the polymorphism−8CG can lead to increased efflux of iron from the macrophages and thereby increased oxidative stress and cell death. FX111 has a pleiotropic function that results from its transglutaminase and proangiogenic function. The insights from a patient with FXIII deficiency showed the role of this factor in timely wound closure. Tognazzo et al. investigated different FXIII-A gene polymorphisms and were able to show that the different gene variants coincide with different venous ulcer size. In relation to genotype distribution between the control group and the patient group, a significant difference was not observed. Nevertheless, they were able to show that the ulcer size was inversely related with the presence of the Leu34 or Leu564 allele in an independent fashion and without synergistic effects. Because of this finding, the authors suggested that the FXIII gene variants can act as a "modulator factor" in lesion progression and extension. The protective role of the two gene variants can be explained by the higher FXIII expression and an increase of the incorporation rate of fibrinolysis inhibitors in fibrin provisional matrix [248].

In the past decade a new pathophysiological mechanism involved in the occurrence of CVUs has been discovered. A major pathological mechanism of the disease is the chronic inflammation that disables the physiological wound closure. In terms of understanding the cause of chronic inflammation, gene expression and gene polymorphisms are increasingly becoming the focus of research. Many changes in gene expression have been identified in nonhealing leg ulcers. Based on the literature review it can be claimed that genes that regulate the inflammation response, genes that code for extracellular peptides, and genes that code for different cellular pathways have a different expression in patients with CVUs when compared to healthy individuals [249]. In addition, different polymorphisms in single genes that promote the appearance of CVUs have been identified. Further insights from different studies show that the precise mechanism responsible for the disease is still unknown and may be much more complex than anticipated in the past [250]. Based on our limited understanding of the underlying ulcer formation mechanisms, at the time of writing it is not possible to precisely predict which patient will experience a prolonged healing process. To the best of our knowledge, genome-wide association studies (GWASs) performed looking for SNPs associated with the risk of venous leg ulcer have not been performed. GWASs remain the gold standard in genetics of complex diseases and would be beneficial in better understanding the genetic basis of CVU.

5.18 Epidemiology

The worldwide incidence of VVs is generally described as 10%–15% for males and about 20%–25% females. However, in practice, these figures appear to be inaccurate. The incidence is almost double. *In our studies on more than 8000 patients, we found a 45% incidence in males and 55% in females.* This is mainly because:

1. The majority of asymptomatic patients do not go for any treatment;
2. Patients are ignorant about the disease;
3. Failure to identifying the disease, even by the treating medical professionals who have incorrect and/or incomplete knowledge of the condition;
4. Some patients resort to other systems of medicine.

CHAPTER 6

Etiology of varicose veins

The predisposing factors of varicose vein are usually ascribed to the following:

1. *Age:* As age advances the chances of developing varicose veins (VV) increase.
2. *Sex:* The female hormones, estrogen and progesterone, increase susceptibility to varicosity.
3. *Pregnancy:* Most commonly, the manifestation of VV is seen in females during the later stages of pregnancy and is usually bilateral.
4. *Profession:* Nurses, traffic policemen, teachers, barmen, shopkeepers, and sentry staff, who have to stand continuously for long hours, claim the development of veins due to their profession.
5. *Race:* In some races there is an increased incidence of VV.
6. *Familial:* A history of many members in the same family carrying the disease.
7. *Deep vein thrombosis:* A block in a deep vein anywhere from the leg to IVC (Inferior Vena Cava).
8. *Pelvic tumors:* Large ovarian cysts, fibroid, or any pelvic pathology obstructing the pelvic vein or in the inguinal region can predispose to VV.
9. *Orthopedic*: Prolonged immobilization in plaster cast following fractures of lower limb, open reduction, and compound fractures.
10. *CABG*: When the long saphenous vein is used for a coronary artery bypass graft, VV develop in later years.
11. *Klippel—Trenaunay syndrome*: Congenital aplasia of pelvic veins.

All the factors mentioned above, in the strictest sense, can be considered only as *predisposing/precipitating factors* in the *etiogenesis* of VV. There is vital factor in general for the causation of this condition. A very common finding is the familial incidence of VV in close relations. A study of the genomic pattern clearly shows that more than 60% of patients give a clear association among family members. More than 30% are unaware of the pattern of VV when they are of finer types, especially when the involvement is above the knee, which usually remains symptom-free. When the involvement is around the gluteal region and situated on the posterior aspect, it is very seldom noticed. Less than

Genesis, Pathophysiology and Management of Venous and Lymphatic Disorders
DOI: https://doi.org/10.1016/B978-0-323-88433-4.00022-X

10% are totally unaware, or the pathology remains dormant. There is a definite genetic factor behind the pathogenesis of VV.

6.1 Genetic associations

Though there are various studies ongoing to identify the genetic association it remains inconclusive. Congenital and genetic disorders are sometimes associated with varicosities. Patients with Klippel–Trenaunay syndrome (KTS) present with VV, limb hypertrophy, and dermal capillary hemangioma. Venous abnormalities such as agenesis, atresia, valvular incompetence, and venous aneurysms are seen in KTS also. These are all the result of structural abnormalities in the venous wall or valves. Another genetic mutation associated with VV includes FOXC2, which is described in patients with lymphedema-distichiasis. While these results are all inconclusive, brighter days are hoped for in the future when with exhaustive genomic studies the causative gene will be identified, which will be a breakthrough for the elimination of this disease for future generations.

6.2 Primary changes in the vein walls/valves

While the clinical progress in the assessment and management of venous diseases remains sluggish and unremarkable, studies into the pathological changes in the vein wall have been intensive and impressive. On one side there are intensive genetic studies and on the other are the pathologies associated with the structural, functional, and biochemical changes to the venous valves and walls (Figs. 6.1–6.3).

6.3 Normal structure of veins

An understanding of the structure of veins is important in learning about changes in the vein wall. Fig. 6.4

The vein wall consists of the following three layers:

1. *Tunica adventitia*: It is the strong outer covering of veins. It is composed of connective tissue, collagen, and elastic fibers. These fibers allow veins to stretch to prevent overexpansion due to the pressure that is exerted on the walls by blood flow.
2. *Tunica media*: It is the middle layer. It is composed of smooth muscle and elastic fibers, and is thinner in veins. The smooth muscles are arranged in a circular fashion centrally and longitudinally on either

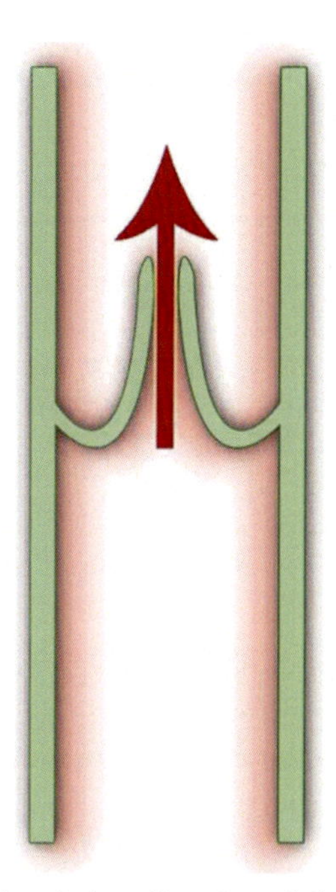

Figure 6.1 Normal venous valve during dilatation of the heart.

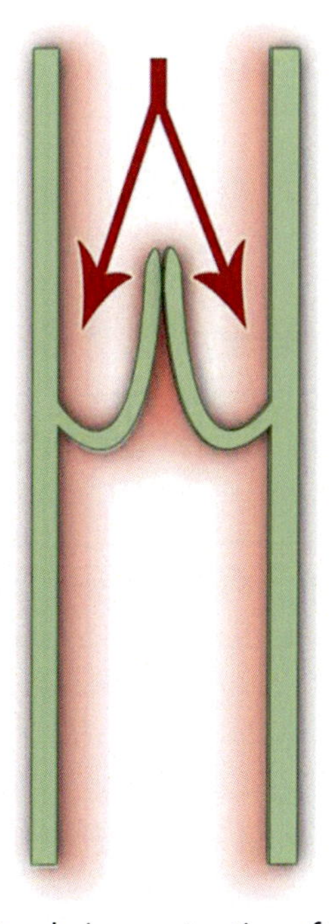

Figure 6.2 Normal venous valve during contraction of the heart.

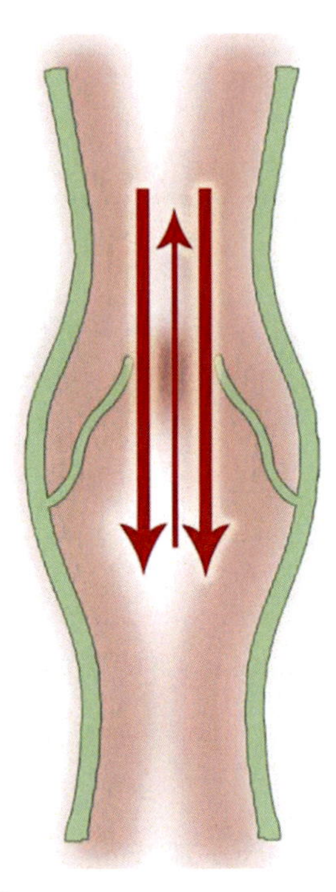

Figure 6.3 Incompetent valve.

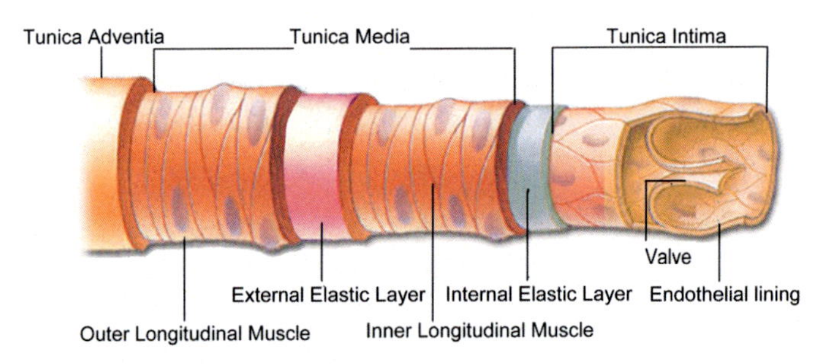

Figure 6.4 Structure of vein.

side. Smooth muscle cells (SMCs) traverse through the central circular layer from the outer longitudinal layer to the inner longitudinal layer.

3. *Tunica intima*: It is the inner layer. In arteries this layer is composed of a basement membrane and endothelium. Veins do not contain the elastic membrane lining that is found in arteries. *In leg veins the tunica intima layer is modified to form valves to keep blood flowing in a single direction.*

In the smaller veins the tunica media and tunica adventitia are very thin, and in larger veins, the tunica adventitia is very thick and the tunica media is very thin.

In the smallest veins the valves are arranged very closely as compared with larger veins. The valves fade off above the level of the umbilicus.

The integrity and tone of the vessel wall depends on the cellular and extracellular matrix (ECM) components. Any degradation of these factors can lead to a reduction in the tensile strength of SMCs of the wall and changes in the collagen content.

Numerous changes have been demonstrated in the cellular and ECM components of the vein wall in varicosity. These changes do not occur in a continuous pattern; it is usually scattered or distributed unevenly. Intimal changes include endothelial proliferation and subintimal plaque formation. The media shows SMC migration, ECM disintegration, or degradation. The adventitial layer shows fibroblastic and SMC proliferation in atrophied connective tissue and loss of vasa vasora.

6.4 Extracellular matrix degradation

The ECM is the most important factor in maintaining the integrity and homeostasis of the vein through interactions with cellular components such as endothelium and SMCs, because the ECM provides a structural framework of collagen, proteoglycans, elastins, glycoproteins, and fibronectins in which the cellular components are embedded. Therefore any degradation of ECM will produce weakening and distortion of the vessel wall. Fragmentation of elastic lamina and thickening of collagen fibers are usually found in VV. The total elastin content is also seen to be reduced, leading to loss of elasticity and a reduction of tensile strength.

6.5 Enzymatic dysregulation

A group of enzymes are known to regulate and maintain the homeostasis of ECM, these include matrix metalloproteinases (MMPs) and endogenous tissue-inhibitory metalloproteinases (TIMPs). MMP–TIMP imbalances have been identified in VV.

6.6 Activation of endothelium

Endothelial cells play a significant role in regulating vascular inflammation, remodeling, and tone. Endothelial cells of VV are seen to be desquamated and degenerated. Such injured cells are activated and are known to release various types of inflammatory mediators and growth factors. Increased expression of inflammatory factors such as vascular cell adhesion molecule-I, intercellular adhesion molecule-I, and von Willebrand factor by the endothelium of VV compared with non-VV has been recorded. There are also increased mast cells, macrophages, and monocytes in VV. Activated leukocytes also release large amounts of superoxide anions and proteases that are able to degrade ECM. The release of growth factors including basic fibroblast growth factor (bFBGF) and platelet-derived growth factor (PDGF) may induce SMC proliferation.

6.7 Smooth muscle cell proliferation and migration

Areas of SMC hypertrophy and proliferation occur in the tunica media of VV although dystrophy may also be present. Rearrangement and migration of SMCs are also seen. SMCs in varicosities appear disorganized and dedifferentiated from contractile to synthetic and proliferative phenotypes. They demonstrate vacuolization and phagocytosis. The growth factors (cell mediators) PDGF and bFBGF may also contribute to some changes of SMCs. Estrogen and progesterone levels may also have an influence to some extent. The contractile strength and tone of SMCs are also impaired.

6.8 Dysregulated SMC apoptosis

Dysregulation of SMC apoptosis also occurs. *Apoptosis* is a type of programmed cell death characterized by zeiosis (membrane blebbing), cytoplasmic condensation, and activation of endogenous endonucleases and specific proteases, and is an important physiological process in normal development, morphogenesis, and tissue regulation. When improperly regulated, apoptosis leads to pathological conditions involving faulty repair of the DNA mechanism and cell cycle disruption, which happens in VV.

CHAPTER 7
Clinical manifestations

The varied presentations according to the stage of venous disease make it somewhat mysterious. The clinical manifestations can remain undiscovered for almost a quarter century, which makes it very differ to most other human diseases. It can remain dormant for a long time and can present with an acute life-threatening disease in a limited number of patients. It can result in a very painful, persistent ulcer, which can refuse to heal over a period of years with any attempted treatments, resulting in extreme suffering and misery for the majority of patients. The difficult nature of the disease can place clinicians and patients in a dogmatic dilemma.

7.1 Clinical complications

The clinical complications of venous disease can be classified into two groups:

1. *Acute*
2. *Chronic*

   The acute disease includes three types:

1. *Acute bleeding:* Thinning of the vein wall and high pressure predispose to rupture of the vein, which can happen in both large and small veins. The gravity of the condition depends on the site and size of the vein; at a lower level, because of the high hydrostatic pressure, the bleeding is usually in a jet. If the bleed occurs at a higher level, due to the low venous pressure, the bleeding is comparatively sluggish. Therefore bleeding from a larger vein at a lower level carries a very high mortality rate.
2. *Acute thrombophlebitis:* When blood flow is sluggish or stagnant, it has a natural tendency to clot. Thus the formed thrombus can become infected, resulting in pain, swelling, induration over the infected site, and the patient becoming febrile. It can proceed to cellulitis and abscess formation.

Genesis, Pathophysiology and Management of Venous and Lymphatic Disorders
DOI: https://doi.org/10.1016/B978-0-323-88433-4.00023-1

3. *Acute cellulitis:* At one stage of the disease, there is lymphedema on the affected leg, which can become infected. Extensive cellulitis may ensue and tissue necrosis and abscess formation may develop.

In *chronic* disease there is a sequence of events that take place over a period of time:

1. Lymphedema of the foot and leg on prolonged standing, which disappears on keeping the leg elevated;
2. Itching, pigmentation, eczematous changes, lipodermatosclerosis, or *atrophie blanche*;
3. All the above, plus recurrent ulceration;
4. Chronic nonhealing ulcers;
5. Chronic ulcer that turns malignant.

7.2 Evaluation of patients

Evaluation of the patient is a threefold process, including:

1. Presenting symptoms
2. Clinical findings
3. Investigations

The patient presents with a long history of:

1. Swollen, tortuous, large, long/short veins with no other complaints;
2. Very small intradermal veins which remain scattered throughout with no other complaints;
3. Combinations of points (1) and (2) in varying proportions;
4. Points (1), (2), or (3) with a complaint of pain/ache in the leg, thigh, or both;
5. All of the above, plus swelling (edema) of the foot in the evening or when standing for prolonged periods, which subsides in the morning after sleep or by keeping the leg elevated;
6. Edema of the leg and foot with no obvious veins visible;
7. Itching (pruritus) of the leg and any of the above;
8. Pigmentation around the ankle region with/without progression to the foot and/or lower leg;
9. Eczematous change around the ankle region with/without progression to the foot and/or lower leg;
10. Thick leathery skin with elephantoid changes;
11. Healed ulceration with any of the above;
12. Active ulceration/nonhealing ulcers;
13. Active bleeding or h/o (history of) bleeding;

14. Large veins appearing hard, swollen, reddish, and painful, with/without constitutional symptoms;
15. Swollen, red, painful lower extremity with constitutional symptoms;
16. H/o previous surgeries for venous disease;
17. H/o of repeated skin grafting.

Clinical findings include:

1. Typical anatomical involvement of lower limb veins;
2. Atypical scattered veins;
3. Combination of both (1) and (2);
4. Only small veins/only large veins/in combination;
5. Above knee involvement/below knee involvement of both;
6. Unilateral/bilateral leg involvement;
7. Skin changes, especially around and adjacent to the ankle joint;
8. Any signs of healed ulceration;
9. Any active ulceration;
10. Any marks of surgery.

The patient should be *specifically asked about*

1. Any family h/o venous disease;
2. Any h/o repeated surgery for varicose vein;
3. Any h/o treatment for deep vein thrombosis (DVT);
4. Any h/o spondylosis/arthritis;
5. Any h/o rheumatoid arthritis;
6. Any h/o neurological problems affecting the lower limb;
7. Any h/o intermittent claudication or numbness of the lower limb;
8. Whether the patient has attended for cosmetic reasons;
9. Whether attending for treatment due to disqualification during a medical examination.

Remember!

Pain and discomfort in the lower limb can be due to the following etiologies:

1. Venous
2. Arterial
3. Neurogenic
4. Orthopedic

Tests include:

1. Trendelenburg's test
2. Perthes' test
3. Homans' sign

All these tests are absolutely useless and have no relevance in the present day management of venous diseases.

7.3 Investigations

1. *General*:
   a. BP (Blood pressure), height, weight
   b. Routine blood counts and Hb (Hemoglobin)
   c. Blood sugar (RBS (Random Blood Sugar))
   d. HIV, HbsAg (Hepatitis B surface antigen), VDRL (Venereal Disease Research Laboratory test)
2. *Special investigations*:
   a. Blood: liver function test, renal function test, lipid profile, HbA1C;
   b. ECG (Electrocardiogram) examinations
   c. Color Doppler scan (CDS) of the venous system of the lower limb up to IVC (Inferior Vena Cava)

      This type of scan is the most important test and should be carried out before any surgical procedure for varicose veins to exclude DVT. Also, the differentiation between primary and secondary varicose veins is confirmed with this investigation. Assessment of valvular function and also identification of perforators can be well studied using this scanning method. Any thrombophlebitis in superficial veins, venous malformations, and arteriovenous fistula can be identified with this technique. The evolution of the CDS has almost totally replaced the old clinical tests such as the Trendelenburg's and Perthes' tests for the identification of valvular incompetence and demonstration of Homans' sign for DVT.
   d. Color Doppler study of the arterial system of the lower limb is used if it is found necessary to exclude vascular claudication due to arterial insufficiency;
   e. X-ray of the thoracolumbar spine to exclude neurogenic claudication due to degenerative changes in the spine;
   f. Biopsy of chronic nonhealing ulcers, if present, to exclude malignant changes;
   g. Other relevant tests and investigations required specifically for evaluation of the physical condition of the patient.

CHAPTER 8

Staging of chronic venous disease

8.1 Development of the classification of chronic venous disease for proper evaluation

Chronic venous disease (CVD) is a very common disease in developed and developing countries. Due to the inadequacy of scientific data recording and nonavailability of proper research work and proper reporting accurate evaluation of the rate of prevalence in developing and underdeveloped countries is not available. However, it is probably one of the most common diseases affecting humans. Many people with the disease remain undetected, and are unaware of it. Others do not seek medical advice or treatment, having noticed the disease but not having any notable clinical problems. For centuries, this disease and its treatment have been an enigma. In is only in the past few years that there has been an increasing enthusiasm among clinicians and pathologists to evaluate the disease from different angles. The social prevalence of the disease, individual predisposition, site of involvement, distribution, extent, and severity of the disease, the type and style of management, response to treatment, recurrence after treatment, and cost of treatment need to be extensively evaluated. Moreover, the etiological analysis, pathophysiological studies regarding the venous system, cellular changes at the site of involvement, and the impact of lymphatic system are all points of great interest.

Until recently, there was no standardized method of description of the disease and defined method of treatment. This created confusion and improper understanding. The necessity for a standardized method for evaluation and treatment thus became an urgent necessity. A common language acceptable to all, worldwide, had to be created. This marked the evolution of the presently followed standardized modality, the *CEAP* (*C*linical—*E*tiological—*A*natomical—*P*athophysiological) classification. Credit goes to the American Venous Forum (AVF) for taking the initiative to form an international ad hoc committee for formulating a *Common Language* available and understandable to all. This was essential for an accurate classification and interinstitutional communication of the disease

Genesis, Pathophysiology and Management of Venous and Lymphatic Disorders
DOI: https://doi.org/10.1016/B978-0-323-88433-4.00020-6

on a worldwide level. This system of classification was established at the sixth annual meeting of the AVF in February 1994 in Maui, Hawaii. The noninvasive method of examination for evaluating the venous system by color Doppler scanning resulted in a revolution in achieving precise identification of the disease process and greater accuracy in classification. In 2004, the recommendations were accepted. The CEAP classification is now accepted as the *gold standard* of classification for CVDs. This was followed by the formation of REVAS (REcurrence after VAricose vein Surgery) in 1998 in Paris and VCSS (Venous Clinical Severity Score) in 2000. VEIN-TERM consensus document was the latest update of terminology for CVD and was published in the *Journal of Vascular Surgery* in 2009. All these are meant only to complement the CEAP classification.

The noninvasive method of examination for evaluating the venous system by the color Doppler scan brought in a revolution to achieve precise identification of the disease process and maintain more accuracy in classification.

8.2 The revised CEAP classification: summary

8.2.1 Clinical classification

C0: No visible or palpable signs of venous disease
C1: Telangiectasia or reticular veins
C2: Varicose veins
C3: Edema
C4a: Pigmentation and/or eczema
C4b: Lipodermatosclerosis and/or atrophie blanche
C5: Healed venous ulcer
C6: Active venous ulcer
S: Symptoms including ache, pain, tightness, skin irritation, heaviness, muscle cramps, as well as other complaints attributable to venous dysfunction
A: Asymptomatic

8.2.2 Etiologic classification

Ec: Congenital
Ep: Primary
Es: Secondary (postthrombotic)
En: No venous etiology identified

8.2.3 Anatomic classification

As: Superficial veins
Ap: Perforator veins
Ad: Deep veins
An: No venous location identified

8.3 Pathophysiologic classification

Basic CEAP

Pr: Reflux
Po: Obstruction
Pr,o: Reflux and obstruction
Pn: No venous pathophysiology identifiable

Advanced CEAP

Same as basic, but allows any of the 18 named venous segments to be used as locators for venous pathology.
The superficial veins include:

1. Telangiectasia/reticular veins
2. Great saphenous vein (GSV) above knee
3. GSV below knee
4. Small saphenous vein
5. Nonsaphenous veins

The deep veins include:

1. Inferior vena cava
2. Common iliac vein
3. Internal iliac vein
4. External iliac vein
5. Pelvic: gonadal, broad ligament veins, other
6. Common femoral vein
7. Deep femoral vein
8. Femoral vein
9. Popliteal vein
10. Crural: anterior tibial, posterior tibial, peroneal veins (all paired)
11. Muscular: gastrocnemial, soleal veins, other
12. Perforating veins

13. Thigh
14. Calf

Example:

A patient presents with painful swelling of the leg and varicose veins, lipodermatosclerosis, and active ulceration. Duplex scanning showed axial reflux of the GSV above and below the knee, incompetent calf perforators, and axial reflux in the femoral and popliteal veins. No signs of post-thrombotic obstruction.

Classification according to basic CEAP: C6,S; Ep; As,p,d; Pr.

Classification according to advanced CEAP: C2,3,4b,6,S, Ep, As,p,d, Pr2,3,18,13,14−17, LII.

8.4 Comment

Despite of all these modifications in the CEAP classification, it is not able to explain severity based on the etiopathogenesis of the disease. Reticular veins and telangiectasia are considered as harmless and of cosmetic importance only, whereas, actually the severity of the disease has to be attributed to the size and position of the small vein as per the observation. *Smaller tributaries (less than 3 mm) involved at the lower end of the leg, ankle region, and foot are more pathogenic than larger tributaries and truncal veins*. The pathogenicity of the veins increases when the size of the vein decreases. The terminologies relating to corona plebetica, spider veins, and telangiectasia in the vicinity of the ankle joint, even in the absence of major tributaries and truncal veins, predisposes very early for a class 6 venous disease. Instead of using the terminologies: *corona plebetica*, *spider veins*, and *telangiectasia* it can be made more meaningful if it is called *radicular veins* (radicle means rootlet).

8.5 Terminology and new definitions in CVD

The CEAP classification describes all forms of CVD, in the full spectrum of its morphologic and functional abnormalities. They include descriptions from the earliest form of the dormant disease up to the final stage of nonhealing ulceration. The letter "C" in CEAP is the most important one for clinicians in their day-to-day practice. Such a vivid and self-explanatory classification can be further elucidated by the following description of the terminologies usually used alongside.

8.5.1 Atrophie blanche

This is also called "white atrophy." It is a localized, circular area of atrophic skin surrounded by very fine veins. There can be pigmentation in the area, and this may sometimes be mistaken for healed ulceration with pigmentation as in "C5," but is easily differentiated from the history.

8.5.2 Corona phlebetica

This condition is characterized by the fan-shaped distribution of intradermal veins, seen on either side of the ankle region. It is also described as *malleolar flare* and *ankle flare.*

8.5.3 Eczema

Erythematous dermatitis with itching, which slowly advances to scaling and blister formation, is usually mistaken for allergic eczema. The presence of veins in the vicinity can give a clue to its differentiation.

8.5.4 Edema

Perceptible edema, confined to the skin and subcutaneous tissue over the foot, ankle, and lower part of the leg is a matter of concern.

8.5.5 Lipodermatosclerosis

This is characterized by recurrent inflammation and subsequent fibrosis of the skin and subcutaneous tissues of the lower part of the leg. Diffuse scarring around the lower leg sometimes gives the leg a *champagne bottle* shape. Usually this area has a very darkish appearance.

8.5.6 Pigmentation

Pigmentation is the most common characteristic of "C4," and is due to extravasation of blood cells into the tissue spaces. Red blood cells are destroyed and release hemosiderin pigments, which in turn result in itching and discoloration. In the earlier phases it is brownish in color but it later progresses to a blackish discoloration.

8.5.7 Reticular vein

The dilated bluish subdermal veins of size 1 mm to less than 3 mm diameter are usually called reticular veins, but are also called *blue veins, subdermal varices,* and *venulectasis.*

8.5.8 Telangiectasia

Dilated intradermal venules of less than *1 mm in diameter*, which remain interconnected, are referred to as telangiectasia. Spider veins and thread veins are synonyms.

8.5.9 Varicose veins

Varicose veins are subcutaneous dilated veins of 3 mm diameter or greater, in the upright position. The saphenous veins or their tributaries or nonsaphenous superficial veins may be involved. In the early stages, that is, in C1, they appear straight and tubular, but in C2, they become tortuous. Varix, varices, and varicosities are synonyms.

8.5.10 Venous ulcer

Full-thickness ulceration occurring in or around the ankle region is usually described as venous ulcer. They do not heal spontaneously, and usually continue to expand and erode (Tables 8.1–8.3).

Table 8.1 Clinical venous terms.[2,3,5,8–13]

Previous definitions	VEIN-TERM update[9]
Chronic venous disorders: include all clinical abnormalities (symptoms or signs) resulting from disease of the lower limb veins and progressing chronically.[8]	*Chronic venous disorders*: this term includes the full spectrum of morphological and functional abnormalities of the venous system.
Chronic venous disease: defined as an abnormally functioning venous system caused by venous valvular incompetence with or without associated venous outflow obstruction, which may affect the superficial venous system, the deep venous system, or both.[10]	*Chronic venous disease*: morphological and functional abnormalities of the venous system of long duration manifested either by symptoms and/or signs indicating the need for investigation and/or care.
Chronic venous insufficiency: implies a functional abnormality of the venous system, and is usually reserved for more advanced disease, including edema (C3), skin changes (C4), or venous ulcers (C5–C6).[8]	*Chronic venous insufficiency* (C3[a]–C6): a term reserved for advanced chronic venous disorders, which is applied to functional abnormalities of the venous system producing edema,[a] skin changes, or venous ulcers (C3[a]: moderate or severe edema, as stratified by Rutherford et al.[3]).
Venous symptoms: may be associated with telangiectasia, reticular, or varicose veins and include lower extremity aching, pain, and skin irritation.[10]	*Venous symptoms*[a]: complaints related to venous disease, which may include tingling, aching, burning, pain, muscle cramps, swelling, sensations of throbbing or heaviness, itching skin, restless legs, and leg tiredness and/or fatigue. Although not pathognomonic, these may be suggestive of chronic venous disease, particularly if they are exacerbated by heat or dependency in the day's course, and relieved with leg rest and/or elevation.
Venous signs: described in the "C" of the CEAP classification.[5,10]	*Venous signs*: visible manifestations of venous disorders, which include dilated veins (telangiectasia, reticular veins, varicose veins), leg edema, skin changes, and ulcers, as included in the CEAP classification.[8]

(*Continued*)

Table 8.1 (Continued)

Previous definitions	VEIN-TERM update[9]
Recurrent varices: the presence of varicose veins in a lower limb previously operated on for varices (with or without adjuvant therapies).[2]	*Recurrent varices*: reappearance of varicose veins in an area previously treated successfully.
Persisting or residual varices: original varicosities that may persist so that the failure of treatment is apparent from an early stage after surgery.[11]	*Residual varices*: varicose veins remaining after treatment.
No previous definition.	*New acronym PREVAIT*: this acronym stands for: PREsence of Varices (residual or recurrent) After operative Treatment.
Postthrombotic syndrome: the term may be used if the patient has experienced an objectively documented prior episode of deep vein thrombosis.[10]	*Postthrombotic syndrome*: chronic venous symptoms and/or signs secondary to deep vein thrombosis.
Pelvic congestion syndrome: characterized by chronic pelvic pain in the setting of pelvic venous varicosities. The syndrome has been shown to be the result of engorgement of the pelvis due to gross dilatation and incompetence of one or both ovarian veins.[12]	*Pelvic congestion syndrome*: chronic symptoms, which may include pelvic pain, perineal heaviness, urgency of micturition, and postcoital pain, caused by ovarian and/or pelvic vein reflux and/or obstruction, and which may be associated with vulvar, perineal, and/or lower extremity varices.
No venous literature definition.	*Varicocele*: presence of scrotal varicose veins.
Venous aneurysm: the diameter above which a vein is considered to be aneurysmal is debated: it is generally accepted that its diameter must be twice that of the normal vein. Aneurysms are classified into saccular and fusiform.[13]	*Venous aneurysm*: localized saccular or fusiform dilatation of a venous segment with a caliber at least 50% greater than the normal trunk.

[a]In the original article, the definition is followed by the sentence: "Existing venous signs and/or (noninvasive) laboratory evidence are crucial in associating these symptoms with chronic venous disorder," which conflicts with the acknowledged existence of the clinical CEAP category COs En An Pn, corresponding to patients complaining of leg symptoms, but presenting with no visible signs and without detectable pathophysiological abnormalities identifiable by routine investigations. This is why we removed this paragraph from the present text.

Table 8.2 Physiological venous terms.[9,13–20]

Previous definitions	VEIN-TERM update[9]
Venous valvular incompetence: abnormal functioning of the veins of the lower extremities is recognized clinically as venous dysfunction.[14]	*Venous valvular incompetence*: venous valve dysfunction resulting in retrograde venous flow of abnormal duration.
Venous reflux: reversal of flow in a segment of vein following its dilatation and/or anatomical or functional incompetence of its valves.[15]	*Venous reflux*: retrograde venous flow of abnormal duration in any venous segment.
Primary valve dysfunction: absence of complete closure of the valves.[15]	*Primary*: caused by idiopathic venous valve dysfunction.
Secondary valve dysfunction: valves irreversibly damaged by the thrombotic process.[15]	*Secondary*: caused by thrombosis, trauma, or mechanical, thermal, or chemical etiologies
Congenital valve dysfunction: atrophy or absence of valve.[15]	*Congenital*: caused by the absence or abnormal development of venous valves.
No previous definition.	*Axial reflux*: uninterrupted retrograde venous flow from the groin to the calf.
Superficial: reflux in the entire great saphenous vein to below the knee.[16]	*Superficial*: confined to the superficial venous system.
Deep: superficial femoral vein and popliteal vein or the deep femoral vein and popliteal vein when those are connected.[17]	*Deep*: confined to the deep venous system.
No precise previous definition.	*Combined*: involving any combination of the three venous systems (superficial, deep, perforating).
No precise previous definition.	*Segmental reflux*: localized retrograde flow in venous segments of any of the three venous systems (superficial, deep, perforating) in any combination in the thigh and/or the calf, but *not* in continuity from the groin to calf.
Perforator incompetence: retrograde (outward) outflow lasting greater than 0.3 s or longer than antegrade flow during the relaxation phase after release of manual compression.[18]	*Perforator incompetence*: perforating veins with outward flow of abnormal duration.

(*Continued*)

Table 8.2 (Continued)

Previous definitions	VEIN-TERM update[9]
Neovascularization: recurrence of varices after vein transection restored by growth of new vessels in the surrounding tissue and vein wall.[19]	*Neovascularization*: presence of multiple new, small tortuous veins in anatomic proximity to a previous venous intervention.
No precise previous definition.	*Venous occlusion*: total obliteration of the venous lumen.
No precise previous definition.	*Venous obstruction*: partial or total blockage of venous flow.
Venous compression: compression by external structures.[20]	*Venous compression*: narrowing or occlusion of the venous lumen as a result of extraluminal pressure.
No precise previous definition.	*Recanalization*: development of a new lumen in a previously obstructed vein.
No precise previous definition.	*Iliac vein obstruction syndrome*: venous symptoms and signs caused by narrowing or occlusion of the common or external iliac vein.
May–Thurner syndrome: compression of the left common iliac vein by vascular bone entrapment. The anterior surface entrapment is the common iliac artery, the posterior is formed by vertebral column.[13]	*May–Thurner syndrome*: venous symptoms and signs caused by obstruction of the left common iliac vein due to external compression at its crossing posterior to the right common iliac artery.

Table 8.3 Descriptive venous terms.[9,21–23]

Previous definitions	VEIN-TERM update[9]
High ligation and division: ligation and division of the long saphenous vein and its tributaries at the saphenofemoral junction.[21]	*High ligation and division*: ligation and division of the great saphenous vein (GSV) at its confluence with the common femoral vein, including interruption of all upper GSV tributaries.
Stripping: removal of the saphenous vein.[22]	*Stripping*: removal of a long vein segment, usually most of the GSV or the small saphenous vein by means of a device.
No precise previous definition.	*Venous ablation*: removal or destruction of a vein by mechanical, thermal, or chemical means.
No precise previous definition.	*Perforating vein interruption*: disconnection of a perforating vein by mechanical, chemical, or thermal means.
No precise previous definition.	*Perforating vein ligation*: interruption of a perforating vein by mechanical means.
No precise previous definition.	*Perforating vein ablation*: disconnection or destruction of a perforating vein by mechanical, chemical, or thermal means.
No precise previous definition.	*Miniphlebectomy*: removal of a vein segment through a small skin incision.
No precise previous definition.	*Sclerotherapy*: obliteration of a vein by introduction of a chemical (liquid or foam).
Endophlebectomy: surgical disobliteration of chronically obstructed venous segment.[23]	*Endophlebectomy*: removal of postthrombotic residue from the venous lumen.

CHAPTER 9

Management

The management of venous disease has to be viewed from different perspectives and depends on many factors, including:

1. The nature of the disease;
2. The stage of the disease;
3. The type of problem.

9.1 The nature of the disease

The disease may be any of the following:

1. Congenital venous malformation;
2. Cavernous hemagioma;
3. Lymphangioma;
4. AV malformation;
5. Primary varicose vein;
6. Secondary varicose vein;
7. Deep vein thrombosis (DVT).

9.2 The stage of the disease

Treatment varies according to the various stages of the disease as assessed by clinical, etiological, anatomical and pathophysiological (CEAP) classification. Classes I and II are uncomplicated and hence the treatment is very simple when compared to the other four stages, which are more time consuming as they includes three phases of treatment:

1. Preparation for the definitive surgical procedure;
2. The planned surgical procedure;
3. Postprocedural corrections and management to minimize morbidity.

9.3 The type of problem

1. *Acute*: This can present as:
   a. Bleeding;
   b. Superficial thrombophlebitis;

Genesis, Pathophysiology and Management of Venous and Lymphatic Disorders
DOI: https://doi.org/10.1016/B978-0-323-88433-4.00015-2

   c. Cellulitis;
   d. Acute DVT.
2. *Chronic*: Aching and discomfort in the lower limb, postural edema, dermatological changes, or even ulcerations develop over varying periods. Though the sequence of development of problems is defined, the timing of appearance varies from person to person. Chronic DVT manifests as varicose veins and can be mistaken for primary varicose veins as it mimics all the features of chronic venous disease.
3. *Cosmetic*: Disfigurement caused to the legs.

Management of venous diseases can be discussed conveniently under three sections:
1. Management of congenital venous disorders;
2. Management of primary varicose veins;
3. Management of deep vein thrombosis.

9.4 Management of congenital venous disorders

Congenital venous malformations, capillary hemangiomas, cavernous hemangiomas, and arteriovenous malformations any where on the body can be treated preferably by sclerotherapy. Attempting to excise them will cause profuse bleeding and is very risky. It may sometimes require several sittings to effect complete sclerosis. After achieving complete sclerosis, large or unsightly lesions can be refashioned for better aesthetic appearance.

9.5 Management of primary varicose veins

The majority of venous disease comes under this category, with almost 98% of venous disease being attributable to primary varicose veins. Hence this requires a very exhaustive description of the management modalities. Varicose veins are conventionally treated in two ways:
1. Conservative management;
2. Surgical management.

9.6 Conservative management

9.6.1 Medical treatment

Varicose veins are traditionally treated in different *systems of medicines* with herbal drugs, chemical drugs, and by supportive measures such as the

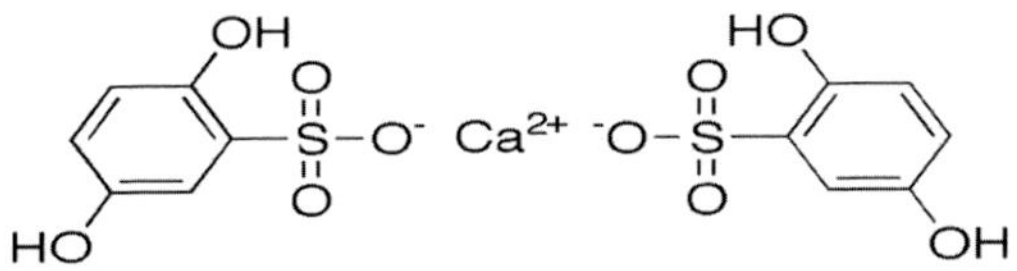

Figure 9.1 Structure of calcium dobesilate.

usage of compression bandages and leech therapy. In the Ayurvedic system of medicines, although there is no specific treatment prescribed in any literature, some Ayurvedic physicians prescribe internal medicines in the form of kashaya and tablets or local application of oils or herbal juices. Homeopathic physicians prescribe internal medicines. In modern medicine also, the drug used most widely is *calcium dobesilate* (2,5-dihydroxy-benzene sulfonate) tablet, with the chemical structure illustrated in Fig. 9.1.

9.7 Calcium dobesilate

Calcium dobesilate is generally considered to be a venotonic drug. Although it has no specific effect on varicose veins, calcium dobesilate significantly improves night cramps and discomfort in some patients. Current evidence suggests that calcium dobesilate is more effective than placebo in improving some chronic venous insufficiency (CVI) symptoms, that there is higher efficacy in more severe disease, and that a dose of 1000 mg/day is as effective and safe as 1500 mg/day. Further adequately powered trials are needed to further evaluate these hypotheses. Some studies reveal its ability to decrease capillary permeability, as well as platelet aggregation and blood viscosity. Furthermore, recent data show that calcium dobesilate increases endothelium-dependent relaxation owing to an increase in nitric oxide synthesis. There is no drug currently available for the specific treatment of varicose veins; these drugs in fact delay the definitive treatment.

Calcium dobesilate (2,5-dihydroxy-benzenesulfonate) is a synthetic drug with vasoprotective and antithrombotic properties that has been widely used in the treatment of diabetic retinopathy and chronic venous disease (CVD). It may act at the vascular endothelium, as it has been shown to reduce capillary hyperpermeability associated with diabetes mellitus, inhibit platelet aggregation, and reduce whole blood viscosity. Consistent with these effects, rat endothelial cell preparations incubated with calcium dobesilate in vitro exhibited enhanced nitric oxide-synthase activity, which is important for vascular homeostasis. Calcium dobesilate was also shown to inhibit microsomal prostaglandin synthesis in vitro, and to reduce the viscoelasticity of whole blood and plasma in patients with ischemic disease after 14 days of treatment [183]. In three studies that measured lower leg volume in CVD patients following calcium dobesilate treatment for 4–8 weeks, leg edema was significantly reduced compared to patients treated with placebo. Pooled analysis of these clinical results revealed that calcium dobesilate treatment was associated with significant improvements in objective measurements of leg edema and in subjective symptoms of leg pain, restless legs, and leg swelling. However, in a recent well-designed randomized, parallel double-blind, placebo-controlled clinical trial in over 500 CVD patients, calcium dobesilate treatment for 3 months elicited no significant improvements compared to placebo in leg edema, CVD symptoms, or quality of life (QoL). Taken together, these results suggest that calcium dobesilate may have beneficial effects in CVD but that further studies are needed to establish a definitive role for this treatment [251].

9.7.1 Daflon

The role of Daflon 500 mg in the treatment of symptomatic patients with varicose veins includes Daflon 500 mg having the ability to inhibit leukocyte activation.

It has been observed that firm leukocyte attachment to the endothelial wall and subsequent migration of leukocytes into the interstitium is a mechanism of tissue damage during inflammation, and the attenuation of this phenomenon during ischemia–reperfusion could explain the positive effects of Daflon 500 mg on clinical edema. In an ischemia–reperfusion model by Korthius, Daflon 500 mg significantly inhibited leukocyte adhesion and migration through the venous endothelium as well as the protein leakage observed in this model. During restoration of venous blood flow, the number of rolling, adherent, and migrating leukocytes as well as cells exhibiting apoptosis in the parenchyma significantly decreased in animals pretreated with Daflon 500 mg compared with control subjects. The molecular mechanism in leukocyte adhesion and activation in CVD patients involves the increased expression of several types of cell adhesion molecules at the leukocyte surface, particularly L-selectins (CD62L) and integrins (CD11b) [252]. The expression of these leukocyte adhesion molecules were substantially decreased on monocytes and neutrophils after a 60-day treatment with Daflon 500 mg in patients with CVD (C2–C4). This implies that Daflon 500 mg would prevent the inflammatory process in CVD. In patients with CVI as well as after prolonged standing, plasma concentrations of endothelial adhesion molecules, vascular cell adhesion molecule (VCAM), and interstitial cell adhesion molecule (ICAM) are increased. In the study by Shoab, the plasma level of VCAM-1 and ICAM-1 significantly decreased in C2–C4 patients pretreated with Daflon 500 mg. This reflects the ability of Daflon 500 mg to prevent the interaction between the endothelium and the leukocytes, which is at the core of CVD progression [253].

9.7.2 Daflon 500 mg: a protective effect against valve damage

Acquired valve dysfunction can occur due to inflammation, as evidenced by monocyte infiltration. It seems possible that activated leukocytes can migrate into the endothelium of proximal surfaces of the vein valves as well as proximal vein walls and promote destruction of supporting structures and remodeling of the valves with consequent valvular insufficiency. Immunohistochemical studies have demonstrated monocyte/macrophage infiltration into the valve leaflets and venous wall of patients with varicose veins. In a recent study by Schmid Schonbein et al. using a pharmacological model of hypertension, Daflon 500 mg was shown to limit the leukocyte infiltration into the vein valve and to inhibit the expression of leukocyte adhesion molecules. As a result, Daflon 500 mg attenuated the vein and valve destruction and significantly reduced the reflux rate in a dose-dependent manner. By acting at the core of the disease, that is, the leukocyte–endothelium interaction, Daflon 500 mg is the only phlebotropic drug with a protective effect on the valve endothelium. Daflon 500 mg delays the reflux appearance and is thus likely to prevent evolution of the CVD toward complications [254].

9.7.3 Effects of Daflon 500 mg on chronic venous disease symptoms

In the reflux assessment and quality of life improvement, with micronized flavonoids (RELIEF) study ($n = 4527$, intention-to-treat population), patients receiving two tablets of Daflon 500 mg daily showed progressive improvement in the CVI symptoms. After 6 months, patients in the per-protocol population showed significant improvement from baseline in the study outcome measure (ankle circumference, pain, leg heaviness, cramps, sensation of swelling). In the 1-year trial of two tablets of Daflon 500 mg daily in 170 patients, a significant reduction from baseline values in physician-assessed clinical symptoms (functional discomfort, cramps, evening edema), ankle and calf circumference, and patient overall assessment of symptom severity was demonstrated at each 2-month evaluation ($P < .001$). The rapid reductions observed during the first 2 months of treatment represented approximately 50% of the total improvement ultimately observed after 1 year of treatment. The reduction in leg edema with Daflon 500 mg, two tablets daily, was demonstrated in the Blume study in which the reduction was assessed by volumetric measurements after 6 weeks of treatment. In this study, the reduction in the mean volume was 263 mL (8%) in all patients, and 392 mL (12%) in patients with leg edema associated with varicose veins. In both trials, the reduction in leg volume was highly significant ($P < .001$).

During the course of a 6-month period of treatment with Daflon 500 mg, changes in the QoL scores were comparable across the different CEAP subgroups. Only patients with symptoms had greater improvement in QoL than patients without symptoms. These changes in QoL scores resulted mainly from the alleviation of symptoms: improvement between D180 versus D0 in pain, heaviness, swelling, and cramps was significant ($P < .001$ for all symptoms). However, the associated signs (telangiectasias, varices, edema, or skin changes) did not show such a direct impact on QoL. The dramatic improvement of QoL observed in patients treated with Daflon 500 mg can be interpreted as a result of symptom alleviation [255].

The preliminary result of a meta-analysis of five comparable, prospective, randomized, controlled studies identified from medical literature databases and from the files of the manufacturer in which 723 patients were pooled [23] confirm that venous ulcer healing is accelerated by adding Daflon 500 mg, two tablets daily, to conventional treatments: the rate of complete ulcer healing at 6 months was in favor of the Daflon 500 mg group (61.3% vs 47.7%; OR = 2.02; CI = 1.05–3.89; $P = .035$).

9.8 Micronized purified flavonoid fraction (Daflon)

Micronized purified flavonoid fraction (MPFF; e.g., Daflon) consists of 90% diosmin and 10% other active concomitant flavonoids (hesperidin, diosmetin, linarin, and isorhoifolin), and is currently one of the most widely available and prescribed venoactive drugs, and the best studied. Diosmin is synthesized from hesperidin, which is extracted from a particular type of small immature orange, and the mixture is micronized to particles of $<2\ \mu m$ diameter to improve bioavailability. The effects of MPFF have been demonstrated in both clinical and nonclinical studies, with reported improvements in venous tone and contractility, microcirculation, trophic disorders, and venous ulcer healing, and reductions in edema, inflammation, leukocyte adhesion and activation, and inflammatory mediator production [256]. A recent meta-analysis of randomized, double-blind, placebo-controlled clinical trials investigating the effectiveness of MPFF treatment in improving the symptoms, signs, and QoL in CVD patients identified seven studies involving 1692 patients [65]. The following results were observed: Most of leg pain, heaviness, and feeling of swelling [risk factors (RR) of 0.35–0.53; $P < .0001$]; cramps (RR 0.51, $P = .02$); paresthesia (RR 0.45, $P = .03$); skin changes (RR 0.18, $p < .0001$); and burning sensation (standard mean difference −0.59, 95% CI −1.15 to −0.02).

Clinical and nonclinical studies support the notion that MPFF generally improves venous tone and contractility. Nonclinical studies in isolated rat veins found that diosmin directly enhanced sympathetic-mediated venous contractility and increased calcium sensitivity and contractility. In similar studies in varicose human saphenous veins, the reported mechanisms of action were different, although with similar results, since diosmin potentiated dose-dependent norepinephrine-induced contractility. Regarding the clinical efficacy of MPFF in improving venous tone, two controlled clinical trials are particularly important. Barb et al., in a study of women with various grades of CVD, demonstrated that MPFF treatment was associated with improvements in venous distension, capacitance, and tone [236]. In addition to its venotonic effects, MPFF has demonstrated various antiinflammatory properties. In several animal models, MPFF treatment was shown to reduce leukocyte adhesion to vascular endothelium. In a hamster ischemia–reperfusion model, neutrophil adhesion in postcapillary venules was lower in animals pretreated with MPFF than in control animals. Similar findings were obtained in two rat ischemia–reperfusion models employing the cremaster muscle and mesentery vein [183].

One mechanism by which MPFF may prevent leukocyte adhesion to damaged epithelium is by inhibiting the production of the surface molecules that mediate adhesion and activation. In CVD patients, MPFF was found to selectively reduce the expression of L-selectin/CD62-L on monocytes and neutrophils after treatment for 60 days. In a rat model of chronic venous hypertension initiated by local venous occlusion followed by reperfusion (which also results in leukocyte adhesion and activation), MPFF treatment significantly mitigated these inflammatory processes and CD62-L expression in neutrophils in a dose-dependent manner. In a novel and recent model of chronic venous hypertension and disease in the hamster, iliac vein ligature induced a progressive increase in hindlimb venous pressure over a period of 6–10 weeks that resulted in increased numbers of rolling and adherent lymphocytes, decreased functional capillary density, and enlarged venules [257].

In this model, MPFF treatment twice daily for 8 weeks significantly and effectively prevented these pathological consequences and was superior to either diosmin or the active concomitant flavonoid combination alone, suggesting a synergistic effect. MPFF has been shown to reduce inflammatory marker concentrations locally and in the circulation. MPFF treatment for 2 weeks prior to scleropathy and 2 months after the procedure in patients with mild CVD (C1) resulted in lower levels of histamine, C-reactive protein, IL-1, TNF-α, and vascular endothelial growth factor (VEGF) in blood taken from treated veins. MPFF treatment for 12 weeks was also associated with significant reductions in systemic blood concentrations of endothelin-1 and TNF-α and with significantly increased antioxidant enzyme ratios in women with CVD [258]. This latter effect is particularly important since excess oxidative stress (low antioxidant enzyme ratios) leads to excessive production of radical oxygen species and has been reported in CVD patient. This may contribute to leukocyte activation, inflammation, and vein wall injury, whereas higher antioxidant enzyme ratios may have protective effects. Coupled with its inhibitory effects on cells of the immune systems, MPFF treatment may also reduce endothelial cell activity in CVD, which is an important trigger of inflammation and thrombosis. In patients with C2–C5 CVD, MPFF treatment for 60 days reduced plasma ICAM-1 concentrations by 32% and plasma VCAM concentrations by 29%. In another study in CVD patients with skin changes (C4), MPFF treatment for 60 days significantly reduced plasma VEGF concentrations by 42% ($P < .02$) [82]. These findings suggest that MPFF may help maintain endothelium quiescence in veins, preventing the binding and activation of leukocytes through these cell adhesion molecules.

MPFF treatment was also found to significantly reduce ankle edema compared to placebo in a meta-analysis of studies encompassing 463 CVD patients and to reduce evening great saphenous vein diameters in patients with transient venous reflux due to daily orthostatic loading. In patients with cyclic edema, MPFF treatment for 6 weeks significantly improved capillary hyperpermeability ($P = .02$) compared to the placebo treatment, and this improvement was accompanied by significant weight loss ($P < .05$) and a significant reduction ($P < .05$) in the sensation of swelling [88]. After 3 months of MPFF treatment (500–2000 mg/day) in CVD patients, transcutaneous oxygen pressure significantly increased ($P < .001$), transcutaneous carbon dioxide pressure significantly decreased ($P < .001$), and disease-related clinical symptoms (subjective manifestations and edema) were improved with no differences across the dosage groups [183].

In patients with symptomatic capillary fragility, as evidenced by spontaneous bruising, frequent nosebleeds, petachiae, and conjunctival hemorrhage, MPFF treatment for 6 weeks significantly increased capillary resistance to rupture by negative suction cup pressure relative to placebo treatment. In the hamster cheek pouch, immediate pretreatment with 10 or 30 mg/kg MPFF, or with equivalent doses of any of its constituent flavonoids, significantly reduced the number of leaky capillary sites produced by ischemia–reperfusion. In a rabbit model of scleropathy, MPFF treatment for 21 days prior to initiation of inflammation prevented increases in venular diameter, preserved functional capillary density, and reduced capillary leakage ($P < .001$). MPFF may also have beneficial effects on ulcer healing, but this outcome has not been extensively investigated in clinical trials. Guilhou et al., in a randomized, double-blind, placebo-controlled study in 105 CVD patients with venous ulcers who also wore compression stockings, demonstrated that complete ulcer healing rates for all patients were 26.5% in the MPFF treatment group (2 months) and 11.5% in the placebo group, although no differences were observed in the risk ratio of not healing [259]. However, in patients with smaller ulcers (<10 cm in diameter), healing rates were also higher in the MPFF treatment group (31.8% vs 12.8%) and the risk ratio of not healing (RR 0.83, 95% CI 0.62–0.98) was statistically significant. Two nonblinded similar studies compared ulcer healing rates in CVD patients receiving compression treatment with MPFF for 6 months versus patients receiving compression treatment alone. After 6 months, ulcer healing rates were 46.5% in treated patients versus 27.5% in untreated patients in one study, and 64.6% versus 41.2%, respectively, in the other study.

Together, these findings suggest that MPFF improves ulcer healing with compression therapy, but additional clinical investigation is needed. The effect of MPFF treatment on skin trophic disorders has been investigated in several randomized controlled trials, although the reported evidence is conflicting. In two studies encompassing a total of 75 CVD patients treated with MPFF, skin trophic improvement was not significantly better than in patients treated with placebo. On the other hand, in two larger trials in 160 and 200 CVD patients, skin trophic changes persisted in significantly fewer patients treated with MPFF (82.5% and 86%, respectively) than in those treated with placebo (95.0% and 96%, respectively) [230]. In a pooled analysis of all four studies, the RR of 0.87 (95% CI 0.81–0.94) for persistence of the skin trophic disorder with MPFF treatment versus placebo, indicated a statistically significant benefit with MPFF.

9.9 Rutosides

Rutosides or rutins represent another class of venoactive bioflavonoids and have been reported to have antiinflammatory properties and to improve CVD signs and symptoms. Rutoside (as pentahydroxyflavone glycoside) was found to be a potent inhibitor of inflammation-related gene expression in activated human macrophages cultured in vitro, and to inhibit the release of nitric oxide, TNF-α, IL-1, and IL-6 from these cells. In adjuvant-induced arthritic rats, rutosides reduced the clinical signs of arthritis, which correlated with the inhibition of inflammatory cytokine production measured in rat sera and in human macrophages. Rutin was also found to significantly inhibit nitric oxide and TNF-α production, as well as myeloperoxidase activity in human peripheral blood neutrophils. In several clinical studies in CVD patients, rutoside preparations were reported to significantly reduce edema and leg volume, and reduced severity of lower leg pain, leg cramps, heaviness, and itching. In a pooled analysis of randomized controlled clinical trials in CVD patients, these venoactive compounds were found to have significant benefits for edema, leg volume, lower leg pain, and leg heaviness [260].

9.10 Sulodexide

As integrity of the glycocalyx is critical for maintaining vein homeostasis, treatments that prevent glycocalyx damage or breakdown and/or promote its repair could help protect the endothelium and reduce the inflammatory cascades stemming from the activation of endothelial cells. Sulodexide is a highly purified glycosaminoglycan mixture, consisting of low-molecular-weight heparin (80%) and dermatan sulfate (20%), both of which are components of the glycoproteins that form the glycocalyx. In patients with type II diabetes, which is commonly associated with disrupted glycocalyx and elevated vascular permeability, sulodexide treatment for 2 months was associated with partial restoration of glycocalyx thickness toward control values and by a reduction in the transcapillary permeability to albumin. Sulodexide has also been shown to have antiinflammatory and antiapoptotic activities, to prevent leukocyte adhesion to endothelium, and to have a protective effect on the endothelial cell lining and vascular wall in the microcirculation. Specifically, in nonclinical and clinical studies, it lowered plasma concentrations of IL-1β and IL-8 in dialysis patients; and inhibited release of monocyte chemoattractant protein (MCP)-1 and interleukin (IL)-6 and formation of free radicals in human umbilical vein endothelial cell cultures, neo-angiogenesis in a rat model of peritoneal perfusion, plasma tumor necrosis factor (TNF)-α release in rat model of peritonitis, and release of a wide range of proinflammatory cytokines, chemokines, and colony-stimulating factors from activated human macrophages [261]. More recently, sulodexide was shown to inhibit the release of IL-2, IL-2 (p70), IL-10, and VEGF from human monocyte/macrophage cells (THP)-1 monocytes stimulated in vitro with wound fluid collected from CVD patients with venous leg ulcers. Sulodexide treatment may therefore help stem the proinflammatory cascades present in venous ulcers. Consistent with its protective effects on the vasculature, sulodexide significantly reduced secretion of pro- and complexed matix-metalloproteinases (MMP)-9 from cultured blood leukocytes and of MMP-1, MMP-9, and MMP-12 from THP-1 monocytes stimulated with wound fluid. These results suggest additional mechanisms by which sulodexide treatment may prevent degradation of vascular extracellular matrix and collagen [262].

In CVD patients, sulodexide treatment was associated with reduced plasma concentrations of MMP-9, IL-6, and MCP-1 compared to pre-treatment concentrations. In an uncontrolled study in 476 CVD patients that compared 60 days of treatment with different doses of sulodexide (50—100 mg/day), statistically significant and clinically relevant improvements in peripheral venous pressure were observed, along with significant reductions in other clinical signs and symptoms. A recent open uncontrolled observational study in 450 CVD patients reported that 3 months of treatment with sulodexide significantly improved objective signs (erythema, skin temperature, induration) and all subjective symptoms of CVD ($P < .0001$) and significantly improved patient-assessed QoL scores in the CIVIQ questionnaire ($P < .0001$). Sulodexide treatment has also shown benefits in venous ulcer healing. In 235 patients with venous ulcers, those treated for 3 months with sulodexide exhibited significantly higher rates of complete ulcer healing and significantly greater reductions in ulcer surface area over time, compared to patients treated with placebo. Similar benefits of this venotonic on ulcer healing in CVD patients have been reported for other studies, including one in which complete ulcer healing with local treatment was found to be significantly faster with a combined sulodexide plus MPFF treatment than with MPFF treatment alone. A recent meta-analysis of three randomized controlled clinical trials investigating the effects of sulodexide on ulcer healing found that, although the quality of the evidence was low, the proportion of completely healed ulcers may be greater with local treatment alone (49% vs 30%; RR 1.66, 95% CI 1.30–2.12) [263].

Venoactive drugs present pharmacological and clinical profiles that can be explained by their action at specific levels of CVD pathophysiology. MPFF increases venous contractility in isolated vein preparations and increases venous tone in persons at risk for varicose veins, whereas these drugs as a class all exhibit broad antiinflammatory activity. Several have been shown to reduce leukocyte adhesion and activation in different models of venous inflammation induced either by ischemia–reperfusion or by transient venous hypertension. In humans, venoactive drugs reduce the concentration of plasma markers of inflammation (TNF-α and other cytokines), endothelial activation (ICAM–VCAM), vascular hypertrophy, and angiogenesis (VEGF), and also the release of proteases involved in the breakdown of extracellular matrix and venous tissue remodeling. Most also appear to improve capillary resistance and decrease excess vascular permeability. The pharmacological actions of these drugs are the basis behind their demonstrated clinical benefits, including significant reductions in leg edema, skin trophic disorders, patient-reported symptoms, and ulcer healing time [230]. Inappropriate endothelial activation in CVD is a mechanism common to other cardiovascular diseases and diabetes. Drugs that specifically target endothelial dysfunction in these diseases may also be beneficial in CVD and some, such as angiotensin-converting enzyme inhibitors, statins, and agents that target endothelial NO synthase coupling, may be worth investigating as adjunct therapies in future studies.

The pharmacological profiles and demonstrated clinical benefits of venoactive drugs provide the rationale behind their use in the treatment of CVD. MPFF and sulodexide are recommended in international guidelines for the treatment of CVD and venous ulcers but MPFF is the only one to obtain a grade 1B in the relief of symptoms associated with C0–C6. MPFF and sulodexide are also recommended in the clinical practice guidelines issued by the Society for Vascular Surgery and the American Venous Forum. Further experience and controlled clinical investigation of venoactive drugs is likely to provide additional clinical evidence to better define the therapeutic benefits of these important treatments to CVD patients.

9.11 Implications for treatment

In patients presenting with varicose veins or early-stage CVD, conservative treatment with compression stockings is generally the standard treatment option used to reduce symptoms and prevent disease progression. Although effective, with general improvement of all stages of CVD, as well as ulcer healing, compression stockings have poor treatment compliance, which limits their effectiveness. For that reason, many CVD patients warrant pharmacological treatment. Fortunately, several phlebotonics, or venoactive drugs, are currently available that intervene with many of the pathophysiological mechanisms of CVD. These drugs offer promising pharmacological efficacy and safety profiles, and most are derived from natural flavonoids extracted from plants. A recent Cochrane systematic review reported a meta-analysis of randomized, double-blind, placebo-controlled clinical trials investigating the efficacy of several venoactive drugs (such as rutosides, hidrosmin, diosmin, and calcium dobesilate) in the treatment of CVD [264].

9.12 Compression bandages

Use of elastocrepe bandage is very common and helpful in varicose vein management, and the same role is met by varicose vein stockings. This is also a supportive treatment and not a preventive treatment for varicose veins. One has to be very careful while applying and selecting the stockings. The following points should be observed. The stockings selected should be of the correct size for the limb and should not be too tight or too loose. When applying the bandage, keep the leg elevated above the horizontal level, start bandaging from just proximal to the toes upwards in a criss-cross fashion maintaining uniform pressure throughout.

9.13 Surgical management

Surgical treatment is undoubtedly the definitive treatment for varicose veins. However, the technique used is not always satisfactory and hence does not meet with universal approval. This is a topic which is totally misunderstood and can be confusing to surgeons and dermatologists throughout the world. A brief insight into the history of varicose vein surgery is very much interesting, informative, and thought provoking for surgeons interested in this field. An overall view into this history (from 3500 BCE to the present) enhances understanding of the stages of development of venous surgery, the contributors, and the numerous experts from the past, who should be remembered and applauded for their work.

CHAPTER 10

History of the surgical management of varicose veins

"The future is the past rewritten"

(var′ə kōs′) **Latin *varicōsus*, from *varix*, *varic-*, swollen**
Origin: L varicosus < varix (gen. varicis)

10.1 Introduction

The history of the treatment of varicose veins dates back to before 3500 BC, as evidenced by the descriptions in Ebers Papyrus from Ancient Egypt. Despite earnest efforts at improving the surgical techniques over the past 56 centuries, there seems to have been no significant improvement in the surgical treatment of varicose veins; the reason for this being mainly attributable to two factors. First, everyone tried to follow the practices of their predecessors without analyzing the real causes of the poor outcomes following surgical treatment, which even now, in principle, remains unchanged. Second, the *methodology* in the technique of the surgical procedure was periodically modified from *invasive to minimally invasive*, but the *surgical principle behind the procedure remained unchanged*. When we speak of different methods in operative procedures and good results are claimed in specialties, a keen unprejudiced search of the history combined with a thorough understanding of the biophysics and etiopathogenesis based on human anatomy and physiology, will prove that it is high time to revolutionize the present understanding and set up of treatment, based on newer outlooks and principles.

The history of surgery of varicose veins is vast. Contributions to this surgical field were initiated, as in so many cases, by *the Father of Medicine*, Hippocrates (460–377 BCE). This was followed by contributions others from different countries, especially the Greeks, Romans, Arabs, Germans, French, and British. The endless innovations and research continue to aim at the great goal of offering a permanent cure for this distressing disease. The contributions are discussed briefly in this chapter in chronological order.

Genesis, Pathophysiology and Management of Venous and Lymphatic Disorders
DOI: https://doi.org/10.1016/B978-0-323-88433-4.00018-8

10.2 Egypt

The Egyptians are credited with initiating studies into the surgical correction of varicose veins. They described them as "serpentine windings." After the bitter experiences of attempting surgery on these veins, they concluded that "anyone attempting to operate on these serpentine windings would make the patient touch the ground."

10.3 Greece

10.3.1 Hippocrates (460–375 BCE)

Hippocrates was the first to write a medical literature on varicose veins. He used the word "varicose" in medical terminology for the time, considering the appearance of varicose veins as "grapelike." He did not advise any radical procedures for treating the condition. In the *Hippocratic Medical Treatises* (460 BCE), he advised compression bandaging after making multiple punctures on the veins. It is worth remembering his quotation, "what cannot be cured by medicaments can be cured by knife, what cannot be cured by knife can be cured by the searing iron and that whatever cannot be cured by the searing iron can be considered incurable" (Fig. 10.1).

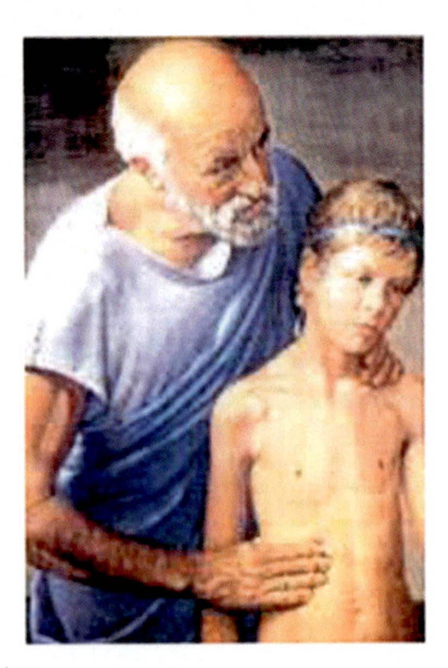

Figure 10.1 Hippocrates.

10.3.2 Paulus Aegineta (CE 625–690)

Paulus Aegineta was a 7th-century Byzantine Greek physician best known for writing the medical encyclopedia *Medical Compendium in Seven Books*, which contained the sum of all Western medical knowledge and was unrivaled in its accuracy and completeness. He also identified and stressed the importance of ligation and excision of the great saphenous vein long before Trendelenburg. He wrote: "Varices of the leg may be operated upon in a manner similar to that for varicocele, making the attempt upon those in the inner parts of the thigh where they gradually arise, for below this they are divided into many ramifications. A tourniquet is placed upon the thigh, and the patient walks. When the vein becomes distended a mark is made with writing ink or collyrium. Having placed the man in a reclining position with his leg extended we apply another tourniquet above the knee, and where the vein is distended we make an incision through the skin. The vein is freed and tourniquets are removed. A double thread is introduced under the vein and so cut as to make two ligatures, and the vein is opened in the middle, and as much blood as is required is evacuated. The wound is dressed with a pledget in it and with an oblong compress soaked in wine and oil. It is then bandaged" (Fig. 10.2).

Figure 10.2 Paulus Aegineta.

10.4 Rome

10.4.1 Celcus (53 BCE–CE 7)

Aurelius Cornelius Celcus, the Roman physician, in the medical treatise *De Medicina* described the ligation and excision of veins and the complications thereof. He was better known as a medical historian than a physician. He exposed the affected vein, made incisions on the vein at a distance of 10 cm without a ligature, applied cautery at the depth of vein, extracted the vein as much as possible, and applied pressure to control bleeding (Fig. 10.3).

Figure 10.3 Celcus.

10.4.2 Galen (CE 131–201)

Claudius Galen, the founder of experimental physiology, gave a description of disconnection of the artery and vein at the affected limb in the hope of reducing pain and gangrene. He was known as the "prince of physicians." He was the most distinguished practitioner in Rome and was known also as the "wonder worker" and "medical Pope" (Fig. 10.4).

Celcus and Galen can be considered the original promoters of the modern *phlebectomies*.

Oribasius from *Pergamum* (CE 325–403), who studied medicine at the famous School of Alexandria, was the personal physician of Emperor Julian the Apostate and was the most eminent early Byzantine physician, who practiced multiple phlebectomies along the length of the great saphenous vein.

Figure 10.4 Galen.

10.5 Arabia

Abu I-Qasim Khalaf ibn al-Abbas al-Zahrawi (CE 936–1013), known in the west as "Abulcasis" and "*Albucasis*" was a celebrated Arabic physician, surgeon, scientist, and writer. His most important work is at-Tasrif, a 30-volume collection of medical knowledge with chapters on surgery. He borrowed heavily from the writings of Aegineta and favored a form of ligation and stripping or dissection of superficial veins. By the 11th century the major principle of treatment of varicose veins was outlined. It was the first textbook of surgery with illustrations of instruments used in surgery ever published. It gained such fame that it became the standard textbook of surgery in prestigious universities in the west and was very widely read. He emphasized that knowledge of anatomy and physiology was essential prior to undertaking any surgery. The book was used in Europe until the 17th century. Al-Zahrawi described the ligature of arteries long before Ambrois Paré. Al-Zahrawi also used cautery to control bleeding. He showed evidence of great experience from details of clinical pictures and surgical procedures, for example, his description of varicose vein stripping, even after 10 centuries, is very similar to modern surgery. "... Have the leg shaved if it is much hairy. The patient gets a bath and his leg is kept in hot water until it becomes red and the veins dilate; or he exercises vigorously. Incise the skin opposite the varicose vein longitudinally either at the ankle or at the knee. Keep the skin opened by hooks. Expose, dissect, and separate the vein. Introduce a spatula underneath it.

Figure 10.5 Abulcasis.

When the vein is elevated above the skin level, hang it with a blunt rounded hook. Repeat the procedure about three fingers from the previous site and hang the vein with another hook as previously done. Repeat the procedure at as many sites along the varicose vein as necessary. At the ankle, ligate and strip it by pulling it from the incision just above. When it reaches there, repeat at the higher incision until all of it is stripped. Ligate the vein and then excise it. If difficulty is encountered in pulling it, ligate its terminal part with a string and pass it under the spatula and dissect it further. Pull gently and avoid its tearing because if it does, it becomes difficult to strip all of it and can cause harm to the patient. When you have stripped it all, put alcohol sponges at the sites of the skin incisions and take care of the incisions until they heal. If the varicose vein is tortuous, you have to incise the skin more frequently, at each change of direction. Dissect it and hang it with the hooks and strip it as previously described. Do not tear the vein or injure it. If this happens, it becomes difficult to strip it. The hooks used should be blunt, eyeless, and rounded; otherwise it can injure the vein" (Fig. 10.5).

10.6 The era of barber surgeons

The 13th century witnessed great progress in surgical procedures, and were led by European physicians described as master surgeons. However, during the era of the barber surgeons (CE 1500–1850), for about 350 years the progress of science and surgery was stalled. The famous French barber surgeon, Ambrois Paré (CE 1510–1590) is an exception. He modified the techniques of Aegineta and Albucasis, favoring ligation at the site of varices and sometimes cautery as Celcus had advised (Fig. 10.6).

Figure 10.6 Ambrois Paré.

10.7 The Renaissance

Jean Louis Petit (1674–1750) wrote extensively regarding radical excision of varices. He stated that anything that obstructed the rising of the blood in the vein was the principal cause of varicosities. Lorenz Heister, a leading German surgeon, a century later (CE 1768) wrote about the crudeness of past procedures to treat varicose veins and recommended bleeding, a diet which amounted to near starvation, and the application of bandages to the legs, so that the walls of the vein might be strengthened. He also taught longitudinal incisions on large varices (phlebotomy) between two percutaneous ligatures put around the vein and to allow letting out one pint of blood and apply compression bandage. Sir Everard Home (1756–1832), brother-in-law of John Hunter, worked on the treatment of leg ulcers and in 1811 described the effect of ligation of the long saphenous vein above the knee.

Morgnani, a physician-scientist, considering that varicose veins seem to affect only upright humans, wrote: "without doubt, it was not very easy for the blood to pass through a liver. But why, then, you will say, did it not stagnate equally in the other veins which go to the trunk of the vena portarum? And for this very reason it was that I said you would immediately understand it, or at least in part. Add therefore, to omit other things, the very great length, which is peculiar to this one vein among others, so that it is much more difficult for the blood to be carried upwards, from this vein, than from the others, especially as the situation of the human body requires it, which without doubt is one of the reasons why other animals are not subject to piles. And if you ask why, in those bodies in which there is any impediment to the quick motion of the blood upwards, the veins of the legs in particular are dilated into varices, you will find the same thing to be the cause of them chiefly which we assign for the piles."

10.8 The era of sclerotherapy

Christopher Ubren (CE 1656) and his associates are reputed to have been the first to introduce drugs intravenously. Using a metal tube they introduced opium into the veins of a dog. This was followed by *D. Zollikofer* in Switzerland in 1682 who injected an acid into a vein. Similar injections were given to a human a few years later by J.D. Major and Casper Scotus [265].

Francis Rynd in 1845 and Parvez in 1851 were known to be the first to introduce the hypodermic syringe. However, Howard Jones pointed out that there was no true "invention" of the hypodermic syringe. The subcutaneous route was adopted by Wood of Edinburgh, and for this purpose he adapted the syringe made by Fergusson in 1855. This opened new avenues of approach to the treatment of varicose veins. It is interesting to note that Sir Everard Home, the brother-in-law of John Hunter, was the first to observe that dilation of a vein made valves incompetent. However, the implications of this fact, together with the observations by John Hunter on the venous circulation, were not taken into full account by many of those who treat varicose veins (Fig. 10.7).

In the early 1850s, the treatment of varices by injection began to attract attention. Cassaignac [265], and also Debout in 1853, used injections of perchloride of iron and reported some success. Desgranges used injections of iodotanin. Soule [265] noted the development of inflammation and suppuration following perchloride of iron injections, and advised the use of compression to prevent dilatation of the veins after injection. Muller [265] reported four cases successfully treated by injections of iron chloride in the 1860s, but Corbiu [265] reported severe phlebitis and sloughing following injections of persulfate of iron. A solution of iodotannin was used by Panas [265], who reported suppuration in both of his patients, and gangrene of the skin in one of them.

Figure 10.7 Sir Everard Home.

In 1876, Weinlechner [265] reported the healing of a varicose ulcer by the injection, with iron perchloride, of varicose veins in the region of the ulcer. The popularity of the injection treatment was now accelerating, and Burroughs [265] reported a series of 60 patients, with successful results in those who completed the treatment. Weber [265] reported using carbolic acid in one patient with success, and Stevenson [265] repeated this in eight cases. However, at the surgical congress of Lyon in 1894 [265], the injection treatment of varicose veins was much discussed, and it was finally decided, in view of the complications which all too frequently developed, that this treatment should be abandoned.

In 1904, Tavel reviewed the subject before the Congress of the Swiss Medical Association and advised the injection of 5% phenol solution into varicose segments of veins. This treatment was not widely accepted because of the severity of the reaction.

Schiassi of Bologna, in 1908, reported his results with high ligation and subsequent injection, and in 1913 he reported improvements of this method with further success. In 1911, Sicard of the Paris University noted the obliteration of veins following injections with luargol solution. The reaction was too florid, and he changed to injecting sodium salicylate in 20%, 30%, and 40% solutions. By 1919, Sicard changed his technique again, this time using sodium carbonate. Kausche, in 1917, introduced quinine and urethane. By 1930, sodium morrhuate had become popular and its use was advocated by Rogers and Winchester [265]. This popularity grew with the increased mortality of operative procedures at that time.

However, the popularity was followed by discredit as, in 1933, Faxon [265] published the results of a follow-up survey of injection treatment, which showed a recurrence rate of 63%. Two years previously, Howard, Jackson, and Mahon [265] had published a recurrence rate of 98% following injection treatment, and Ochsner and Mahorner [265] reported a 60% recurrence rate. With the advance of surgical techniques and the discredit of injection therapy, treatment returned to the principles of Hippocrates, Galen, and Aegineta.

However, most books written on the subject of the treatment of varicose veins at that time still described various injection techniques and sclerosing substances. Ochsner and Mahorner [265] injected varicosities with the patient standing up, and advocated compressing the limb with bandages for 4–6 weeks. Barrow [265], who also injected the patient standing up, gave an injection into the varices, followed by bandaging, and gave no further injection until the first one had produced fibrosis of the varices. McPheeters and Anderson [265] applied tourniquets before injecting, and advocated the necessity for immediate ambulation of the patient. They described better results following multiple injections.

The use of air and a sclerosing drug in combination was described by Orbach in 1944: *the air block technique*. Pratt [265] stated that a tourniquet should not be used when injecting. He advocated the use of the fingers to diagnose the areas to be injected, and injected the patient standing up. Like McPheeters, he advocated ambulation of the patient. In 1995s Cabrera et al. and in 1997s Monfreux published papers about their attempts to make a stable sclerosing foam by mixing air with the solution in a glass syringe. In 1997s Henriet tried this in minor varicosities. In 1999s Cavezzi and Frullini reported their results using the Monfreux technique in major veins and recurrent veins after surgery. In 1999, Tissari described another technique of mixing sclerosant with air using two plastic syringes and a three-way stopcock, which is even now followed.

Linton, in 1938, was responsible for a major advance in the treatment of chronic venous insufficiency. He described in detail the anatomy of the perforators of the leg and an operative technique for their ligation when they became incompetent.

Many of the basic principles of treatment suggested in this monograph have been mentioned by other authors in the past: injection of a sclerosant, compression of the leg, ambulation of the patient, dietary measures, and operative approaches. All these procedures performed, appeared to have shown a lack of attention to the pathophysiology of the underlying cause.

Despite many incarnations of surgery for varicose veins it took until the early 1900s for the crossectomy technique (high saphenofemoral ligation) to be established.

10.9 Back to invasive procedures

10.9.1 Benjamin Brodie (1783–1862)

Sir Benjamin Collins Brodie, who was born in Winterslow, Wiltshire, England, requires special mention. His father, a clergyman, was a classical scholar, who had his son educated at home rather than at one of the English public schools and universities. In 1801, the 18-year-old Benjamin was ready to take up his medical studies in London, where he attended the anatomy lectures of John Abernethy (1764–1831), a pupil of John Hunter (1728–93), at St. Bartholomew's Hospital, and later at the Great Windmill Street School of Medicine established by the older Hunter brother, William (1718–83). Two years later Brodie became the trainee of Sir Everard Home (1756–1832), the brother-in-law of John Hunter, at St. George's Hospital, London, where he graduated in 1805. It was Sir Everard Home to whom Brodie dedicated his famous book on *Pathological and surgical observations on diseases of the joints.*

Brodie's eponym is still in vogue in relation to a mass of inflammed anal mucosa at the lower end of a fissure in ano, "Brodie's pile," and "Brodie's tumor" of the breast (giant fibroadenoma). Brodie first operated on a patient with varicose veins in 1814. He was not only a versatile surgeon but also a great writer, thinker, philosopher, a good administrator, and above all a philanthropist. He was the first to observe the reverse flow of blood in the varicose veins of the saphenous system. He also gave the first description of intermittent claudication, and the test for incompetence of the valves of the saphenous veins in 1846, now credited to Friedrich Trendelenburg (1844–1924), Professor of Surgery in Leipzig, half a century later. Trendelenburg's observation was a verbatim translation of Benjamin Brodie's observations. His views on the surgical treatment of varicose veins were clouded by the nonantiseptic era with its attendant mortality. For this very reason he thought it unadvisable to submit these patients to surgery. He abandoned the subcutaneous excision of varicose veins by stating "I always observed that, if I cured one cluster of veins,

Figure 10.8 Benjamin Brodie.

two smaller ones appeared on either sides and that, ultimately, I left the patient no better than I found him." (Fig. 10.8).

Davat [265] and Schede [265] described the percutaneous ligation of veins, but the recurrence rate was very high.

10.9.2 Friedrich Trendelenburg (1844–1924)

The advent of anesthesia and antiseptic surgery advanced the treatment of varicose veins with great pace. *Friedrich Trendelenburg*, perhaps one of the most well-known venous surgeons, popularized mid-thigh ligation of the greater saphenous vein (GSV). He made a 3 cm transverse incision at the junction of the middle and upper third of the thigh and ligated the vein in situ. He believed that groin dissection was unnecessary as blood would flow through perforators alleviating back pressure, perhaps similar to the theories applied by CHIVA today. Twelve patients were hospitalized for 5 weeks and he claimed he could do the operation so fast that no anesthetic was required. Trendelenburg published recurrence rates of *22% at 4 years*. This procedure was later modified by Trendelenburg's student Perthes, who advocated a groin incision and a saphenofemoral ligature. Although called the *Trendelenburg procedure* (1895), the mid-thigh ligation procedure was actually performed as early as the seventh century. Through the ages, the vast majority of patients who have been treated for varicose disease have undergone some variant of this procedure. Perthes, the student of Trendelenburg, published recurrence rates of 18% in 41 procedures (Figs. 10.9 and 10.10).

Figure 10.9 Friedrich Trendelenburg.

Figure 10.10 Perthes.

10.9.3 Otto Wilhelm Madelung (1846–1926)

Madelung's operation (1884) a radical procedure consisting of the complete extirpation of the long saphenous vein and its main tributaries through

Figure 10.11 Otto Wilhelm Madelung.

Figure 10.12 Madelung's operation.

lengthy incisions in the thigh and lower leg, later fell out of favor due to high incidence of pulmonary embolism (Figs. 10.11 and 10.12).

10.9.4 Peterson

Peterson's operation (1893), also known as *Schede's operation*, was done by making a circular incision through the skin and subcutaneous tissue. The object was to interrupt the superficial circulation in the veins at that level (Fig. 10.13).

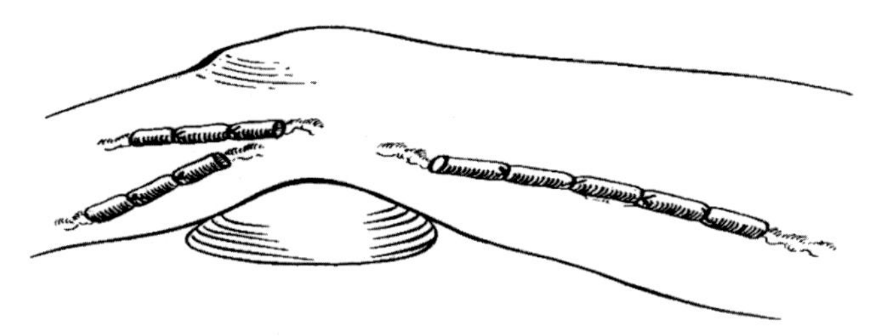

Figure 10.13 Schede's operation.

Figure 10.14 Mayo.

Keller [265], introduced a twisted wire through the varicose vein along the lumen and the withdrawal of the wire inverted the vein, which was pulled out through its own lumen with satisfactory results. Several new approaches to stripping the GSV were introduced in the first few years of the 20th century. The Mayo [265] stripper is an extraluminal ring that cuts the tributaries as it passes along the vein. Mayo's original "ring" stripper, which was passed outside the vein, has now been largely replaced by the intraluminar stripper designed by Myers (see page 72 figure), or by one of its modifications. The Babcock [265] device is an intraluminal stripper with an acorn-shaped head that pleats up the vein as it pulls the vessel loose from its attachments. The Keller device is an internal wire used to pull the vein through itself, as is done today with perforation–invagination strippers (Fig. 10.14).

Many historical surgical approaches were unpalatable to patients. The Rindfleisch–Friedel method [265] involved cutting a deep (i.e., to the level of the deep fascia) spiral gutter that wrapped around the leg six

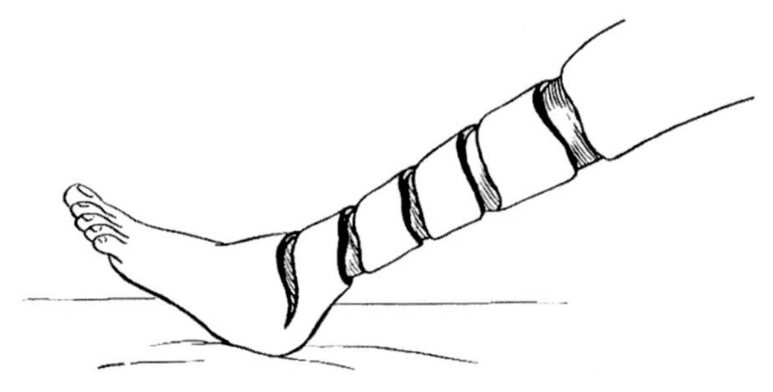

Figure 10.15 Friedel operation.

times, bringing into view a large number of superficial veins, each of which was ligated. This wound was left open to heal by granulation. The Linton procedure, introduced in the late 1930s, also used an open approach for removal of incompetent vessels and subfascial interruption of perforating veins, and this procedure also led to cosmetically undesirable outcomes. Oschner described this operation as "one of the most atrocious crimes a heavy-handed surgeon ever inflicted on the patients" (Fig. 10.15).

Thelwall Thomas, a Liverpool surgeon, described ligation of the long saphenous vein at the saphenous opening, however, credit for this procedure went to the San Franciscan John Homans in 1916. The second paper was published in Australia by William Moore from Melbourne [265]. He described a crossectomy under local anesthetic as an outpatient. The slow speed of communication between the northern and southern hemispheres at the turn of the century meant that the credit for this went to Geza de Takats of Chicago [265]. In fact, Homans credits the crossectomy procedure to Madelung who reported full excision of the GSV. Homans' paper describes the use of a transverse incision several inches long made in the groin about an inch below Poupart's ligament. Through this incision the GSV is divided at the saphenous opening. At the same time any other veins that parallel it, or enter it from above are found and divided, "in order to do away with any vessel capable of reestablishing a large single collateral trunk." Homans, like Langenbeck and Perthes before him, recognized the importance of preventing new vessel formation (neovascularization) and stated as his first requisite that the GSV should be eradicated in such a way that there was no possibility of the reformation of its channel or the formation of a similar channel.

Today, crossectomy forms part of the gold standard procedure for the treatment of primary great saphenous varicose veins. A small incision is made in the groin crease medial to the pulse emitted by the femoral artery. The saphenofemoral junction is identified with careful dissection. Once the GSV has been identified it can be divided between dips. All of the tributaries should be identified, and rather than simply tying them adjacent to the saphenofemoral junction, they should be dissected back to their first branch before ligation. The authors use a technique of diathermy avulsion of the tributaries back to, and beyond, the first branch, which is both fast and efficient. The ligation of the saphenofemoral junction itself can be performed a number of different ways and as yet there is no good evidence to suggest one technique is better than another. Surgeons may use single/double ligation, transfixion, or a method of oversewing the junction flush to the common femoral vein. It has become apparent that failure of varicose vein surgery (and the development of recurrent varicose veins) is most common at the level of the groin. Most recurrent GSVs are due to recurrent saphenofemoral incompetence, particularly when the original surgery is done by a vascular specialist. Previously, when the operation was done by a general surgeon or a surgical trainee, inadequate groin dissection was common. There is considerable controversy about whether the groin recurrence is due to growth of new vessels at the ligated junction (angiogenesis, or neovascularization), or whether it is due to dilatation of existing collateral also.

The idea of using endovenous electrosurgical devices for venous wall collagen denaturation is not new. Over the past few decades, monopolar electrosurgical desiccation has been used sporadically [266]. Endovenous obliteration with radiofrequency resistive heating is a more advanced method, including precise heating, feedback controlled by the venous wall temperature, and impedance. The idea behind using this was to shrink the vein so as to make the valves competent again, but on the contrary, obliteration of the vein was observed. This intention to reduce vein luminal diameter in order to eliminate vein reflux using controlled collagen-denaturation contraction was not efficient enough, and it has been abandoned [267].

Endovenous obliteration with radiofrequency resistive heating has turned out to be a feasible method for the treatment of superficial venous reflux. The two sizes of catheters allow for obliteration of veins from 2 to

12 mm in diameter and they are not too tortuous for catheter passage. The technique is minimally invasive, but still able to provide satisfactory immediate and long-term results [268].

A number of techniques including cryostripping [265], endovenous laser obliteration [265], saphenous valvuloplasty (EV-SFJ (external valve-plasties of the sapheno-femoral junction) and CHIVA (Conservatrice Hémodynamique de l'Insuffisance Veineuse)) [265], *which survived only for a limited period of time*, angioscopic techniques [265], transposition of a competent tributary vein [265] and echo-sclerotherapy using a sclerosant foam [265], have been proposed to minimize the trauma of chronic venous disease treatment.

Crossectomy and stripping have been the standard of care for primary great saphenous varicose veins since the high failure rates of sclerotherapy became apparent in the 1970s. As the specialty of venous surgery has evolved, a number of clinical trials have established the optimal methods of surgical treatment, and the clinical benefit of routine stripping. Long-term trials, however, have uncovered a high recurrence rate after varicose vein surgery that approaches 70% after 10 years. There is much debate about whether this is the result of the dilatation of existing tributaries in the groin or the growth of new veins as a result of angiogenesis that follows surgical treatment and healing (neovascularization). The addition of barrier technology to current crossectomy has the potential to improve the results of surgery in the future. In the meanwhile, new techniques are evolving to obliterate the great saphenous vein, including endovenous laser, radiofrequency ablation, and foam sclerotherapy. Randomized clinical trials are urgently required to compare these new treatments against standard surgery, and they will need to focus on whether the short-term gains in reduced convalescence and morbidity are balanced by durable long-term results.

Now, in this 21st century, due to the unsatisfactory outcomes following invasive and noninvasive procedures, every surgeon chooses his own technique based on his skill and convictions. While the majority still continues to follow the Trendelenburg operation and stripping/multiple ligation/microphlebectomy, some resort to the minimally invasive techniques such as radiofrequency ablation, endovenous laser treatment, sclerotherapy (plain/microfoam/USG-guided/endovenous videoscopic sclerotherapy), or combinations thereof.

10.10 Comments

From my experience of having treated around 14,000 patients, in order to get a better outcome from any procedure, one has to have a very clear knowledge of the anatomy and physiology of the circulatory system, the biophysics, and the etiopathogenesis behind the development of this chronic venous disease, rather than trying to copy a technique postulated or advocated by someone centuries ago. In the past, as well in current surgical practice, it shows an ascending curve from below the knee toward the inguinal region. The current practice is nothing but copying previous techniques without changing the principle of surgery with a newer outlook on the etiopathogenesis. Only the knife has been replaced by a laser,

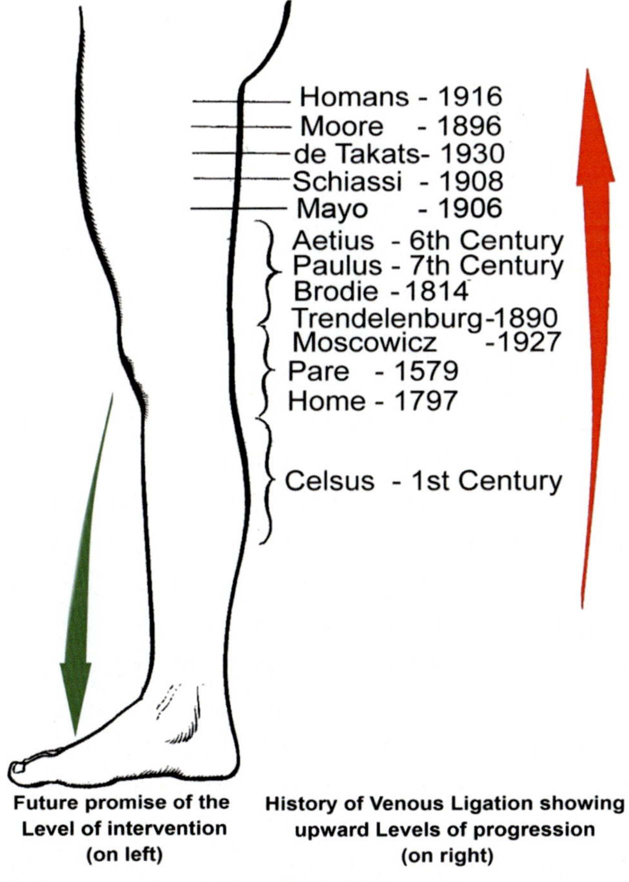

Figure 10.16 Past, present, and future surgical interventions.

radiofrequency probes, or chemical sclerosants. The day is not too far off for revision of the principle of this technique with a newer outlook. Then we may have to look down toward the ankle and foot, rather than look up, for a promising effective future technique. This would be the chemical weapon, the sclerosants, in effect chemical *bombing* of the venous network of the diseased superficial venous system (Fig. 10.16).

CHAPTER 11

Contemporary surgical procedures

There are two different types of surgical management:

1. Open/operative procedures
2. Closed/minimally invasive/endovenous procedures

11.1 Open/operative procedures

1. Trendenlenburg's operation
2. Excision of short saphenous veins
3. Saphenectomy/vein stripping and ligation
4. Microphlebectomy
5. Transilluminated powered phlebectomy

11.2 Closed/minimally invasive/endovenous procedures

1. Endovenous laser ablation/treatment
2. Endovenous radiofrequency ablation
3. Sclerotherapy: various techniques
4. Endovenous steam ablation
5. MOCA (ClariVein)
6. Cryo-stripping
7. Glue (VenaSeal)
8. SEPS (subfascial endoscopic perforator surgery)
9. Echo-therapy (Sonovein)

11.3 Trendenlenburg's operation

11.3.1 Friedrich Trendelenburg

Friedrich Trendelenburg was a German surgeon, born on May 24, 1844, in Berlin, and who died on December 15, 1924, in Nikolassee near Berlin. He was the son of the a well-known philosopher Friedrich Adolf Trendelenburg (1802–72). Although he was born in Berlin, at his time a

Genesis, Pathophysiology and Management of Venous and Lymphatic Disorders
DOI: https://doi.org/10.1016/B978-0-323-88433-4.00019-X

world center of medical science and teaching, he chose to study medicine in Scotland, at the universities of Glasgow [as a pupil of Allen Thomson (1809–84)] and Edinburgh. He completed his studies in Berlin under Bernhard Rudolf Konrad von Langenbeck (1810–87) and received his medical doctorate in 1866 with a dissertation on surgery in ancient India. He was later made doctor of honor at the University of Aberdeen, Scotland.

During the years 1868–74, Trendelenburg worked as an assistant at Langenbeck's clinic, and in 1874 he received a senior position as medical director of the surgical station of the Friedrichshain hospital in Berlin. The following year he left Berlin to enter a position as Professor of Surgery in Rostock. Seven years later, in 1882, he assumed the same office in Bonn, and in 1895 succeeded Karl Thiersch (1822–95) in Leipzig. In Leipzig he was entrusted with the position of Surgeon-in-Chief at the University Clinic. He remained in Leipzig until his retirement in 1911.

It was during this time as assistant to Langenbeck that Trendelenburg worked on stricture of the trachea. In Rostock he introduced gastrostomy in the treatment of esophageal stricture, and in 1878 he was the first surgeon to suture the patella in Germany. He first used his position for operating on viscera in 1881.

Trendelenburg introduced an operation for varicose veins and, in 1907, made an attempt at surgical removal of a thrombosis in a patient with pulmonary embolism. He lived to see his pupil, Martin Kirschner (1879–1942), perform the first successful embolectomy in 1924. He is also connected with a cannula/needle used for preventing patients from swallowing blood during operations of the larynx.

Friedrich Trendelenburg founded the German Surgical Society in 1872, and was greatly interested in surgical history. He wrote an account of ancient Indian surgery as well as an autobiography.

Trendelenburg spent his last years in Nikolassee near Berlin. His son Wilhelm Trendelenburg (born 1877) was Professor of Physiology at the University of Innsbruck.

The advent of anesthesia and antiseptic surgery advanced the treatment of varicose veins with great pace. Friedrich Trendelenburg (May 24, 1844–December 15, 1924), perhaps one of the most well-known venous surgeons, popularized mid-thigh ligation of the great saphenous vein (GSV). He made a 3 cm transverse incision at the junction of the middle and upper third of the thigh and ligated the vein in situ. He believed that groin dissection was unnecessary as blood would flow through perforators alleviating back pressure; perhaps similar to the theories applied by Conservatrice Hémodynamique de l'Insuffisance Veineuse (CHIVA) today. Twelve patients were hospitalized for 5 weeks, and he claimed he could do the operation so fast that no anesthetic was required. Trendelenburg published recurrence rates of 22% at 4 years. This procedure was later modified by Trendenlenburg's student Perthes, who advocated a groin incision and a saphenofemoral ligature. Although called the *Trendelenburg procedure*, the mid-thigh ligation procedure was actually performed as early as the seventh century. Through the ages, the vast majority of patients who have been treated for varicose disease have undergone some variant of this procedure. Perthes published recurrence rates of 18% in 41 procedures.

The currently practiced technique is a repeatedly modified one and, although still described as Trendenlenburg's operation, it has no resemblance to what Trendelenburg did. Even before Trendelenburg, as evidenced through the history of varicose vein surgery, the current procedure has evolved over a period of centuries, starting from *Paulus Aegineta* (CE 625–690) till today. The presently followed open procedure is described as a *crossectomy* combined with stripping of the long saphenous vein or multiple subfascial ligation or excision of varices.

11.4 Procedure in brief

Indications

1. Symptomatic veins
2. Cosmetic
3. Varicose ulceration
4. Lipodermatosclerosis

Preoperative workup

1. Doppler scan to confirm superficial reflux;
2. Venous duplex imaging with or without junction marking—demonstrates incompetent perforators and deep veins;
3. Mark veins preoperatively with indelible marker (Fig. 11.1).

11.5 Tributaries ligated in the procedure

- Superficial inferior epigastric
- Superficial circumflex iliac
- Superficial/deep external pudendal
- Lateral/anterior cutaneous vein of thigh

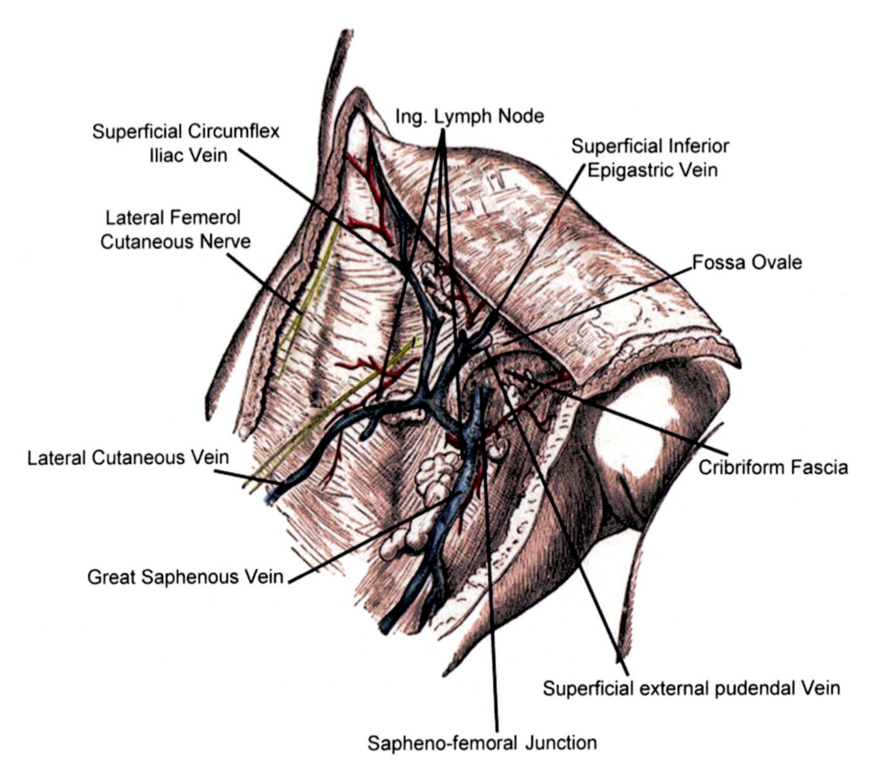

Figure 11.1 Trendenlenburg's operation.

11.6 Risk factors

Risk factors include any cause of obstruction, including: deep venous thrombosis (DVT), pregnancy, running, malignancy, and smoking.

11.7 Trendelenburg operation

1. General/spinal anesthesia;
2. Supine position;
3. 1.5 cm incision lateral and below pubic tubercle (site of SFJ)—4 cm in groin crease;
4. Dissect tributaries of SFJ (superficial inferior epigastric, superficial circumflex iliac, deep/superficial external pudendal);
5. Ligate and divide tributaries;
6. Ligate SFJ;

7. Pass stripper down long saphenous vein (LSV) to knee, preferably up to the level of medial malleolus;
8. Stab incision over stripper and deliver;
9. Strip vein back to groin;
10. Close incision;
11. Avulse local varicosities in lower calf;
12. Apply compression bandage.

11.8 Complications

- Hematoma
- Recurrence (approximately 20%)
- Saphenous nerve injury—loss of sensation of medial thigh
- Rarely DVT (Fig. 11.2).

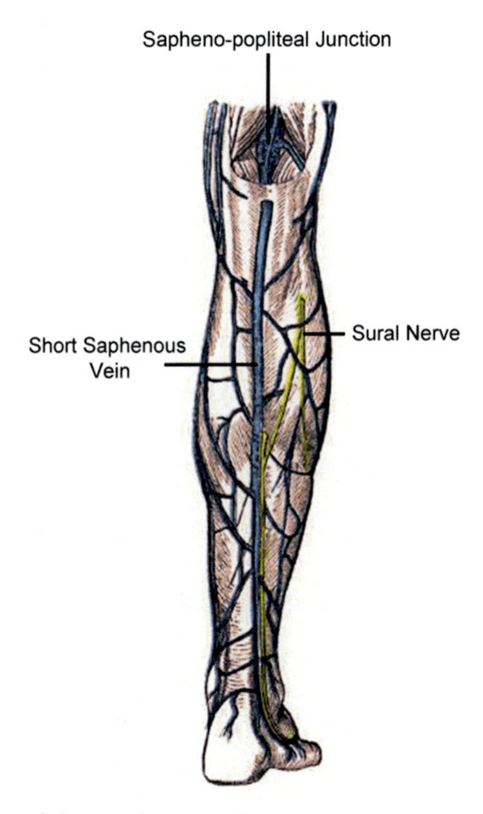

Figure 11.2 Course of short saphenous vein.

11.9 Excision of the short saphenous vein

1. Transverse skin crease incision;
2. Dissect down to short saphenous vein junction (SSVJ);
3. Tie off the junction at the point where SSV enters the popliteal vein (avoid sural nerve which is lateral to Sapheno-Popletial Junction (SPJ)).

11.10 Varicose vein stripping with conventional ligation

11.10.1 Saphenectomy: vein stripping with ligation, avulsion, or ablation

Vein stripping is done to remove varicose veins in the legs. Mayo's original "ring" stripper, which was passed outside the vein, has now been largely replaced by the intraluminar stripper designed by Myers, or by one of its modifications. This stripper consists of a coiled spring with a wire core, 35 in. (85–90 cm) long; to one end is fitted an acorn-shaped head, and to the other, an olivary tip of smaller size. The stripper, with its smaller olivary-tipped end leading, is passed along the lumen of the vein—usually in a proximal direction, so that it is less likely to be impeded by valves, or to deviate from the main trunk into tributaries. It is passed up the vein to as high a level as possible, where it is brought out through a second incision. The vein is then divided at the lower incision, and the upper cut end is ligatured around the stripper. Traction in an upward direction then avulses the vein, which is telescoped against the acorn-head. Before commencement of the procedure, the varicosities should be marked out on the skin while the patient is standing up. A decision is provisionally made regarding any areas where stripping is unlikely to be successful, and where excision of varices through a direct exposure will probably be required.

11.11 Technique of stripping of LSV

For stripping of the long saphenous vein, it is advised that the Trendelenburg operation is first carried out at the groin, by exposing the vein, and all tributaries entering its upper 3 in. (7–5 cm) are divided. The vein is then ligatured at its junction with the femoral, and is divided below the ligature. The distal cut end is occluded by a rubber band held on forceps. The lower end of the long saphenous vein is then exposed through a small incision just above the medial malleolus; the stripper is introduced, and is passed upwards until it emerges through the ligature of strong silk that is then tied firmly around the vein containing the stripper, so that the vessel is anchored against the acorn-head. The ends of this ligature are left long enough to follow the stripper to the groin, so that they remain as a guide to the lower end of the vein, and a means of delivering it again at the lower wound, if this is found to be necessary. The vein is then divided below the ligature, the lower cut end is tied-off, and the skin wound is sutured, at this stage. The leg is then elevated to reduce bleeding, and traction is applied to the stripper in an upward direction, not continuously, but in a series of short sharp jerks. In this way the vein is avulsed in stages, and, in the ideal case where no difficulties are encountered, it will eventually emerge complete on the stripper, telescoped against the acorn-head. Smaller tributaries are torn off from the main trunk without undue difficulty during the stripping process, but the resultant bleeding which occurs into the subcutaneous track must be controlled by firm bandaging from below upwards. This is best done after each stage in the stripping process; the bandage is applied up to the head of the instrument, which can be easily felt beneath the skin. Finally, after delivering out the stripper, the groin incision is closed and the bandaging is completed.

Very few stripping operations, however, will be completed in this simple manner, mainly because of tortuosity, thrombosis, or other causes. It may be quite impossible to pass the stripper for more than a short distance along the vein. Frequently it is arrested at knee level. In such cases it is necessary to cut open over the tip of the instrument and deal with the venous situation in that area under direct vision. Usually it will be possible to strip the segment of vein already traversed and to excise the local varicosities after tying off the tributaries. The remaining segment of the vein may then be avulsed after reinsertion of the stripper, in one or more stages as required. Tributaries may be dealt with in a similar manner. Marked resistance to the actual stripping process usually indicates the presence of a large tributary or perforator at the level of arrest. It is then advisable to desist from traction. This can help to avoid troublesome hemorrhage. The local condition is dealt with through a direct exposure, made by cutting down upon the acorn-head of the stripper. Most patients receive either general or spinal anesthesia for surgery. Vein stripping takes about 1 hour, depending on the surgeon's experience (Figs. 11.3—11.5).

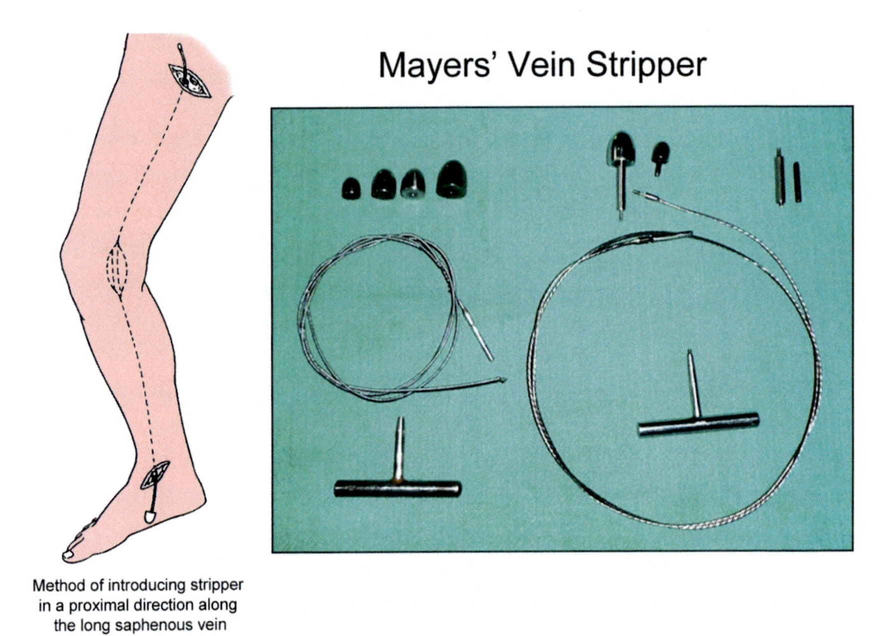

Figure 11.3 Mayer's vein stripper.

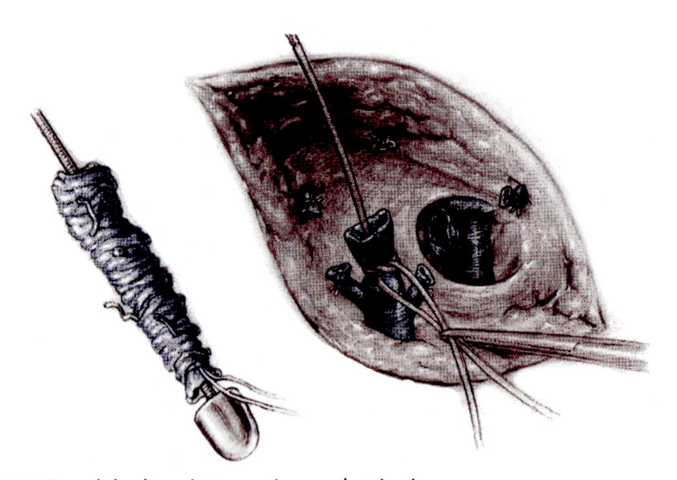

Figure 11.4 Trendelenburg's operation and stripping.

11.12 Modified vein-stripping technique

Until recently, vein ligation and stripping was the standard treatment for large or perforated varicose veins. If the majority of the valves in the vein

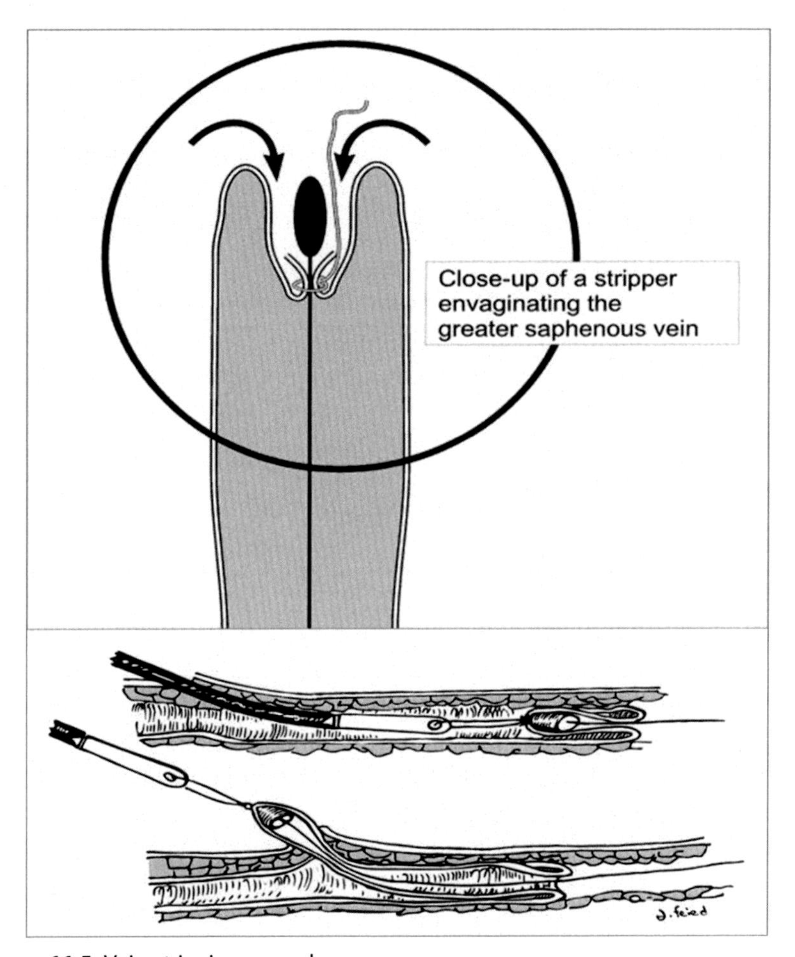

Figure 11.5 Vein stripping procedure.

are healthy, ligation can be used to isolate the faulty valves and the rest of the vein may be left in place to continue circulating blood. If the vein is badly damaged, however, it is usually removed or stripped.

To strip a varicose vein, an incision is made above and below the damaged vein. The standard procedure is total removal of the GSV by two incisions, one in the groin and the other on the medial part of the ankle joint just above the medial malleolus. The stripper is introduced usually from the upper incision where the GSV is exposed and brought out through the lower incision. The acorn head of the stripper is tied firmly to the lower end of the vein and pulled up, and the vein is fully stripped out. As the stripper is pulled up a compression bandage is used from below upwards to prevent bleeding. Most people are able to return to normal daily and recreational activities within a few weeks of the procedure.

11.13 Common side effects of vein ligation and stripping

Common side effects include temporary pain or discomfort, bruising, hematoma, numbness, and (less frequently) wound infection. Vein stripping is an invasive procedure and should not be performed on older patients because surgery may pose risks due to other existing medical conditions. It is also not appropriate for pregnant women or people with poor leg circulation, skin infections, blood-clotting defects, or artero venous fistula (Fig. 11.6).

11.14 Microphlebectomy

Microphlebectomy microextraction is a minor surgical procedure which allows for the removal of large surface varicosities through very small incisions that need no stitches. The procedure is normally done as a day-care procedure using a local anesthetic.

The procedure normally takes less than 1 hour. There is very little discomfort with the procedure and a shorter recovery time. Most patients are able to go return to work the day following the procedure, although many patients take the day off to ensure they are fully recovered.

After cleansing and anesthetizing the skin, minute skin incisions are made with an instrument that limits the size of the microincision to 1.5 mm. Another instrument is inserted under the surface of the skin to extract a length of the varicose vein through this microincision.

With this procedure, a compression bandage is worn for 5–7 days until the patient is seen in the doctor's office for follow-up. After this, a support stocking is worn until all bruising resolves, usually in about 1 more week. The compression bandage and support stocking minimize swelling and discomfort and allow for proper healing.

11.15 The basic steps of the microphlebectomy procedure

Some patients take a mild sedative about 1 hour before the procedure to help them relax. This is not required and very few patients opt for a

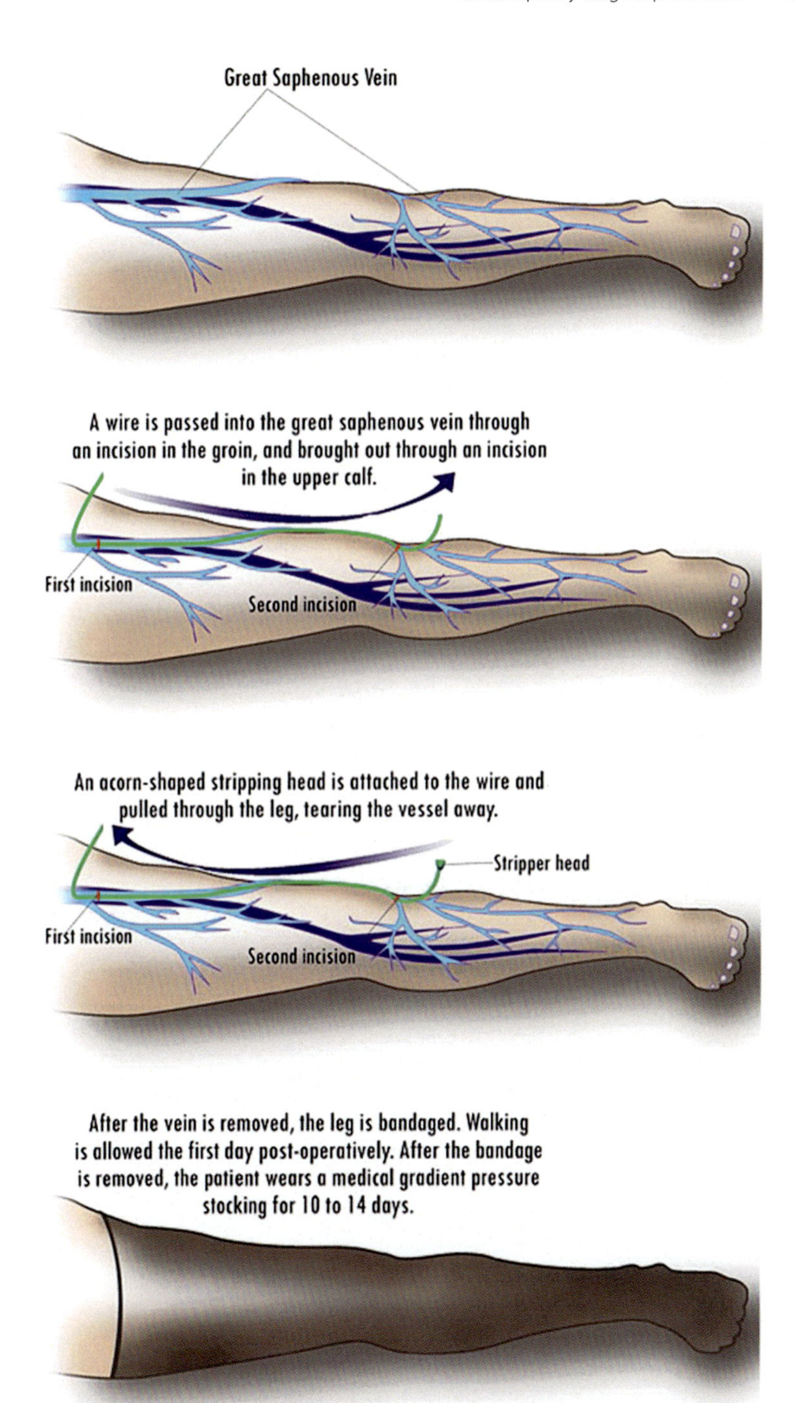

Figure 11.6 Diagrammatic representation of stripping.

sedative. The areas to be treated are marked with an indelible marker while the patient is standing.

The patient then lies down on a table and the area of the leg to be treated is cleansed with a sterile disinfectant. Sterile towels are then put in place to cover the nonsterile parts of the skin. A small area of skin is then anesthetized with a local anesthetic, immediately over one of the varicose veins. Through this anesthetized skin, a 1.5 mm microincision is made, and through this microincision, additional anesthetic is infiltrated into the tissue around the varicose vein. This helps to loosen the varicose vein and make its extraction easier. It also helps to anesthetize the deeper tissues so that the patient remains comfortable during the procedure. Through the microincision, fine surgical instruments are used to extract as great a length of the varicose vein as can be reached from the microincision. Normally, this is a length of about 1 in. in either direction, so about 2 in. of vein can be extracted through one microincision. Another small area of skin is anesthetized with a local anesthetic, usually about 2 in. from the first microincision, and these steps are repeated until the entire length of the varicose vein is removed. The surgeon then places some butterfly tapes over the microincisions, the leg is comfortably wrapped with a compression wrap, and the patient is able to walk out of the operating room with a healthier leg.

11.16 The benefits and expected results

1. A very simple procedure;
2. Performed under local anesthesia;
3. Cost effective; Minimally invasive, so minimal risk of scarring or postoperative infection;
4. Takes less treatment time and rapid recovery with reduced postoperative pain;
5. Normal activities can be resumed immediately;
6. Very good clinical and cosmetic results.

11.17 Transilluminated powered phlebectomy

Transilluminated powered phlebectomy (TIPP) is a newer technique that avoids the disadvantages of stab avulsion phlebectomy, which include long operating times, the risk of scar formation, and a relatively high risk of infection developing in the microincisions. TIPP is performed with an illuminator and a motorized resector. After the patient has been anesthetized with general or spinal anesthesia, the surgeon makes only two small incisions: one for the illuminating device and the other for the resector. After making the first incision and introducing the illuminator, the surgeon uses a technique called tumescent anesthesia to plump up the tissues around the veins and make the veins easier to remove. Tumescent anesthesia was originally developed for liposuction. This involves the injection of large quantities of a dilute anesthetic into the tissues surrounding the veins until they become firm and swollen.

After the tumescent anesthesia has been completed, the surgeon makes a second incision to insert the resector, which draws the vein by suction toward an inner blade. The suction then removes the tiny pieces of venous tissue left by the blade. After all the clusters of varicose veins have been treated, the surgeon closes the two small incisions with a single stitch or Steri-Strips. The incisions are covered with a gauze dressing and the leg is wrapped in a sterile compression dressing.

11.18 Risks

Vein ligation and stripping carries the same risks as other surgical procedures under general anesthesia, such as bleeding, infection of the incision, and an adverse reaction to the anesthetic. Patients with leg ulcers or fungal infections of the foot are at increased risk of developing infections in the incisions following surgical treatment of varicose veins.

Specific risks associated with vascular surgery include:

- DVT;
- Bruising: Bruising is the most common complication of phlebectomies, but heals itself within a few days or weeks;
- Scar formation: Phlebectomy has been found to produce permanent leg scars;
- Injury to the saphenous nerve: This complication results in numbness, tingling, or burning sensations in the area around the ankle. It usually disappears without further treatment within 6–12 months;
- Seromas: A seroma is a collection of uninfected serum or lymphatic fluid in the tissues. Seromas usually resolve without further treatment, but can be drained by the surgeon, if necessary;
- Injury to the arteries in the thigh and groin area: This complication is extremely rare, but it can have serious consequences and may result in limb amputation;

- Leg swelling: This complication is caused by disruption of the lymphatic system during surgery. It lasts about 2–3 weeks and can be managed by wearing compression stockings.
- Recurrence of smaller varicose veins.

11.19 Normal results

Normal results of vein ligation and stripping, or ambulatory phlebectomy, include reduction in the size and number of varicose veins in the leg. About 95% of patients also experience significant pain relief.

11.20 Morbidity and mortality rates

The mortality rate following vein ligation and stripping has been reported to be 1 in 30,000. The incidence of DVT following vascular surgery is estimated to be 0.6%.

CHAPTER 12

Endovenous laser treatment

12.1 Introduction

In endovenous laser treatment (EVLT) and radiofrequency ablation (RFA), the principles and techniques are almost the same; both produce thermal burns to the vein wall, however the technology is different.

Since the introduction of endovenous lasers (EVLs) in the early 2000s, procedure methodologies, as well as device technologies, have evolved extensively in the endeavor to improve treatment outcomes. As each novel parameter has been studied, new data have enabled the venous ablation community to acquire an enhanced understanding of the laser's mechanism of action. The primary subject matter in EVL studies has routinely included one or a combination of the following: laser fiber vein-wall contact, *linear endovenous energy density* (LEED), laser power settings, variable laser wavelengths, and most recently, covered laser fibers. The following is a detailed review of the major topics, from EVL inception to the latest methodologies employed.

12.2 Evolution of endovenous laser technology

12.2.1 The shift from pulsed energy to continuous laser energy

The initial investigators into EVLs employed methodologies that involved laser fiber vein-wall contact and bare-tip (BT) fibers to deliver pulsed energy. Users combined manual compression with a slow pullback of the fiber. At this early juncture, it was believed that the primary mechanism of action for vein obliteration was direct contact with the vessel wall. *The pulsed method with applied compression produced several perforations at the site of contact of the BT fiber with the vessel wall, resulting in high rates of postoperative pain and bruising.* Considering these early adverse findings, investigators began utilizing continuous energy instead of pulsed energy, and discontinued the use of manual compression (Fig. 12.1).

Other early researchers postulated that laser-induced steam bubble formation, similar to direct fiber-tip contact, caused perforations of adjacent

Genesis, Pathophysiology and Management of Venous and Lymphatic Disorders
DOI: https://doi.org/10.1016/B978-0-323-88433-4.00009-7

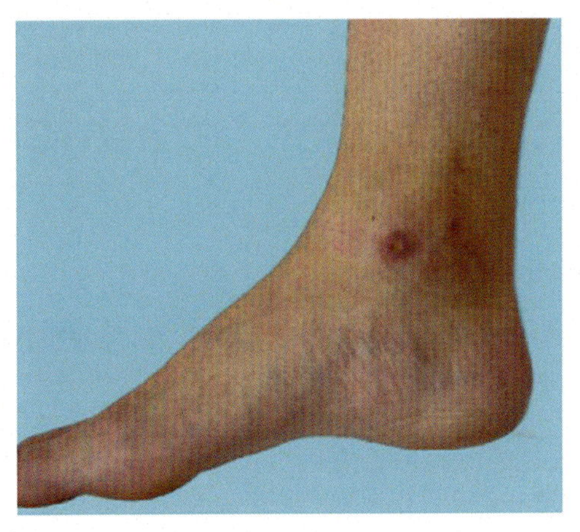

Figure 12.1 Laser burn.

wall areas. Proebstle and Perkowski proposed that the primary mechanism of action for 940 nm EVL was the formation of steam bubbles via delivery of laser energy, causing thermal injury to the vein endothelium, resulting in thrombotic occlusion. This mechanism was further defined by Proebstle et al. using in vitro generation of steam bubbles with 810-, 940-, and 980-nm lasers. Each laser was examined in saline, plasma, and hemolytic blood. None of the lasers was able to produce steam bubbles in saline or plasma alone, but they did create perforations at sites of direct laser-tip contact. However, all lasers did produce steam bubbles in hemolytic blood, indicating that hemoglobin plays a key role in inflicting thermal damage to the vein wall. Original data from these studies helped to form the opinion that vein-wall perforations and extravasation of blood into surrounding tissues are the culprits in causing EVL postoperative pain and bruising. Procedurally, it is now largely accepted that the use of manual compression to achieve direct laser fiber-tip contact actually exacerbates the incidence of perforation and extravasation, and hence the incidence of pain and bruising.

12.3 Endovenous laser ablation

Saphenous vein reflux is the underlying primary abnormality in the majority of cases of superficial venous insufficiency. Thus, approaches to dealing with saphenofemoral junction and saphenous truncal incompetence have dominated the thinking of phlebologists. Trendelenburg described saphenofemoral junction ligation alone, without stripping of the incompetent saphenous vein, in the 1890s. The advantages of ligation alone over ligation and stripping, which are still extolled today, include preservation of the saphenous trunk for possible future use as a bypass graft and avoidance of saphenous nerve injury. High ligation by itself is less invasive, quicker and simpler to perform, and associated with an easier recovery when compared to vein stripping. While it is true it routinely "spares" the saphenous trunk, the use of a diseased saphenous vein as a conduit has been associated with an increased risk of graft failure [269].

Most importantly, there is no longer any question that high ligation alone usually results in persistent reflux in the saphenous trunk. Varicose recurrence is significantly reduced and the reoperation rate is 60%–70% less if the saphenous vein is stripped compared with ligation alone. Also, after ligation alone, recurrence or residual communication with the junction in the groin was found in 80% of patients, while 34% also had mid-thigh perforator incompetence via the unstrapped great saphenous vein (GSV). As Neglen concluded, stripping of the GSV of the thigh is essential to minimize the recurrence that is caused by redevelopment of incompetent communication with the saphenofemoral confluence, and due to thigh perforator incompetence. Simply put, the shortcomings of ligation alone outweigh its advantages [270].

It is important to note that recurrence is common even after ligation and stripping of the saphenous vein. While inadequate surgery of the saphenofemoral junction and progression of disease are mechanisms that explain some cases of recurrence, another important mechanism is neovascularization around the junction after venous surgery. In fact, neovascularization has been reported as the principal cause of recurrence, with neovascular channels of variable size, number, and tortuosity accounting for the reflux to recurrent varicosities in the majority of cases. Though some have expressed doubt as to the veracity of true neovascularization, there is clear histological evidence that neovascularization is a cause of recurrent varicose veins. Early reports suggest, in contrast, that endovenous ablation techniques are associated with a very low incidence of neovascularization. It may be that the development of neovascularization is largely prevented by avoiding groin dissection and by preserving venous drainage in normal junctional tributaries [271].

Endovenous laser ablation (EVLA), like RFA and foam sclerotherapy, is a less invasive alternative to vein stripping. EVLA is indicated in an ambulatory patient with great, small, or accessory saphenous vein reflux with surface varices and/or symptoms or complications related to superficial venous insufficiency. EVLA is routinely performed using dilute local anesthesia, and short- and mid-term studies of EVLA, regardless of wavelength used, seem remarkably consistent, typically reporting ablation of refluxing saphenous veins in 90% or more of cases. EVLA of the saphenous vein has been shown to correct or significantly improve the hemodynamic abnormality and clinical symptoms of chronic venous insufficiency in clinical, etiological, anatomical, pathophysiological (CEAP) clinical class 3–6b patients with superficial venous reflux [272]. Outcomes seem equal to or better than those of stripping, with better quality of life scores in the postoperative period compared to stripping. High patient satisfaction rates have been reported. The total cost of endovenous procedures is likely equal to or lower than that of surgery.

Early data on treatment of the GSV with 810 and 940 nm devices suggest treatment failure is uncommon in patients treated with >70 J/cm [30,31]. A withdrawal rate of 2 mm/s at 14 W delivers 70 J/cm.

12.3.1 Mechanism of action

The following wavelengths are in current use for EVLA: 810, 940, 980, 1064, 1319, 1320, and 2068 nm. It has been postulated that vein wall injury is mediated both by direct effect and indirectly via laser-induced steam generated by heating of small amounts of blood within the vein. Some have suggested that the choice of wavelength greatly impacts results. The main chromophore of 1320 and 2078 nm lasers, at least initially, is water, while other wavelengths used for EVLA primarily target hemoglobin. Obviously, it is imperative to thermally damage the vein wall adequately in order to obtain effective ablation. Some heating may occur to obtain effective ablation. Some heating may occur by direct absorption of photon energy (radiation) by the vein wall as well as by convection from steam bubbles and conduction from heated blood. However, it is unlikely that these latter mechanisms account for the majority of impact on the vein. The maximum temperature of blood is 100°C [273].

Laser treatment has been found to produce carbonization of the vein wall. Carbonization of the laser tip, which occurs at about 300°C, is noted following EVLA, and seems to occur regardless of the wavelength used. Carbonization of the laser fiber tip creates a point heat source and essentially reduces light penetration into tissue to zero. Mordon et al. stated

"The steam produced by absorption of laser energy by the blood is a tiny fraction of the energy necessary to damage the vein wall and cannot be the primary mechanism of injury to the vein with EVL." The carbonization and tract within the vein walls seen by histology following EVL can only be the result of direct contact between the laser fiber tip and the vein wall. Dr Rox Anderson, director of the Wellman Centre for Photomedicine at Massachusetts General Hospital, reported that carbon appears to be a secondary but key chromophore that is probably independent of wavelength. Note that fiber tip and shape may impact the development of carbonization [274].

12.3.2 Tumuscent anesthesia

EVLA should be performed under local anesthesia using large volumes of a dilute solution of lidocaine and epinephrine (average volume of 200–400 mL of 0.1% lidocaine with 1:1,000,000 epinephrine) that is buffered with sodium bicarbonate. This solution should be delivered either manually or with an infusion pump under ultrasound guidance, so the vein is surrounded with the anesthetic fluid along the entire length of the segment to be treated. The benefits of tumescent anesthesia for endovenous ablation include:

- Anesthesia;
- Separation of vein to be treated from surrounding structures;
- Thermal sink, which reduces peak temperatures in perivenous tissues;
- Vein compression, which maximizes the effect of treatment on the vein wall.

Although the maximum safe dosage of lidocaine using the tumescent technique for venous procedures is not well studied, a dosage of 35 mg/kg is a reasonable estimate.

12.3.3 Contraindications to endovenous laser ablation

- Allergy to local anesthetic
- Hypercoagulable states
- Infection of the leg to be treated
- Lymphedema
- Nonambulatory patient
- Peripheral arterial insufficiency
- Poor general health
- Pregnancy

- Recent/active venous thromboembolism
- Thrombus or synechiae in the vein to be treated
- Tortuous GSV (it may be difficult to place the laser fiber)

12.3.4 Adverse sequelae

Short-term pain and ecchymoses have been commonly observed after EVLA. Intermittent-pulsed laser fiber pullback has been reported, in a retrospective review, to cause significantly greater levels of postoperative pain and bruising, compared with a continuous pullback protocol. Adding a short-stretch bandage for 3 days following intermittent-mode EVLA substantially reduced patient-reported bruising and pain. Employing continuous-mode pullback further reduced the severity of pain and bruising to such an extent that levels were similar to those reported by patients treated with RFA. Preliminary reports suggest there may be some differences in postoperative course depending on the wavelength used to perform EVLA. However, this is based on sparse data with short-term follow-up [275].

12.3.5 Perivenous thermal injury

Mean peak intravascular temperatures during EVLA (goat jugular vein, 12 W, 1-second pulses, 1-second intervals), measured flush with the laser tip, averaged 729°C, while those 4 mm distal to the tip averaged 93°C. However, the risk of collateral thermal injury depends on perivenous tissue heating, not intravascular temperature. Collagen has been noted to contract at about 50°C, while necrosis occurs between 70°C and 100°C. The extent of thermal injury to tissue is strongly dependent on the amount and duration of heat the tissue is exposed to. Henriques and Moritz investigated the time–temperature response for tissue exposed to up to 70°C. They found that skin could withstand temperature rises for very short exposure times, and that the response appears to be logarithmic as the exposure times become shorter. For example, an increase in body temperature to 58°C will produce cell destruction if the exposure is longer than 10 seconds. Tissues, however, can withstand temperatures up to 70°C if the duration of the exposure is maintained for less than 1 second. Li et al. reported that heating endothelial cells to 48°C for 10 minutes did not induce cell death. They also found that osteoblasts, after exposure for 10 minutes or less at 45°C, underwent transient and reversible changes [276].

12.3.6 Major complications

Major complications following EVLA have been reported rarely. Rates of deep venous thrombosis, pooled from multiple series, are much lower than 1%. One group reported an incidence of thrombus extension into the femoral vein of 7.7%. However, in that study EVLA was done under general or spinal anesthesia. The fact that patients were not able to ambulate immediately postoperatively may have contributed to the high incidence of thrombus extension. There is a single report of an arteriovenous fistula that developed following EVLA of the short saphenous vein. One patient developed septic thrombophlebitis following EVLA combined with open ligation of perforators and stab phlebectomy. This resolved with antibiotic treatment and debridement [277].

12.3.7 Alternative approaches

EVLA and RFA [54,55] both appear to be effective treatments for saphenous incompetence. Advantages of EVLA over RFA include shorter procedure times and lower per treatment cost. Reported occlusion rates of EVLA generally are slightly higher than those obtained with RFA [56]. Disadvantages of EVLA may include more bruising and discomfort in the early postoperative period, although this may be technique dependent. Both techniques continue to undergo refinement, which will improve results. Both procedures, when performed using tumescent anesthesia, are associated with low complication rates.

12.3.8 Instrument to perform endovenous laser ablation

Brand name: Dimedlaser
Model number: CHERYLAS
Maximum power: 15/30/60 W
Control mode: True color touchscreen
Laser type: GaA1As diode laser
Wavelength: 810/980 nm

The underlying goal for all thermal ablation procedures is to deliver sufficient thermal energy to the wall of an incompetent vein segment to produce irreversible occlusion, fibrosis, and ultimately disappearance of the vein (Fig. 12.2). The mechanism of vein wall injury after EVLA is controversial. It has been postulated to be mediated both by direct effect and indirectly via laser-induced steam generated by the small amounts of blood within the vein. Adequately damaging the vein wall with thermal

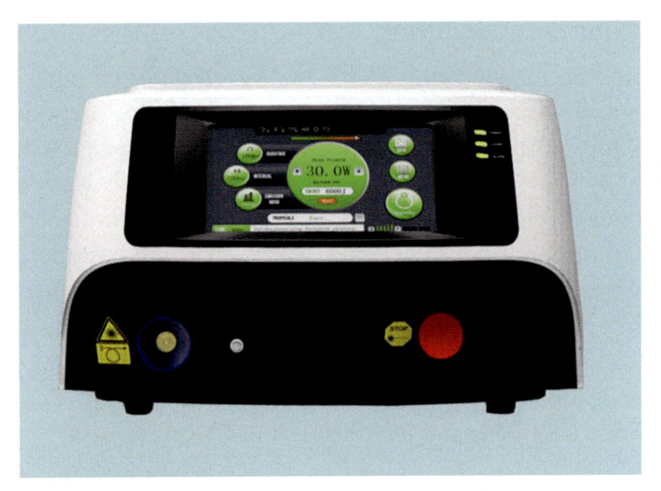

Figure 12.2 Laser Equipment.

energy is imperative to obtain effective ablation. Some heating may occur by direct absorption of photon energy (radiation) by the vein wall, as well as by convection from steam bubbles and conduction from heated blood. However, these later mechanisms are unlikely to account for most of the impact on the vein.

12.3.9 Diode laser advantages

Diode lasers are most commonly used for EVLA. Laser generators exist with multiple wavelengths, including lower wavelengths that are considered hemoglobin-specific and include 810, 940, 980, and 1064 nm. Higher wavelengths are considered water-specific and include 1320 and 1470 nm. Although it is not yet definitely established in the literature, some authors suggest that the higher wavelength lasers produce similar efficacy at lower power settings with less postprocedure symptoms [278].

12.3.10 Linear endovenous energy density

Subsequent to the steam bubble mechanism of action premise, several researchers began to evaluate LEED for its effect on treatment outcomes. *LEED is best defined as the number of joules delivered per centimeter of the target vein during an EVL procedure.* Efficacy has been the primary endpoint of LEED studies, evaluating low LEED versus high LEED.

EVLT is an efficient method to treat incompetent saphenous veins with high occlusion rates. Major side effects reported with 810 and

980 nm diode lasers are postoperative pain and bruising. Recently, laser systems with higher wavelengths (WSLWs), associated with new energy delivery devices, seem to reduce some side effects previously reported. The purpose of various experimental studies was to assess the adequate parameters required for vein wall destruction and to evaluate the role of fiber pullback velocity on vessel wall degradation. Varicose vein segments were treated with 1470 nm diode laser with 3–9.5 W power. The fiber moved through the vein at a velocity of 0.7 or 1.5 mm/s; the applied LEED was 40–95 J/cm. The temperature of the vein surface in the course of laser irradiation was controlled by infrared thermography. The intact collagen in treated vein specimens was studied by differential scanning calorimetry. The increase in the surface temperature with applied energy was found to be about three times slower for the pullback velocity of 0.7 mm/s than that of 1.5 mm/s. The collagen in the tissue was totally denatured in the case of the surface temperature of about 91°C. The critical values of LEED that ensured complete degradation of vein wall were 53 and 71.5 J/cm for velocities of 1.5 and 0.7 mm/s, respectively. An experimental study conducted by Moscow State University supports the conception that it is laser power and pullback velocity rather than LEED value that determine the temperature as well the collagen framework degradation and therefore the thermal response of the procedure.

The EVLA procedure is currently a well-established treatment modality for the entire incompetent GSV segment. Laser energy is used to destroy the GSV in situ in combination with tumescent anesthesia. The energy is administered with pullback of the laser fiber. Thermal destruction of the venous wall is generally accepted to be responsible for vein occlusion in endovenous treatments. However, not only is the exact mechanism of EVLA-induced temperature rise not well documented, but there is no standardized protocol for the EVLA procedure yet either. In particular, the large variety of laser power level and pullback velocity is used in actual clinical practice, as well as in experimental models. The ratio of laser power and pullback velocity (termed LEED) is commonly used to report the dose administered to the vein, to compare and discuss the clinical or experimental results [279]. This parameter was believed to be a major determinant of treatment outcome. A more recent theoretical study predicted that the highest power setting and fastest pullback velocity would result in the highest vein wall temperature. Therefore the thermal efficacy of EVLA is clearly driven by the laser power, rather than by the laser energy. There are not many experimental studies comparing the

EVLA results at the constant LEED value by combining of various powers and pullback velocities. Early investigators of EVLA focused on power (watts) and rate of fiber pullback to achieve vein obliteration. The approach regarding power settings has developed over time, with higher power originally deemed as one of the key components to procedural success. Further studies revealed that high power outputs were associated with increased pain and bruising [280].

In addition, the rate of fiber pullback was measured initially in mm/s, but LEED became the measurement of choice because it was a more accurate metric of energy delivered to a segment of treated vein. LEED has been described in the literature as the delivered energy per centimeter (J/cm) of a treated vein. Similar to early studies that utilized high power settings, studies with higher LEEDs in excess of 80 J/cm were reported as highly efficacious [2,7,8]. However, LEEDs greater than 100 J/cm were later shown to cause increased pain, bruising, and paresthesia. Currently available laser wavelengths include 810, 940, 980, 1064, 1320, 1470, and 1500 nm. Wavelengths on the shorter end of the spectrum target hemoglobin as the chromophore; whereas longer wavelengths have an increased affinity for water as the chromophore. Correspondingly, the 1064 nm and shorter wavelengths are categorized as hemoglobin-specific laser wavelengths, and the 1320, 1470, and 1500 nm lasers are categorized as water-specific laser wavelengths [281].

The specificity that hemoglobin-specific laser wavelengths have for hemoglobin in red blood cells leads to steam bubble formation, which secondarily induces thermal damage to the intima. Although this mechanism is effective in vein obliteration, it results in increased carbonization and perforation of the vein wall. Perforations have been associated with increased postprocedural pain and bruising, largely due to extravasation. Conversely, the water-specific laser wavelengths appear to target the interstitial water present in the endothelial cells and red blood cells of the vein wall, leading to collagen contraction, eventual fibrosis with increased efficiency, resulting in lower energies utilized and producing fewer symptoms. Regardless of wavelengths, there is still a risk for vessel damage from direct contact between the fiber tip and the vein wall. Hence, the second facet of EVLA, which has evolved, is fiber design. A jacket-tip (JT) fiber design has been devised to sheath the distal tip of the laser fiber. By covering the energy-emitting portion of the fiber, perforation of tissue by contact is essentially eliminated. In addition, a JT fiber significantly expands the emitting surface of the fiber [282].

This enables the delivered energy to be diffused over a greater surface area with a lower power density, similar to the technology involved with radial-emitting fibers. The following clinical study was conceived on the premise that the origin of postoperative pain and bruising is directly correlated to tissue perforation during laser treatment. In order to analyze this relationship, 810-, 980-, and 1470-nm wavelengths with the use of BT and JT fibers were evaluated for the treatment of incompetent GSVs. Given the hypothesis that postoperative pain and bruising are correlated with increased laser fiber/vein wall contact, the aim of this study was to determine if fiber type had a greater effect on postprocedural pain and bruising as compared to laser wavelength. To further evaluate the relationship between fiber type, laser wavelength, and tissue perforation, an in vitro study was conducted that compared the 810- and 1470-nm lasers using both BT and JT fibers.

12.3.11 Prospective study

From 2001 to 2004 and from 2006 to 2014, 213 limbs in 213 patients were treated for venous insufficiency of the GSV at the Vein Institute of New Jersey and at the New York Vein Center, New York University Langone Medical Center. The interventionists who performed the treatments in this study comprised eight vascular surgeons, all of whom had significant experience performing EVLAs. Surgeon preference dictated the laser/fiber combination that was used. Patients received isolated treatments with an 810, 980, or 1470 nm laser (Delta 810, Venacure 980, or 1470; Angiodynamics, Latham, NY), using either a BT or JT fiber (Venacure or nevertouch; Angiodynamics). There were no concomitant treatments performed, including microphlebectomies or sclerotherapy. Data were collected prospectively in a nonrandomized fashion, up to a prescribed number in each group and for whom adequate data were available. The institutional review board approved the protocol for a retrospective review of the data. No institutional review board consent was required for evaluation of the deidentified data. EVLAs were performed under ultrasound guidance (LOGIC E; GE Healthcare, Milwaukee, Wisconsin) and using the same technique. In patients with CEAP clinical classification C2 or C3, the vessel was entered at or just below the knee.

For patients presenting with scores ranging from C4 to C6, the vessel was accessed at the lowest point of reflux. In all procedures, care was taken to access the GSV deep to the saphenous fascia and greater than 1 cm from the skin. Using ultrasound, the tip of the sheath was positioned 2–3 cm peripheral to the saphenofemoral junction. After removal of the guidance, either a 600-μm BT or JT fiber was advanced into the sheath. The sheath was withdrawn and secured to the fiber, exposing 2.5 cm of the fiber tip outside of the sheath while maintaining the 2–3 cm distance peripheral to the saphenofemoral junction. Tumescent anesthesia was delivered to ensure a 1 cm ring of tumescent solution surrounded the treatment vein.

For the ablation, laser energy was delivered according to the general consensus of what the optimal power and LEED were at the time the procedures were performed. Fiber pullback was continuous, at a speed that achieved the target LEED. The 810 nm BT fiber group was treated at 14 W, with a target LEED of 80 J/cm. The 810-nm JT fiber group was treated with similar settings. In the 980-nm cohort, both the BT and JT group had a lower power setting of 6 W and a target LEED of 42 J/cm.

There is a paucity of data comparing BT fibers to other fiber types. Initial studies of different fiber types were first reported in a randomized trial between RFA and the 980 nm laser with both BT and JT fibers [24]. JT and BT fibers were set to a power of 12 W and averaged 82.3 and 86.2 J/cm, respectively. At 72 hours, the JT group reported less pain (0.96) than the BT group (1.87), with scores similar to that of RFA (0.80). Doganic and Demirkilic [25] evaluated the 980 nm laser with a BT fiber in comparison to the 1470 nm laser with a radial-tip fiber, both at 15 W of power and a target LEED of 90 J/cm. Pain was measured according to duration, yielding 3.2 days for the 980 nm group and 2.2 days for the 1470 nm group, with only two patients exhibiting ecchymosis versus 13 patients in the 980 nm group ($P > .001$) [275].

In the pain analysis, none of the wavelength comparisons using the same fiber-tip demonstrated significance, except for the 810-BT versus the 980-BT group. There was not a substantial difference between any of the wavelengths that used the JT fiber, again implying that the fiber type may have a more profound and overarching impact on perforation as compared to the wavelength. These results were corroborated in vitro, which allowed for the measurement of tissue thermal injury depth for varying combinations of power, LEED, wavelengths, and fiber type. When the same wavelength was used in order to compare BT or JT, the JT group demonstrated diminished tissue injury depth as compared to the BT group. The same was observed with different wavelengths paired against different fibers, with the 810-BT versus 1470-JT revealing the largest gap in thermal injury depth.

Analogously, when different wavelengths paired with the same fiber type were contrasted, the mean difference in depth favored the 1470 nm wavelength, but there was a suggestion that the effect was less significant. Both the 810-BT versus 1470-BT and the 810-JT versus the 1470-JT cohorts did exhibit statistical significance, with the JT comparison demonstrating the least variability of any group. When aligned with the prospective clinical comparisons, the BT groups exhibited the deepest tissue thermal injury depth, as well as the highest pain and bruising scores. These findings reaffirm the clinical results, demonstrating that thermal depth, pain, and bruising are likely greater in the shorter 810-nm wavelength in comparison to the longer 1470-nm wavelength. A suggestion of an even more compelling outcome was the depth disparity between the BT and JT fibers, mirroring our findings of significantly less pain and bruising in all JT groups.

The JT fiber was able to eliminate or significantly reduce perforation, likely by significantly reducing the possibility of the emitting portion of the fiber tip from achieving direct tissue contact with the vein wall. The glass ferrule construction expands the emitting surface of the fiber, thereby dispersing the laser power over a much greater surface area as compared to the BT fiber. This results in more efficient energy delivery. Increased efficiency enables use of the laser at a considerably lower power and LEED. The advantage is that lower power and LEED utilization result in reduced tissue penetration and fewer side effects. Secondarily, the absorption properties of the longer wavelengths achieved decidedly selective destruction of the vein wall, while using less energy than wavelengths with a hemoglobin affinity [283]. Water-specific wavelengths have specificity to the vein wall at least five times that of the lower hemoglobin-specific wavelengths. Additionally, water as a chromophore is 40 times more efficient for energy absorption as compared to hemoglobin. Ultimately, the goal is to obtain a treatment with as low a side-effect profile as can be reasonably achieved while also maintaining long-term efficacy for treatment. The design of this study precluded long-term evaluation; however, there are data in the literature that demonstrate the mid- and/or long-term efficacy of the laser wavelengths utilized in this study [284].

As a result of the limited number of limbs in each clinical cohort, the lack of data on certain patient characteristics such as body habitus and body mass index, there are limitations to this prospective, nonrandomized study. In addition, the CEAP classifications were not recorded prospectively, and therefore, they were omitted from this study. First, the 810-nm laser pain score was originally measured on a scale of 0–5 using a visual analog score. Because the other laser wavelengths in the study were based on the 0–10 pain visual analog score, the 810-nm wavelengths pain score was multiplied by a factor of 2 [285].

Finally, the in vitro study represents a novel methodology, and this warrants further validation. With regard to specific methodology, the in vitro study did not include use of the 980-nm laser. Overall, the data within this study support conclusions regarding the dominant effect of JT fibers, and this is corroborated by the clinical as well as the in vitro study. However, any conclusions drawn remain hypothesis generating and warrant additional evaluation in a randomized trial setting, and they merit clinical corroboration with quality of life data. Finally, the short-term effects of pain and bruising need to be correlated with long-term outcomes, and this warrants additional study. There are data to suggest that the mid- and long-term efficacy of laser ablation using the hemoglobin-specific as well as the water-specific wavelengths results in excellent long-term outcomes, with recurrence rates that are comparable to RFA and to high ligation with saphenectomy, provided that the appropriate power settings are utilized to minimize short-term complications [286].

The surgical procedure, postprocedure care, advantages and disadvantages, complications and follow-up are almost the same as for RFA therapy.

This procedure can be combined with crossectomy or sclerotherapy.

CHAPTER 13
Radiofrequency ablation

In the historical surgical approach, ligation and division of the saphenous trunk and all proximal tributaries are followed either by stripping of the vein or by avulsion phlebectomy. Proximal ligation requires a substantial incision at the groin crease. Stripping of the vein requires additional incisions at the knee or below and is associated with a high incidence of minor surgical complications. Avulsion phlebectomy requires multiple 2–3-mm incisions along the course of the vein and can cause damage to adjacent nerves and lymphatic vessels.

Endovenous ablation has replaced stripping and ligation as the technique for elimination of saphenous vein reflux. One of the endovenous techniques is a radiofrequency-based procedure. Newer methods of delivery of radiofrequency were introduced in 2007. Endovenous procedures are much less invasive than surgery and have lower complication rates. The procedure is well tolerated by patients, and produces good cosmetic results. Excellent clinical results are seen at 4–5 years, and the long-term efficacy of the procedure is now known due to having 10 years of experience. The original radiofrequency endovenous procedure was cleared by the United States Food and Drug Administration (FDA) in March 1999.

Endovenous techniques [endovenous laser therapy (EVLT), radiofrequency ablation (RFA) and endovenous foam sclerotherapy] are clearly less invasive and associated with fewer complications.

13.1 Endovenous radiofrequency ablation for the treatment of varicose veins

Varicose veins affect approximately 26% of the adult population and are a frequent cause of discomfort, loss of productivity, and deterioration in health-related quality of life. Numerous therapies have been developed for the treatment of this condition. Conventional open surgical interventions include ligation of the great saphenous vein (GSV) at the saphenofemoral junction and stripping. Smaller veins have also been treated with phlebectomies. More recently, less invasive modalities, such as foam sclerotherapy, EVLT, and endovenous RFA, have also been used. While

Genesis, Pathophysiology and Management of Venous and Lymphatic Disorders
DOI: https://doi.org/10.1016/B978-0-323-88433-4.00004-8

endovenous approaches are associated with fewer postoperative complications, such as hematoma, pain, or saphenous nerve injury, there is currently no strong evidence to suggest an overall advantage for any particular treatment approach [287].

The RFA procedure involves using a catheter electrode to deliver a high-frequency alternating radiofrequency current that leads to venous spasm, collagen shrinkage, and physical contraction. The patient's leg is prepped with antiseptic solution and draped in a sterile fashion. With ultrasound guidance, the vein is cannulated, and local tumescent anesthetic is then injected around the target venous segment. The catheter is then introduced through a sheath. The radiofrequency current is then delivered, resulting in circular homogeneous denaturation of the venous collagen matrix and endothelial destruction at a temperature of 110°C–120°C. Venous segments 3–7 cm in length are treated in 20-second cycles. Patients are instructed to wear 20–30 mm Hg graduated elastic compression stockings for at least 14 days. Compared with conventional open surgery, RFA can be performed in the outpatient setting without the requirement for hospital admission or general anesthesia [287].

However, the procedure is not feasible in tortuous or very small or large veins, and it may be less cost-effective than open surgery because of the cost of the catheters. The RFA technique is performed from the venous lumen without anatomic excision of the pathophysiological level. The radiofrequency technique uses a radiofrequency catheter guided by medical imaging. The catheter heats the vein wall by means of thermal energy delivered by a generator. The rise in temperature causes destruction of the intima and the media with contraction and thickening of the collagen fibers. These phenomena lead to fibrous changes, gradually leading to remote occlusion of the venous lumen. Tumescent local anesthesia is highly recommended for this procedure because, in addition to reducing the burning sensation, it produces a compression effect on the vein (physical hydrostatic compression and compression by spasms induced by the product used), which maximizes the ablative effects of RFA on the vein wall. Compared to the first-generation catheter used (Closureplus), the second-generation catheter (ClosureFast), introduced to the market in 2006, produces more heat (85°C vs 120°C) and involves a segmental approach with 20-second cycles. The segmental approach can speed up the procedure and reduce the variability of the heat dosage delivered [288].

13.1.1 Basic principles

The goal of RFA is to induce thermal injury to the tissue through electromagnetic energy deposition. The term RFA applies to coagulation induced by all electromagnetic energy sources with frequencies less than 900 kHz. The term RF refers not to the emitted wave but rather to the alternating electric current that oscillates in this frequency range. In monopolar RFA, the patient is part of a closed-loop circuit that includes an RF generator, an electrode needle, and a large dispersive electrode (ground pads). An alternating electric field is created within the tissue of the patient. Because of the relatively high electrical resistance of tissue in comparison with the metal electrodes, there is marked agitation of the ions present in the target tissue that surrounds the electrode, since the tissue ions attempt to follow the changes in direction of the alternating electric current. The agitation results in frictional heat around the electrode. The discrepancy between the small surface area of the needle electrode and the large area of the ground pads causes the generated heat to be focused and concentrated around the needle electrode. The thermal damage caused by RF heating is dependent on both the tissue temperature achieved and the duration of heating. Heating of tissue at 50°C–55°C for 4–6 min produces irreversible cellular damage [289].

At temperatures between 60°C and 100°C near-immediate protein coagulation is induced, with irreversible damage to mitochondrial and cytosolic enzymes as well as nucleic acid–histone protein complexes. Cells experiencing this extent of thermal damage most often, but not always, undergo coagulative necrosis over the course of several days. In fact, the zone of coagulation, while predominantly comprising coagulative necrosis, often lacks the classic well-defined histologic appearance of coagulative necrosis in the acute postablation period or even within some zones of adequately ablated tissue for many months after ablation. Indeed, in many cases, specialized stains are required to confirm that cellular death has been achieved after thermal ablation [290].

In the early experiences with RF treatment, a major limitation of the technique was the small volume of ablation created by conventional monopolar electrodes. These devices were capable of producing cylindrical ablation zones not greater than 1.6 cm in the short axis. Therefore multiple electrode insertions were necessary to treat all but the smallest lesions. Subsequently, several strategies for increasing the ablation zone achieved with RF treatment have been used. Heat efficacy is defined as the difference between the amount of heat produced and the amount of heat lost. Therefore effective ablation can be achieved by optimizing heat production and minimizing heat loss within the area to be ablated. The relationship between these factors has been characterized as the "bio-heat equation." The "bio-heat" equation governing RF-induced heat transfer through tissue has been previously described by Pennes, with this equation simplified to a first approximation by Goldberg et al. as follows:

$$\text{Coagulation} = \text{energy deposited} \times \text{local tissue interactions} - \text{heat loss}$$

Heat production is correlated with the intensity and duration of the RF energy deposited. On the other hand, heat conduction or diffusion is usually explained as a factor of heat loss in regard to the electrode tip. Heat is lost mainly through convection by means of the blood circulation. Therefore most investigators have devoted their attention to strategies that increase the energy deposited into the tissues and several corporations have manufactured new RFA devices based on technologic advances that increase heating efficacy. To accomplish this increase, the RF output of all commercially available generators has been increased to 150–200 W, which may potentially increase the intensity of the RF current deposited at the tissue.

13.2 Latest medical devices used in radiofrequency ablation

13.2.1 Venefit (VNUS closure) procedure

The Venefit procedure, previously known as the VNUS closure procedure, offers patients an efficient, nonsurgical alternative for treating varicose veins to improve the skin's appearance and to reduce health risks related to large varicose veins. The Venefit procedure utilizes RF energy to heat and contract the collagen found inside the vein walls, causing the vein to collapse and seal shut. The main treatment alternative is to reroute blood flow through healthy veins. Traditionally, this has been done by surgically removing (stripping) the troublesome vein from the leg. Closure provides a less invasive alternative to vein stripping by closing the diseased vein instead (Fig. 13.1).

The closure procedure can be performed on an outpatient basis using either local or general anesthesia in which the physician numbs the leg before treatment. Currently, it is predominantly performed in a hospital

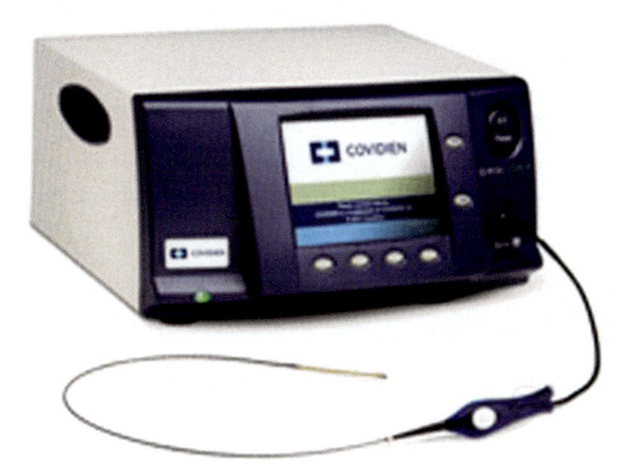

Figure 13.1 Radio Frequency Machine.

setting, although it may also be performed in a physician's office. The procedure consists of four principle steps:

1. *Map the saphenous vein:* A typical procedure begins with noninvasive ultrasound imaging of the diseased vein to trace its location. This allows the physician to determine the site where the closure catheter will be inserted and to mark the desired position of the catheter tip to begin treatment.
2. *Insert the closure catheter:* After the physician accesses the saphenous vein, the closure catheter is inserted into the vein and advanced to the uppermost segment of the vein. The physician then typically injects a volume of dilute anesthetic fluid into the area surrounding the vein. This numbs the leg, helps squeeze blood out of the vein and provides a fluid layer outside the vein to protect surrounding tissue from heat once the catheter starts delivering RF energy. Saline is then slowly infused into the vein from the tip of the catheter to further create a near-bloodless field inside the vein, allowing the catheter to preferentially heat the vein wall, rather than the blood.
3. *Deliver RF energy and withdraw catheter:* Noninvasive ultrasound is used to confirm the catheter tip position and the physician then activates the RF generator, causing the electrodes at the tip of the catheter to heat the vein wall to a target temperature of typically 85°C or 185°F. As the vein wall is heated, the vein shrinks and the catheter is gradually withdrawn. During catheter pullback, which typically occurs over 15–18 minutes, the RF generator regularly adjusts the power level to

maintain the target temperature to effectively shrink collagen in the vein wall and close the vein over an extended length.

4. *Confirm closing of vein:* After treatment, ultrasound imaging is used to confirm closing of the vein. If a portion of the vein is not closed, the catheter can be reinserted and energy reapplied. After the procedure, the narrowed vein gradually becomes fibrous, sealing the interior of the vein walls and naturally redirecting blood flow to healthy veins. Experienced physicians often complete the procedure in 45–60 minutes.

Of patients who have undergone the closure procedure, 98% are willing to recommend it to a friend or family member. The physician generally instructs patients to walk regularly for several days after the closure procedure and return within 72 hours for an ultrasound examination. Physicians may prescribe compression stockings to be worn for several days or weeks after the procedure. Compression stockings are prescribed as a routine item for vein procedures with the goal of enhancing patient comfort in the initial days after treatment.

13.2.2 Preprocedure

Duplex ultrasonography is used to confirm and map all areas of reflux and to trace the path of the refluxing great saphenous trunk from the saphenofemoral junction down the leg to the lower thigh or upper part of the calf. The vein, the saphenofemoral junction, and the anticipated entry point are marked in some way on the skin. An appropriate entry point is selected just above or below the knee, at a point permitting cannulation of the vessel with a 16-gauge needle introducer.

13.2.3 Procedure

The leg is prepared and draped, and a superficial local anesthetic agent is used to anesthetize the site of cannulation. Needle puncture of the vessel is guided by Duplex ultrasonography. The Seldinger technique is used to place a guidewire into the vessel, and an introducer sheath is passed over the guidewire, which is removed. The ClosureFast catheter is passed through the sheath, and the tip is advanced to 2 cm below the saphenofemoral junction under duplex ultrasonographic visualization.

With ultrasonographic guidance, a local anesthetic agent is injected into the tissues surrounding the GSV above and within its fascial sheath. The anesthetic is injected along the entire course of the vein from the catheter insertion point to the saphenofemoral junction. In most patients, 200–400 mL of lidocaine (0.1%) is sufficient to anesthetize and compress the vessel. Note the importance of delivering the anesthetic agent in the correct intrafascial location, with a volume sufficient to compress the vein and dissect it away from other structures, such as nerves, along its entire length. Duplex ultrasonography is used to position the catheter tip 2 cm below the level of the terminal valve of the saphenofemoral junction. The catheter must not extend into the femoral vein because injury to the femoral vein may cause *deep vein thrombosis* (DVT).

In the previous RFA system, when the console is switched on and the test mode activated, the baseline impedance should be 250–300 Ω and the baseline temperature should be 32°C–37°C. When radiofrequency energy is applied, the thermocouple temperature should rise to 80°C–85°C within 10–15 seconds. In the new system, when the radiofrequency is activated, the catheter core temperature should rapidly rise to 120°C and should be sustained for 15 seconds of the 20-second pulse cycle. If the temperature does not rise quickly, a malpositioned catheter tip should be strongly suspected.

In the previous system, after the temperature reaches 85°C and remains constant for 15 seconds, the catheter tip is slowly withdrawn at a rate of approximately 1 cm per minute (1 mm every 6 seconds). In the new system, two 20-second cycles are performed in the proximal section, after which the catheter is withdrawn 7 cm as per catheter markings. The next 20-second cycle is repeated once, and, if 120°C is maintained, the catheter is then withdrawn another 7 cm until the entire vein is treated.

When proper tumescent anesthesia is applied, the patient should never experience a sudden heat sensation. If this happens, more anesthesia is injected.

13.2.4 After procedure

Postprocedural duplex ultrasonography confirms the contraction of the vessel and the absence of flow along the entire length of the treated vessel. In the previous system, if persistent flow is observed, the procedure may be repeated immediately, provided the catheter can still be easily passed along the vessel to the desired site of treatment. However, in the new ClosureFast system, the procedure is not repeated because the targeted vessel typically shows no flow.

Schematic images of the radiofrequency endovenous process are illustrated in Fig. 13.2.

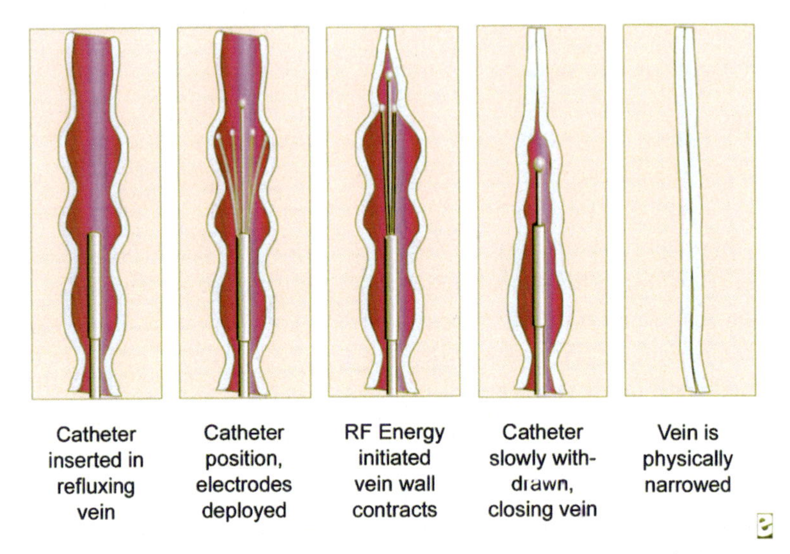

Figure 13.2 Schematic images of the radiofrequency endovenous occlusion process.

13.2.5 Follow-up

Compression is of vital importance after any venous procedure. Compression is effective in reducing postoperative bruising and tenderness, and it can also reduce the risk of DVT and venous thromboembolism in both the treated leg and the untreated leg.

A class II (30–40-mm Hg gradient) compression stocking is applied to the treated leg, and if the patient is willing, it is also applied to the untreated leg. Bed rest and lifting of heavy objects are forbidden, and *normal activity is encouraged.*

The patient is reevaluated 3–7 days after the procedure, at which time duplex ultrasonography should demonstrate a closed greater saphenous vein and no evidence of thrombus in the femoral, popliteal, or deep veins of the calf.

At 6 weeks, an examination should reveal clinical resolution of truncal varices, and an ultrasonographic evaluation should demonstrate a completely closed vessel and no remaining reflux. If any residual open segments are noted, sclerotherapy is performed under ultrasonographic guidance.

13.2.6 Complications

Reported complications of the procedure are rare.

1. Local paresthesia can occur from perivenous nerve injury but is usually temporary.

2. Thermal injury to the skin was reported in clinical trials when the volume of local anesthetic was not sufficient to provide a buffer between the skin and a particularly superficial vessel, especially below the knee.
3. Progression of thrombus from local superficial phlebitis has occasionally been observed when compression was not used. The greatest current area of concern is DVT, with one 2004 study documenting deep vein thrombus requiring anticoagulation in 16% of 73 limbs treated with an RFA procedure.

13.2.7 Final outcome

Published results show a high early success rate with a very low subsequent recurrence rate up to 10 years after treatment. Early and midrange results are comparable to those obtained with other endovenous ablation techniques. The overall experience has a 90% success rate, with rare incidents, where patients required a repeat procedure in 6–12 months. Overall efficacy and lower morbidity have resulted in endovenous ablation techniques replacing surgical stripping. Patient satisfaction is high and downtime is minimal, with 95% of patients reporting that they would recommend the procedure to a friend.

In the early 2000s, the most common treatment for varicose veins was a very painful surgery called "vein stripping," which was often performed in the operating room (OR), and was a technique for removing the problem veins entirely. This remained the gold standard until when minimally invasive thermal technology emerged as an alternative standard of care. Now techniques such as ClariVein, RFA, Varithena, Venaseal, sclerotherapy, and EVLT are the new gold standards. Here, we discuss ClariVein in depth.

13.3 ClariVein

13.3.1 Clinical aspects

ClariVein is an infusion catheter with a rotating wire tip. During the treatment, the end of the catheter rotates while dispensing special chemical medication. This gets the vein closing medication to the targeted treatment area. No thermal energy is used during this procedure, eliminating the need for multiple needle-stick injections of anesthesia (pain-numbing) medication along the length of the treated area. When using ClariVein, the patient is guaranteed a faster and simpler procedure for vein treatment;

there has been research that shows patients who choose ClariVein experience almost 80% less pain than patients who choose other minimally invasive peripheral vascular treatments. Overall, ClariVein patients benefit from minimal to no pain treatment, along with eliminating any bruising after the procedure (bruising is a common consequence of procedures such as laser ablation treatment) [291].

For ClariVein to work, an incision the size of a tiny pin is made as an entrance point for a specially designed slim catheter (tube) that is temporarily inserted into the vein. Since the ClariVein is several times smaller than other devices used in peripheral vascular treatments, this allows the entrance point to be smaller also. All vein procedures have the same outcome in mind. Removing or closing the varicose vein, which will allow blood to naturally take an alternative path to healthy and nonaffected veins in the legs; leading to improved circulation back from the legs to the heart.

Varicose veins are a type of medical condition that can cause unpleasant symptoms but recent medical advancements have made varicose vein treatment somewhat easier and more effective than previously. ClariVein ablation is a current advancement in treating varicose veins and offers numerous benefits. Veins in contrast to arteries have valves that operate in a one-way basis, which transfers oxygen-depleted blood to the heart. Arteries bring blood rich in oxygen to the heart. In the case that the valves of the veins do not work well, blood does not flow efficiently. In the case that the valves leak or fail, blood will flow back through the veins resulting to a condition referred to as reflux or venous deficiency. The ClariVein technique is an effective treatment for clearing varicose veins.

This is an outpatient procedure that does not need an injection of local anesthetic or incisions. Sclerotherapy makes use of sclerosing that destroys the inside part of the vein causing scarring and the closure of the veins. One aspect about sclerotherapy is that it works best with smaller veins but not larger veins such as the saphenous veins. Sclerotherapy has been used for decades to treat vein problems of all kinds, the ClariVein included. The ClariVein endovascular procedure originated in the United States and has been endorsed by leading surgeons all over the world. ClariVein treatment is healthier and moves beyond the original success of laser treatments because it offers nonthermal vein ablation that can avoid any damage to the surrounding tissues. Thermal ablation can cause some slight nerve damage as well, which is avoided with ClariVein. Studies have shown that ClariVein is faster and simple, and statistics reveal that it is 74%

less painful that other techniques that offer minimal invasive treatments [292] (Fig. 13.3).

Unlike any other technique, ClariVein does not require the use of heat. ClariVein ablation makes use of a unique patented catheter with a rotating tip. The rotating tip allows much greater application of the sclerosing chemical to seal the affected vein. ClariVein is the only disposable, single-use catheter treatment that combines the use of a mechanical head with specific vein closing (sclerosing) medications to close off the diseased vein [293]. The physician uses ultrasound therapy to obtain a better look at the varicose vein from the surface of the skin and, since the treatment is minimally invasive, there are no incisions. ClariVein catheter is a conduit that is put into a target vein through a small cut in the skin. The peak of the catheter has a wire that rotates at a 360 degrees angle to coat the surface of the vein with medication. The 360 degrees spin by the wire enables the absorption and medication to work effectively. The schedule takes about 15 minutes. The damage caused by the medication and the spinning wire causes the vein to block as the veins become wider and are reabsorbed by the body. When this happens, blood begins to flow to other healthier and nearby veins. This procedure is referred to as mechanochemical ablation (MOCA). The MOCA techniques use the principle of double injury. The MOCA catheter uses ultrasound guidance to position itself close to the saphenopopliteal or saphenofemoral junction, and a revolving catheter tip creates the mechanical tension of the endothelium. Most research indicates that ClariVein is a safer procedure, but this has not been proven concerning the treated veins being closed permanently.

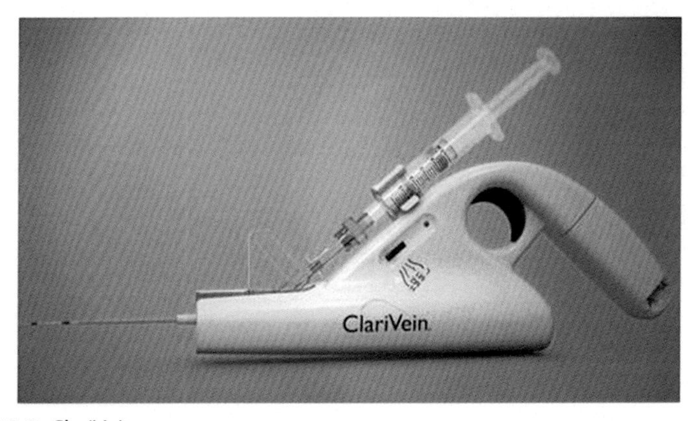

Figure 13.3 ClariVein.

Witte et al. noted that endovenous mechanochemical occlusion using the ClariVein catheter is a new technique combining mechanical injury to the venous endothelium coupled with simultaneous catheter-guided infusion of a liquid sclerosant. This produces irreversible damage to the endothelium, resulting in fibrosis of the vein. The technique is related to a low complication rate and a success rate of 96% at 2 years with sustained quality-of-life improvement. This closure rate is comparable to endothermal techniques, but significantly less postoperative pain and earlier return to normal activities and work have been reported with endovenous mechanochemical occlusion. The authors concluded that mechanochemical occlusion using ClariVein has proven to be safe and effective and has several advantages compared to endothermal techniques. The possibility of retrograde ablation of distal short saphenous vein (SSV) insufficiency in C6 ulceration is considered a significant advantage. Moreover, they stated that randomized comparative studies with long-term follow-up will continue to define the definite place of mechanochemical occlusion. Deijen et al. stated that mechanochemical endovenous ablation is a novel technique for the treatment of GSV and SSV incompetence which combines mechanical injury of the endothelium with simultaneous infusion of liquid sclerosant. The main objective of this study was to evaluate early occlusion [294].

All consecutive patients who were eligible for the treatment with mechanochemical endovenous ablation were included. The inclusion period was from the introduction of the device in hospitals (September 2011 and December 2011) until December 2012. A total of 449 patients were included, representing 570 incompetent veins. In 506 treated veins, duplex ultrasonographies were performed at follow-up: 457 veins (90%) were occluded at a follow-up of 6–12 weeks. In univariate and multivariate analysis, failure of treated GSV was associated with saphenofemoral junction incompetence [odds ratio (Or) 4; 95% confidence interval (CI): 1.0–17.1, $P = .049$]. The authors concluded that the ClariVein device appeared to be safe and had a high short-term technical effectiveness. Long-term clinical outcomes are needed to ascertain the clinical value of the ClariVein. In a randomized controlled trial (RCT), Lam et al. identified the ideal polidocanol dosage and form for MOCA in order to occlude the GSV. When adhering to safe dosage levels, sclerosants with higher concentrations potentially limit the extent of treatment. It has been demonstrated that this problem may be overcome by using polidocanol as microfoam. This chapter has been established on the findings of a preliminary analysis. The initial study was a single-blinded multicenter RCT where patients were allocated to three treatment arms. Group 1 consisted of MOCA + 2% polidocanol liquid, group 2: MOCA + 3% polidocanol liquid, and group 3: MOCA + 1% polidocanol foam. A total of 87 [34 males and 53 females (60.9%)] patients with a mean age of 55 years [Standard deviation, SD 16.0 (range 24–84 years)] were enrolled in the study. The treatment length was 30 cm (range 10–30) for 95.2% of the patients. The mean operating time was 16 minutes (range 5–70). The mean saphenofemoral junction diameter (7.7 mm) was similar in all three groups. At 6 weeks posttreatment duplex ultrasound showed that 25 of 25 (100%), 27 of 28 (96.4%), and 13 of 23 (56.5%) were occluded in the MOCA + 2% polidocanol liquid, MOCA + 3% polidocanol liquid, and MOCA + 1% polidocanol microfoam, respectively ($P < .001$). However, stricter scrutiny showed that the anatomical success rate defined as occlusion of at least 85% of the treated length to be 88.0%, 85.7%, and 30.4%, respectively ($P < .001$). The authors concluded that MOCA using ClariVein combined with 1% polidocanol microfoam is significantly less effective and should not be considered as a treatment option of incompetent truncal veins.

They stated that further investigation to determine the ideal polidocanol liquid dosage with MOCA is advocated and is being conducted accordingly. Vun and colleagues [295] evaluated the effectiveness of the ClariVein system of MOCA for superficial vein incompetence. ClariVein treatment uses a micropuncture technique and a 4-Fr sheath to allow a catheter to be placed 1.5 cm from the saphenofemoral junction. Unlike EVLT or RFA, no tumescence is required. The technique depends on a wire rotating at 3500 r/minute, causing endothelial damage while liquid sclerosant (1.5% sodium tetradecyl sulfate) is infused. The wire is pulled back while continuously infusing sclerosant along the target vessel's length. Initially, 8 mL of dilute sclerosant was used, but this was subsequently increased to 12 mL. No routine postoperative analgesia was prescribed and specifically no NSAIDs. Procedure times and pain scores (visual analog scale) were recorded and compared to EVLT and RFA. All patients were invited for duplex postprocedure. A total of 51 GSV and six SSV were treated and followed-up with duplex in the 10 months from July 2011. No major complications or DVT were reported. Duplex showed patency of three treated veins with two more veins having only a short length of occlusion, giving a technical success rate of 91%. Comparison with 50 RFAs and 40 EVLTs showed procedure times were significantly less for ClariVein (23.0 ± 8.3 minute) than for either RFA (37.9 ± 8.3 minute) or EVLT (44.1 ± 11.4 minute). Median pain scores were significantly lower for ClariVein than RFA and EVLT (1 vs 5 vs 6, $P < .01$). The authors concluded that MOCA with the ClariVein system is safe and effective. After some initial failures, the use of 12 mL of dilute sclerosant resulted in a very high technical success rate that was greater than 90%, which accorded with the limited published literature; and procedure times and pain scores were significantly better than for RFA and EVLT. These researchers stated that they were awaiting the long-term clinical outcomes. Bootun and associates [296] noted that endovenous techniques are, at present, the recommended choice for truncal vein treatment. However, the thermal techniques require tumescent anesthesias, which can be uncomfortable during administration. Nontumescent, nonthermal techniques would, therefore, have potential benefits. In an RCT, these investigators compared the degree of pain that patients experience while receiving MOCA or RFA. The early results of this RCT were reported here. Patients attending for the treatment of primary varicose veins were randomized to receive MOCA (ClariVein) or RFA (Covidien Venefit).

13.4 Cryostripping

13.4.1 Modern miniphlebectomy

Cryosurgery is the best method for the elimination of varicose veins, which meets patients' and doctors' expectations. It allows cure of the disease with minimal damage and results in the removal of perforators without leaving scars (miniphlebectomy technique). It is also unnecessary for patients to stay at hospital and undergo anesthesia. This technique is based on insertion of the probe through a very small skin incision. The tip of the probe decreases the temperature to −80°C, resulting in a cryoapplication effect, that is, sticking of the surrounding tissues to the probe.

13.4.2 Cryostripping method

Cryostripping is bases on the Babcock method, which involves extirpation of varicose veins by freezing. In order to remove a varicose vein, CryoProbe is introduced into the lumen of a vessel, or applied externally on the vessel wall. Cryostripping is introduced into the lumen of a vessel or applied externally on the vessel wall. Cryostripping allows removal of varices after previous sclerotherapy, that is, injection of obliterating substances. A low temperature exerts an analgesic effect with simultaneous occlusion of small vessels, resulting in prevention of intraoperative bleeding. The design of the probe allows the surgeon to obtain an approach to all varicose veins without the necessity for additional skin incisions. This method provides an ideal method for the treatment of varicose veins and complications of varicose disease. In these cases, this technique is very competitive with the classic, vary drastic Linton operation (skin incision from the knee to the ankle), or a novel, but very expensive method, of endoscopic ligation of perforators [297].

13.4.3 Advantages of cryostripping

- Small skin incisions
- Short time of treatment
- Local anesthesia
- No complications
- Quick recovery
- Good cosmetic results
- Low cosmetic damage
- Low costs of treatment
- No need for hospitalization
- Cryostripping is the most advanced form of mechanical stripping
- Cryostripping is a popular miniphlebectomy method

Cryostripping involves the removal of a sufficient vein segment by introducing the special probe into the vein or by its external application, and it can reach a low temperature in a few seconds. Taking advantage of the adhesion phenomenon (freezing of the probe to the vessel wall) and using different sizes of probe, from a few skin punctures varicose veins of different size and shape are removed (enhanced miniphlebectomy). This applies both to primary and secondary varicose veins in chronic venous insufficiency (CVI).

Cryostripping is the best method for the removal of varicose veins after a previous obliterative treatment and a history of thrombophlebitis. Cryostripping has also proved to be the most convenient method for removing meandering varicose plexuses.

13.4.4 Cryosurgery in venous ulcer treatment

Excellent results have been recently obtained also in the removal of perforators in surgical venous ulcer treatment. Perforator vein cryosurgery involves introduction of the probe in an area of healthy skin (over the skin–muscle scleroderma). The skin incision after probe introduction does not exceed 3–4 mm. Then in the epifascial layer, after freezing, all the veins near the ulcer area, connecting the deep venous system to the superficial system, are removed (stripping).

This is the technique of extirpating a saphenous vein using a metal probe which has been cooled with liquid nitrogen. Although this is still above a cryogenic temperature (−180°C or −150°C), it fulfills its purpose as a saphenous vein extractor. There are many advantages to using cold temperatures to ablate saphenous veins rather than mechanical stripping or endothermal heating with laser, steam, or radiofrequency treatments.

The advantages of cryostripping include a reduction in bleeding because many of the tributaries come out with the probe and the remainder of the detached tributaries are sealed during the freezing process. Nerve injury is reduced because the trunk is removed from within. This is in contrast to stripping using external stripper heads and endothermal techniques which may interfere with the saphenous nerves. Furthermore, the freezing process may reduce the incidence of neovascularization through reduced tissue trauma and the cold temperature may provide temporary pain relief. From an economic point of view this method is very affordable because the freezing probes can be resterilized and reused. Although patients require general or regional anesthesia, cryostripping can be performed in an ambulatory setting with a rapid return to work and a fast recovery [298] (Fig. 13.4).

13.4.5 Cryostripping equipment

13.4.5.1 Cryo-s Electric II

This state-of-the-art cryosurgical device manufactured by Metrum Cryoflex is the next generation of apparatus used by many experts in the field since 1992. Cryo-s Electric II is controlled by a microprocessor, and all parameters are displayed on an Liquid Crystal Display (LCD) screen.

The Cryo-S Electric II device is a smart technology and the new standard in cryosurgery—What is new?

- Mode selection, cleaning the probe, and freezing can be performed automatically using a foot switch or touchscreen, keeping the procedure site under sterile conditions.

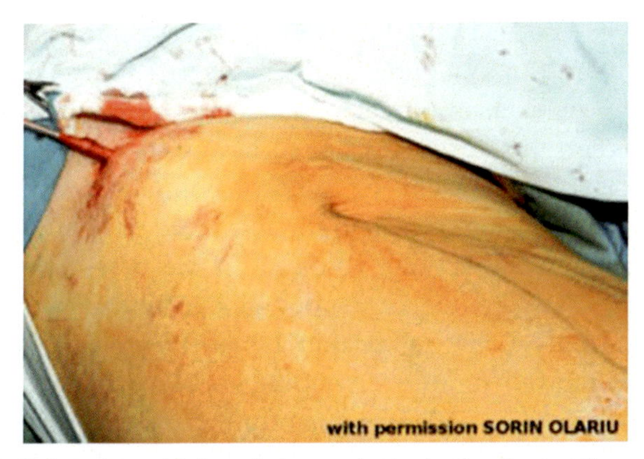

Figure 13.4 Cryoprobe withdrawal (cryostripping). Gentle traction removes the saphenous vein from the patient.

- Electronic communication (chip system) between the main device and the connected cryoprobe. The unit recognizes optimal operating parameters and autoconfigures to probe characteristics.
- Pressure and gas flow are set automatically, and manual adjustment is unnecessary.
- It contains an automatic probe-cleaning system and two freezing modes (continuous and discontinuous).
- The temperature of the probe, pressure, gas flow, and time of the procedure are displayed during freezing.

13.4.6 Cooling medium

The working medium for Cryo-S series devices is nitrous suboxide (N_2O) or carbon dioxide (CO_2), which are very efficient and easy-to-use gases. A 10-L cylinder serves approximately 60–75 patients. In contrast to liquid nitrogen, both gases can be stored safely in steel cylinders and without losses.

13.5 The evolution of varicose vein treatment: from 19th century ligation to noninvasive therapy (cyanoacrylate glue)

In the 1890s, Friedrich Trendelenburg introduced the idea that visible varicose veins were caused by underlying truncal valve dysfunction and, to cure them, he introduced the Trendelenburg ligation. Previous to this, treatments had been aimed at the visible varices alone. The following century showed little advances in varicose vein treatments, apart from studies showing that stripping the GSV was superior to ligation alone, and that the treatment of incompetent perforators helped to heal venous leg ulcers. In contrast, the last 20 years or so have seen a massive increase in the advancements in treatments for varicose veins and venous reflux disease. Although most people point to the introduction of endovenous surgery as being the major turning point, all of the new advances actually stem from the development of venous duplex ultrasonography in the mid-1980s and early 1990s. It was only because this noninvasive imaging modality that allowed us to see venous function in real time became widely available that our understanding of venous disease leapt forward.

The ability to identify venous reflux, different reflux patterns, the differences between passive (diastolic) reflux that everyone understands and active (systolic) reflux that many doctors struggle with, and the size of target veins, has shaped many of the new approaches to varicose vein treatment. Moreover, the identification of venous reflux in leg varicose veins arising from pelvic veins has revolutionized the concept that varicose veins can be thought of as a problem isolated to the lower limb. However, to think that everyone has reached the same conclusions from the advent of venous duplex ultrasound would be wrong [299]. Whereas some have used the technique to improve understanding and hence results, most doctors simply use it to identify which truncal vein they are going to treat, ignoring complex patterns, perforators, or pelvic venous reflux. It is not surprising that randomized trials of different modalities are not conclusive, if they only concentrate on treating incompetent truncal veins and ignore the other causes of varicose veins.

The introduction of venous duplex ultrasonography split the venous world into two main factions. Unfortunately, a great many doctors who "do" varicose veins as a job and do not attend conferences or read around the subject are unaware of this huge divide. In the English-speaking world, most doctors practice ablative surgery. Some still ligate and strip, although most now have moved on to some form of thermal ablation. Other nonthermal ablative techniques are becoming more widely used. Nevertheless, a large number of doctors remain unaware of the hemodynamic approach to varicose veins and venous reflux disease, championed by the conservative hemodynamic correction of venous insufficiency method (CHIVA) [299]. Often called "saphenous sparing surgery," advocates of this approach present series where the results have been demonstrated to be comparable with stripping. Whereas those of us who treat all of the reflux pathways would regard such results as suboptimal, the randomized studies that have been performed, where surgeons ignore perforator vein reflux and pelvic vein reflux, appear to show stripping as equivalent to the thermal ablation techniques. As such, the hemodynamic approach may begin to gain some traction.

13.5.1 The endovenous revolution

Following on from venous duplex ultrasonography, the greatest revolution in the treatment of varicose veins was the invention of successful endovenous thermal ablation. At the end of the 1990s, catheter-based RFA and endovenous laser ablation proved to be successful, causing endovenous surgery to take off. Not only did these endovenous thermal techniques destroy the vein, the catheters were introduced under ultrasound control into the distal vein passed proximally, without the need for open surgery in the groin. This minimally invasive approach allowed the development of tumescent anesthesia. With truncal ablation and phlebectomies being possible under tumescent anesthesia, true "walk-in, walk-out" ambulatory surgery became possible for the treatment of varicose veins. More than just a new technique of treating veins, this allowed vein centers to be set up outside hospitals that could concentrate on ambulatory venous surgery [300]. In fast succession, treatment of incompetent perforators was developed in 2001 using the transluminal occlusion of perforator technique, steam vein sclerosis, and several different radiofrequency and laser devices became available for leg veins.

All of these thermal ablation devices require tumescence because of the heat generated during treatment. This has led to the investigation and development of nonthermal, and therefore nontumuscent, ablation techniques. In 1985, a patent was granted to allow detergent sclerotherapy fluids to be mixed with gas to make foam. This was then taken on by a British company hoping to replace surgery with a chemical ablation technique. Although it has become clear in the last decade that foam sclerotherapy works well in small veins within walls, medium- and long-term results have been poor in truncal veins, which are larger and have thicker walls. As such, although foam sclerotherapy is an essential technique to be used by any doctor providing venous treatments, it has been shown to have comparatively poor results when used as a sole treatment modality.

To improve sclerotherapy results, the endovenous MOCA catheter (MOCA, ClariVein) was developed to traumatize the venous wall mechanically and to allow sclerosant to penetrate deeper. Research has shown that this increases the cell death within the vein wall, improving the long-term ablation over foam sclerotherapy alone. Finally, in the nonthermal nontumuscent area, cyanoacrylate glue has been injected intravenously with good results in medium-term studies. This appears to use a different mechanism, as the vein wall is not ablated in the same way as it is with the previously described techniques. However, patient satisfaction is high and clinical results are very good (Fig. 13.5).

Cyanoacrylate tissue adhesive (e.g., the VariClose Vein Sealing System and the VenaSeal Closure System; Sapheon Inc., Morrisville, NC) is a minimally invasive, nontumescent, nonthermal, and nonsclerosant procedure that uses a medical adhesive to close the diseased vein in patients with symptomatic venous reflux disease. Unlike other treatments, the VenaSeal Closure System does not require tumescent anesthesia, allowing patients to return to their normal activities following the procedure; it also eliminates the risk of nerve or other heat-related injuries associated with thermal-based procedures, and thus may reduce the need for compression stockings postprocedure. Toonder et al. noted that percutaneous thermo-ablation techniques are still being used today and seem more effective than nonthermal techniques. However, thermal techniques require anesthesia and potentially may cause inadvertent damage to surrounding tissues such as nerves. Cyanoacrylate adhesive has a proven record, but not for the treatment of chronic venous disease of the leg.

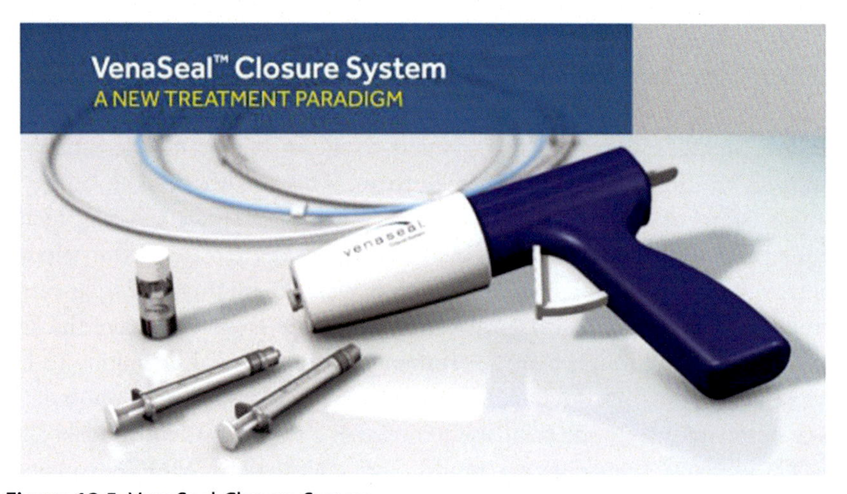

Figure 13.5 VenaSeal Closure System.

Innovation has led to the development of the VenaSeal Sapheon Closure System, which has been designed to use modified cyanoacrylate glue as a new therapy for truncal vein incompetence. These researchers examined the feasibility of ultrasound-guided cyanoacrylate adhesive perforator embolization (CAPE).

The authors stated that the results of this feasibility study showed a 76% occlusion rate of incompetent perforating veins (IPVs) without serious complications; further investigation with a dedicated delivery device in a larger patient population is needed. McHugh and Leahy stated that endothermal treatment of the GSV has become the first line of treatment for superficial venous reflux. Newer treatments, especially nonthermal ablation, have potential benefits both for patient acceptability and decreased risk of nerve injury. These researchers described the current nonthermal options available, including advantages and disadvantages. Ultrasound-guided foam sclerotherapy avoids the risk of nerve injury; however, it is not as effective as endothermal ablation. Mechanochemical endovenous ablation combines mechanical endothelial damage using a rotating wire, with the infusion of a liquid sclerosant (the ClariVein System). Reports have suggested that this system is safe and effective, eliminating the need for tumescent anesthesia with no reported case of nerve injury.

Finally, the VenaSeal Closure System comprises the endovenous delivery of cyanoacrylate tissue adhesive to the vein causing fibrosis. Perioperative discomfort seems to be minimal, but the complication of thrombophlebitis has been reported in up to 15% of patients. The authors concluded that nonthermal options promise comparable treatment efficacy without the added morbidity associated with high thermal energies. They stated that the potential of treating venous reflux without the risk of nerve damage may change how surgeons approach venous disease. On February 20, 2015, the FDA granted premarket approval of the VenaSeal Closure System to treat superficial varicosities of the legs through endovascular embolization and is intended for adults with clinically symptomatic venous reflux diagnosed by duplex ultrasound. The FDA approval was based on a multicenter RCT by Morrison et al. [301], who noted that preliminary evidence suggests that CAPE may be effective in the treatment of incompetent GSVs. These investigators reported early results of a RCT of CAPE versus RFA for the treatment of symptomatic incompetent GSVs [302].

13.6 Subfascial endoscopic perforator surgery in perforator vein insufficiency

Patients with CVI and venous ulcers were surgically correlated using long incisions through diseased skin and subcutaneous tissues already compromised by venous hypertension. This procedure involved ligation of incompetent perforator veins described by Linton, Cockett, and Dodd, this technique was often complicated by wound infections and poor healing. However, in 1985, Hauer demonstrated a new surgical technique where incompetent perforator veins were directly visualized using an endoscope in the subfascial space. This seminal contribution marked the advent of subfascial endoscopic perforator vein surgery (SEPS). The idea to use this approach was based on the possibility of creating, using laparoscopic instruments, a virtual space, and this was of great interest as it offered the possibility to avoid further damage to the scarred tissues surrounding the ulcer and thus eliminating the wound complications that affected the Linton technique.

13.6.1 Subfascial endoscopic perforator vein surgery equipment

Most of the instruments used in this procedure are used for laparoscopic cholecystectomy, and these include:

- Insufflator for introducing carbon dioxide to maintain the working space;
- A rigid 5 or 10 mm endoscope;
- A three-chip video camera, preferably with a xenon light source;
- A TV monitor;
- A 10 mm cannula, with which the rigid endoscope is introduced into the subfascial working space;
- A 5 mm cannula is used for all other equipment.

Other additional instruments important for the successful performance of the operation include a balloon dissector. Although dissection of the subfascial plane can be created via endoscopic instruments manually, the balloon dissector significantly expedites the dissection process and helps to create a large, operative working space. The balloon dissector used in this technique has a capacity of 300 mL balloons with a protective removable cover, and a guide rod to aid in the introduction and placement, and also a 10 mm laparoscopic cannula with a skin seal. A second important but optional instrument is the 5 mm reticulating endograsper, where the tip articulates and rotates, offering a high degree of maneuverability. The 5 mm clip applier needs a 5 mm port. Its small size also offers a high degree of maneuverability and visibility when working in a small endoscopic space. The applier delivers an 8 mm long (medium/large) clip in a convenient and multifire configuration.

13.6.2 Preoperative preparation

Preoperative evaluation includes color Doppler scanning, which can be used to document superficial, deep, or perforator incompetence and guide the operative intervention. The incompetent perforators on the skin are accurately mapped and marked, which is mandatory as this assists the surgeon during surgery. Ultrasonographers can help by marking the sites of incompetent perforators and also that of an incompetent saphenopopliteal junction with the help of a skin marker.

13.6.3 Operative technique

The SEPS procedure is performed under general/spinal anesthesia with the patient supine and in the Trendelenburg position with the knee slightly flexed and elevated. In anticipation of concomitant stripping of superficial veins, the entire extremity is prepared circumferentially.

A 10 mm incision is made through the skin, which is 4 cm medial to the tibia and 10–12 cm below the popliteal crease. Subcutaneous tissue is dissected, the posterior compartment is identified, and a 10 mm transverse incision is made into the fascia. The subfascial space is identified and retractors are placed to keep it open. The balloon dissector is introduced into the fascial incision and directed toward the medial malleolus. After removal of the balloon cover sheath, the dissection balloon is inflated with 200–300 mL saline. The balloon is designed in such a way that initial radial expansion occurs, followed by distal expansion toward the malleolus, as the balloon everts distally. Dissection occurs along planes of least resistance by the balloon; hence, the perforating veins are not disrupted in the dissection process [303].

The balloon is deflated and removed once the dissection is accomplished; the rotating seal of 10 mm is secured to the fascial incision. The cannula is introduced into the space dissected, and the guide rod and obturator are removed. The skin seal is rotated into the fascial incision to provide a gas seal. CO_2 is then insufflated at a pressure of 15 mm Hg to create the working space. A 0-degree 10 mm rigid laparoscope with attached video camera and light cable are introduced and the subfascial space is visualized on a video monitor. A working 5 mm laparoscopic port is then inserted in the mid-calf under direct endoscopic guidance [304]. This trocar is placed as posteriorly as possible to make a wide working axis. This arrangement of the trocar aids visualization of the working instrument and facilitates instrument manipulation in the confines of the calf. The perforating veins may be visible immediately or may require some amount of blunt dissection and exploration. Skin markings carried out with the help of duplex venous studies are useful in guiding the surgeon to the location of the perforators. Once identified, each perforating vein is double clipped with the 8 mm titanium clips with a 5 mm clip applier. Generally, all perforating veins which can be identified are clipped.

As the perforator continuity is interrupted by the clips, the veins are usually not divided. However, division of the perforator between the clips can be performed, when desired, with endoscopic shears to facilitate distal exposure. When interruption and/or division of the perforators is complete, the trocars are removed, and the skin incisions are closed with interrupted mattress stitches using monofilament sutures. Superficial ligation and stripping can be performed in the standard fashion in patients with superficial venous insufficiency, a nonadherent dressing is used to cover all wounds, and the operated leg is wrapped with a compression bandage extending from the forefoot to the upper calf or thigh. Usually, patients are discharged on the day of surgery and advised for routine follow-up in the outpatient department 1 week after surgery [305].

13.6.4 Postoperative management

Once the effect of the anesthetic wears off, patients are encouraged to ambulate and are discharged on the same day or the day after surgery. Patients receive two postoperative doses of antibiotics in addition to the intraoperative intravenous antibiotic. For the first 24 hours after surgery, they are provided with adequate parenteral analgesia, which is changed to oral analgesia upon discharge. Postoperative instructions stress the need for active ambulation, elevation of the operated limb, and regular maintenance of the elastic bandage. Patients are seen for removal of skin sutures in the outpatient department a week to 10 days after surgery. Those patients with an active ulcer need further regular dressings until the ulcer heals. Class 2 graduated compressive stockings are prescribed to all patients on a long-term basis.

SEPS has gained a great deal of attention around the world. Many controlled trials have been conducted, with many in favor of SEPS. The goal of this review was to ascertain if the SEPS procedure for perforator incompetence is superior to the conventional open surgical procedure, and if so what the benefits are and how it could be more widely instituted. There is a great deal of diversity in RCTs. The main variables in these trials are:

- Number of patients in trial;
- Withdrawal of cases;
- Blinding;
- Intention to treat analyses;
- Publication biases;
- Local practice variations;
- Prophylaxis antibiotic used;
- Follow-up failure.

Without proper details on all these parameters, it is difficult to draw a conclusion. One should always think of SEPS and the conventional open procedure as being complementary to each other.

A successful outcome requires greater skill of the operating surgeon with adequate training in the field of minimal-access surgery. SEPS requires different skills and technological knowledge. In fact, many studies have shown that the outcome of SEPS was influenced by the experience and technique of the operator. In a study by Anjay Kumar, 21 varicose veins patients with an incompetent perforator underwent SEPS using a harmonic scalpel. Various parameters were studied. The result of their study was that all ulcers healed in 8 weeks with no recurrence in the 11.9-month follow-up period. There was one case of wound infection and one saphenous nerve neuropraxia as complications were noted postoperatively. They concluded that using an ultrasonic scalpel in SEPS is technically feasible, causing less tissue damage as the thermal effect it generates is very low, and also the study was associated with minimal morbidity [306].

In another study by Luebke and Brunkwall, a meta-analysis of SEPS for the treatment of CVI was carried out. Here, a multiple health database search was performed, including Medline, Embase, Ovid, Cochrane Database of Systematic Reviews, and Cochrane Database of Abstracts of Reviews of Effectiveness, on all studies published between 1985 and 2008, that reported on health outcomes in patients with CVI treated with SEPS and comparing this therapy with the conventional Linton procedure. Three studies, which compared SEPS with conventional surgery, were included in the meta-analyses. The results of the study found that between the SEPS and Linton groups, there was a significantly lower rate of wound infections in the SEPS group [OR 0.06 (95% CI: 0.02−0.25)] and a significantly reduced hospital stay for SEPS [OR 8.96 (95% CI: 11/62 to −6.30)]. In addition, there was a significantly reduced rate of recurrent ulcers in the SEPS group. There was no significant difference between the groups in the following dimensions: death at 6 months [OR 3.00 (95% CI: 0.11−78.27)], rate of hospital readmission [OR 0.21 (95% CI: 0.03−1.310)], healing rate of ulcer at 4 months [OR 0.44 (95% CI: 0.09−2.12)], and the rate of DVT [OR 0.35 (95% CI: 0.01−8.85)]. The conclusion drawn from this study was that when SEPS was used as a part of a treatment regimen for severe CVI it benefits most patients in the short term regarding ulcer healing and also the prevention of ulcer recurrence. Also, SEPS, if safely performed, has less early postoperative complications compared with the Linton procedure. However, further prospective randomized trials are required to define the long-term benefits of SEPS.

In a randomized study by Kianifard et al., the effect of adding SEPS to standard GSV stripping was studied. The authors studied the fate of IPVs in patients undergoing standard varicose vein surgery versus those treated with standard varicose vein surgery and SEPS. Patients were included in this study if they were undergoing surgery for varicose veins and also had venous reflux (0.5 seconds) in the GSV. All patients in the study also had IPVs. Patients were randomly allocated to standard surgery (saphenofemoral ligation, stripping, and phlebectomies alone) or standard surgery with the addition of SEPS. Patients were excluded from the study if they had recurrent varicose veins, deep venous reflux, deep venous thrombosis, ulceration, or saphenopopliteal reflux. Using duplex ultrasound, IPVs were determined preoperatively, and at 1 week, 6 weeks, 6 months, and 1 year after surgery. Visual analog scores for pain and quality-of-life questionnaires were obtained at the same time periods. There were 34 patients in the no SEPS group and 38 patients in the SEPS group. During the follow-up period, the groups did not differ with respect to quality-of-life scores, pain, or mobility, but at 1 year, there was a higher proportion in the no SEPS versus SEPS group that had IPVs. The conclusion drawn was that SEPS when used as an adjunct to standard varicose vein surgery reduces the number of IPVss at 1 year but has no effect on quality of life or recurrence of varicose veins at 1 year.

Florian Roka et al., in their study, investigated the mid-term (mean, 3.7 years) clinical results and the results of duplex Doppler sonographic examinations of EPS in all patients with mild to severe CVI (clinical classes 2–6) and also assessed the factors associated with the recurrence of IPVs. Around 80 patients with mild to severe CVI undergoing SEPS were evaluated, and duplex findings as well as clinical severity were compared. Those patients with prior DVT (<6 months) or prior to SEPS procedure were excluded from their study.

The conclusion of the experimental model and statistical analysis revealed that SEPS is a feasible, safe, and effective treatment of IPVs in patients with advanced CVI. In this review it has been found that SEPS is a promising technique for the treatment of patients with perforator incompetence. It may be optimally utilized in cases with failure of conservative therapy or those with advanced CVI. The favorable ulcer-healing rate and improvement in clinical symptoms suggest that SEPS plays a considerable role in correcting the underlying pathology in CVI caused by IPVs.

CHAPTER 14
Sclerotherapy

14.1 Introduction

Christopher Ubren (1656) and his associates are reputed to have been the first to introduce drugs intravenously. Using a metal tube they introduced opium into the veins of a dog.

Treatment of varicose veins (VVs) using a sclerosant is not new. Different types of sclerosants and techniques were tried as far back as the 17th century. *In 1682, Zollikofer in Switzerland was the first to initiate sclerotherapy by injecting an acid into a vein.* The inefficacy and the toxicity of the drugs used and the complications after the procedures discouraged other surgeons from continuing with this technique. After a long silence of nearly three centuries Schiassi (1908) combined high ligation and subsequent injection. Refinements in technique and the search for safer drugs continued and now an era has evolved which could replace all other surgical techniques that are currently being used. This is because of the extreme safety and simplicity of the technique of sclerotherapy. Most of the failures of the procedure still discussed are only due to the improper understanding of the science behind it and improper technique still practiced all over the world. In all the current techniques used for VV, everyone is focusing on the elimination of visible veins to prevent so-called "venous reflux," which is always overemphasized. Sclerotherapy is still used as an alternative to stripping, radiofrequency ablation (RFA), or endovenous laser therapy (EVLT) and is never a total substitute for them. *In our practice the principle is totally different. It is aimed at the elimination of mainly radicular veins. It can be used to efficiently eliminate the truncal veins also. It is "chemical bombing" for the total eradication of both truncal and radicular veins.*

A word about our institute: This is an exclusive institute solely dedicated to the treatment of venous diseases. We are carrying out a lot of research work on venous diseases including genomic analysis and molecular studies, in collaboration with the Rajiv Gandhi Centre for Biotechnology (Dept. of Science & Technology, Govt. of India), Thiruvananthapuram, Kerala. I am privileged to have carried out around 34,000 sclerotherapies myself.

Genesis, Pathophysiology and Management of Venous and Lymphatic Disorders
DOI: https://doi.org/10.1016/B978-0-323-88433-4.00001-2

Now we offer the procedure to more than 3000 patients annually, with most of the patients belonging to classes C4, C5, and C6. Many of the patients present to us after the failure of repeated surgeries, EVLT, or RFA in highly reputed centers. The procedure details given in this chapter are those followed in our institute.

14.2 Definition

Sclerotherapy is a procedure in which a safe drug is injected into the veins or lymphatic channels to produce a chemical inflammation in the endothelial lining of the vessel with subsequent scarring and thrombosis, resulting in permanent luminal occlusion.

The historical background to this has already been discussed.

14.3 Evolution of the methodology of sclerotherapy

The evolution of sclerotherapy followed the following methods:

Plain sclerotherapy
Foam sclerotherapy
Ultrasound-guided sclerotherapy
Endovenoscopic sclerotherapy
Microfoam sclerotherapy (MFST)
Modified microfoam sclerotherapy (MMFST) or chemical bombing

14.4 Principles of sclerotherapy

The Greek word *sclero* means *hardening*. The scientific principle behind this technique is that no instrument can be passed through a vein freely beyond a particular limit and therefore the maneuverability of such an instrument is only up to a particular size of vein. Only a solution can pass through any type of vein, whether it be too small and invisible or too large. It is also capable of selectively producing a chemical burn on the endothelial lining of veins, whereby an inflammatory process is induced resulting in fibrosis and scarring. The interconnectivity of the veins between deep veins and superficial veins are maximal at the tributary level and not at the truncal level. The perforators are only the interconnecting channels between the deep and superficial main channels and are very few. In a patient with a VV, the venous return, despite the valvular dysfunction, is facilitated by the overflowing mechanism from the superficial vein bearing very high venous pressure compared to the deep vein which maintains a negative pressure. By simply obstructing the outlets of the superficial systems there is compartmentalization of the superficial system and the back pressure progressively increases within the system and, if the tributaries are involved in the disease process, the pressure is reflected at the cellular level and leads to progressive worsening of the symptoms of the patient. If the tributaries are not involved, no adverse situation appears in the patient as the increasing pressure in the truncal vein helps the healthy tributaries to bypass the blood to the deep system. *Any method of surgical block induced by any technique proximal to the tributaries will be detrimental to the tissues, provided the tributaries are involved in the disease process.*

In Western countries, the treatment of VV is mainly a cosmetic procedure. In India, however, this is not the case, and it is mainly done for acute or chronic disease complications.

14.5 Sclerosant drugs

Different types of sclerosants have been tried over the centuries, and are traceable from 1682, as already described in detail. The 20th century witnessed the use of the following:

1. Phenol in almond oil
2. Sodium morrhuate
3. Hypertonic saline
4. Sodium tetradecyl sulfate (STS)
5. Polidocanol (POL)
6. Chromated glycerine

Of these, only the last three are currently in use.

14.6 Historical evolution of sclerosing agents

1840 Absolute alcohol (Monteggio, Leroy D'Etiolles)
1851–1853 Ferric chloride (Pravaz)
1855 Iodo tannic liquer (Desgranges)
1880 "Chloral" (Negretti)
1904 5% Phenol solution (Tavel)
1906 Potassium iodo-iodine (Tavel)
1910 "Sublime" (Scharf)
1917 Hypertonic glucose/50%–60% calorose (Kausch)
1919 30% Sodium salicylate (Sicard and Gaugier)
1919 Sodium bicarbonate (Sicard and Gaugier)
1920 1% Bichloride of mercury (Wolf)

1922 12% Quinine sulfate with 6% urethane (Genevrier)
1922 Bi-iodine of mercury (Lacroix, Bazelis)
1926 Hypertonic saline with procaine (Linser)
1927 50% Grape sugar (Doeriffel)
1929 Sodium citrate (Kern and Angel)
1929 20%–30% Hypertonic saline (Kern and Angel)
1930 Sodium morrhuate (Higgins and Kittel)
1933 Chromated glycerin (Jausion)
1937 Ethanolamine oleate (Biegeleisen)
1946 STS (Reiner)
1949 Phenolated mercury and ammonium (Tournay and Wallois)
1959 Stabilized polyiodinated ions (Imhoff and Sigg)
1966 POL (Henschel and Eichenberg)
1969 Hypertonic saline/dextrose

14.7 Hypertonic saline

Hypertonic saline was used for some time due to its comparative safety of administration, as sodium chloride is a natural constituent of the body. It has been approved by the US Food and Drug Administration (FDA). When used in a concentration of 23.4% it is a good sclerosant for medium-sized veins; but a lower strength saline (half strength) is used for very tiny veins. The drawback of saline is its poor efficacy in producing a clinical effect in the maximum loadable dose. When extravasated it produces excruciating pain and tissue necrosis. It also produces hemosiderin skin staining. All these side effects made the drug unacceptable to patient and surgeon alike, and hence it is no longer in use now.

14.8 Sodium tetradecyl sulfate

14.8.1 $C_{12}H_{25}(OCH_2CH_2)_n \cdot OH$

14.8.1.1 Polyethylene glycol monododecyl ether

An ideal, safe, reliable, and effective synthetic surfactant (soap) was the only FDA-approved sclerosant in the United States until early 2010. It is commercially available in concentrations of 1% and 3% solutions. This will also produce postsclerosing hemosiderin staining. When extravasated, it produces severe pain and tissue necrosis. It is used in concentrations of 0.1–0.4% depending on the size of the vein to be sclerosed. Very rarely it may produce anaphylaxis. Compared to POL it is very painful for the patient but the effectiveness is almost equal (Fig. 14.1).

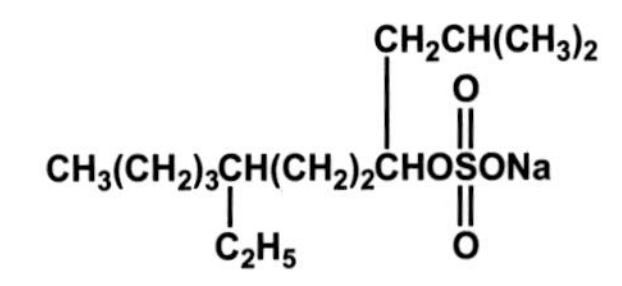

Figure 14.1 Structure of sodium tetradecyl sulfate.

14.9 Polidocanol

POL is a liquid surfactant that when infused into the vein produces a selective endotheliolytic property. It is very widely accepted as the sclerosant of choice in the treatment of VV and congenital venous disorders.

POL was used to sclerose leg VV in more than 30,000 patients from 1995 to 2019, in varying dilutions, as a plain solution and microfoam, using the usual techniques and also MMFST in different sizes of veins. The direction of the flow of drug was visually observed and videographed, particularly in tributaries and smaller veins. The results were compared with STS and POL, using plain and foam versions in varying strengths and different methods. All the available published papers were reviewed. Excellent promising results were obtained in the study. The specific retrograde flow and centrifugal distribution of the drug are noteworthy. This feature adds to its efficacy and safety over other sclerosants. The optimal sclerosant effect is observed at a lower universal dilution of 0.2% and not with increasing concentrations proportionate to the size of veins [307].

14.9.1 Overview

Chronic venous disease (CVD) of the lower extremity is the most common of all venous diseases. The history of VV is traceable from BCE 3500, when the Egyptians cautioned on the risk of surgery on veins. It was in the early 1st century AD, that Celcus initiated surgical corrections in Rome. Until the 19th century no substantial improvements in treatment took place. The medieval period witnessed the most brutal methods of treatment, and it was only after middle of 20th century that a new horizon arose. The 21st century brought promising results with minimally invasive techniques like EVLT, RFA, and sclerotherapy [308].

It was in 1682 in Switzerland that Zollikofer first tried sclerotherapy by injecting an acid into a VV. However, the final outcome was rather disastrous. After a long pause, the search for a better method was taken over by surgeons with differing techniques and using various chemicals. Phenol in almond oil was used in 1950, but the sclerosant effect was transient and the recurrences were very frequent. This procedure was done usually along with crossectomy to avoid the crude method of vein stripping. Prior to this, ferric and iodine compounds, tannic acid, and hypertonic saline were used in the 19th century but were abandoned due to their several limitations. Following the inception of two new detergents, POL and STS, after 1930, progress in the methodology of treatment was increasing. STS was approved by the FDA in 2004 for use in small veins, but POL only gained approval in March 2010. Even before its approval, POL had been used extensively by physicians in the United States and Europe [6] because of its clinical efficacy. The main drawback of STS was the pain during and for a few hours following injection. The foaming capacity is far less when compared to POL. Thus POL has proved to be superior to all the other sclerosants, in all respects [309].

The present practice of sclerotherapy is as a substitute to open surgical procedures like crossectomy and stripping, minimally invasive procedures like EVLT and RFA, or in their combinations. All these procedures are basically set on the currently followed concepts; the refluxes from the deep venous system (DVS) to the superficial venous system due to the incompetence of the venal valves, specifically at the saphenofemoral (SF) and saphenopopliteal (SP) junctions and perforators. Moreover, too much emphasis was given to the erect posture, age, gender, family relation, obesity, etc., but they are only predisposing factors. Failures of the major venous valves, due to pathological changes in the vein wall and/or valves have been well established. Recent studies on microscopic venous valves (MVVs) have proved the importance of MVVs in the development of CVD by impediment of the microcirculation. Failure of valves at the smallest tributaries and smallest veins closer to the capillaries can only induce tissue changes in CVD. Moreover, recent genetic studies have proved that *FOX-C2-Promotor Variant c.-512C<T* is associated with increased susceptibility to CVD. This newer finding supports a newer concept on the etiopathogenesis of VV and subsequent CVD changes. The evolution of the disease is not primarily dependent on the etiological factors mentioned above but on the genetic factors. Hydrostatic pressure also contributes to a greater extent. If the smaller veins are healthy no CVD changes will occur. The distance from the foot also matters a lot, and the greater the height from the foot the more the morbid changes are minimized. Therefore also, the smaller the vein, the greater the morbidity.

It can thus be concluded that the extent of morbid changes in CVD is inversely proportional to the size and height of the involved segment of vein. Hence it is clear, for the successful treatment of CVD, that elimination of the smallest veins at the lowermost part of the leg is pivotal. This illustrates the importance of a good sclerosant solution and sclerotherapy, over all other modalities of treatment for VV.

14.9.2 History of the evolution of polidocanol

Soon after the development of POL in 1931, it made its initial appearance as a local anesthetic, but this did not last long due to its very poor performance in this field. Following this, it reappeared in the field of cosmetics, such as shampoos, hair conditioners, body lotions, and facial creams from 1970 because of its remarkable nonionizing, emulsifying, and surfactant qualities. It was used in concentrations of 3–4%. This helped the product to undergo extensive evaluation regarding the sensitivity reactions on skin, toxicity, and bodily absorption. Endovenously it was used in the late 1960s, but it was only in the 1990s that it attained popularity in this area. Though a plain solution was used initially for endovenous ablations, by the beginning of 21st century various manual techniques were introduced and attained new area of use in the field of sclerotherapy [310]. On March 30, 2010, the FDA announced the approval of the POL product, Asclera Injection, manufactured by Chemische Fabrik Kreussler & Company of Wiesbaden, Germany. Following the approval of POL by the FDA various manufacturers started producing automatic foam-generating mechanisms (Verisolve, Varithena). The most recent addition is Clarivein (2014).

14.9.3 Physical properties

It is a viscous liquid at room temperature, with a melting point of 15–21°C, and it is miscible in water. The pH is 6.0–8.0 (1% solution in water) and it has a density of 0.97 g/cm^3 at room temperature, which is very close to that of water. It is a surfactant and when agitated with normal atmospheric air, CO_2, O_2, or N_2O it becomes a very stable microfoam. Topical applications in different forms have proved that it is nonirritant and does not cause any sensitization.

14.9.4 Chemical properties

POL is alkyl polyglycol ether of lauryl alcohol. It is also known as "Laureth-9." It is chemically an alcohol ethoxylate with an average alkyl

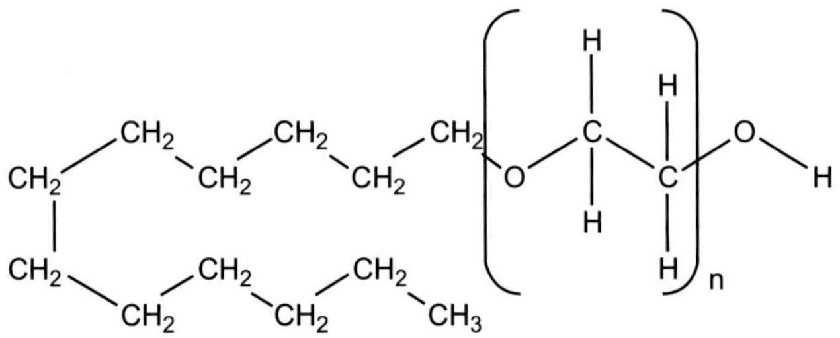

Figure 14.2 Chemical structure of polidocanol.

chain of 12–14 carbon atoms (C_{12}–C_{14}) and an ethylene oxide chain of nine ethylene oxides. Its average empirical formula is C_{30} H_{62} O_{10}, and its structural formula is as shown in Fig. 14.2.

14.9.5 Mechanism of action

The selective endotheliolytic action of POL on the inner lining of veins is a specific characteristic and this property qualifies it for its uniqueness. Cellular calcium signaling and nitric oxide pathways become activated and cellular death occurs. The timing of endothelial cell death is predictable and is based on the sclerosant concentration. There is a potential danger of blood cell destruction in concentrations above 0.5% [311]. However, in lower concentrations it is very safe. There is a homogeneous distribution of the drug within the vein lumen. The sluggishness of blood flow in the vein, along with the gravitational backward thrust, allows better contact in a very short time on the inner wall of the vein. When the microfoam is injected forcibly into the vein it is mixed with plasma. With albumin there is a further accentuation of foam formation and increased stability of the foam which further potentiates the sclerosant action.

The drug owes the peculiarity of a centrifugal retrograde flow within the veins and the superficial venous plexuses which allows spontaneous infusion into the smallest veins of $>3\ mm^2$. This remains the greatest advantage for treatment of CVD. Even in ultrasound guided foam sclerotherapy (UGFS), while injecting the drug through the cannula, the risk of POL entering the deep vein and development of subsequent deep vein thrombosis (DVT) and venous thromboembolism (VTE) is uncommon due to this unique characteristic.

With regard to concentration of the drug for infusion, an optimum universal concentration is observed at 0.2%. As against the current belief, there is absolutely no necessity to change the concentration according to the size of vein. This helps to cover larger areas, lower dosage, and increased safety, and also helps to avoid repeated sessions in extensive varicosities. The same concentration is equally effective in treating congenital venous malformations, AV fistulae, and Klippel–Trénaunay syndrome (KTS).

14.9.6 Metabolism

Once POL comes into contact with the endothelial layer of vein, a chemical reaction is set up very quickly, which results in cellular destruction and splitting of the ether bondage as well as oxidation of the alkyl chain. Lower molecular weight polyethylene glycol-like compounds, carbon dioxide, and water are thus formed. The polyethylene glycol-like substances are excreted by the kidneys and a large volume of CO_2 through the lungs. This metabolic process does not produce any known biologically active compounds. Almost 75% of the metabolites are eliminated through urine, 5% through feces, and 4% through the lungs within 24 hours. Within a week almost 99% is cleared. The rapid inactivation of the POL in circulating blood adds to its high safety profile [312].

14.9.7 Foam creation

POL was initially used as a plain solution in variable strengths of 0.5%, 1.00%, and 3%. The disadvantage of a plain solution was mainly dilution with the blood, whereby the sclerosing efficiency became unpredictable. Larger doses of drug were therefore required for a predictable response, which may lead to toxicity. A safe dosage would require several weekly interventions, which is most uncomfortable for patient and doctor alike.

Because POL is a detergent, studies were routed to make foamy solutions, so that the volume can be expanded enormously. The principle of surface tension is made use of by mixing a surfactant to water and a gas. Soap bubbles have very large surface areas with very little mass. Bubbles in pure water are unstable. Therefore, as soon as this mixture of POL is injected into the vein, instead of getting diluted with blood these bubbles are attracted to the cell wall and an immediate chemical burn is produced on the intimal surface of the vein. This chemical reaction is stable enough to produce fibrosis later. Moreover, because of the specific affinity of POL to the vein intima, there is minimal exposure of the drug to blood cells within the lumen. Therefore no remarkable changes happen to the blood. There is no change in the pH of the blood. These are additional advantages of the drug mixture. The pH of the POL solution and water match, and the diluted solution remains isotonic with blood.

Water is added to POL to obtain the desired dilution. To make the foam, different gases are used. Normal atmospheric air is commonly used. Other gases currently in use are CO_2, O_2, and N_2O. Different methods are also used to create the foam, some are manual and some are with the help of mechanical devices. Lots of papers on different gases, methods of foam formation, and concentration of the POL mixture have been published over the last few years. All the above mentioned methods are author specific and their clinical experiences are very limited.

14.9.8 Method of administration

14.9.8.1 Manual methods

For nearly 70 years, several contributors worked hard to obtain an efficient foam. The first attempt at making sclerosant foam was in 1944 by Orbach, by mixing a detergent sclerosant and air in a glass syringe and shaking. The foam thus created was larger in size and hence it had poor performance. However, it was several times more effective than plain sclerotherapy, and is known as the air block technique. Subsequently, Cabrera et al., in 1997, made the foam using CO_2 and it was known as a tensio-active agent. In 1997, Monfreux described the Monfreux Ultrasound Guided Sclerotherapy (MUS) method that generated simple foam with air by means of a glass syringe. Mingo-Garcia developed a special device to produce foam with compressed air in 1998. Benigni and Sadoun published a method to produce very short-lasting foam in a plastic syringe, and in 1999 Lorenzo Tessari presented an original method of foam production with two disposable syringes and a three-way tap. In 2000 Frullini, following on from Tissari, developed a different method that generated foam of sclerosing solution in a vial with a rubber cap. Tissari's technique is the one of the most popular, and is a very simple and effective method. Normal atmospheric air is commonly used and offers excellent sclerosant effect and has minimal complications, however CO_2 is preferred by some phlebologists [313].

14.9.8.2 Mechanical methods

Following the approval of POL by the FDA, newer gadgets made their appearance.

Varithena is a proprietary foam-generating device containing POL 1%, and was the first FDA-approved foam for great saphenous vein (GSV) system incompetence. The ratio of gas mixed is $O_2:CO_2$ (65:35) gas mixture with <0.8% nitrogen. It delivers reliably small bubbles (median diameter <100 μm; all ≤500 μm). The concentration of delivered sclerosant is 0.13% (1.3 mg POL per mL foam). This cohesive foam fills the lumen for circumferential contact. It is designed to displace blood to facilitate efficient sclerosis.

Varisolve POL microfoam is a pharmaceutical form of microfoam that emulates the foam in a standardized way. Foam is generated and dispensed using a proprietary pressurized canister mechanism. The system contains the liquid agent and a gas mixture of oxygen and carbon dioxide with only a trace (0.01%–0.08%) amount of nitrogen present. Passage of the gas and liquid under pressure through a microfoam-producing mechanism yields a microstructurally consistent 1% POL microfoam with reproducible rheological properties. Foam properties including stability are in general sensitive to the materials and techniques used to produce foam. In particular, the bubble size for Varisolve foam is appreciably smaller than that resulting from manual foam production techniques, and the absence of nitrogen facilitates more rapid absorption of bubbles within the body. Both these considerations are important to the safety profile, given the potential for gas embolism to occur with any type of foam sclerosant therapy.

Clarivein is a new addition, which helps the administration SF more comfortably without much pain anesthesia with the help of a mechanical device as a substitute to ultrasound guided (USG) which is gaining popularity among phlebologists for the ablation of the GSV.

An investigation into the influence of various gases and concentrations of sclerosants on foam stability has proved that the half-life varied with different gases used for foam formation according to the sclerosant concentration when atmospheric air, O_2, or a mixture of CO_2 and O_2 was used, but not when CO_2 was used. Room air foam is more than three times as stable as CO_2 foam and 1.5 times as stable as a mixture of CO_2 and O_2. The foam half-lives for atmospheric air and O_2 are similar at low concentrations of STS but differ at higher concentrations [314].

A review of the published literature has focused on a few variables of the material and methods used to produce SF with Tissari's method and to inject SF. In SF production, differences in gas components, liquid-to-gas ratio, as well in disposable material can variably influence the resulting SF. Similarly, SF injection through ultrasound guidance, with a needle or with short/long catheter may exhibit different foam behaviors according to the variable material and techniques which are employed. More recently the introduction of long catheters, possibly together with hook phlebectomy, seems to potentiate the short-/mid-term outcomes of foam sclerotherapy.

14.9.8.3 Strength of sclerosant

To suit the preferences of various users, POL is available in varying concentrations from 0.25% to 3%. It is available plain (Asclera, Asklerol) and as a foam (Varithena, Varisolve). Most phlebologists prefer using variable increasing strengths proportionate to the size of veins. The higher the concentration the proportionately greater the chance of VTE due to damage to blood cells and subsequent microparticle formation.

14.9.9 Indications

It has a wide spectrum of indications in addition to VV of the lower extremities. It is equally effective in treating hemorrhoids, esophageal varices, congenital venous disorders, arteriovenous fistula, and hemangiomas and lymphangiomas in different parts or organs of the body, KTS, vulvovaginal varicosities, varicoceles of the testes, and ganglions.

14.9.10 Safety, efficacy, and tolerability

POL is a gifted drug. Many studies and reviews have appeared recently to evaluate its safety, efficacy, and tolerability. All these studies conclude with excellent remarks with very limited limitations. It has taken the upper hand over all other known sclerosants. Though there are different modalities of treatment for management of CVD, sclerotherapy has attained newer indications to ablate the GSV, especially with the evolution of Clarivein [307].

POL is extremely well tolerated in aqueous as well as foamed forms. There are absolute and relative contraindications. The absolute contraindications are known allergy, presence of acute superficial or DVT, advanced peripheral arterial occlusive disease, first trimester and late pregnancy, and prolonged immobility. Relative contraindications include longstanding history of leg edema, hypercoagulable state, late-stage diabetic complications, and asthma. Most of these are not specific to POL but common to many other situations. The overall profile of POL qualifies it as a wonder drug for the management of CVD.

There are certain complications which can arise immediately during the procedure or later. The treating physician has to be well aware of these possibilities, most of which are very trivial. Though extremely rare, serious complications do occur.

Complications can be immediate or late. The immediate complications are mainly pain during infusion of POL of variable nature and are mostly subjective. Pain is experienced when injecting into the smaller veins and never in the larger veins, and is tolerable and transient. Neurogenic shock may occur and is seen in younger patients due to apprehension or pain. Some may rarely have a sense of oppression in the chest or coughing or visual disturbances which are also transient and self-limiting. Migraine is also occasionally reported. Prophylaxis with aminaphtone, an oral antiendothelin drug, has been found to be very effective. This is more common in patients who had a history of migraine. Hypersensitivity to the drug is extremely rare, although the possibility should not be underestimated. Urticarial rashes may appear due to histamine release due to tissue injury, but is theoretical. Extravasation and intraarterial injections are procedural complications and not related to the drug [315].

Late complications include cutaneous skin necrosis, particularly over the lowermost part of the leg, ankle, or foot where there is a conglomeration of very tiny veins present. Patchy areas of skin loss can appear around the ankle region and foot. These will heal over a few weeks. There will be induration over the veins which can last for a few weeks or months and then settle down spontaneously. There will be pigmentation and occasionally itching over the line of the vein due to deposition of hemosiderin which will also subside over a period of time. Hypertrichosis along the line of the sclerosed vein is not too uncommon. Air embolism is quite uncommon but neurological problems have been reported in patients with patent ductus arteriosus (PDA). Fatal air embolism can occur only if more than 100 mL of air is injected into the venous system at rates greater than 100 mL/s. The most important of all the complications is VTE, but it is the rarest [315].

Most of the complications can be controlled by regulating the dosage, volume, concentration, and also good know-how.

All the reviewed reports added to my vast experience clearly support the excellent profile of POL. No drug can match POL in safety and tolerability. There were no remarkable complications or fatal outcomes. POL is a wonder drug and, if used ingeniously, there is no drug or modality that can replace foam sclerotherapy (FS) with POL. The specific retrograde flow and the centrifugal distribution of the drug are noteworthy. This feature adds to its efficacy and safety over all other sclerosants. The optimal sclerosant effect is observed at a lower universal dilution of 0.2% and not on increasing concentrations proportionate to the size of veins.

14.10 Chromated glycerine (Sclermo)

This is a new sclerosant which is very popular in Europe. However, it has not yet been approved by the FDA. The sclerosant effect is comparatively less and it is hence selectively used only for small veins. It does not produce tissue staining or tissue necrosis when extravasated. However, because of its high viscosity it is very painful for the patient. It is highly allergenic and there are incidences reported of having produced ureteric colic and hematuria. Hence it is not at all a preferred drug for the patient or surgeon.

14.10.1 Selection of technique

Different methodologies are being adopted by different surgeons who claim different success rates. However, all the evaluation studies so far verified basically lack one thing; in addition to the inadequacy of a sufficient period of clinical follow-up, they lack reasonable experience and expertise. Most of the surgeons still use this method as a pure substitute for the other surgical techniques. They consider and categorize this as one of the minimally invasive techniques, however this is not the case in reality.

14.11 Planning for surgery

Before any surgical procedure it is always ideal and reasonable to observe the following steps before offering the procedure:

1. Proper evaluation of the patient with respect to the venous disease and other coexisting diseases;
2. A proper data compilation in a specified format;
3. Cary out relevant investigations, especially a color Doppler study of the venous system.
4. An awareness program for the patient and their close relatives to make them understand their disease, the complications, the various methods of treatment available, the details of the technique followed at the center, the advantages and disadvantages, if any, of the technique to be adopted, possible risks involved in the procedure, and the immediate postprocedural care to be observed by the patient.

5. It is very important to stress the importance of the follow-up which is essential to evaluate and institute further treatment. Preferably a follow-up schedule should be worked out and given to the patient.
6. According to the stage of disease, hospital visits will vary. C2 patients will require scheduled follow-up, whereas C6 patients require frequent visits to the hospital or require admission for grafting.

14.12 Preparation of the patient

1. After clinical and laboratory evaluations, the patient is prepared for the procedure.
2. Usually it is done as a day-care procedure and hence the patient has to report to the hospital at 8 a.m. after a good bath on the day specified.
3. If the patient is on medication, allow him to continue it as usual. Anticoagulants or antiplatelet drugs need not be withdrawn.
4. The procedure is usually done only on one limb. It is preferable to carry out the procedure on the other leg after 3 months.
5. The proposed leg is washed with soap, cleaned, and shaved from the inguinal region to the foot.
6. In the case that there is mild edema of the foot, the leg should be kept elevated until the patient is taken to theater for the procedure.
7. A skin test should be carried out with the sclerosant to rule out any hypersensitivity to the drug, although it is extremely rare.

14.13 Materials for the procedure

This is the least expensive procedure, with no costly equipment required. All that is required is a set of four 5-mL plastic syringes, a three-way connector, a set of 26 gauge hypodermic needles, and the sclerosant. The choice of sclerosant and technique varies according to the preference of the surgeon. Here we describe our time-tested preferred technique and sclerosant. We use only POL (Asklerol) and occasionally STS. The disadvantage of STS is that it is very painful.

14.14 Technique

In various literature sources, it is found that this procedure can be done in the doctor's office. However, this is not advisable. The safety of the patient is the first concern in any surgical procedure. Hence it is always preferred that the procedure should be carried out in a well-equipped operation theater under the supervision of a qualified anesthesiologist in order to monitor and tackle any problems that may arise during the procedure. No anesthesia is required. Adverse reactions are extremely rare, yet one has to be cautious. The technique is described in three stages. The whole procedure may take about 10–15 minutes.

Step 1: Positioning of the patient

The most convenient position for the procedure is to have the patient sit on the operating table with their legs hanging freely down. The limb is very well cleaned with spirit and the feet are placed on a sterile towel kept on the lap of the surgeon.

Step 2: Reassessment of veins and the procedure (Figs. 14.3–14.5)

The surgeon reviews the distribution of veins over the foot, leg, and thigh. This allows him to evaluate the areas to administer the sclerosant. There is no hard and fast rule that one should start working from above or below. This is decided by the overall distribution of veins and their size. Two staff nurses make the foam with air and sclerosant using *Tissari's technique*. The dilution of the solution is variable and is decided by the surgeon based on the size and site of veins. The smaller the veins the greater the dilution is (0.5%), and vice versa. The quantity of sclerosant to be injected is assessed by the appearance of the vein as well as the pressure felt on the injecting fingers of the surgeon. As with any other procedures,

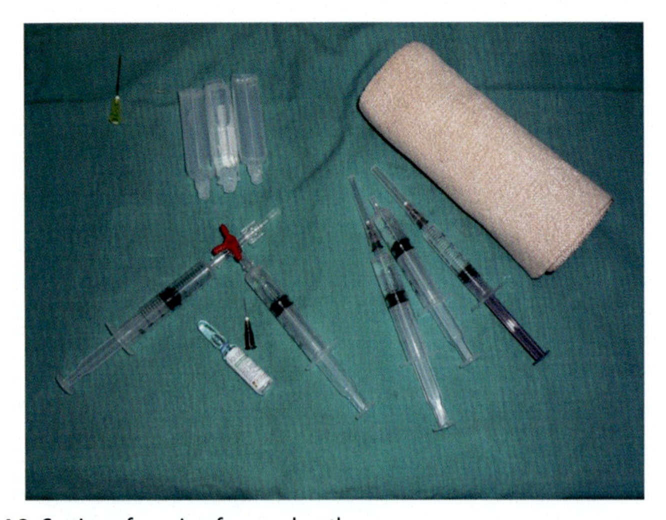

Figure 14.3 Settings for microfoam sclerotherapy.

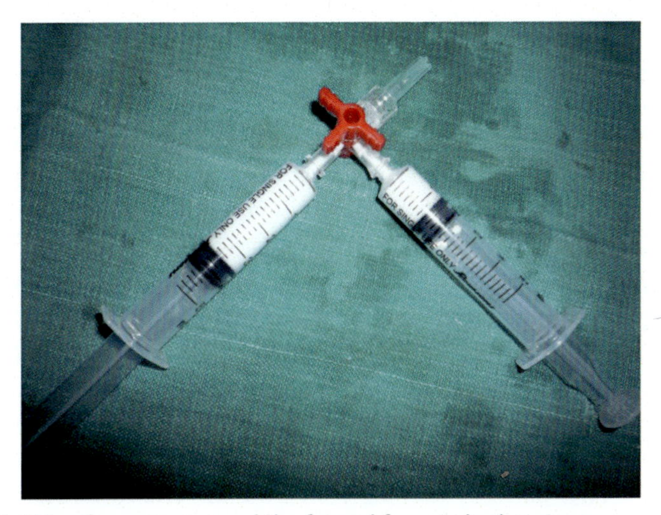

Figure 14.4 The three-way tap and the formed foam in both syringes.

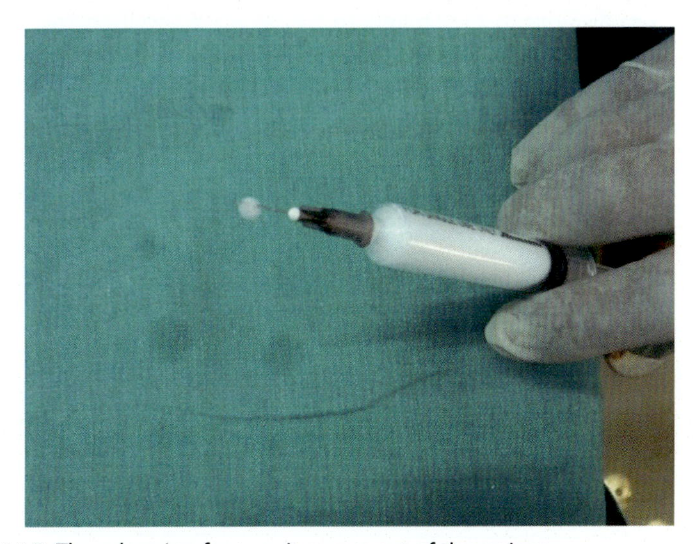

Figure 14.5 The sclerosing foam as it comes out of the syringe.

the surgeon requires good observation, skill, and experience for a very successful outcome. Four nurses are required to keep the puncture sites compressed.

As regards the method of administration of the sclerosant and site of administration, the usual practice is to inject the sclerosant into medium- and large-sized truncal veins and at perforator sites, mimicking the principle of standard surgical procedures such as vein stripping, junctional blocks, or perforator blocks, carried out either by knife, laser, or radiofrequency. Some claim to have achieved greater sophistication with ultrasound-guided foam sclerotherapy and endovenous videoscopic sclerotherapy. With my vast experience in successfully performing MFST in around 34,000 cases, I feel that the current surgical principle is absolutely wrong and it is high time to review and revise the principles of sclerotherapy with a newer outlook about the disease. I perform sclerotherapy in the opposite way, mainly concentrating on the smallest tributaries, especially toward the lowermost levels of the leg, ankle, and foot, and work up toward the larger veins, as far as possible avoiding the administration of the sclerosant too close to the perforators and SF/SP junctions. With my technique the risk of DVT and pulmonary embolism can also be avoided by keeping away from the vicinity of the deep vein, and the therapeutic effectiveness is greatly enhanced to near total perfection when the venous flow is blocked at the smallest tributary level, whereby the output of venous flow to the saphenous system is very well controlled at the initial point. Once the procedure is completed, a compression bandage is applied from below upwards, starting from the foot, excluding toes, toward the thigh to help the vein remain collapsed during the process of endovascular fibrosis achieved by the chemical burns. The bandage should be applied for at least 3 weeks. The foot has to be kept raised above the body pelvic level with the knee fully extended (Figs. 14.6 and 14.7).

Even in extensive venous distributions of very large and small veins, all can be eradicated usually in a single sitting, if properly planned.

Step 3: Immediate postprocedural care

Patients usually remain stable and are shifted to the postoperative recovery room. They are kept under observation for 2 hours. Thereafter they are either discharged or sent back to their respective rooms.

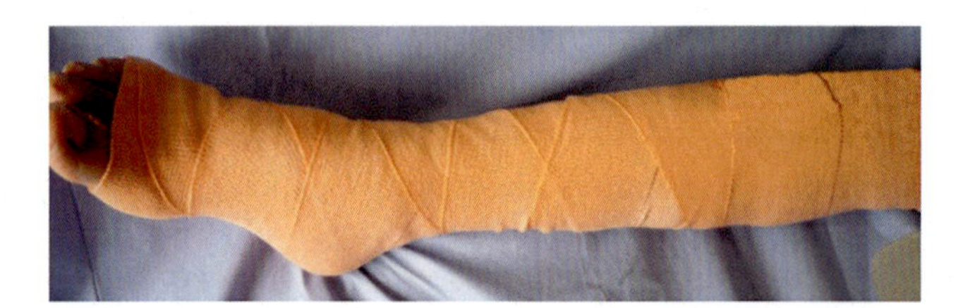

Figure 14.6 Application of the compression bandage after the procedure.

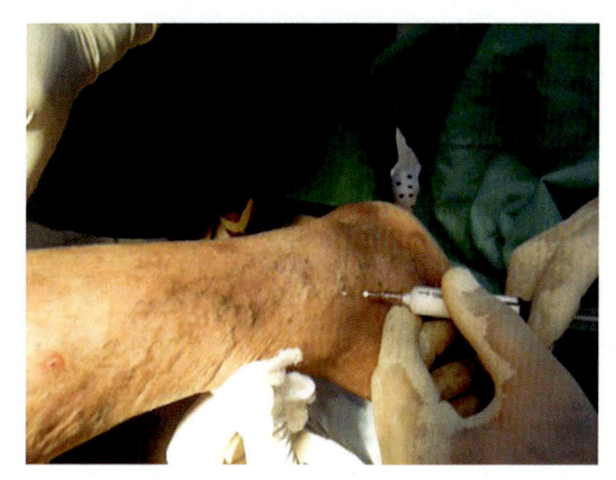

Figure 14.7 Microfoam sclerotherapy procedure.

Step 4: Late postprocedural care

At the time of discharge the patient is provided with a printed discharge summary along with an instruction booklet containing descriptive outcomes following the procedure for the subsequent 3 months, including when they should contact the doctor or when to appear for reviews. A *review schedule* worked out in the hospital is also given to the patients. All the leaflets are printed in the local language as well as in English. Depending on the stage of disease, additional instructions are given to the patients and their relatives regarding any additional care they have to observe by trained nurses. They are also given a scheduled time to personally speak to the doctor on a cell phone and also the email address for any required assistance.

14.15 Retreatment

The main reasons for repeated sclerotherapy are attributable to the following:

1. Neovascularization;
2. Recanalization;
3. Inadequacy of periodic follow-up;
4. Very large number of truncal veins, where the dosage limitation restricts full administration of sclerosant.

The majority of patients have full correction from the first attempt. In a very small group of patients recanalization of the sclerosed veins can happen. In the process of neovascularization some veins may make a superficial appearance which can be controlled easily if the patient reports for review as per the prescribed schedule. Any corrections should be done preferably 3 months after the procedure. In our practice, not more than two sessions are required after the first procedure.

14.16 Complications

Sclerotherapy is definitely superior to any other techniques and is quite safe. However, occasionally, as in any surgical interventions, some complications may occur. These can include the following:

1. Allergy or hypersensitivity to the drugs used
2. Skin necrosis
3. Hyperpigmentation
4. Blister formation
5. Clot retention in large veins
6. Induration over the sclerosed veins
7. DVT
8. Local hirsutism
9. Microembolism

14.16.1 Minor allergic reactions

Very minor allergic reactions are quite common. For example, localized urticaria and edema may occur secondary to histamine release. In the vast majority of cases, such reactions are self-limited, typically resolving in less than 1 hour. Itching often accompanies this response, but it usually resolves by the time the patient leaves the office. Should reactions persist, oral antihistamines or, on rare occasions, steroids may be required.

With the sclerosants used today, anaphylactic reactions are extraordinarily rare but can be life-threatening. The incidence of anaphylaxis with STS or POL is not known accurately but is certainly very low. The reaction is usually mediated by immunoglobulin-E and occurs within minutes of exposure. In some places, the sclerotherapy is performed in the doctor's office or a small clinic. Appropriate emergency measures must be undertaken immediately, including subcutaneous administration of epinephrine, delivery of supplemental oxygen, and securing of the airway. The patient should then be given antihistamines and transferred to an emergency department for continued evaluation and treatment. As noted, a properly stocked emergency response cart, including endotracheal intubation supplies and medications, is essential in any location where sclerotherapy is performed. Periodic review of procedures with staff and maintenance of the emergency medications and supplies is imperative.

14.16.2 Skin necrosis

Skin necrosis occurs with 0.2%–1.2% of sclerotherapy injections. It is a potentially devastating complication and is often unpreventable. Depending on the extent of necrosis, healing may take months. The main causes of necrosis are extravasation of the sclerosant into subcutaneous tissue, inadvertent injection into an arteriole, and vasospasm. Extravasation of the sclerosant can destroy tissue, with the degree of damage determined by the type, concentration, and amount of sclerosant used. Necrosis is rare, when small amounts of dilute (<0.25%) STS or POL are given, but extensive skin and soft-tissue necrosis has been observed when higher concentrations of STS/POL (3%) are administered to treat VV. Inadvertent injections into the arteriole feeding the telangiectasia are impossible to prevent and probably occur frequently. Blanching of the skin often occurs with intraarteriolar injections. Skin massage or, if spasm persists, application of nitroglycerin ointment to the skin may increase microcirculation. Why ulcerations develop in some patients but not others is unknown. The question of whether it is related to injection pressure or injectate volume also remains unanswered.

14.16.3 Hyperpigmentation

This is quite common, occurring in a significant percentage of patients along the course of the vein. It is due to the release of hemosiderin pigments derived from the clotted blood inside the vein which remains stagnant. This hyperpigmentation usually resolves with time, but the process may take months. An unsightly permanent scar is no doubt worse than short-term pigmentation!

14.16.4 Deep vein thrombosis

The precise incidence of DVT after sclerotherapy is unknown but appears to be extremely low overall. The risk is somewhat higher when more concentrated solutions are used or larger volumes administered. Injecting in very close proximity to the deep vein, especially very close to the SF/SP junctions and into the perforators, will greatly increase the incidence of DVT and subsequent pulmonary embolism.

14.16.5 Blister formation

Though relatively uncommon, this happens when sclerosants are injected into clusters of very small veins, especially located over the foot, ankle, and lower leg. This is due to extreme thinning of the vein wall which becomes necrosed when sclerosants are injected. It is very painful on most occasions and takes a few weeks for recovery.

14.16.6 Clot retention

Following sclerotherapy, there will be a lot of blood clots in the larger veins. Usually they remain symptom-free, but occasionally they may become infected. Then there will be severe discomfort and redness over the site. This may necessitate puncture or even evacuation.

14.16.7 Induration

Some pain and swelling over the vein is expected for sclerotherapy in the postprocedural period. The word *sclero* in Greek means *hardening*. This is part of the normal response of the body showing the signs of inflammation. When this is associated with retention of clots and infection, evacuation is required. The induration will usually progress for about 3 months and then will slowly start regressing and over a period of months will fully subside. The smaller ones disappear faster than the larger ones.

14.16.8 Local hirsutism

It has been occasionally observed that some patients develop growth of hair at the area where the sclerosant was administered, however the incidence is comparatively low (Fig. 14.8).

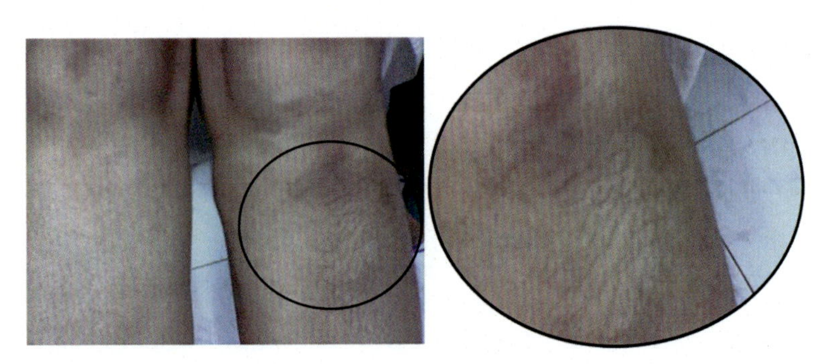

Figure 14.8 Local hirsutism.

14.16.9 Microembolism

Researchers have reported that microembolism of foam particles is common during foam sclerotherapy for VV, and among patients with PFO. These microembolisms could be linked to neurologic symptoms. They present with transient scotomas or a migraine attack immediately after the initiation of foam sclerotherapy. This is particularly noted in patients with PFO and the incidence is reported to be less than 2%. However, they suggest that caution be exercised when foam sclerotherapy is performed in patients with a known PFO, and that patients with overt neurologic symptoms may undergo additional echocardiographic examination to identify a known PFO. Further prospective studies are needed to evaluate and confirm these observations.

14.17 Follow-up

Follow-up after the procedure is very important. Though the procedure is the same for all patients, they usually belong to different classes (clinical, etiological, anatomical, pathophysiological). Because of this, a follow-up schedule has to be worked out for each patient depending on the stage of their disease. The follow-up regime ideally should have two sessions:

1. *Immediate*: In the first 3 months following the procedure, a monthly review can help the patient a great deal. Postprocedure complications such as intravenous thrombosis in the sclerosed veins, which produce some pain, discomfort, and anxiety, can be identified and solved. Therefore any necrosis or bleb formation can be made out early and treated also.
2. *Late*: Following the immediate review sessions, patients should be advised to appear for review as per the schedule given below, to facilitate planning for any residual therapy. Skipped veins can be recanalized or newly opened up veins and newly formed veins by neovascularization can be identified and treated. This would also help to evaluate the improvement after the therapy. Improvement is assessed by progress in the healing of the wound, return of normal color of the skin, and the absence of edema and symptomatic relief of discomfort in the leg. Moreover, these evaluations help to keep clinical records for future improvement and evaluation of the technique.

   The *follow-up schedule* is as follows:

   Once a month × 3 months

   Once in 3 months × 1 year

   Once in 6 months × 2 years

   Once in a year × 3 years

14.18 Ultrasound-guided sclerotherapy and endovenous videoscopic sclerotherapy

Both these techniques aim at sclerosing the perforators, SF and SP junctions as invasive procedures.

14.19 Comments

I would like to add a few comments about USG sclerotherapy and endovenous videoscopic sclerotherapy. These procedures are carried out to mimic EVLT and RFA. Injecting sclerosant very close to the valves or perforator very close to the deep vein will increase the risk of DVT. A retrograde administration is always advisable and never a forward one. The use of a catheter and endovenoscopes is likely to produce more complications, and will also make the treatment cost very expensive.

The following is a summary of the results of the "European Consensus Conference on Foam Sclerotherapy," April 4–6, 2003, Tegernsee, Germany:

1. Foam sclerotherapy is an appropriate method for the treatment of VV. It is a refinement of sclerotherapy with liquid sclerosants featuring better control and a stronger effect.
2. Due to the greater risks in the case of improper use, foam sclerotherapy should be performed only by those who are experienced in sclerotherapy. The patient must be informed about the details of foam sclerotherapy regarding use, efficacy, and side effects.
3. Indications and contraindications basically do not differ from sclerotherapy with liquid sclerosants.
4. In the case of large VV or recurrent VV, the results of foam sclerotherapy are better than those with liquid sclerotherapy. Promising results are also reported in venous malformations. In the case of known symptomatic PFO, special caution should be used.
5. Sclerotherapy of spider veins can be performed sufficiently with liquid sclerosants.
6. When treating large VV, a rather viscous foam should be used. In large VV elevation of the leg is recommended (the "lighter" foam moves "upward," and elevation of the extremity prevents rapid entry into the DVS).
7. Foam sclerotherapy requires fewer injections per session, with larger distances between them. When treating large VV, a single puncture site is often sufficient for administration.

8. Puncture should always be made at the safest and most easily accessible site. The puncture site in the case of truncal VV should be at least 10 cm from the saphenous junction.
9. When sclerosing large VV, irrespective of the concentration, a total amount of 6–8 mL foam per session in the double syringe system technique of Tessari or 4 mL per session in the Monfreux technique should not be exceeded.
10. As foam possesses a more intense sclerosing effect, the treatment goal can be reached using sclerosant of lower concentration as in liquid sclerotherapy.
11. After foam sclerotherapy of large VV, vasospasm occurs in a higher percentage in the sclerosed vein.
12. There is a positive correlation between spasm and good therapeutic results.
13. When performing foam sclerotherapy on the saphenous veins, in the groin, and popliteal cavity, and for the treatment of recurrent VV and of perforating veins, a duplex-guided approach is greatly recommended.
14. Before applying compression therapy, one should wait several minutes to prevent premature displacement of the sclerosing foam into other regions.
15. Side effects of foam sclerotherapy are comparable with those of liquid sclerotherapy.

14.20 Other sclerotherapy indications

Other sclerotherapy indications include:
1. Esophageal varices
2. Hemorrhoids
3. Varicocele
4. Ganglion
5. Hemangioma liver
6. Congenital arterial/venous/lymphatic malformations

CHAPTER 15

Modified microfoam sclerotherapy

- Varicose refers to dilated, elongated, and tortuous veins.
- It can affect the superficial venous system (SVS) anywhere in the body.
- It includes congenital malformations and genetic vein-valve weakness.
- Varicose veins are most commonly formed in the lower extremities.
- These dilated veins usually remain harmless throughout life without producing complications, however, rarely the vein may rupture due to extreme weakness of the wall resulting in severe hemorrhaging, and it can be fatal when it occurs in very large veins.
- Large vein varicosity can occur when the blood flow is very weak and as a result the blood can become clotted and later infected. When described as thrombophlebitis, the blood vessel ruptures, which is the only acute complication related to larger veins.

15.1 Acute complications related to small veins

- When there is edema of the leg and foot it may become infected and produce severe infection, which is called cellulitis. It may extend from the foot as high as the thigh. If not properly treated it may result in severe septicemia, toxemia, and be fatal.
- However, when this complication occurs in very small veins it is never fatal—the blood jets out as if from a syringe with a small needle and the subject becomes frightened and rushes for medical treatment.
- The sequelae of varicose vein of the lower limb include the serious chronic complications which pass through seven stages, clinically classified as 1–6. The description of these is given below. These stages are related to the skin and subcutaneous tissue over and surrounding the ankle joint.
- From the ankle joint it may extend toward the foot and upto the knee, distally toward the foot and proximally toward the lower two-thirds of the leg.
- The upper third of the leg and the thigh are usually spared.

Genesis, Pathophysiology and Management of Venous and Lymphatic Disorders
DOI: https://doi.org/10.1016/B978-0-323-88433-4.00016-4

- The various stages of progression of the disease are attributable to the small veins of the lower leg, ankle, and foot, and not directly attributable to the large veins of the leg including the thigh.

15.2 Relevance of the size of veins in chronic venous disease

The incidence of venous disease globally is estimated at 10%–20% based on racial variations, but this is based only on visible veins, whether large or small. These microscopic veins are labeled as problem creators. The size of the veins and the height from the level of the foot have an important relationship with the morbidity. As the level and size increase, the lower the morbid. In contrast, the smallest veins at the lower most part of the leg have the greatest morbidity. It is surprising that the microscopic veins which forms the major part of the venous networking system are never taken into account in the involvement of chronic venous disease (CVD). Microscopic veins make up more than 85% of the venous system, and remain in the intracellular space upto the capillary level. The size varies from less than 0.008 mm upto 1 mm. Those minor tributaries which are the thickness of cotton wool or thin hair with a light pink color include millions of venules and venulets.

15.3 Nonreturn valves

There are nonreturn valves in the veins from the head to the feet in infancy. However, during the transitional phase from infancy to adulthood, the venous valves get atrophied from head to heart, as the flow is in accordance with the gravitational force. No non-return valves are necessary for the blood to flow from above downwards. From the level of the heart to the feet, the valves are able to withstand gravitational force and the weight of the blood column. The microscopic valves in and around the ankle region are arranged in very close proximity. The number of valves increases progressively toward the foot to withstand the increasing hydrostatic pressure in the veins and to prevent the backflow toward the capillaries.

In the macroscopic veins from the ankle toward the heart, the number of valves is reduced. The major valves in the perforators and junctions are given too much importance in the treatment of venous disease as they do not have any direct relation to the capillaries. If there are many macroscopic veins distributed throughout the lower limb from the feet upto the

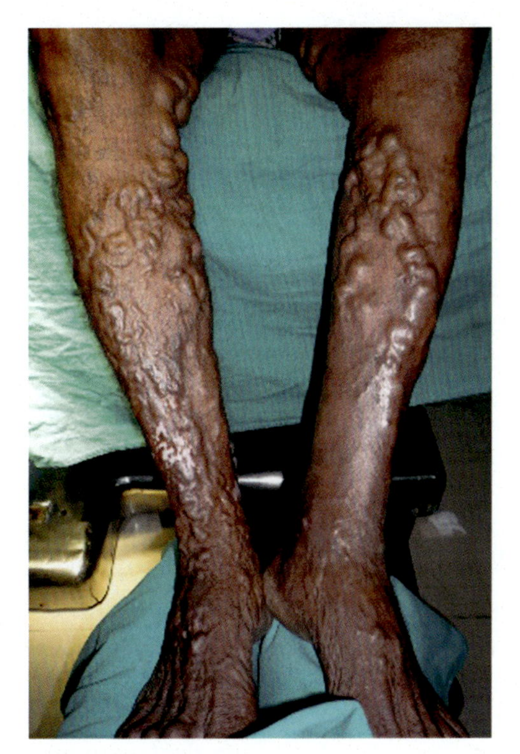

Figure 15.1 Lower limb with large veins.

uppermost level of the thigh, complications of chronic venous disease usually do not develop. The association of tiny veins with large veins may contribute to some extent to the development of CVD by indirectly helping the thrust towards the capillaries due to gravity and incompetent valves. Patients with no macroscopic veins develop severe complications of CVD, even at very young age, supports the importance of microscopic venous valves in the lower leg. The old statement that the gravity of the disease increases as age advances is incorrect—patients with nonhealing ulcers even in their early 20s are proof of this. Again very aged patients with innumerable large veins throughout the lower limb do not have complications, as shown in Fig. 15.1.

15.4 The junctional valves and perforators

Too much stress is placed on the importance of incompetence at the major veins, specifically at the sapheno-femoral, sapheno-popliteal, and a dozen perforators below the knee and half a dozen perforators above the knee. The ablation of these veins at the junctions and perforator levels followed by ablation of truncal veins are considered as the global standard of management. These larger veins are traditionally considered as having severe morbidity and hence all the available contemporary treatments are focused on these veins, whether by cutting, stripping, thermal burns (endovenous laser treatment, radiofrequency ablation), or chemical burns (sclerotherapy). In contrast, the so-called spider veins, reticular veins, and telangiectasia are considered cosmetic, and fortunately these are the start of a large network underneath the skin which is currently considered as the marker of an enormous venous network which creates the real problem of CVD by noting the progressive skin damage from edema, pigmentation, and ulceration, as shown in Figs. 15.2–15.4.

15.5 Relevance of microscopic veins

CVD is a disease affecting the SVS of the lower limbs. The deep venous system is the main drainage system of the lower limb, and is not usually involved except in very rare conditions such as deep vein thrombosis (DVT), deep venous reflux syndrome, or agenesis of venous valves. It is the deep vein that carries the blood from the feet toward the heart in the case of a blockage in the deep vein, and the circulation is maintained through the SVS and its drainage area is confined only to the skin and subcutaneous fat. To be brief, the deep vein drains all the areas below the deep fascia of the leg and the superficial system drains the thin layer of skin and subcutaneous fat outside the deep fascia. There are innumerable interconnecting veins between these two systems and all tissue drainage has access to both systems. The deep vein is supported by the calf muscles in addition to the nonreturn valves. As the heart dilates, there is suction pressure (negative) in the deep vein to pull the blood from the heart. However, the SVS is entirely dependent on the integrity of the vein wall and valves. The high suction pressure in the deep vein sucks the blood into the deep vein from the SVS. If there is any incompetence of the junctional valves, perforators, or truncal veins, the backthrust of blood will not be reflected at the capillary level and tissue function of the integrity of the venules and the venulets is sufficient and no CVD complications manifest. If only microscopic and macroscopic veins are diseased then the large veins give indirect thrust toward the capillaries through microscopic veins.

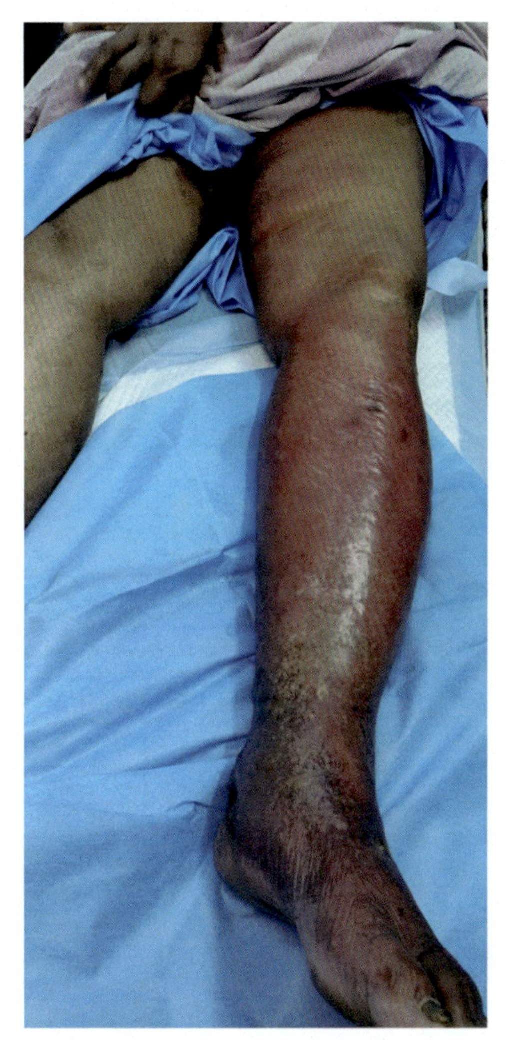

Figure 15.2 Edema.

The above-mentioned newer outlook has changed the treatment modality for the correction of CVD by eliminating the major veins. By this method, a vein with a lower velocity upward flow is completely obstructed and increased venous pressure is reflected in the microscopic veins whereby the so-called cosmetic veins become severely morbid; obstructing the flow in the major superficial veins is nonphysiological as those veins with partial flow are completely obstructed. Partial flow can be resumed without interference. The SVS flows from below upwards

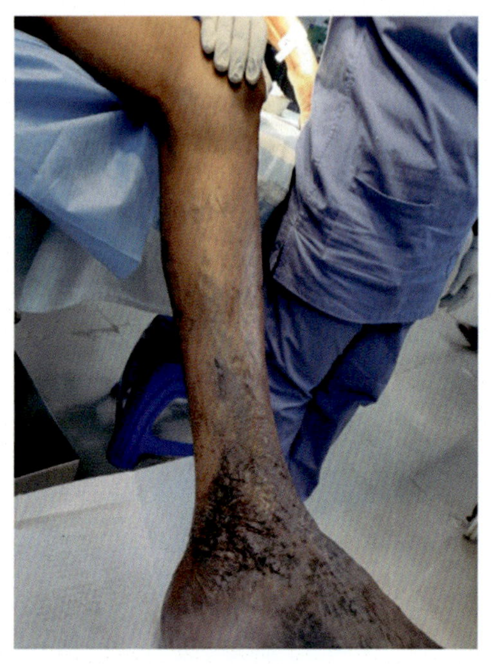

Figure 15.3 Hyperpigmentation.

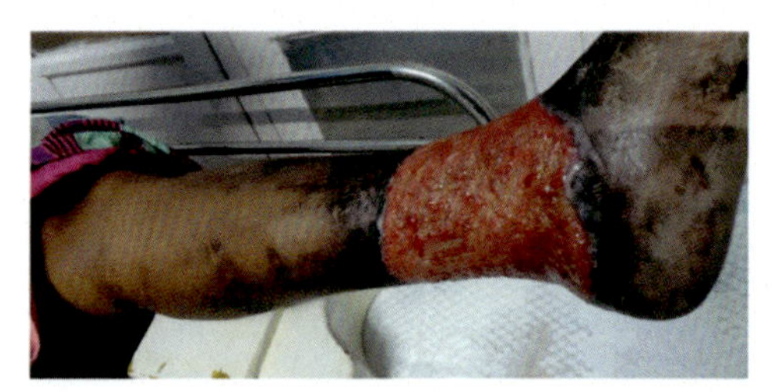

Figure 15.4 Ulcer.

without a proper supportive mechanism, unlike the deep venous system. It is solely dependent on venous valves and the suction pressure of the deep venous system. It is from the feet that these venulets, venules, and tributaries carry blood toward the heart. If the venous plexus at the lower most part of the leg is involved in the disease process, starts eliminating the smallest veins and hidden venous plexuses. This cannot be achieved using any of the conventional techniques. These venous plexuses can only be eliminated with the help of a modified sclerotherapy technique.

15.6 Modified microfoam sclerotherapy—principle

Modified microfoam sclerotherapy (MMFST) is a new technique based on knowledge of the microscopic veins and microcirculatory impediment. In order to prevent backflow toward the capillaries through the SVS, the collecting venous system at the tissue level has to be eliminated by a suitable sclerosant—polidocanol (POL).

The treatment is initiated by elimination of the SVS from the feet toward the leg by injecting microfoam POL through the very fine tributaries. The specific property of POL is made use of in eliminating the minute collecting system. This drug has a centrifugal retrograde distribution from larger to smaller and from smaller to the smallest veins, unlike any other intravenous injections which promptly go from smaller to larger blood vessels, and then to the heart. The details of the drug are available in Chapter 14.

15.6.1 Procedure

The procedure for a leg with microscopic (truncal veins, major and minor tributaries) vessels from the foot to the upper thigh is as follows:

- The procedure starts from the uppermost part of the truncal vein.
- The sclerosant is injected and the point compressed with gauze, and pressure is used by fingers to prevent the drug flowing toward the iliac veins.
- Compression is maintained until the whole procedure is finished.
- One or two additional injections maybe given at the proximal large veins.
- Maintain finger pressure with gauze at the injected site.
- Proceed to sclerose the small tributaries on the foot, followed by around the ankle, and work upward proximally toward the knee.
- Eliminate, all the veins especially the smaller ones.
- Avoid injecting at blow-outs at perforator sites and junctions as it may predispose to DVT.
- Any tampering at junctions and perforator levels will lead to an untoward outcome, depending on the availability of drug.
- Proceed to eliminate those in the thigh.
- If this is not possible, just leave it, as involvement of those veins does not produce any problems for the patient.
- These can be eliminated later if there is recurrence in the leg.

15.6.2 Drug dilution

Twelve hours after intravenous application, about 90% of the POL administered will have been eliminated from the blood. No accumulation is expected, even after repeated doses of POL at intervals, which is normal for sclerotherapy. The daily dose should not exceed 2 mg/kg body weight.

The dose and concentration of the solution in the conventional sclerotherapy technique depend on the size of the varicose vein. The smallest varices are preferably treated with 0.5%–1.0% solution, medium-sized varices with 1.0%–2.0% solution, and large varices with 3.0%–5.0% solution. It is available in the market in 2 mL ampules with each mL containing 30 mg of POL.

In MMFST the dilution is totally different. Variable concentrations based on vein size appear to be unnecessary. In more than 30,000 patients treated we have observed that the concentration does not increase the potency, and a standard dilution of 0.25% POL is very effective in sclerosing small and large veins. This would help to cover a large area without risk, and help to avoid subsequent repeated interventions.

For a normal healthy adult with an average weight of 70 kg, the total quantity of the sclerosant (POL) is 140 mg. The level of drug administered is below the standard quantity, and so the safety margin is very high.

We use 0.25% (universal dilution) irrespective of the size of vein, so that the coverage is very large, and in one sitting in a majority of cases the whole leg can be treated. While completing 30,000 cases, no notable or remarkable complications were observed. A crepe bandage only is applied immediately after sclerotherapy to keep the veins compressed to help the vein lumen adhere during the healing process. This bandage application is maintained for 3 weeks.

The wearing of knee caps, socks, and tight bandages must be avoided to prevent proximal lympho-venous obstruction. The leg has to be kept elevated continuously during the day time, with frequent dorsi-flexion exercises. Elimination of truncal veins alone is not sufficient for the treatment of chronic venous disease but there is elimination of the so-called varicose vein. No equipment or instrument currently available can reach the venous network (minor tributaries, venules, and venulets). There is no treatment for below the tibial malleolus, and the foot is always spared maximum intervention over the veins of the thigh and lower half of the leg.

MMFST is the direct opposite of the conventional method of treatment so far described and practiced throughout the world. MMFST gives excellent results in nonhealing ulcers, especially after various repeated surgical interventions and those left to their own fate. The safety and efficacy are exceptionally high.

CHAPTER 16

Deep vein thrombosis

16.1 Introduction

Deep vein thrombosis (DVT) is a condition in which a blood clot (thrombus) forms in one or more of the deep veins in the body, commonly in the legs (Fig. 16.1). Usually DVT causes pain in the leg, but it can be silent or symptom free and hence is also called the *silent killer*.

Sitting still for a long time, such as when traveling in an aircraft or prolonged immobilization in a plaster cast after orthopedic procedures, or prolonged recumbency following chronic illnesses or surgery, or some conditions that affect the blood coagulation mechanism can contribute to DVT. DVT is a very serious condition as the thrombus that has formed in the deep vein can break loose and be carried to the lungs. However, most of the time it can remain unnoticed and be absorbed spontaneously. Venous thromboembolism (VTE), a multifactorial disease, occurs due to the interactions between the risk factors and predisposing genetic (thrombophilias) or acquired conditions. VTE presents either as clots forming in the deep veins of the limbs, DVT, or as clots embolizing to the lungs and pulmonary embolism (PE). The VTE incidence rate varies according to age. The overall age-adjusted annual incidence rate globally is 130 per 100,000 in males and 110 per 100,000 in females. Incidence rates are much higher in women of childbearing age as compared to men, but in individuals >45 years of age, the scenario is reversed. Although it is

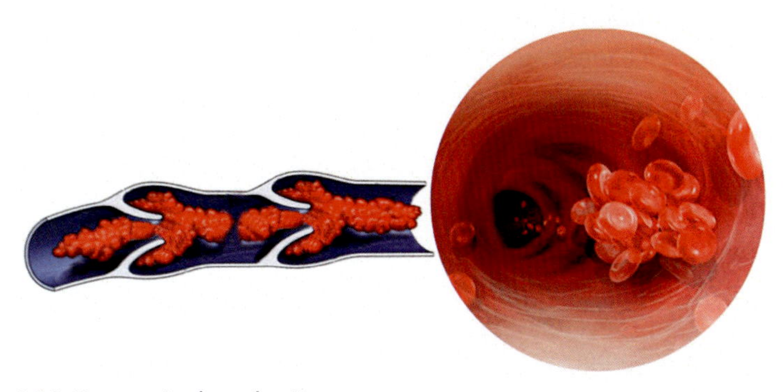

Figure 16.1 Deep vein thrombosis.

Genesis, Pathophysiology and Management of Venous and Lymphatic Disorders
DOI: https://doi.org/10.1016/B978-0-323-88433-4.00005-X

primarily a disease of the elderly, its prevention and management in all other age groups are equally important [316].

16.2 Epidemiology

About 2 million new cases of DVT and 600,000 new cases of PE occur every year in the United States. In India, traditionally, VTE was considered a rarity. However, in the recent past, due to increasing awareness and the availability of modern diagnostic technology, the incidence of VTE has risen. A study spanning 19 Asian centers revealed that DVT occurred in 41% of patients undergoing major joint surgery without thromboprophylaxis. In an epidemiological study, 52% (42% medical and 64% surgical) of 68,183 (55% medical and 45% surgical) inpatients in 358 hospitals across 32 countries were found to be at risk for developing VTE. India contributed 2058 patients (46% medical and 54% surgical), where 54% (45% medical and 61% surgical) of hospitalized patients had risk factors for VTE. Autopsy studies have reported the incidence of VTE in hospitalized patients to be as high as 34.7%, with fatal PE in 9.4% of cases. In a study conducted in South India, it was observed that over a period of 10 years, the incidence of VTE increased from 8 per 10,000 population to 24 per 10,000 populations, with overall incidences being 17.46 per 10,000 population [317]. Similarly, the incidence of DVT in patients undergoing lower limb surgeries in North India was estimated to be 18.13% (33 in 182), while that in rural India was 5.94% (12 in 202) of patients undergoing lower limb orthopedic surgery.

16.3 Pathogenesis

The Virchow's triad explains the relation between stasis, changes in the vessel wall, or changes in the blood, leading to venous thrombosis. Thrombosis is believed to begin at the venous valves; the areas of hypoxia, stasis, and a raised hematocrit providing a procoagulant environment. Abnormalities of the valves further add to the risk.

The antithrombotic proteins, including thrombomodulin and endothelial protein C receptor, are present on the valves and are very sensitive to the environment. Hypoxia and/or inflammation lead to their downregulation and contribute to the initiation of thrombosis. Hypoxia, itself, can lead to the upregulation of various procoagulant factors including tissue factor on the endothelium. Upregulation of P-selectin by hypoxia leads to the recruitment of leukocytes or leukocyte microparticles, containing tissue factor, serving as the nidus for the thrombotic response. Risk of thrombosis increases with increased levels of coagulation factors, particularly factor VIII, von Willebrand factor, factor VII, and prothrombin (factor II). Factor VIII is inherently unstable following activation; hence the need for replenishment in order to obtain a stable thrombus. Prothrombin, besides increasing thrombin generation, acts as an inhibitor of activated protein C anticoagulant activity [318]. The vessel wall damage forms the third and final component of the triad which can occur as a consequence of surgery, trauma, and/or burns. Postinjury, the anticoagulant effect provided by the heparin-like glycoaminoglycans is replaced by a procoagulant state leading to the release of tissue factor and activation of the intrinsic pathway and later the extrinsic pathway and formation of a fibrin clot (Fig. 16.2).

16.4 Symptoms

In a large number of cases DVT may remain symptom free. When signs and symptoms of DVT occur, they include:

1. Swelling in the affected leg, especially in the calf region;

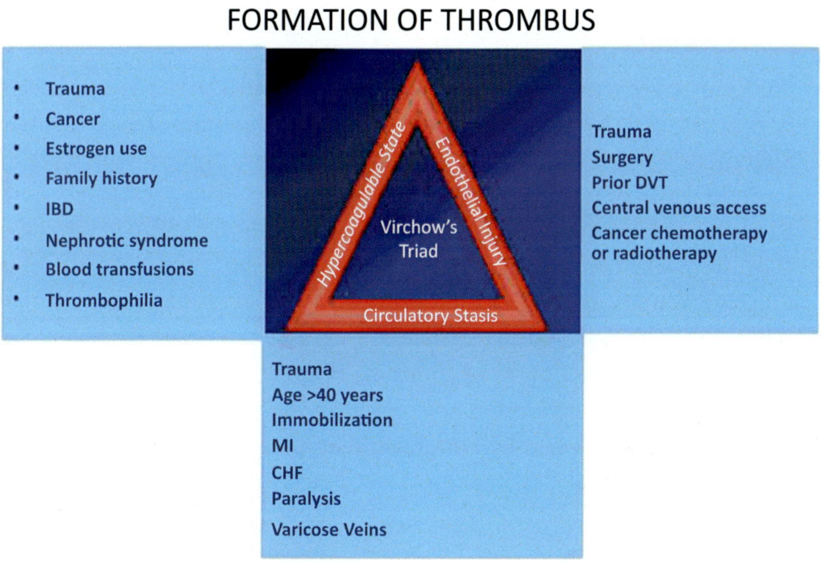

Figure 16.2 Formation of thrombus.

2. Pain in the leg; this can include pain in your ankle and foot. This pain often starts in the calf and may feel like cramping;
3. Redness and warmth over the affected area;
4. Turgid feel to the calf, which is the commonest site;
5. Acute severe chest discomfort and breathlessness (due to PE).

16.5 Signs

In symptomatic cases the signs vary according to the condition, whether acute or chronic. Chronic DVT may be silent or symptomatic.

16.5.1 Acute

Homans sign: Discomfort in the calf muscles on forced dorsiflexion of the foot with the knee straight has been a time-honored sign of DVT. However, Homans sign is neither sensitive nor specific; it is present in less than one-third of patients with confirmed DVT, and is found in more than 50% of patients without DVT. *Pratt's sign* is another clinical test. This is done by applying pressure on the calf muscle. If the calf is turgid and painful, it is indicative of acute DVT. The significance of these tests is becoming reduced due to the availability of MRI and color Doppler scanning.

16.5.2 Chronic

1. Turgidity over the calf when soleal sinuses are affected;
2. Secondary varicose veins of lower limb and sometimes over the lower abdomen;
3. Lymphedema of leg;
4. Pigmentation of the foot and leg;
5. Recurrent ulceration;
6. Chronic venous ulcers.

The warning signs of a PE include:

1. Unexplained sudden onset of shortness of breath;
2. Severe chest pain, discomfort, or oppression in chest;
3. Breathlessness worsened by cough or deep breath;
4. Coughing up blood;
5. Cyanosis;
6. Severe apprehension;
7. Collapse.

16.6 Wells clinical score for deep vein thrombosis

The Wells clinical prediction guide quantifies the pretest probability of DVT. The model enables physicians to reliably stratify their patients into high-, moderate-, or low-risk categories. Combining this with the results of objective testing greatly simplifies the clinical workup of patients with suspected DVT. The Wells clinical prediction guide incorporates risk factors, clinical signs, and the presence or absence of alternative diagnoses.

Clinical parameter score	Score
Active cancer (treatment ongoing, or within 6 months or palliative)	+1
Paralysis or recent plaster immobilization of the lower extremities	+1
Recently bedridden for >3 days or major surgery <4 weeks	+1
Localized tenderness along the distribution of the deep venous system	+1
Entire leg swelling	+1
Calf swelling >3 cm compared with the asymptomatic leg	+1
Pitting edema (greater in the symptomatic leg)	+1
Previous deep vein thrombosis (DVT) documented	+1
Collateral superficial veins (nonvaricose)	+1
Alternative diagnosis (as likely or greater than that of DVT)	−2
Total of above score	
High probability	>3
Moderate probability	1 or 2
Low probability	≤0

Source: Adapted from Ioannidis, J. P. A., Cappelleri, J. C., & Lau, J. (1998). Issues in Comparisons Between Meta-analyses and Large Trials. *JAMA*, *279*(14), 1089–1093. https://doi.org/10.1001/jama.279.14.1089.

16.7 Predisposing factors

Many factors influence the risk of developing DVT, including:

1. *Immobility*: Sitting for long periods of time, such as when driving or flying, may predispose to DVT. When the legs remain still for long periods, the calf muscles fail to contract, which normally helps blood to circulate. Blood clots can form in the calf muscles, especially in the soleal sinuses. Although sitting for long periods is a risk factor, the chance of developing DVT while flying or driving is relatively low. It is sarcastically described as "economy class syndrome."
2. *Recumbency*: Prolonged bed rest, such as during a long hospital stay, or paralysis. When the legs remain still for long periods, the calf muscles do not contract to help the blood circulate, which can result in blood clot development.

3. *Trauma*: Injury to the veins or surgery can slow blood flow, increasing the risk of blood clots. General or spinal anesthesia used during surgery can make the veins dilate, which can increase the risk of stasis of blood and, where there is stasis, thrombosis follows.
4. *Pregnancy*. Pregnancy increases the pressure in the veins in the pelvis and legs. Women with an inherited clotting disorder are especially at risk. The risk of blood clots from pregnancy can continue for up to 6 weeks after delivery.
5. *Malignancy*: Some forms of cancer treatment also increase the risk of blood clots due to changes in coagulation.
6. *Cardiac problems*: People with congestive heart failure (CHF) and acute myocardial infarction are at risk of DVT because a damaged heart does not pump blood as effectively as a normal heart does. This increases the chance of stasis in deep veins and predisposes to clot formation.
7. *Hormones*: Oral contraceptives and hormone replacement therapy can increase the risk.
8. *Intravenous catheters*. These can irritate the blood vessel wall and decrease blood flow, promoting thrombus formation.
9. *Recurrent DVT*: A history of DVT may increase the risk in the future.
10. *Hematologic*: Polycythemia rubra vera, thrombocytosis, inherited disorders of coagulation or fibrinolysis, antithrombin-III deficiency, protein C deficiency, protein S deficiency, prothrombin 20210A mutation, factor V Leiden, dysfibrinogenemias, and disorders of plasminogen activation.
11. *Vasculitis*: Systemic lupus erythematosus and the lupus anticoagulant, Behçet syndrome, homocystinuria.
12. *Hereditary*: With a family history of DVT or PE, the risk of developing DVT is increased.
13. *Obesity*: Being overweight increases the pressure in the veins in the pelvis and legs.
14. *Smoking*: Smoking affects blood clotting and circulation, which can increase the risk of DVT.
15. *Old age*: The chances of DVT are comparatively high after the age of 60.
16. *Varicose veins*: Clots may form due to sluggish flow through the superficial veins.
17. *Venous surgical procedures*: There is a chance of DVT but it is extremely rare.

18. *Heparin therapy*: Heparin-induced thrombocytopenia (HIT) is not infrequent. In this condition, platelet aggregation induced by heparin may trigger venous or arterial thrombosis with significant morbidity and mortality. Unfortunately, it is not possible to predict which subset of patients will develop thrombosis.

16.8 Other provoking risk factors

The provoking risk factors include malignancy, surgery, trauma, fracture, immobilization, pregnancy and puerperium, prolonged travel, hospitalization, catheterization, and acute infection [319].

16.8.1 Measures to prevent bleeding

- Risk categorization: low, moderate or high [319];
- Extremes of age: higher risk of bleeding;
- Liver or renal diseases: higher risk of bleeding;
- Gastrointestinal (GI) bleeding;
- Acute upper GI bleeding is common and potentially life-threatening;
- All patients need to undergo endoscopy to diagnose, assess, and treat any underlying lesion;
- Prevention: drink plenty of water; review use of aspirin; avoid straining at stool (use laxatives); evaluate for polyps; avoid movement for an hour after food to prevent acid reflux;
- Give proton pump inhibitors for ulcers, treatment for *Helicobacter pylori*, avoid concomitant salicylates and/or nonsteroidal anti-inflammatory drugs (if absolutely necessary, add Proton pump inhibitors).

16.8.2 Methods to control bleeding from life-threatening traumatic wounds

- Cold water rinse;
- Pressure point compression;
- Manual pressure;
- Tourniquets;
- Hemostatic dressing;
- Elastic adhesive dressing;
- Suturing;
- Electrocautery;
- Topical thrombin powder;
- Tranexamic acid mouth rinse 5%;

Table 16.1 Thrombus locations.

Location	%
• Distal veins	40%
• Popliteal veins	16%
• Femoral veins	20%
• Common femoral veins	20%
• Iliac veins	4%

- Aminocaproic acid mouth rinse 5%;
- Avoiding additional risk factors for 24 hours;
- Diagnosis of VTE (Table 16.1).

16.9 Complications

16.9.1 Pulmonary embolism

The most serious complication of DVT is PE. A PE occurs when the pulmonary artery is blocked by a blood clot (thrombus) which becomes dislodged and travels to the lungs from another part of the body, usually from the veins of the legs, and is highly fatal. The mortality from DVT is very unpredictable as the patient may not have sufficient time for any emergency care to be instituted.

As PE is highly fatal, it is important to be on the lookout for signs and symptoms of a PE which requires emergency treatment. Signs and symptoms of a PE include:

- Sudden unexplained shortness of breath.
- Chest pain or discomfort. This pain or discomfort is usually aggravated when taking a deep breath or when the patient coughs.
- Feeling light-headed or dizzy, or fainting.
- Coughing up blood.
- Restlessness.
- Cyanosis.
- Collapse.

16.9.2 Postthrombotic syndrome

A common complication that can occur after DVT is a condition characterized by persistent calf symptoms called postthrombotic syndrome (PTS), which occurs in less than 10% of patients. This syndrome is used to describe a collection of signs and symptoms, including:

1. Chronic leg pain;

2. Chronic lymphedema of legs;
3. Skin pigmentation and discoloration;
4. Secondary varicose veins, especially over the abdominal wall along the leg varicosities;
5. Chronic ulceration of leg and foot.

PTS is a common complication of DVT. After an episode of proximal DVT, 20%–50% of patients develop PTS and 5%–10% of patients develop severe PTS within the first 2 years. The more proximal the DVT, the higher the risk of PTS, and the risk increases with the recurrence of ipsilateral DVT. The other risk factors include obesity, lack of physical activity, and inherent thrombophilia. Numerous systems have been introduced to aid diagnosis and to grade severity; including the clinical, etiology, anatomic, and pathophysiology classification; the Widmer classification; the venous clinical severity; Brandjes scale and the Prandoni–Villalta score. The Prandoni–Villalta scale correlates well with the other classification systems, has a good interobserver reproducibility, correlates with the quality-of-life scores, and is the most frequently used aid [320]. The diagnosis, however, is made clinically on the basis of the presence and severity of the characteristic signs and symptoms of venous insufficiency in the leg ipsilateral to the DVT. Management of PTS is primarily prophylactic. The key strategies are to prevent the occurrence of DVT by using appropriate and adequate thromboprophylaxis in surgical and medical patients at high risk; to prevent the occurrence of PTS post-DVT by the use of compression stockings, exercises, and in select cases, catheter-directed thrombolysis (CDT) of liofemoral DVT, if the risk of bleeding is low; and to prevent ipsilateral DVT recurrence by ensuring an optimal duration of anticoagulation. A recent metaanalysis suggested that asymptomatic DVT was associated with a 58% increased risk of developing late PTS [RR = 1.58; 95% confidence interval (CI), 1.24–2.02; $P < .0005$] various surgical methods, namely, ablation of superficial venous reflux, valve reconstruction, repair, transportation, or transplantation, have been suggested. However, their long-term efficacy is questioned [321].

16.10 Investigations

In addition to the routine investigations and coagulation assessments, the following tests have to be carried out.

1. *Color Doppler scan*: The most important specific investigation is color Doppler scanning of the leg. This will give the correct information of the site as well as the extent of the clot.

2. *CT or MRI scans*: These can provide visual images of the veins and may show the clot. Sometimes a clot is found when these scans are performed for other reasons.
3. *Venogram*: This test is used much less frequently because currently less invasive studies can usually confirm the diagnosis.
4. *Blood test*: Almost all people who develop severe DVT have an elevated blood level of a clot-dissolving substance called *D-dimer*. This is a rapid, low-cost, and simple peripheral blood test with good accuracy and reproducibility providing a valuable tool for physicians to rule out the presence of DVT, as well as reducing the need for more time-consuming and expensive testing. Testing for plasma *D-dimer* in the ambulatory care or emergency setting has emerged as an excellent noninvasive triage test for patients with suspected DVT. The test has a high negative predictive value for DVT when used properly.
   The test for increased *D-dimer* in the blood is most useful for ruling out DVT or for identifying people at risk of recurrence. Unfortunately, the test can be positive in a number of conditions, such as recent surgery or pregnancy. Hence a positive test is never reliable, but a negative test can definitely rule out the incidence of DVT, a PE, or any VTE.
5. *Factor V Leiden*: Some people inherit a disorder that makes their blood clot very quickly. This inherited condition may not cause problems unless combined with one or more other risk factors. Factor V Leiden (sometimes written as *factor* V_{Leiden}) is the name given to a variant of human factor V that causes hypercoagulability.
6. *Other blood tests*: Protein S, protein C, antithrombin III, factor V Leiden, prothrombin 20210A mutation, antiphospholipid antibodies, and homocysteine levels can be measured.
7. *Nuclear medicine imaging studies*: Nuclear medicine studies with I^{125}-labeled fibrinogen are no longer recommended for patients in the emergency department. They are relatively insensitive for proximal vein thrombosis and take longer than 24 hours to obtain results. I^{125}-labeled fibrinogen is no longer available.
8. *Ventilation–perfusion (V/Q) scan*: A ventilation–perfusion lung scan, also called a V/Q scan, is a type of medical imaging using scintigraphy and medical isotopes to evaluate the circulation of air and blood within a patient's lungs, in order to determine the ventilation/perfusion ratio. The ventilation part of the test looks at the ability of air to reach all parts of the lungs, while the perfusion part evaluates how

well blood circulates within the lungs. The test is most commonly done in order to check for the presence of a blood clot or abnormal blood flow inside the lungs (such as PE), although computed tomography with radiocontrast is now more commonly used for this purpose. The V/Q scan may be used in some circumstances where radiocontrast would be inappropriate, as in renal failure. A V/Q lung scan may be performed in the case of serious lung disorders such as chronic obstructive pulmonary disease or pneumonia, as well as a lung performance quantification tool pre- and postlung lobectomy surgery.

16.11 Differential diagnosis

1. Achilles tendonitis or myofascitis;
2. Arterial insufficiency (ischemic pain due to thromboangitis obliterans, atherosclerosis, or other occlusive arterial diseases);
3. Arthritis;
4. Asymmetric peripheral edema secondary to CHF, liver disease, renal failure, or nephrotic syndrome;
5. Cellulitis/lymphangitis;
6. Extrinsic compression of iliac vein secondary to tumor, hematoma, or abscess;
7. Hematoma;
8. Lymphedema;
9. Muscle or soft-tissue injury;
10. Neurogenic pain;
11. Postphlebitic syndrome;
12. Stress fractures or other bony lesions;
13. Superficial thrombophlebitis in association with or without varicose veins.

16.12 Treatment

The goal of treatment is to achieve the following:

1. To prevent mortality and morbidity;
2. To prevent the clot spreading up the vein and growing;
3. To prevent a large embolus breaking off and producing PE;
4. To reduce the risk of development of PTS;
5. To reduce the risk of a further DVT in the future.

As it is a medical emergency, it requires immediate attention in an emergency medical care unit to prevent PE by initiating anticoagulation therapy and other supportive measures.

The current guidelines recommend short-term anticoagulation with low molecular—weight heparin (LMWH) SC (Grade 1A), unfractionated heparin (UFH) IV (Grade 1A), fixed-dose UFH SC (Grade 1A), or Fondaparinux SC (Grade 1A). Initial treatment with LMWH, UFH, or Fondaparinux should continue for at least 5 days and until the international normalized ratio (INR) is >2 for 24 hours (Grade 1C). A vitamin K antagonist (VKA) such as warfarin should be initiated together with LMWH, UFH, or Fondaparinux on the first treatment day (Grade 1A).

16.13 Anticoagulation therapy (thrombolysis)

16.13.1 Thrombolysis

Despite effective anticoagulation, only 20%, 39%, and 58% of the patients have a normal compressed ultrasonography (USG) at 3 months, 6 months, and 1 year, respectively. Residual thrombus has been implicated in VTE recurrence and the occurrence of PTS. Evidence suggests that systemic thrombolysis can reduce PTS but at the expense of an increased risk of major bleeding. Hence, thrombolysis is not recommended, various endovascular interventions have been tested in patients with iliofemoral DVT, including CDT and/or thrombectomy. It has been observed that CDT reduced PTS at 24 months with a number needed to treat of 7 in a randomized trial of 209 patients with ileofemoral DVT. At present, CDT is referred for patients with ileofemoral DVT and/or limb-threatening circulatory compromise, acute or subacute symptoms, and a low risk of bleeding [322].

Endovascular thrombectomy and stenting have demonstrated a reduced recurrence rate and PTS at 30 months compared with anticoagulation in a randomized trial involving 169 patients with proximal DVT [323], whether pharmacomechanical CDT in reality reduces recurrence and PTS in patients with proximal DVT is yet to be answered.

Systemic thrombolysis is recommended for high-risk PE patients presenting with hypotension and with a low bleeding risk. Utilizing a low-dose CDT with or without USG guidance minimizes the risk of bleeding with systemic thrombolysis [324]. CDT is recommended in acute iliofemoral DVT, in patients with symptoms of <14 days, reasonable life expectancy, and good functional status, provided the expertise and resources are available. Once successful, the same intensity and duration of anticoagulant therapy as for comparable patients who do not undergo CDT is recommended.

Removal of clots using a mechanical thrombectomy device may be considered as an option in patients with ileofemoral DVT with symptoms for <7 days, a good functional status, life expectancy of >1 year, and subject to the availability of resources and expertise.

16.14 Venous thromboembolism and management in special populations

The management of anticoagulation in patients undergoing surgical procedures is challenging because interrupting anticoagulation for a procedure increases the risk of thromboembolism. At the same time, surgery and invasive procedures have associated bleeding risks that are increased by the anticoagulants to prevent thromboembolism.

16.15 Management of venous thromboembolism in surgical patients

According to the 8th and 9th American College of Chest Physicians (ACCP) evidence-based clinical practice guidelines:

- *High risk for thromboembolism*: bridging anticoagulation with parental anticoagulants.
- *Moderate risk for thromboembolism*: bridging anticoagulation, on a case-by-case basis, with parenteral anticoagulants.
- *Low risk for thromboembolism*: low-dose parenteral anticoagulants or no bridging therapy.
- After an acute episode of VTE, defer surgery until patients have received at least 1 month, and preferably 3 months, of anticoagulation.
- If surgery must be performed within 1 month of an acute VTE, administer IV UFH while the INR is <2.
- If surgery must be performed within 2 weeks after an acute VTE, IV heparin may be withheld 6 hours preoperatively and 12 hours postoperatively, if the surgery is brief.
- If the acute event was within 2 weeks of major surgery and/or patients have a high risk of postoperative bleeding, an Inferior Vena Cava (IVC) filter should be inserted preoperatively or intraoperatively.
- Warfarin must be withheld for four doses if an episode of VTE occurred 1–3 months before surgery.

- If the patient was anticoagulated for >3 months, five doses of warfarin should be withheld before surgery.

16.15.1 Preoperatively

- Subcutaneous UFH or LMWH should be administered in immobilized inpatients with an INR of <1.8.
- Fondaparinux sodium must be stopped at least 5 days prior to surgery, and much earlier with renal impairment.

16.15.2 Postoperatively

- If the patient had an episode of VTE within 3 months before surgery, IV UFH is recommended until the INR is >2.
- In patients with an IVC filter, IV heparin can be avoided in the early postoperative period.
- If no episodes of VTE occurred within 3 months, postoperative IV heparin is not indicated; subcutaneous heparin is recommended.
- Fondaparinux can be restarted 6 hours after surgical closure, provided hemostasis has been established and there is no risk of bleeding (e.g., active bleeding and acquired bleeding disorders such as acute liver failure).
- Patients with a bare metal coronary stent who require surgery within 6 weeks of stent placement: continue aspirin and clopidogrel in the perioperative period.
- Patients with a drug-eluting coronary stent who require surgery within 12 months of stent placement: continue aspirin and clopidogrel in the perioperative period.
- Patients on VKAs who are undergoing minor dental procedures: continue the VKAs around the time of the procedure as well as coadminister an oral prohemostatic agent.
- Patient on VKAs who are undergoing minor dermatologic procedures or cataract removal: continue the VKAs perioperatively.

16.15.3 Venous thromboembolism in pregnancy

The overall incidence of pregnancy-associated VTE is around 200 events per 100,000 women-years. Virchow's triad of hypercoagulation, vascular damage, and venous stasis all occur in pregnancy (RR = 4.3, 95% CI 3.5–5.2) for VTE in pregnant or postpartum women compared with nonpregnant women. Prior superficial vein thrombosis is an independent risk factor.

Additional risk factors include age >35 years, obesity, grand multiparity, varicose veins, urinary tract infection, preexisting diabetes mellitus, stillbirth, obstetric hemorrhage, preterm history of VTE, or thrombophlebitis.

16.15.4 Diagnosis

DVT occurs with equal frequency in all trimesters, mostly involving the left leg and the left ileofemoral vein. Typical symptoms are unilateral leg pain and swelling. Pain with foot dorsiflexion (Homans sign) may be seen but is not diagnostic.

PE occurs more commonly during the puerperium than during pregnancy (RR = 15.0; 95% CI, 5.1–43.9), usually after a cesarean delivery. Clinical presentation can vary from mild dyspnea and tachypnea to cardiopulmonary collapse.

16.16 Diagnosis of pulmonary embolism in pregnancy

16.16.1 Treatment

Airway, breathing, and circulation are to be addressed immediately. In life-threatening PE, thrombolytic therapy, percutaneous catheter thrombus fragmentation, or surgical embolectomy may be used, depending on local resources. Evidence of safety and efficacy of thrombolytic therapy is lacking. Empiric anticoagulation may be started if clinical suspicion is high, then discontinued if VTE is excluded. Warfarin should be avoided in pregnancy as it crosses the placenta and may cause fetal hemorrhage, but is compatible with breast feeding. Nonvitamin K antagonist oral anticoagulant (NOACs) have demonstrated a potential to cross the placenta; the outcomes of pregnancy in 85 women were 31 live births (36.5%); 21 miscarriages/abortions (24.7%); 27 elective pregnancy interruptions (31.8%), while six pregnancy (7.1%) were still ongoing at the time of assessment. Fondaparinux, on the other hand, has been found to be safe in pregnancy; the sample size, however, was very small for coming to a definitive conclusion [325].

16.17 Intrapartum management

Heparin should be discontinued at the onset of labor in order to prevent anticoagulant complications during delivery. For a cesarean section, heparin should be discontinued 24 hours prior to surgery. For high-risk patients, such as those with mechanical heart valves or recent VTE, the

American College of Obstetricians and Gynecologists (ACOG) recommends switching to IV heparin at the onset of labor. To minimize spinal and epidural hematoma risk, the ACOG and the American Society of Regional Anesthesia advise avoiding regional anesthesia for 24 hours after the last LMWH dose for women on twice-daily therapeutic doses of enoxaparin (Lovenox), and for 12 hours after the last dose of LMWH for women receiving daily prophylactic dosing [326].

16.17.1 Anticoagulation in the obese

Obesity appears to be associated with an increased risk for VTE. Studies have reported the risk of VTE as being twice as common in the obese as compared to the general population. Abdominal obesity has been intimately linked to the development of idiopathic VTE both in men (OR = 2.31; 95% CI, 1.48–3.62) and women (OR = 1.84; 95% CI, 1.19–2.84) [327]. Similarly, increasing Body Mass Index (BMI) values are associated with a greater risk of VTE. The frequency of recurrent VTE was 9.3% (95% CI, 6.0%–12.7%) among patients with a normal BMI and 17.5% (95% CI, 13.0%–22.0%) among patients who were obese [326].

With the high prevalence of obesity and the potential interplay of obesity with other risk factors for VTE, implementation of effective, safe, and practical prevention strategies is critical.

The ACCP guidelines do not recommend using mechanical prophylaxis alone for VTE prevention among patients with morbid obesity unless a high bleeding risk precludes the use of pharmacological prophylaxis. The ACCP guidelines suggest weight-based dosing of anticoagulants for VTE prophylaxis. However, the ACCP and the National Institute for Health and Clinical Excellence underscore the importance of individualized prophylaxis according to the estimated risk for VTE. Therefore the education of the individuals at high risk and healthcare providers is essential.

16.17.2 Anticoagulation in renal impairments

As LMWH is partially cleared by renal excretion and metabolism, drug accumulation is expected to occur with long-term use, especially in those with significant renal insufficiency. A greater risk of bleeding is observed in patients with renal impairment. Guidelines recommend the adjustment of the dose based on antifactor Xa monitoring; VKAs are considered to be a safer option for long-term management of these patients. Smaller doses of warfarin are required in renal impairment and the risk of bleeding increases with decreasing renal function [328]. On the other hand, lower rates of stroke or systemic embolism with dabigatran 150 mg twice daily versus warfarin across all creatinine clearance categories (>80, 50 to <80, and <50 mL/min) have been observed, along with lower rates of major bleeding in participants with creatinine clearance >80 mL/min.

16.18 Venous thromboembolism in cancer

Individuals with cancer are at a greater risk of VTE. More than 1% of cancer patients develop VTE annually, which is a fourfold increased risk in cancer patients (incidence rate ratio + 3.96; 95% CI, 3.68–4.27). Among average-risk patients, the overall risk of VTE was estimated to be 13 per 1000 person-years (95% CI, 7–23), with the highest risk among patients with cancers of the pancreas, brain, and lung. Amongst patients at high risk (due to metastasis or receipt of high-risk treatments), the risk of VTE was 68 per 1000 person-years (95% CI, 48–96), with the highest risk among patients with brain cancer (200 per 1000 person-years; 95% CI, 162–247) [329].

The NOACs are the attractive measures to manage VTE in cancer patients as they are administered orally. They have very few drug interactions and do not require monitoring. They are noninferior to warfarin in preventing recurrent VTE without increasing the risk of bleeding and are safe for long-term secondary prophylaxis of VTE. In a subgroup analysis of patients with active cancer participating in the RE-COVER trial ($n = 121$), recurrence of VTE was observed in 5.3% of cancer patients treated with dabigatran versus 3.1% of patients treated with warfarin ($P = .49$), whereas in the RE-MEDY study, the recurrence of VTE was 1.8% with dabigatran and 1.3% with warfarin at 6–36 months of treatment. The EINSTEIN-DVT study included 207 patients with active cancer and the EINSTEIN-PE study included 223. Pooled analysis of these studies showed that treatment with rivaroxaban or LMWH followed by VKA had similar rates of recurrent VTE in patients with active cancer [330]. VTE recurrence occurred in 3.9% of patients treated with NOACs and in 6.0% treated with VKAs (OR 0.63; 95% CI, 0.37–1.10). Clinically relevant bleeding occurred in 14.5% of patients treated with NOACs and in 16.5% of patients treated with VKA Odds Ratio (OR) 0.85; 95% CI, 0.62–1.18).

Recurrence rates of up to 32% have been reported in patients with cancer treated with IVC filters. Incidences of fatal PE are also on record. In the Prevention of Recurrent Pulmonary Embolism by Vena Cava Interruption (PREPIC) trial, the use of permanent IVC filters in conjunction with anticoagulation resulted in a decreased incidence of PE at the expense of an increased risk for recurrent DVT and without a reduction in overall mortality during 8 years of follow-up. As a result, their use is restricted to patients with acute VTE and contraindications to anticoagulation [329].

16.19 Management of upper extremity deep vein thrombosis

Upper extremity deep vein thrombosis (UEDVT) accounts for 5%–10% of all cases of DVT, with increasing incidence due to the higher frequency of intravenous catheter use. The axillary and subclavian veins are the most common sites of UEDVT; the brachial vein may also be involved. Secondary UEDVT, also known as Paget–Schroetter syndrome, is encountered more frequently than the primary disease. Catheter-associated UEDVT is responsible for 93% of all cases of UEDVT.

The diagnosis is by medical history and clinical findings, and radiographic imaging aids diagnosis. Venous duplex ultrasonography is the most frequently used imaging technique. CT and MRI can be used if duplex ultrasonography is indeterminate.

Compared to lower extremity DVT, UEDVT has a lower risk of PE; 5%–8% of patients compared with UEDVT with a mortality of 0.7%. Subclinical PE is far more common, up to 36%, while PTS is seen in up to 13% of patients. In those with central venous catheters, UEDVT may result in the inability to draw from or infuse into the catheter as well as a long-term complication of loss of venous accessibility.

Unless contraindicated, the cornerstone of treatment is anticoagulation. The aim is to obtain early venous recanalization and restore vein patency. Subsequently mechanical thrombectomy can be utilized. The ACCP guidelines recommend anticoagulant therapy alone over thrombolysis. However, thrombolysis can be considered in patients in whom there are severe symptoms, extent of thrombus from the subclavian to axillary vein, symptoms <14 days, good performance status, life expectancy >1 year, and low risk for bleeding. Surgical decompression may be needed.

In catheter-associated UEDVT, the ACCP recommends the removal of the offending catheter, if the catheter is no longer needed or no longer working; with overlap anticoagulation prior to catheter removal followed by a minimum of 3 months of anticoagulation. If the catheter is not removed, anticoagulation should be continued for as long as the CVC remains in place and continued for 3 months after its removal. CDT may be considered in those with severe symptoms and who require continued use of the CVC. The use of superior vena cava filters should only be considered in rare cases in those patients with contraindications for anticoagulant and PE.

16.19.1 Cerebral venous sinuses thrombosis

Cerebral venous sinuses thrombosis [331] is rare, with an incidence rate of 0.22–1.32/100,000/year. It is more common in the young and females. The principal pathology is thrombosis of cerebral veins, at the junction of cerebral veins and larger sinuses. CT scan, MRI, and magnetic resonance venography aid diagnosis. Treatment is predominantly anticoagulation in order to prevent thrombus extension and thus reduce the chance of an embolism. The American Heart Association and American Stroke Association (AHA/ASA) and the European Federation of Neurological Societies guidelines recommend that the patients be anticoagulated even in the presence of hemorrhage; although it is unclear as to whether LMWH or UFH should be used. If anticoagulation is contraindicated, or severe CVT is not responding to anticoagulants, endovascular thrombolysis or mechanical thrombectomy might be an option, although evidence to support this approach is currently lacking. Following the immediate management, long-term VKAs with a target INR of 2–3 should be used. The AHA/ASA guidelines recommend anticoagulation for 3–6 months in provoked CVT, or CVT associated with severe thrombophilias.

16.19.2 Management of splanchnic vein thrombosis

Splanchnic vein thrombosis, involving the portal, splenic, mesenteric, or hepatic veins, is uncommon in the general population, but occurs at significant rates in individuals with intraabdominal malignancies [332].

In a retrospective cohort of 832 patients with splanchnic vein thrombosis, of which 27% had underlying cancer, warfarin therapy and GI tract varices were independent predictors of bleeding. Furthermore, VTE recurrence was not reduced after anticoagulant therapy. An international registry of 613 patients with splanchnic vein thrombosis reported that most patients were treated with anticoagulation and that the risk of major bleeding was low.

In patients with acute, symptomatic splanchnic vein thrombosis without contraindications to anticoagulation, guidelines recommend the use of anticoagulant therapy. Guidelines are not available for the management of patients with incidentally detected splanchnic vein thrombosis. However, it would be reasonable to withhold anticoagulation if the patient is asymptomatic, especially if radiologic evidence indicates that the thrombus is of a chronic nature. Repeated imaging seems prudent to detect thrombus progression, if anticoagulation is not given [333].

16.19.3 Heparin

Anticoagulation remains the mainstay of the initial treatment for DVT. Regular UFH was the standard of care until the introduction of LMWH products. Heparin prevents extension of the thrombus and has been shown to significantly reduce (but not eliminate) the incidence of fatal and nonfatal pulmonary emboli as well as recurrent thrombosis. The primary reason for the persistent, albeit reduced, risk of PE is the fact that heparin has no effect on preexisting nonadherent thrombus. Heparin does not affect the size of existing thrombus and has no intrinsic thrombolytic activity (Table 16.2).

Heparin therapy is associated with complete lysis in fewer than 10% of patients studied with venography after treatment.

Heparin therapy has little effect on the risk of developing the postphlebitic syndrome. The original thrombus causes venous valvular incompetence and altered venous return, leading to a high incidence of chronic venous insufficiency and postphlebitic syndrome.

The anticoagulant effect of heparin is directly related to its activation of antithrombin III. Antithrombin III, the body's primary anticoagulant, inactivates thrombin and inhibits the activity of activated factor X in the coagulation process.

Table 16.2 Unfractionated heparin.

- Heterogeneous mixture of branched GAG chains (MW 3000–30,000)
- Mechanism: Binds to and potentiates the action of antithrombin III
  - Inactivates factors II, IX, X, XII (intrinsic pathway, prothrombin time)
  - Complex is converted to a rapid thrombin inhibitor

Heparin is a heterogeneous mixture of polysaccharide fragments with varying molecular weights but with similar biological activity. The larger fragments primarily interact with antithrombin III to inhibit thrombin. The low molecular–weight fragments exert their anticoagulant effect by inhibiting the activity of activated factor X. The hemorrhagic complications attributed to heparin are thought to arise from the larger higher molecular-weight fragments.

Warfarin therapy is overlapped with heparin for 4–5 days until the INR is therapeutically elevated to 2–3. Heparin must be overlapped with oral warfarin because of the initial transient hypercoagulable state induced by warfarin. This effect is related to the differential half-lives of protein C, protein S, and the vitamin K-dependent clotting factors II, VII, IX, and X. Long-term anticoagulation is definitely indicated for patients with recurrent venous thrombosis and/or persistent or irreversible risk factors.

When intravenous UFH is initiated for DVT, the goal is to achieve and maintain an elevated *activated partial thromboplastin time* (aPTT) of at least 1.5 times the control. The pharmacokinetics of heparin is complex; the half-life is 60–90 minutes. A protocol for IV heparin use is as follows:

1. Give an initial bolus of 80 U/kg;
2. Initiate a constant maintenance infusion of 18 U/kg;
3. Check the aPTT or heparin activity level 6 hours after the bolus and adjust the infusion rate accordingly;
4. Continue to check the aPTT or heparin activity level every 6 hours until two successive values are therapeutic;
5. Monitor the aPTT or heparin activity level, hematocrit, and platelet count every 24 hours.

Traditionally, heparin has been used only for patients admitted with DVT. Fixed-dose subcutaneous UFH, 333 U/kg of UFH, was administered subcutaneously initially, followed by a fixed dose of 250 U/kg twice daily and was compared to LMWH, Enoxaparin or Dalteparin, 100 IU/kg twice daily. This was overlapped with oral warfarin for 5 days until the INR was considered therapeutic.

16.19.4 Heparin-induced thrombocytopenia

HIT is an immune-mediated reaction caused by the production of antibodies that activate platelets in the presence of heparin [334,335]. Despite thrombocytopenia, bleeding is rare; rather, HIT is strongly associated with thromboembolic complications. Up to 8% of patients receiving heparin develop HIT antibodies, 1%–5% progress to develop HIT with thrombocytopenia and one-third may suffer from arterial and/or venous thrombosis [336]. HIT occurs in approximately 1 in 5000 hospitalized patients, with a large variability among patient populations. The risk is greater in those receiving UFH, than in those receiving LMWH; and it is more so if the patient undergoes a major surgery than a minor surgery. Thrombosis in HIT is associated with a mortality rate of approximately 20%–30%, with an equal percentage of patients becoming permanently disabled by amputation, stroke, or other causes. The thrombocytopenia of HIT is typically of moderate severity, with median platelet counts ranging between $50-80 \times 10^9$/L. Severe thrombocytopenia (platelets $<15 \times 10^9$/L) is unusual. This fall in platelet count typically starts 5–14 days after initiation of heparin treatment, but onset may be rapid or delayed. The platelet count starts to rise within 2–3 days and usually returns to normal within 4–10 days after cessation of heparin treatment, and it takes another 2–3 months for antibodies to disappear. Delayed-onset HIT may, however, be seen after cessation of heparin treatment [337]. A platelet factor 4-heparin should be ordered if signs and symptoms are suggestive of HIT. Treatment of HIT is with discontinuation of heparin and administration of direct thrombin inhibitors like Argatroban (in the United States, Canada, the EU, and Australia) or Dnaparoid (in Canada, the EU, and Australia) as alternative anticoagulants. Platelet transfusion is advocated only in severely thrombocytopenic patients. Fondaparinux and Bivalirudin can also be employed but are not approved for the same. The probability of HIT can be determined by the 4Ts: Thrombocytopenia, Timing of platelet count fall, Thrombosis or other sequelae, and Thrombocytopenia.

Prothrombin complex concentrate (PCC) versus fresh frozen plasma (FFP) in the management of life-threatening hemorrhage due to warfarin.

FFP and PCC are used for increasing the concentration of vitamin K-dependent coagulation factors, factors II, VII, IX, and X to restore the normal levels of the clotting fraction. FFP is human plasma frozen within a specific time period after collection from a donor. PCC is derived from pooled, virus-inactivated human plasma products and provides coagulation factors (three-factored or four-factored) rapidly. PCC has been demonstrated to be more effective and safe in treating overanticoagulation.

In a randomized controlled trial (RCT) of 94 patients with intracranial hemorrhage due to anticoagulants, it was found that four-factored PCC was superior to FFP in normalizing the INR and was associated with a smaller hematoma expansion.

In another RCT in 50 patients with mechanical heart valves, 76% of the patients in the PCC group and only 20% of the patients in the FFP group reached the INR target. Five of 25 in the PCC group received an additional dose of PCC, whereas 17 patients in the FFP group received a further dose of FFP ($P = .001$). With respect to safety, in the FFP group, the INR was high (INR > 2.5) in 86% of the patients. However, hemorrhage was not reported in either group. The study concluded PCC to be better in reversing overanticoagulation than FFP.

In a retrospective study comparing the effectiveness of FFP and PCC in patients with acute bleeding, it was observed that the median time to INR <1.3 was 0.5 (range 0.5–1.5) versus 15.5 (range 5–96) hours for PCC versus FFP, respectively ($P < .001$). Packed red cell transfusion did not differ between the groups ($P = .3$), but more FFP was transfused in the FFP versus PCC group ($P < .001$). The duration of hospital admission was longer in the FFP (median = 7 days, range 1–93) versus PCC patients (median = 6 days, range = 1–35) ($P = .04$), concluding that PCC was better than FFP in the management of anticoagulant-associated hemorrhage [338].

16.19.5 Low molecular–weight heparin

LMWH is prepared by selectively treating UFH to isolate the low molecular–weight (<9000 Da) fragments. Its activity is measured in units of factor X inactivation, and monitoring of the aPTT is not required. The dose is weight adjusted (Table 16.3).

LMWH products are administered subcutaneously, and their half-life permits daily or twice-daily dosing. Its use in the outpatient treatment of DVT and PE has been evaluated in a number of studies.

At the present time, four LMWH preparations, Enoxaparin, Dalteparin, Tinzaparin, and Nadroparin, are available. Enoxaparin, Dalteparin, and Tinzaparin have received US Food and Drug Administration approval for the

Table 16.3 Low molecular–weight heparin.

- Drug: Enoxaparin (clexane), Dalteparin (fragmin)
- Enzymatic depolymerization of unfractionated heparin
  - Molecular weight 4000–5000
  - Lower molecular-weight fragments with reduced binding to proteins
- Mechanism: Specific binding ATIII (factors X and II)
  - More predictable dose–response relationship

NAME	STATUS		DOSE
ENOXAPARIN	FDA approved		1.0 mg/kg twice daily 1.5 mg/kg once daily
DALTEPARIN	Approved/ not in DVT		100 U/kg twice daily or 200 U/kg once daily
NADROPARIN	Not in US		4100U twice daily <50kg 6100U for 50-70kg 9200U for >70kg
REVIPARIN	Not in US		3500U twice daily pts35-45kg 4200U for 46-60kg 6300U for >60kg
TINZAPARIN	FDA approved		175 U/kg OD

Figure 16.3 Dosage schedule of low molecular–weight heparin.

treatment of DVT in the United States. Enoxaparin is approved for inpatient and outpatient treatment of DVT. Nadroparin is approved for DVT treatment in Canada.

The increased bioavailability and prolonged half-life of LMWH allows for outpatient treatment of DVT using once- or twice-daily subcutaneous treatment regimens. Outpatient treatment of acute DVT with LMWH has been successfully evaluated in a number of studies and is currently the treatment of choice if the patient is eligible. In general, outpatient management is not recommended if the patient has proven or suspected concomitant PE, significant comorbidities, extensive ileofemoral DVT, morbid obesity, renal failure, or poor follow-up (Fig. 16.3).

The efficacy and safety of LMWH for the initial treatment of DVT have been well established in several trials.

16.19.6 Fondaparinux, a direct factor Xa inhibitor in acute deep vein thrombosis

Currently, *Enoxaparin* and other LMWH agents are recommended for the treatment of DVT. However, the data on once-daily or twice-daily dosing of Enoxaparin are not clear. Second, the practical issues that surround the administration of a weight-based 1 mg/kg dose from fixed-volume syringes of Enoxaparin may be an issue for some patients. Third, the incidence of HIT, although reduced with Enoxaparin, is not completely eliminated. *Fondaparinux*, a direct selective inhibitor of factor Xa, overcomes many of these disadvantages. Pharmacokinetic studies of Fondaparinux reveal that only a once-daily subcutaneous dose is required. Furthermore, a single dose of 7.5 mg is effective over a wide range of patient weights between 50 and 100 kg. Daily doses of 5 mg or 10 mg are appropriate for patients who weigh less or more than that weight range. HIT has not been reported. Therapeutic monitoring of laboratory parameters such as the prothrombin time or partial thromboplastin time is also not required.

16.19.7 Isolated calf vein deep vein thrombosis

Despite the lower (but not zero) risk of PE and mortality associated with calf vein DVT, current guidelines recommend short-term anticoagulation for 3 months in symptomatic patients. Asymptomatic patients with isolated calf vein DVT do not require anticoagulation, and surveillance ultrasound studies over 10–14 days to detect proximal extension are recommended instead.

16.20 Thrombolytic therapy for deep vein thrombosis

Thrombolytic therapy offers significant advantages over conventional anticoagulant therapy, including the prompt resolution of symptoms, the prevention of PE, the restoration of normal venous circulation, the preservation of venous valvular function, and the prevention of postphlebitic syndrome. Thrombolytic therapy does not prevent clot propagation, rethrombosis, or subsequent embolization. Heparin therapy and oral anticoagulant therapy must always follow a course of thrombolysis.

Unfortunately, most patients with DVT have absolute contraindications to thrombolytic therapy. Thrombolytic therapy is also not effective once the thrombus is adherent and begins to organize. Venous thrombi in the legs are often large and associated with complete venous occlusion. The thrombolytic agent that acts on the surface of the clot may not be able to penetrate and lyse the entire thrombus.

Nevertheless, the data from many published studies indicate that thrombolytic therapy is more effective than heparin in achieving vein patency. The unproven assumption is that the degree of lysis observed on posttreatment venography is predictive of future venous valvular insufficiency and late (5–10 years) development of postphlebitic syndrome. Preliminary evidence suggests that thrombolytic therapy reduces but unfortunately does not entirely eliminate the incidence of postphlebitic syndrome at 3 years.

The hemorrhagic complications of thrombolytic therapy are formidable (~3 times higher) and include the small but potentially fatal risk of intracerebral hemorrhage. The uncertainty regarding thrombolytic therapy is likely to continue. Currently, the ACCP consensus guidelines recommend thrombolytic therapy only for patients with massive ileofemoral vein thrombosis associated with limb ischemia or vascular compromise.

Catheter-directed intrathrombus thrombolysis (CDT) is an image-guided therapy where a thrombolytic agent is administered directly into the thrombus and enhances thrombus removal. A variety of specialized catheters and mechanical devices are used to optimally deliver the drug and mechanically remove the clot. Second, balloon angioplasty and stents may be used at the same time to treat any underlying venous obstruction that predisposes the patient to recurrent DVT. Direct intrathrombus delivery of the thrombolytic agent achieves a higher drug concentration at the site of thrombosis with a lower total dose than would be used by systemic intravenous thrombolytic therapy. This is the suggested mechanism for the lower incidence of systemic and in particular intracranial hemorrhagic complications with CDT.

16.20.1 Filters for deep vein thrombosis

The major indications for vena caval filters are primarily for patients with a contraindication to anticoagulation or for patients with major complications while anticoagulated (hemorrhage or HIT). This filter prevents clots that break loose and lodge in the lungs. The filters are sometimes referred to as "umbrellas" because they look like the wire spokes of an umbrella (Fig. 16.4).

The use of vena caval filters has expanded to include primary VTE prophylaxis in special patient populations such as major trauma patients, major surgery patients, advanced malignancy, and neurological or neurosurgical patients with paralysis or prolonged immobilization. These special

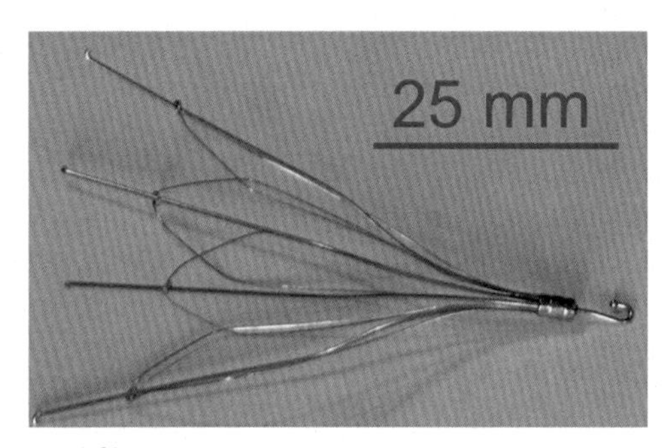

Figure 16.4 Vena caval filter.

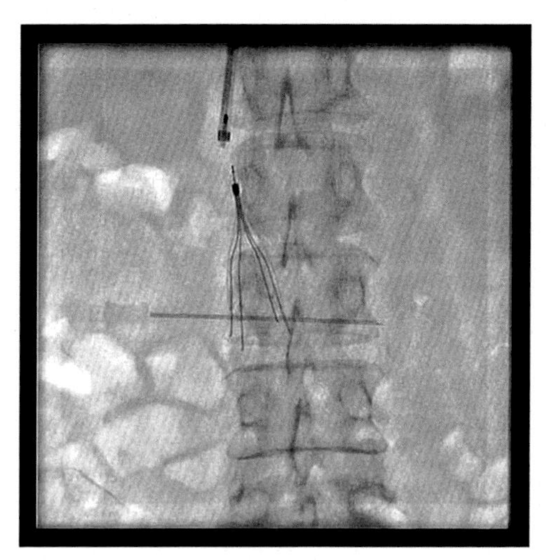

Figure 16.5 Insertion of the vena caval filter.

patient populations are generally characterized by contraindications to anticoagulation, ineffective anticoagulant prophylaxis, hypercoagulable states, or other exceedingly high risks of PE.

Currently, the newer filters are placed under ultrasonographic guidance either by transabdominal or intravascular ultrasonography. The advantage of ultrasonography is that the filters may be placed at the bedside in the ICU or the ED, thereby avoiding the pitfalls and difficulties of transporting the patient to the angiography suite. Transabdominal ultrasonography machines are generally more readily available, do not require a separate femoral venous puncture, and there is more experience with their use. However, the patient's body habitus must provide adequate acoustic windows to permit the transabdominal technique (Fig. 16.5).

There is a need for more studies and data on which filter to use. The temporary optionally retrievable filters have the ultimate advantage, but currently are removed less than half the time and no proven long-term results are available.

16.21 Surgery for deep vein thrombosis

Surgical therapy for DVT may be indicated when anticoagulant therapy is ineffective, unsafe, or contraindicated. The major surgical procedures for

DVT are thrombectomy and partial interruption of the inferior vena cava to prevent PE.

The rationale for thrombectomy is to restore venous patency and valvular function. Thrombectomy alone is not indicated because rethrombosis is frequent. Heparin therapy is a necessary adjunct. Thrombectomy is reserved for patients with massive ileofemoral vein thrombosis (*phlegmasia cerulea dolens*) with vascular compromise when thrombolysis is absolutely contraindicated.

16.21.1 Compression stockings

Compression stockings help prevent swelling associated with DVT. They should be worn on the leg from the foot to about the level of the knee. This pressure helps reduce the chances of blood pooling in the soleal sinuses and thrombosis. The patient is advised to wear these stockings for at least a year (Fig. 16.6).

16.22 Lifestyle

To prevent DVT from worsening or happening again all patients should be given the following advice:

- *Regular checkups* to evaluate by color Doppler scan or modify treatments if need.
- *Advice on diet* as vitamin K can affect drugs such as warfarin, which is a VKA. Foods high in vitamin K include green leafy vegetables and canola and soybean oils.

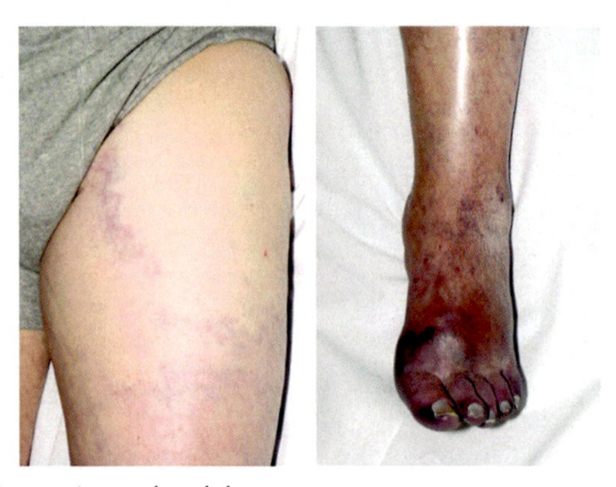

Figure 16.6 Phlegmasia cerulea dolens.

- *Exercise for calf muscles* if sitting a long time. Whenever possible, get up and walk around. If one cannot get up to walk around, try raising and lowering the heels while keeping the toes on the floor, then raising the toes while the heels are on the floor.
- *Movement*: Those on bed rest, because of surgery or other factors, the sooner they start moving, the less chance of DVT.
- *Advice on lifestyle*: Lose weight, quit smoking, and control BP.
- *Wear compression stockings* to help prevent venous stasis.
- *Watch for bleeding*: This can be a side effect of taking medications like warfarin. In such cases, call on the doctor immediately.
- *Blood values to be maintained*: Warfarin 5 mg PO daily is initiated and overlapped for about 5 days until the INR is therapeutic >2 for at least 24 hours. INR to be maintained at around 3 and should never go above the value 4. The platelet count should not fall below 80,000. The prothrombin time (PT or PTT) should also be monitored very frequently (6-hourly during initiation of warfarin therapy and then after stabilization, daily or on alternate days). In acute DVT continue warfarin for 3–6 months, and in recurrent DVT continue for a minimum period of 1 year.

CHAPTER 17

Anticoagulation in the prevention and treatment of venous thromboembolism

17.1 Introduction

Venous thromboembolism (VTE) is an extremely challenging problem to current surgical practice, and deep vein thrombosis (DVT) is considered a silent killer. Every advanced surgical technique carries a potential risk of DVT and resultant pulmonary embolism (PE). It is a challenge to the cardiovascular surgeon, orthopedic surgeon, general surgeon, gynecologist, interventional cardiologist, and oncologist. Terminology and new definitions:

- *Thrombosis*: This is the *formation of a solid mass in circulation* from the constituents of flowing blood; the mass by itself is called *thrombus*.
- *Clot*: Coagulation in vitro.
- *Hematoma*: Extravascular collection of a blood clot.
- *Hemostatic plug*: Clot at the cut end of a normal blood vessel to stop bleeding.
- *Hemostatic plugs are useful to prevent loss of blood, whereas thrombus formation in an undamaged cardiovascular system is life-threatening.*
- *Thrombus in an artery* will lead to a decrease or stoppage of the blood supply to an organ or tissues.
- *Thrombus in a vein* predisposes to VTE.

17.2 Problems caused by venous thromboembolism

VTE, which refers to DVT and/or PE, is thought to be the most preventable cause of death among hospitalized patients.[1] DVT and PE are manifestations of the same disease process, and are often clinically silent. Around 1 in 1000 people per year present with clinical symptoms.[2]

[1] Anderson FA, et al. Arch Intern Med 1991;151:933–8.
[2] SIGN Guideline 36; March 1999.

Genesis, Pathophysiology and Management of Venous and Lymphatic Disorders
DOI: https://doi.org/10.1016/B978-0-323-88433-4.00021-8

According to the American College of Chest Physicians (ACCP), the majority of VTE patients with proximal DVT, when studied carefully, also have PE (symptomatic or asymptomatic) and vice versa. However, as the Scottish Intercollegiate Guidelines Network (SIGN) points out, most patients with proven PE do not have clinically evident DVT.[3] PE contributes significantly to in-hospital mortality rates.[4,5]

Key points:

- *VTE* is the most preventable cause of death among hospitalized patients (see footnote 1).
- *DVT* occurs 1 in 1000 people per year and presents with clinical symptoms (see footnote 2).
- *PE* is a serious complication of DVT → 90% cases result from asymptomatic DVT (see footnote 2).
- *PE* contributes significantly to in-hospital mortality rates (see footnotes 3–5).

17.3 Pathophysiology

The fluid state of the blood is maintained normally by a homeostasis of two phenomena—*coagulation and bleeding*. Injury to blood vessel initiates a hemostatic repair mechanism (thrombogenesis). The primary events leading to thrombus formation are described as Virchow's triad.

17.3.1 Virchow's triad

1. Endothelial damage
2. Changes in blood flow
3. Hypercoagulation of blood

17.4 Role of blood vessels

17.4.1 Blood vessel wall integrity

Endothelium has the following functions:

1. *Protects* the flowing blood from the thrombogenic properties of the subendothelium.

[3] Department of Health, Report on confidential enquires into maternal deaths in the United Kingdom, 1994–96.

[4] Gillies TE, et al. Br J Surg 1996;83:1394–95;

[5] Baglin TP, et al. J Clin Pathol 1997;50:609–10;

2. *Produces* some antithrombotic (thrombo-inhibitory) factors such as:
   a. *Heparin-like substance* which accelerates the action of antithrombin III and other clotting factors.
   b. *Thrombomodulin* (converts thrombin into activator of protein C, an anticoagulant).
   c. *Inhibitors of platelet aggregation* like ADPase (adenosine diphosphate), PGI2 (prostacyclin), or prostocyclin.
   d. *Tissue plasminogen activator* which accelerates the fibrinolytic process.
3. *Releases* some prothrombotic factors with procoagulant properties:
   a. *Tissue thromboplastin*: released from endothelium.
   b. *Von Willebrand factor*: helps platelets *adhere to subendothelium.*
   c. *Platelet-activating factor*: helps platelet aggregation.
   d. *Inhibitor of plasminogen*: suppresses fibrinolysis.

Vascular injury exposes subendothelial connective tissue (collagen, elastin, fibronectin, laminin, and glycosaminoglycans) which are thrombogenic and plays a very important role in initiating hemostasis and thrombosis. Added to this, vasoconstriction also helps to control hemorrhage.

17.5 Role of platelets

Following endothelial cell injury, platelets lead the main role in effecting hemostasis and thrombosis.

1. *Platelet adhesion* (primary aggregation)
2. *Platelet release reaction*

   Alpha granules: fibrinogen, fibronectin, platelet-derived growth factor, platelet factor IV (Intravenous) (an antiheparin), and cationic protein

   Dense bodies: ADP, Ca^{+}, 5-HT *(serotonin, histamine, and epinephrine)*

   This platelet activation and release reaction leads to the intrinsic pathway of coagulation.
3. *Platelet aggregation*

   Release of ADP (a potent platelet aggregating agent) leads to secondary aggregation and the formation of a temporary hemostatic plug. Fibrin, thrombin, and thromboxane A2 help to make the plug very stable.

17.6 Role of the coagulation system

The coagulation mechanism is the conversion of *plasma fibrinogen to solid fibrin mass.* The coagulation system is involved in both the hemostatic process and thrombus formation.

The cascade of events occurs through three pathways:

1. *The intrinsic pathway* (surface contact).
2. *The extrinsic pathway* (tissue damage).
3. *The common pathway* (at the point of convergence of both extrinsic and intrinsic pathways).
   a. *Intrinsic pathway*: contact with abnormal vessel wall activates factor XII with subsequent interactions of factor, XI, IX, VIII, and finally with factor X along with factor IV (Ca^{+}) and platelet factor III.
   b. *Extrinsic pathway*: Tissue damage releases tissue factor (thromboplastin). This interacts with factor VII and activates factor X.
   c. *Common pathway*: both intrinsic and extrinsic pathways meet at one point where factor X gets activated to Xa.

17.7 Alteration of blood flow

1. *Turbulence* in flow, in arteries and heart leads to *thrombosis*.
2. *Sluggishness* of flow or *stasis* in vein leads to *thrombosis* (Tables 17.1 and 17.2).

17.8 Anticoagulants

1. *Vitamin K antagonists (VKAs)*: coumarines (warfarin, acenocoumarol, dindivan)
2. *Heparins*
   a. Unfractionated: heparin*
   b. Fractionated: low-molecular-weight heparin (LMWH)* (enoxaparin, nadroparin, dalteparin)
3. *Factor Xa (FXa) inhibitors*
   a. Indirect**: fondaparinux
   b. Direct***: riveroxaban

Table 17.1 The three pathways of classical blood coagulation.

Prevention of VTE
Treatment of VTE
Treatment of ACS

Table 17.2 Coagulation factors.

Coagulation factors

Coagulation factors and related substances		
Number and/or name	**Function**	**Associated genetic disorders**
I (fibrinogen)	Forms clot (fibrin)	Congenital afibrinogenemia, Familial renal amyloidosis
II (prothrombin)	Its active form (IIa) activates I, V, VII, VIII, XI, XIII, protein C, platelets	Prothrombin G20210A, Thrombophilia
III Tissue factor	Co-factor of VIIa (formerly known as factor III)	
IV Calcium	Required for coagulation factors to bind to phospholipid (formerly known as factor IV)	
V (proaccelerin, labile factor)	Co-factor of X with which it forms the prothrombinase complex	Activated protein C resistance
VI	*Unassigned* - old name of Factor Va	
VII (stable factor, proconvertin)	Activates IX, X	congenital proconvertin/factor VII deficiency
VIII (Antihemophilic factor A)	Co-factor of IX with which it forms the tenase complex	Haemophilia A
IX (Antihemophilic factor B or Christmas factor)	Activates X: forms tenase complex with factor VIII	Haemophilia B
X (Stuart-Prower factor)	Activates II: forms prothrombinase complex with factor V	Congenital Factor X deficiency
XI (plasma thromboplastin antecedent)	Activates IX	Haemophilia C
XII (Hageman factor)	Activates factor XI, VII and prekallikrein	Hereditary angioedema type III
XIII (fibrin-stabilizing factor)	Crosslinks fibrin	Congenital Factor XIIIa/b deficiency
von Willebrand factor	Binds to VIII, mediates platelet adhesion	von Willebrand disease
prekallikrein (Fletcher factor)	Activates XIII and prekallikrein; cleaves HMWK	Prekallikrein/Fletcher Factor deficiency
high-molecular-weight kininogen (HMWK) (Fitzgerald factor)	Supports reciprocal activation of XII, XI, and prekallikrein	Kininogen deficiency
fibronectin	Mediates cell adhesion	Glomerulopathy with fibronectin deposits
antithrombin III	Inhibits IIa, Xa, and other proteases	Antithrombin III deficiency
heparin cofactor II	Inhibits IIa, cofactor for heparin and dermatan sulfate ("minor antit	Heparin cofactor II deficiency
protein C	Inactivates Va and VIIIa	Protein C deficiency
protein S	Cofactor for activated protein C (APC, inactive when bound to C4b-binding protein)	Protein S deficiency
protein Z	Mediates thrombin adhesion to phospholipids and stimulates degradation of factor X by ZPI	Protein Z deficiency
Protein Z-related protease inhibitor (ZPI)	Degrades factors X (in presence of protein Z) and XI (independently)	
plasminogen	Converts to plasmin, lyses fibrin and other proteins	Plasminogen deficiency, type I (ligneous conjunctivitis)
alpha 2-antiplasmin	Inhibits plasmin	Antiplasmin deficiency
tissue plasminogen activator (tPA)	Activates plasminogen	Familial hyperfibrinolysis and thrombophilia
urokinase	Activates plasminogen	Quebec platelet disorder
plasminogen activator inhibitor-1 (PAI1)	Inactivates tPA & urokinase (endothelial PAI)	Plasminogen activator inhibitor-1 deficiency
plasminogen activator inhibitor-2 (PAI2)	Inactivates tPA & urokinase (placental PAI)	
cancer procoagulant	Pathological factor X activator linked to thrombosis in cancer	

4. *Thrombin inhibitors*: dabigatran (oral), hirudin
5. *Antiplatelet drugs* (inhibit platelet adhesion and activation)
   a. Aspirin
   b. Pentoxaphylline (trental)
   c. Thromboxane A_2 receptor antagonists

*Multitargeted activity.

**Fondaparinux: a selective FXa inhibitor through the strong activation of ATIII.

***Rivaroxaban: acts directly on Xa (Oral) (November 2008).

17.9 Selection of anticoagulant

- Efficacy
- Safety
- Economy

17.10 Mechanisms of anticoagulation

- The efficacy/safety ratio for currently available therapies is less than satisfactory due to their ill-defined, multitargeted activity (Table 17.3). New antithrombotic strategies are needed that offer an improved efficacy/safety profile compared with existing antithrombotic agents.
- Currently available antithrombotic agents include the heparins [UFH (Unfractionated Heparin) and lovenox], VKAs (warfarin), and direct thrombin inhibitors (hirudins).
- The most widely used agents, heparins and VKAs, have a range of actions on various components of the coagulation cascade. This contributes to the unpredictable clinical responses associated with these agents.
- Other limitations of currently available antithrombotics include:
  - High incidence of serious adverse effects, particularly bleeding complications;
  - Routine monitoring of coagulation markers may be needed and represents a substantial burden in terms of time and costs;

Table 17.3 Mechanisms of anticoagulation.

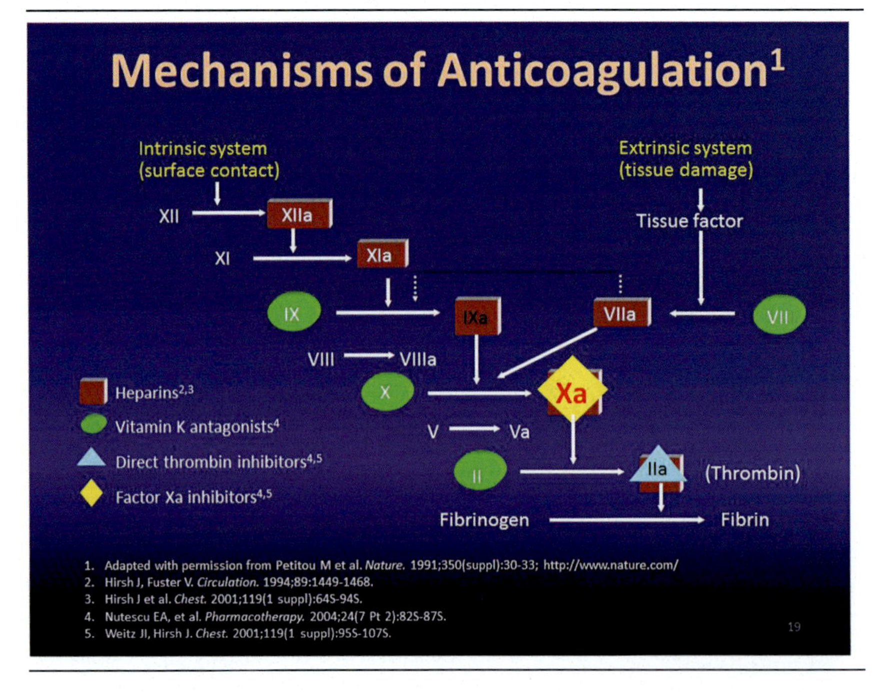

- Narrow therapeutic margin;
- Limited effectiveness in preventing VTE.

- FXa inhibitors are a novel class of antithrombotic agents designed to selectively target only one core step in the coagulation cascade, leading to potent and targeted effectiveness (Table 17.4).

A. Oral anticoagulants:

Vitamin K antagonists:

- Warfarin:

  Most commonly used.
  Plant origin
  Action is very slow
  Requires 2–3 days, for getting the clinical effect
  Bioavailability: almost 100%
  Requires monitoring weekly/biweekly international normalized ratio (INR)
  Unsafe for aged patients (above 80 years)
  Calcification in arteries and cardiac valves

Table 17.4 Site of actions of new anticoagulant drugs.

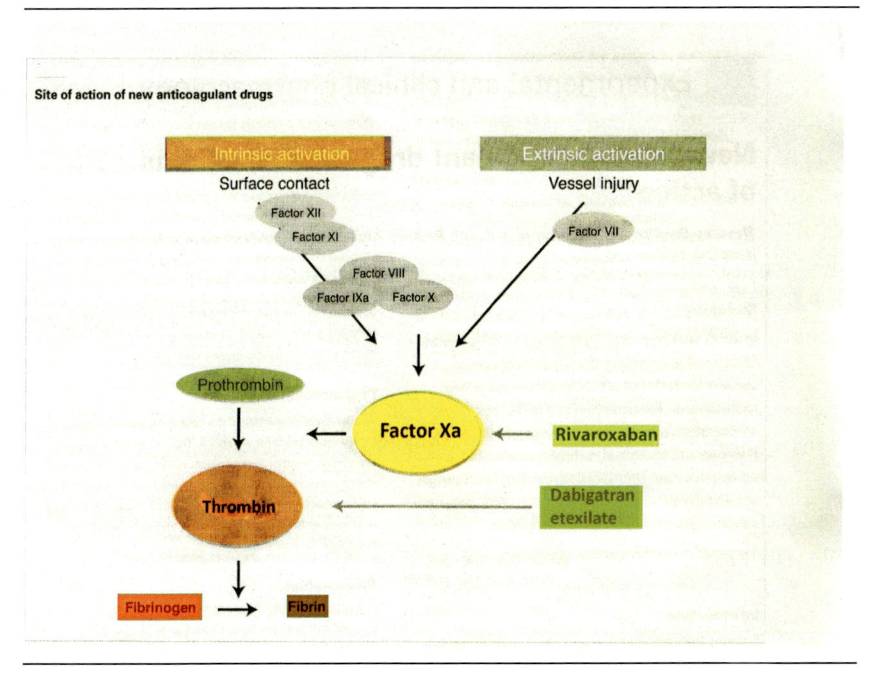

Dosage: 10–20 mg to start with; maintain with 2–10 mg daily (OD)

- *Acenocoumarol (acitrom)*
  Action is faster than warfarin
  Peak action in 24 hours and lasts for 2–3 days
  More stable INR maintenance
  Superior anticoagulant effect than warfarin
  Effect is reversible with vitamin K
  Dose: 10–15 mg to start with and maintain with 2–10 mg OD
  Side effects: gastric upset, ulceration in mouth, dermatitis
- *Phenindione (dindivan)*
  Action starts in 5 hours and lasts for 1–3 days
  Adverse reactions: gastrointestinal (GI) upset, hypersensitivity reactions

1. *Direct Xa inhibitor: Riveroxaban*
   New oral anticoagulant
   This was approved in November 2008, by the Therapeutic Goods Administration, Australia, but is not yet approved by US Food and Drug Administration (FDA). It is a reversible antagonist of FXa. Bioavailability is 80%. Available in 10 mg tablet. Peak plasma concentration in 2.5–4 hours. Half-life is 5–9 hours for young adults, and about 11–13 hours for aged patients. Excreted via kidney and special caution to be observed for patients with renal impairment. Requires once or twice OD dosage.
2. *Direct thrombin inhibitor: dabigatran etexilate*

Dabigatran etixilate is approved in the EU for the prevention of stroke and systemic embolism in patients with nonvalvular atrial fibrillation (NVAF) and one or more risk factors. Dabigatran etixilate is a prodrug of dabigatran, a direct inhibitor of thrombin. In patients with NVAF in the III RE-LY trial, dabigatran etixilate is a prodrug of dabigatran, a direct inhibitor of thrombin. In patients with NVAF in the phase III RE-LY trial, dabigatran etixilate dosages of 110 and 150 mg twice OD were noninferior to warfarin with regard to the risk of stroke or systemic embolism (primary efficacy end point). The higher dosage was associated with a significantly lower risk of stroke or systemic embolism than warfarin, with no significant between-group difference in the risk of major bleeding (primary safety end point). Both dosages of dabigatran etixilate were associated with significantly lower rates of hemorrhagic stroke, intracranial bleeding, and life-threatening major bleeding than warfarin. Dabigatran etixilate was also effective and generally well tolerated across various

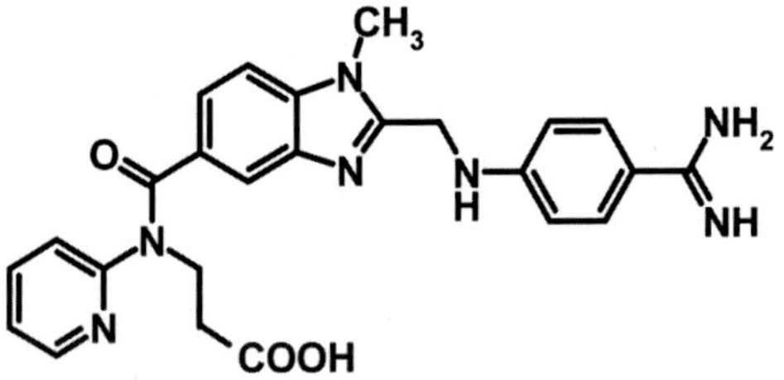

Figure 17.1 Molecular structure of Dabigatran.

patient subgroups. The efficacy and tolerability of dabigatran etixilate were maintained for upto 6.7 years in the RELY-ABLE extension study. Routine anticoagulation monitoring is not required in patients receiving dabigatran etixilate, and it is currently the only non-vitamin K antagonist oral anticoagulant (NOAC) with a specific reversal agent available. Although direct comparisons with other NOACs would be beneficial, dabigatran etixilate is a useful option for the prevention of stroke and systemic embolism in patients with NVAF (Fig. 17.1).

Oral VKAs such as warfarin have long been the mainstay of stroke prevention in patients with atrial fibrillation (AF). However, the use of warfarin in clinical practice is challenging due to problems such as drug–drug and drug–food interactions, a narrow therapeutic index, and unpredictable anticoagulant effects, all of which result in the need for regular laboratory monitoring. Designed to overcome the limitations of warfarin, NOACs have a rapid onset and relatively quick offset of action, fewer drug interactions and no requirement for routine anticoagulation monitoring [339]. Currently available NOACs include the direct FXa inhibitors apixaban, edoxaban, and rivaroxaban, and the direct thrombin (factor IIa) inhibitor dabigatran etixilate. In the EU, dabigatran etixilate is indicated for the prevention of stroke and systemic embolism in patients with NVAF and one or more risk factors. Dabigatran etixilate is also indicated for the primary prevention of VTE after total hip or knee replacement and for the treatment of VTE and prevention of recurrent VTE; discussion of these indications is outside the scope of this review.

17.10.1 Pharmacological properties

The pharmacodynamics profile of dabigatran etixilate is well established and has been previously reviewed. Dabigatran etixilate is a small-molecule prodrug of dabigatran. Dabigatran is a potent, competitive, and reversible direct inhibitor of thrombin, a serine protease responsible for converting fibrinogen to fibrin in the coagulation cascade. Dabigatran inhibits both free and fibrin-bound thrombin, thereby preventing thrombus formation. In vitro, dabigatran inhibited human thrombin in a concentration-dependent manner with an inhibition constant (K_i) of 4.5 nmol/L. Dabigatran was highly selective for thrombin, with selectivity ratios of >700 to >10,000 for thrombin versus most other serine proteases in the coagulation cascade. Dabigatran exhibited potent and dose-dependent inhibition of thrombin-induced platelet aggregation in human gel-filtered platelets [340]. The dabigatran concentration required to achieve 50% of the inhibitory effect (IC_{50} 35 nmol/L) is less than the FXa inhibitors apixaban (IC_{50} 312 nmol/L). Orally administered dabigatran etixilate prolonged standard blood coagulation parameters in healthy volunteers, including activated partial thromboplastin time (aPTT), thrombin time (TT), and ecarin clotting time (ECT), in a dose-dependent manner. There was a linear association between dabigatran plasma concentrations and increased TT and ECT, while a curvilinear relationship was seen between dabigatran plasma concentrations and prolonged aPTT.

The maximum effect (E_{max}) of dabigatran etixilate on clotting parameters occurred at the same time as the maximum plasma concentration (C_{max}), that is, within 2 hours of administration. The pharmacodynamics effects of dabigatran etixilate decreased in parallel with declining plasma concentrations. A rapid initial reduction was observed 4–6 hours after C_{max}, followed by a slow terminal phase. Prolongation of blood coagulation returned to approximately 50% of E_{max} 12 hours after administration, with small residual effects observed 24 hours after cessation of treatment. Although routine anticoagulation monitoring is not required in patients receiving badigatran etixilate, measurement of dabigatran-related anticoagulation may be helpful in identifying patients with an increased bleeding risk caused by excessive dabigatran exposure. TT, ECT, and aPTT may provide useful information, although results should be interpreted with caution given that these tests are not standardized [341].

The diluted TT test provides a quantitative estimation of dabigatran plasma concentration. INR should not be used, as it is unreliable in patients receiving dabigatran etixilate. According to prospective biomarker substudies of the RE-LY trial, dabigatran etixilate 110 and 150 mg twice OD reduced D-dimer and apolipoprotein B (ApoB) levels to a significantly ($P < .05$) greater extent than warfarin in patients with NVAF, with minimal effects on factor VIIa and apolipoprotein A1 levels. There were strong associations between elevated D-dimer levels and an increased risk of stroke or major bleeding, indicating that when used in conjunction with established clinical risk factors and widely scoring systems, D-dimer levels may improve risk prediction in patients with NVAF. Further studies are required to investigate the clinical impact of the potential pleiotropic effect of dabigatran etixilate on ApoB metabolism [342]. Another biomarker substudy of the RE-LY trial demonstrated that a growth differentiation factor of 15 was an independent risk matter for all-cause mortality ($P = .006$) and major bleeding ($P = .0002$), but not stroke or systemic embolism, in patients with NVAF.

17.10.2 Pharmacokinetic profile

Dabigatran etixilate is rapidly absorbed after oral administration and is completely converted to dabigatran via esterase-catalyzed hydrolysis in plasma and in the liver. In healthy volunteers receiving oral dabigatran etixilate 10–400 mg (single doses) or 50–400 mg (three times OD for 6 days), C_{max} values were reached within 2 hours. Dabigatran exhibited linear pharmacokinetics; area under the plasma concentration–time curve (AUC) and C_{max} values increased in a dose-dependent manner. Dabigatran has an absolute oral bioavailability of 6.5%. Dabigatran etixilate is formulated as a hard capsule filled with pellets coated with dabigatran etixilate mesylate [343]. When the pellets are administered without the capsule shell, the oral bioavailability of dabigatran, the time to C_{max} is delayed by 2 hours. The volume of distribution of dabigatran is 60–70 L, indicating moderate tissue distribution. The plasma protein binding of dabigatran is relatively low (35%).

Dabigatran undergoes conjugation to form four pharmacologically active positional isomers of acylglucuronide; each isomers accounts for <10% of total plasma dabigatran. The elimination of dabigatran occurs predominantly via renal excretion of unchanged drug. The renal clearance of dabigatran is 100 mL/minute, which corresponds to the glomerular filtration rate. Following IV administration of a single dose of radiolabeled dabigatran in healthy male volunteers, 85% of the radioactivity was excreted in the urine and 6% in the feces. However, following oral administration, 7% of the dose is eliminated in the urine. In healthy male volunteers receiving a single IV dose of radiolabeled dabigatran, 88%–94% of the administered dose was recovered in 168 hours [344]. The terminal half-life of dabigatran is 12–24 hours after repeated administration. Decreasing renal function increases exposure to dabigatran. For instance, relative to subjects with normal renal function, the AUC from time zero to infinity of dabigatran was 1.5-, 3.2-, and 6.3-fold higher in subjects with mild, moderate, and severe renal impairments, respectively.

Dabigatran exposure is increased in elderly subjects. The AUC of dabigatran is 40%–60% higher and the C_{max} is >25% higher in elderly than in younger healthy subjects. In the RE-LY study, the C_{trough} of dabigatran was 31% higher in patients aged >75 years and 22% lower in patients aged <65 years, compared with patients aged 65–75 years. The pharmacokinetics and pharmacodynamics of dabigatran were not affected by moderate (Child–Pugh class B) hepatic impairment following a single oral dose of dabigatran etixilate of 150 mg. Higher body weight (>100 kg) is associated with a 20% lower C_{trough} of dabigatran than a lower body weight (50–100 kg) [345]. Dabigatran pharmacokinetics do not differ between Caucasian, African–American, Hispanic, Japanese, or Chinese patients. In a pooled analysis of data from 18 phase I and II studies, there were no clinically relevant differences in the pharmacokinetics and pharmacodynamics of dabigatran between Caucasian and Japanese subjects. Dabigatran and dabigatran etixilate are not metabolized by the CYP (Cytochrome P450) system. Coadministration of dabigatran etixilate and the CYP3A4/5 substrate atorvastatin did not alter the pharmacodynamics or pharmacokinetics of either drug. There is no pharmacokinetic interaction between diclofenac (metabolized by CYO2C9) and dabigatran etixilate [346].

Given that dabigatran etixilate is a substrate for P-glycoprotein (P-gp), there is potential for interaction when it is coadministered with P-gp inhibitors or inducers. The EU summary of product characteristics (SPC) recommends close clinical surveillance when dabigatran etixilate is coadministered with strong inhibitors of P-gp; use of some strong P-gp inhibitors is contraindicated (systemic ketoconazole, ciclosporin, dronedarone, and itraconazole) or not recommended (tactolimus). Mild-to-moderate inhibitors of P-gp (amiodarone, posaconazole, quinidine, ticagrelor, and verapamil) and dabigatran etixilate should be coadministered with caution; dosage adjustment of dabigatran etixilate is required when it is coadministered with verapamil, but not with amiodarone or quinidine. Coadministration of dabigatran etixilate with P-gp inducers (e.g., carbamazepine, phenytoin, rifampicin, hypericum) should be avoided. There is no change in dabigatran exposure when dabigatran etixilate is coadministered with the P-gp substrate digoxin.

Coadministration of dabigatran etixilate with a 300 or 600 mg loading dose of clopidogrel is associated with a 30%–40% increase in dabigatran exposure [4]. The hemorrhagic risk of dabigatran etixilate may be increased by the coadministration of aspirin, nonsteroidal antiinflammatory drugs (NSAIDs), clopidogrel, selective serotonin reuptake inhibitors, selective norepinephrine reuptake inhibitors, and other drugs which may impair hemostasis. Dabigatran etixilate should only be coadministered with such drugs if the benefit outweighs the risk. Concomitant treatment with any other anticoagulant (e.g., UFH heparin, LMWH, heparin derivatives, oral anticoagulants) is contraindicated, except when switching anticoagulant therapy or when UFH is given at doses necessary to maintain an open central venous or arterial catheter.

17.10.3 Therapeutic efficacy of dabigatran etexilate

The efficacy of dabigatran etexilate compared with warfarin for the prevention of stroke or systemic embolism in patients with NVAF was assessed in the large, phase III, multicenter RE-LY trial. Subsequent to the initial publication of RE-LY trial results, several additional outcome events were identified. Inclusion of these newly identified events in the RE-LY database did not appreciably alter the results and did not change the conclusion of the study, wherever possible, updated results are used. Patients in RE-LY were aged >18 years with documented NVAF at screening or within the past 6 months and at least one of the following risk factors: prior stroke or transient ischemic attack (TIA); left ventricular ejection fraction <40%; New York Heart Association class II or higher heart failure symptoms within the past 6 months; age >75 years; or age 65–74 years with diabetes, hypertension, or coronary artery disease (CAD). Patients were randomized to receive dabigatran etexilate 110 or 150 mg twice OD, each administered in a blinded fashion, or open-label warfarin. Concomitant use of aspirin (<100 mg/day) or other antiplatelet agents was permitted [347]. The primary end point was the incidence of stroke (defined as the sudden onset of a focal neurological deficit in the area of a major cerebral artery and categorized as ischemic, hemorrhagic, or unspecified) or systemic embolism (defined as an acute vascular occlusion of an extremity or organ) in the intention-to-treat population. The primary analysis tested the noninferiority of both dosages of dabigatran etexilate; two-sided *P*-values were used for all subsequent superiority testing. Dabigatran etexilate 110 or 150 mg twice OD was effective for the prevention of stroke and systemic embolism in patients with NVAF and a risk of stroke. Both dosages of dabigatran etexilate were noninferior to warfarin at preventing stroke or systemic embolism. Dabigatran etexilate 150 mg twice OD was superior to warfarin in the prevention of stroke or systemic embolism [348].

17.10.4 Tolerability of dabigatran etexilate

Dabigatran etexilate was generally well tolerated in patients with NVAF in the RE-LY study. The most commonly reported adverse events occurring in >8% of patients in any treatment group were dyspepsia, dizziness, dyspnea, and peripheral edema. The vast majority of upper GI symptoms (i.e., gastroesophageal reflux, upper abdominal pain, dysmotility-related symptoms, gastroduodenal injury) were of mild-to-moderate intensity. Patients receiving either dosage of dabigatran etexilate were significantly ($P < .001$) more likely than those receiving warfarin to discontinue treatment because of serious adverse events (2.7% and 2.7% vs 1%) [349]. GI symptoms such as pain, vomiting, and diarrhea were the reason for discontinuation in 2.1%, 2.2%, and 0.6% of patients receiving dabigatran etexilate 150 mg twice OD, dabigatran etexilate 110 mg twice OD, and warfarin, respectively. Of the patients permanently discontinuing symptoms, approximately half of the discontinuations occurred during the first 3 months of treatment.

17.10.5 Bleeding outcomes

The primary safety outcome of the RE-LY study was major bleeding, defined as a reduction in the hemoglobin level of >20 g/L, transfusion of >2 units of blood, or symptomatic bleeding in a critical area or organ. The incidence of major bleeding was not significantly different between patients receiving dabigatran etexilate 150 mg twice OD and those receiving warfarin. However, dabigatran etexilate 110 mg twice OD was associated with a significantly lower incidence of major bleeding than warfarin. Rates of life-threatening major bleeding, minor bleeding, major or minor bleeding, and intracranial bleeding were significantly lower with each dabigatran etexilate regimen than with warfarin. Dabigatran etexilate 150 mg twice OD was associated with a significantly higher rate of major GI bleeding than warfarin. There were no significant differences between dabigatran etexilate and warfarin with regard to rates of nonlife-threatening major bleeding or extracranial bleeding [350]. Rates of life-threatening major bleeding, nonlife-threatening major bleeding, intracranial bleeding, and extracranial bleeding were not significantly different between the two dabigatran etexilate groups. For the primary safety outcome of major bleeding, results from several subgroup analyses were consistent with those seen in the overall RE-LY trial. In general, the relative effects of dabigatran etexilate 150 or 110 mg twice OD versus warfarin on the rate of major bleeding were observed across a number of subgroups, regardless of previous TIA or stroke, prior VKA exposure, comorbid heart failure, history of hypertension, prior CAD or M1, valvular heart disease, ethnicity, and use of antiplatelet therapy or digoxin.

The risk of major bleeding increased with increasing $CHADS_2$ score[6] and with increasing age. Exploratory analyses showed that the risk of major bleeding was significantly lower with both dosages of dabigatran etexilate than with warfarin among patients aged <75 years, but not among those aged >75 years. Among patients receiving either dosage of dabigatran etexilate, the risk of GI major bleeding was significantly ($P < .01$) higher in patients with gastroesophageal reflux-related symptoms, dysmotility-related symptoms, or gastroduodenal injury than in patients without these upper GI symptoms [351]. In terms of major bleeding, significant interactions were seen between treatment and baseline renal function, center-based INR control and region. Rates of major bleeding were significantly lower with dabigatran etexilate 150 mg twice OD than warfarin at centers with poor INR control, while in centers with higher INR control the regimens had similar levels of risk. The rate of major bleeding was significantly lower with dabigatran etexilate 150 mg twice OD than with warfarin in Asian patients, but not in non-Asian patients [352].

Besides potential drug interactions, the EU SPC states that factors that may increase the bleeding risk in patients receiving dabigatran etexilate include moderate renal impairment, congenital or acquired coagulation disorders, thrombocytopenia or functional platelet defects, recent biopsy or major trauma, bacterial endocarditis and esophagitis, and gastritis or gastroesophageal reflux. In these settings, dabigatran etexilate should only be administered if the benefit outweighs the bleeding risk. Among patients who interrupted dabigatran etexilate or warfarin for surgery or an invasive procedure ($n = 4591$), there was no significant difference in the rate of periprocedural major bleeding between dabigatran etexilate 150 mg twice OD or dabigatran etexilate 110 mg twice OD and warfarin [353].

In all three treatment groups, the risk of major bleeding was significantly ($P < .001$) higher in patients undergoing an urgent surgery or

[6] It is a score to determine stroke risk in patients having cardiac surgery independent of AF has been advocated.

procedure than in those undergoing an elective surgery or procedure. Data from the real-world setting were generally consistent with those from the RE-LY study. Some studies demonstrated a lower risk of major bleeding in patients receiving dabigatran etexilate compared with those receiving warfarin, while other studies demonstrated no significant difference in the risk of major bleeding between dabigatran etexilate and warfarin. Although two studies demonstrated a significantly ($P < .05$) higher risk of major bleeding with dabigatran etexilate versus warfarin, it should be noted that these studies had a number of methodological limitations. In some studies, dabigatran etexilate was associated with an increased risk of GI bleeding compared with warfarin, while other studies showed a similar or lower risk of GI bleeding with dabigatran etexilate versus warfarin.

17.10.6 Dosage and administration

In the EU, oral dabigatran is approved for use in the prevention of stroke and systemic embolism in patients with NVAF and one or more risk factors, such as prior stroke or TIA, age >75 years, diabetes, hypertension, or heart failure [354]. The recommended dosage of dabigatran etexilate is 150 mg twice OD, taken with or without food. For patients aged >80 years or receiving concomitant verapamil, the recommended dosage is 110 mg twice OD. A dosage of 110 or 150 mg twice OD should be selected based on individual assessment of thromboembolic risk versus the risk of bleeding for patients with the following characteristics: age 75–80 years, moderate renal impairment, gastritis, esophagitis, or gastroesophageal reflux, or other factors associated with an increased risk of bleeding. Dabigatran etexilate should be used with caution in conditions with an increased bleeding risk and in situations with concomitant use of drugs that affect hemostasis via inhibition of platelet aggregation. Patients with an increased risk of bleeding should be monitored closely, and treatment should be interrupted if clinically relevant bleeding occurs. Dabigatran occurs and the source of bleeding should be investigated. The specific reversal agent idarucizumab may be used in the event of life-threatening or uncontrolled bleeding, emergency surgery, or urgent procedures when rapid reversal of the anticoagulant effects of dabigatran etexilate is required.

Dabigatran etexilate is not recommended in patients with liver enzymes >2× ULN (Upper Limit of Normal) and is contraindicated in patients with severe renal impairment [CL (Chloride)$_{CR}$ <30 mL/minute]. Other contraindications include active clinically significant bleeding, lesions or conditions considered a significant risk factor for major bleeding, liver disease or hepatic impairment expected to have an impact on survival, and prosthetic heart valves requiring anticoagulant treatment. In all patients, renal function should be assessed prior to starting treatment and if a decline in renal function is suspected during treatment. In patients aged >75 years or with mild-to-moderate renal impairment, renal function should be monitored at least annually or more frequently as required in specific clinical situations when it is suspected that renal function could decline or deteriorate [355].

17.10.7 Place of dabigatran etexilate in nonvalvular atrial fibrillation

Current European Society of Cardiology treatment guidelines generally consider NOACs as being preferable to VKAs for stroke prevention in most patients with AF. The UK National Institute for Health and Care Excellence recommends dabigatran etexilate and the other approved NOACs (apixaban, edoxaban, and rivaroxaban) as options for the prevention of stroke and systemic embolism in patients with NVAF. In the RE-LY study, both dabigatran etexilate 110 and 150 mg twice OD were noninferior to warfarin with regard to the prevention of stroke and systemic embolism (primary end point) in patients with NVAF, with superiority subsequently shown for the higher dosage of dabigatran etexilate versus warfarin [356]. Dabigatran etexilate was as good as or better than warfarin in terms of most secondary efficacy endpoints. In RELY-ABLE, an extension of the RE-LY trial, the effects of dabigatran etexilate were consistent over an extended time period of upto 6.7 years. Dabigatran etexilate was also effective across various patient populations. Of note, warfarin recipients in the RE-LY trial had a mean time within the therapeutic range of 64%, which is higher than that typically seen in real-world clinical practice. This may have resulted in an underestimation of the benefits of dabigatran etexilate versus warfarin in RE-LY. The adverse event profile of dabigatran etexilate was generally similar to that of warfarin in the RE-LY study, except for the incidence of dyspepsia. Dabigatran etexilate capsules contain dabigatran-coated pellets with a tartaric acid core to enhance absorption of the drug; this acidity may partly explain the increased incidence of dyspepsia with both dosage of dabigatran etexilate [357]. Although dabigatran etexilate was as good as or better than warfarin with regard to most bleeding endpoints, including major bleeding, the dabigatran etexilate 150 mg twice-OD regimen was associated with significantly higher rates of GI major bleeding in patients aged >75 years. Data on bleeding outcomes from various subgroup analyses were generally consistent with those from the RE-LY study.

In an exposure response analysis of the RE-LY trial, the risks of ischemic stroke or systemic embolism and major bleeding correlated with the C_{trough} of dabigatran etexilate. These results indicate that among high-risk groups (i.e., very elderly patients and/or those with renal impairment), adjusting the dosage of dabigatran etexilate to optimize exposure may improve the benefit–risk ration for patients at either extreme of the concentration range. Indeed, in the EU, the lower dosage of dabigatran etexilate (110 mg twice OD) is indicated in patients aged >80 years and is also recommended as an option for patients aged 75–80 years and those with moderate renal impairment. Results of analyses conducted in real-world settings in patients with NVAF have been mixed, although they generally demonstrate that dabigatran etexilate 110 or 150 mg twice OD was as good as or better than warfarin in preventing stroke [342]. In terms of bleeding outcomes, data from real-world studies were generally consistent with those from the RE-LY study. Ongoing, prospective, observational registries such as GARFIELD and GLORIA-AF will provide longitudinal real-life data on the use and uptake of NOACs. The GARFIELD registry is examining outcomes in newly diagnosed patients with NVAF at risk for stroke. Preliminary analyses indicate that increased use of NOACs has improved adherence to guidelines recommending the use of antithrombotic therapy in this patient population. The GLORIA-AF registry is investigating patient characteristics influencing the choice of antithrombotic treatment for stroke prevention in patients with NVAF.

A prespecified interim analysis found that the introduction of NOACs into clinical practice has had a considerable influence on treatment patterns, particularly in Europe and North America; in these regions, overall NOAC use is more common than VKA use. Dabigatran etixilate offers a number of potential advantages over warfarin. While warfarin acts indirectly by inhibiting the synthesis of several vitamin K-dependent coagulation factor (II, VI, IX, and X), dabigatran etexilate works by directly inhibiting the enzymatic activity of thrombin. Dabigatran, the active form of dabigatran etexilate, has high affinity and selectivity for thrombin, with predictable effects on anticoagulation. Unlike warfarin, dabigatran etexilate has a rapid onset of action and predictable pharmacokinetics for routine laboratory testing. Nevertheless, the use of dabigatran etixilate is not without possible limitations [358]. The discontinuation rate during 2 years of follow-up in RE-LY was relatively high (21% for dabigatran etexilate and 17% for warfarin). The discontinuation rate is expected to be higher outside the setting of a controlled clinical trial. Nonadherence is likely to be an issue in real-world clinical practice; particularly as dabigatran etexilate has a short half-life and requires twice-OD dosing. As regular laboratory testing to monitor the anticoagulant effects of dabigatran etexilate is not required, compliance may be an issue with NOACs. Two national cohort studies assessed adherence to dabigatran etexilate in patients with NVAF. During the first year of treatment, the mean proportion of days covered (PDC) was 84%, with the majority of patients demonstrating adequate adherence (e.g., 77% had >80% PDC) [359].

In some clinical settings, it is important to be able to rapidly reverse anticoagulation. The anticoagulant effects of warfarin can be reversed by a number of methods, including dose omission, administration of vitamin K and, if rapid reversal is required, replacement of coagulation factors with prothrombin complex concentrates or recombinant-activated factor VII. Conversely, NOAC reversal is hindered by their competitive inhibition of thrombin or activated factor X. Although several specific antidotes for NOAC reversal are being developed, dabigatran etexilate is currently the only NOAC with an approved reversal agent (idarucizumab). The humanized monoclonal antibody fragment idarucizumab is indicated for patients treated with dabigatran etexilate when reversal of its anticoagulant effects is required for emergency surgery/urgent procedures, or in the event of life-threatening or uncontrolled bleeding [360]. Early pharmacoeconomic analyses showed that dabigatran etexilate 150 mg twice OD was a cost-effective alternative to dose-adjusted warfarin for the prevention of stroke and systemic embolism in patients with NVAF. However, these analyses were conducted prior to the approval of idarucizumab. In two cost analyses, use of idarucizumab lowered costs associated with the use of blood products, although these cost reductions were offset by the acquisition cost of idarucizumab. Analyses taking into account the impact of the introduction of idarucizumab on the cost-effectiveness of dabigatran would be of interest.

Currently, among the NOACs, dabigatran etexilate has the longest term randomized data to support its efficacy and tolerability. To date, no randomized controlled trials (RCTs) have directly compared NOACs in patients with NVAF. A meta-analysis of 71,683 patients included in the RE-LY, ARISTOTLE (apixaban), ENGAGE-AF (edoxaban), and ROCKET-AF (rivaroxaban) trials demonstrated that high-dose NOACs significantly ($P < .0001$) reduced the incidence of stroke or systemic embolism by 19% compared with warfarin; this benefit was mainly driven by a 51% reduction in hemorrhagic stroke. In another systematic review and meta-analysis of 72,720 patients included in the RE-LY, ARISTOTLE, ENGAGE-AF, ROCKET-AF, and J-ROCKET (rivaroxaban) trials, NOACs were associated with a clinically significant 4% reduction in the risk of serious adverse events compared with VKAs. Although network meta-analyses, other indirect comparisons, and real-world studies have demonstrated some apparent minor differences between NOACs, the results of these analyses should be interpreted with caution. RCTs directly comparing dabigatran etexilate with other NOACs are needed [361].

To conclude, in patients with NVAF, dabigatran etexilate 150 mg twice OD is more effective than warfarin for the prevention of stroke and systemic embolism, without an increase in the risk of major bleeding. Dabigatran etexilate has a rapid onset of action, relatively few drug interactions, and no requirement for routine laboratory monitoring. In addition, an approved specific reversal agent is available. Therefore, dabigatran etexilate is a useful option for the prevention of stroke or systemic embolism in patients with NVAF.

17.11 Dabigatran: pharmacological analysis

17.11.1 General dosage information

Conversion from warfarin to dabigatran etixilate mesylate: discontinue warfarin and initiate dabigatran etixilate mesylate when the INR is less than 2.

Conversion from dabigatran etixilate mesylate to warfarin: adjust the starting time of warfarin based on renal function; CrCl 50 mL/minute or greater, initiate warfarin 3 days prior to dabigatran discontinuation; CrCl 30–50 mL/minute, initiate warfarin 2 days prior to dabigatran discontinuation; CrCl 15–30 mL/minute, initiate warfarin 1 day prior to dabigatran discontinuation.

Conversion from parenteral anticoagulant therapy to dabigatran etixilate mesylate: initiate dabigatran etixilate mesylate 0–2 hours before the time of the next scheduled dose of the parenteral anticoagulant or at the same time as discontinuation of parenteral anticoagulant continuous infusion (e.g., IV UFH).

Conversion from dabigatran etixilate mesylate to parenteral anticoagulant therapy: in patients who have a CrCl of 30 mL/minute or greater, postpone initiation of a parenteral anticoagulant for 12 hours after the last dose of dabigatran etixilate mesylate; in patients with a CrCl of less than 30 mL/minute, postpone initiation of parenteral anticoagulant for 24 hours after the dose of dabigatran etixilate mesylate.

Invasive or surgical procedure: discontinue 1–2 days prior (CrCl 50 mL/minute or greater) or 3–5 days prior (CrCl less than 50 mL/minute), because of increased bleeding risk; consider longer time in patients undergoing major surgery, spinal puncture, or placement of a spinal or epidural catheter or port, in whom complete hemostasis may be necessary. If surgery cannot be delayed, weigh the risk of bleeding against the urgency of surgery and use a specific reversal agent.

Invasive or surgical procedure, with a standard risk of bleeding: discontinue 24 hours prior to the day of surgery (CrCl greater than 50 mL/minute); 2 days prior to the day of surgery (CrCl greater than 30–50 mL/minute); 4 days prior to the day of surgery (CrCl 30 mL/minute or less).

Invasive or surgical procedure: resume only after hemostasis and no ongoing bleeding.

Invasive or surgical procedure, minor procedures: resume with a reduced dose of 75 mg the evening of the procedure and increase to the regular dose of 110 or 150 mg twice OD the following morning.

Invasive or surgical procedure, increased risk of bleeding: resume 48–72 hours after surgery starting with the regular dose.

Invasive or surgical procedure, indwelling catheter for neuraxial anesthesia: resume only after 4 hours or longer following catheter removal.

Invasive or surgical procedure, bowel paralysis after major abdominal surgery: resume after intake of oral medications is possible. Bridging with LMWH is recommended.

Invasive or surgical procedure, no oral medication: resume after intake of oral medications is possible. Bridging with LMWH is recommended.

17.11.2 Atrial fibrillation: thromboembolic disorder; prophylaxis

150 mg orally twice OD (FDA dosage).

Following an ischemic stroke or TIA, initiation of therapy within 14 days is reasonable, but initiation may be delayed beyond 14 days in the presence of high risk for hemorrhagic conversion.

17.11.3 Deep venous thrombosis, following parenteral therapy

150 mg orally twice OD.

17.11.4 Deep vein thrombosis, recurrence; prophylaxis

CrCl greater than (30 mL/minute) 150 mg orally twice OD.

17.11.5 Postoperative deep vein thrombosis; prophylaxis; repair of hip

110 mg orally 1–4 hours after surgery and after hemostasis is achieved, then 220 mg orally once OD for 28–35 days.

17.11.6 Pulmonary embolism, following parenteral therapy

CrCl greater than (30 mL/minute) 150 mg orally twice OD.

17.11.7 Pulmonary embolism; prophylaxis; repair of hip

110 mg orally 1–4 hours after surgery and after hemostasis is achieved, then 220 mg orally once OD for 28–35 days.

17.11.8 Contraindications/warnings

Contraindications:

- Active pathological bleeding.
- Mechanical prosthetic heart valve.
- Serious hypersensitivity reaction (e.g., anaphylactic reaction or anaphylactic shock) to dabigatran or any components of the product.

17.11.9 Precautions

Beers criteria: avoid use in elderly patients with CrCl less than 30 mL/minute due to a lack of evidence for efficacy and safety; dose reduction recommended in patients with impaired renal function but CrCl greater than 30 mL/minute who receive concomitant interacting drugs. Use with caution in patients 75 years of age or older due to increased risk of GI bleeding than with warfarin.

Cardiovascular: Acute coronary syndrome risk increases.

Valvular heart disease including presence of bioprosthetic heart valve: use not recommended.

Concomitant use: P-gp inducers such as rifampin; avoid use.

Concomitant use: P-gp inhibitors in patients receiving treatment or prophylaxis for DVT and PE with renal impairments (i.e., CrCl less than 50 mL/minute); avoid use.

Concomitant use: P-gp inhibitors in patients with NVAF with sever renal impairments (i.e., CrCl 15–30 mL/minute); avoid use.

Concomitant use: dronedarone or systemic ketoconazole in patients with NVAF and moderate renal impairment (i.e., CrCl 30–50 mL/minute); dose reduction recommended.

Elderly: older patients are at increased risk of severe bleeding; hemorrhage resulting in hospitalization, disability, or death has been reported.

Hematologic: significant and potentially fatal bleeding has been reported, especially in the very elderly, with labor and delivery, and with concomitant use of drugs that increase bleeding risk such as antiplatelet agents, heparin, fibrinolytic therapy, and chronic NSAID use; monitoring recommended and discontinue with active pathological bleeding.

Renal: renal impairment; risk of increased drug exposure an risk of bleeding; dose adjustment or discontinuation recommended.

Surgery: invasive or surgical procedures increase risk of bleeding; discontinue prior to procedure or weigh benefit versus risk.

Adverse effects

Common [GI]:

Esophagitis, gastritis, gastroesophageal reflux disease (AF, 5.5%), GI hemorrhage (DVT and PE, 0.7%–3.1%; NVAF, 6.1%), GI ulcer, indigestion (DVT and PE, 4.1%–7.5%).

Hematologic:

Hemorrhage (DVT and PE treatment or prophylaxis, 9.7%–12.3%; NVAF, 16.6%).

Serious

Cardiovascular:

Myocardial infarction (DVT and PE, 0.1%–0.66%; NVAF, 0.7%).

GI:

GI hemorrhage, major (DVT and PE, 0.1%–0.6%; NVAF, 1.6%).

Hematologic:

Hemorrhage, major (DVT and PE, 0.3%–2%; NVAF, 3.3%) thrombosis.

Immunologic: anaphylaxis

Neurological: epidural hematoma, intracranial hemorrhage (NVAF, 0.3%; DVT and PE, 0.1%), traumatic spinal subdural hematoma

Respiratory: hemorrhage, alveolar

Generic availability: NO

Toxicology

Clinical effects:

Direct thrombin inhibitors

Uses: direct thrombin inhibitors (DTIs) are used as anticoagulants. Argatroban, bivalirudin, desirudin, and lepirudin are used parenterally and are indicated for adjuvant anticoagulation for percutaneous cardiac interventions or as a substitution for heparin/LMWHs in cases of heparin-induced thrombocytopenia (HIT). Dabigatran etixilate is an oral medication approved for the treatment of VTE and stroke prophylaxis in AF.

17.11.10 Pharmacology

These agents directly inhibit thrombin, leading to inhibition of clot formation and stabilization.

Toxicology

The toxic effects are extensions of the pharmacological effects and primarily include bleeding complications.

Epidemiology

Overdose data are limited. One patient, a 66-year-old man, ingested 9 g dabigatran in a suicide attempt and subsequently developed hypotension, bradycardia, and coagulopathy. Inpatient medication errors may occur in 1%–2% of patients receiving DTIs. The incidence of overdose is likely to increase as new DTIs are approved for in-hospital parenteral use and oral anticoagulation indications.

Overdose: mild to moderate toxicity

Bleeding complications that do not lead to cardiovascular compromise may be considered mild or moderate.

Severe toxicity

Bleeding complications that lead to hypotension or difficulty with oxygen delivery can be considered severe. Intracranial hemorrhage, hemothorax, cardiac tamponade, retroperitoneal hemorrhage, or massive GI hemorrhage may occur even at therapeutic doses of DTIs. Allergic reactions have been reported.

Adverse effects: common

Hemorrhage, dyspepsia, back pain, nausea, vomiting, and diarrhea.

Less common: Anemia, fever, hematomas, hematuria, GI and rectal bleeding, epistaxis, intracranial bleeding, hypotension, cardiac arrest, dyspnea, cardiac dysrhythmias, abnormal hepatic and renal function, and hemothorax.

Rare: Acute allergic reactions and formation of antihirudin antibodies have also been reported.

17.12 Apixaban: pharmacological analysis

Apixaban is used to help prevent strokes or blood clots in people who have AF (a condition in which the heart beats irregularly, increasing the

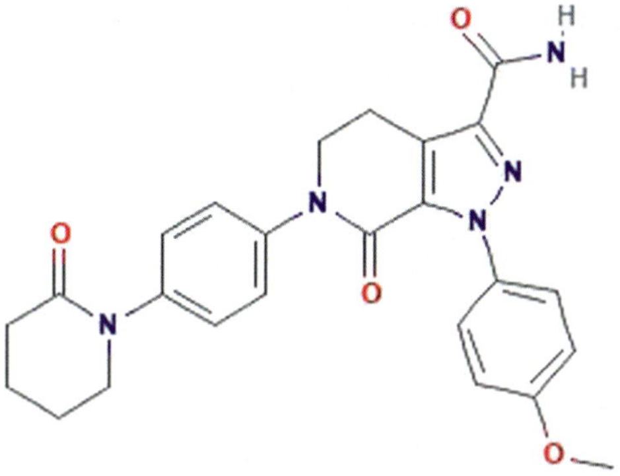

Figure 17.2 Molecular structure of Apixaban.

chance of clots forming in the body and possibly causing strokes) that is not caused by heart valve disease. Apixaban is also used to prevent DVT (a blood clot, usually in the leg) and PE (a blood clot in the lung) in people who are having hip replacement or knee replacement surgery. Apixaban is also used to treat DVT and PE, and may be continued to prevent DVT and PE from happening again after the initial treatment is completed. Apixaban is in a class of medications called FXa inhibitors. It works by blocking the action of a certain natural substance that helps blood clots to form (Fig. 17.2).

Apixaban comes as a tablet to take by mouth. It is usually taken with or without food twice a day. When apixaban is taken to prevent DVT and PE after hip or knee replacement surgery, the first dose should be taken at least 12–24 hours after surgery. Apixaban is usually taken for 35 days after a hip replacement surgery and for 12 days after knee replacement surgery. Apixaban is taken at around the same times every day. The directions on the prescription label should be followed carefully, and the doctor or pharmacist asked to explain any part not understood. Apixaban should be taken exactly as directed. No more or less of it should be taken, nor should it be taken more often than prescribed by the doctor.

Apixaban can cause a very serious blood clot around the spinal cord if undergoing a spinal tap or receiving spinal anesthesia (epidural), especially if the patient has a genetic spinal defect, if they have a spinal catheter in place, if they have a history of spinal surgery or repeated spinal taps, or if they are also using other drugs that can affect blood clotting. This type of blood clot can lead to long-term or permanent paralysis.

17.12.1 Before taking apixaban

The patient should tell their doctor and pharmacist if they are allergic to apixaban, any other medications, or any of the ingredients in apixaban tablets. The pharmacist should be asked or the medication guide checked for a list of the ingredients.

The patient should tell their doctor and pharmacist what other prescription and nonprescription medications, vitamins, and nutritional supplements they are taking or plan to take. They should mention any of the following: carbamazepine (Carbatrol, Epitol, Equetro, Tegretol, Teril); clarithromycin (Biaxin, in Prevpac); itraconazole (Onmel, Sporanox); ketoconazole (Nizoral); phenytoin (Dilantin, Phenytek); rifampin (Rifadin, Rimactane, in Rifadin, in Rifater); ritonavir (Norvir, in Kaletra); selective serotonin reuptake inhibitors (SSRIs) such as citalopram (Celexa), fluoxetine (Prozac, Sarafem, Selfemra, in Symbyax), fluvoxamine (Luvox), paroxetine (Brisdelle, Paxil, Pexeva), and sertraline (Zoloft); and serotonin and norepinephrine reuptake inhibitors (SNRIs) such as duloxetine (Cymbalta), desvenlafaxine (Khedezla, Pristiq), milnacipran (Fetzima, Savella), and venlafaxine (Effexor). The doctor may need to change the doses of medications or monitor the patient carefully for side effects. Many other medications may also interact with apixaban, so the patient must tell their doctor about all the medications they are taking, even those that do not appear on this list.

- The patient should know that apixaban may interact with certain medications that may be used to treat them if they have a stroke or other medical emergency. In the case of an emergency, the patient or a family member should tell the doctor or emergency room staff who treat them that they are taking apixaban.
- If the patient has an artificial heart valve or if they have heavy bleeding anywhere that cannot be stopped. The doctor would probably tell them not to take apixaban.
- If the patient has or has ever had any type of bleeding problem, or kidney or liver disease.
- If they are pregnant, plan to become pregnant, or are breastfeeding. If they become pregnant while taking apixaban, they should inform their doctor.
- If they are having surgery, including dental surgery, they should tell the doctor or dentist that they are taking apixaban. The doctor may

tell them to stop taking apixaban before the surgery or procedure. If they need to stop taking apixaban because they are having surgery, the doctor may prescribe a different medication to prevent blood clots during this time. The doctor will inform them when you should start taking apixaban again after any surgery. These directions should be followed carefully.

17.12.2 What is apixaban?

Apixaban is used to lower the risk of stroke caused by a blood clot in people with a heart rhythm disorder called AF.

Apixaban is also used after hip or knee replacement surgery to prevent a type of blood clot called DVT, which can lead to blood clots in the lungs (PE).

Apixaban is also used to treat DVT or PE, and to lower the risk of having a repeat DVT or PE. Apixaban can cause a very serious blood clot around the spinal cord if undergoing a spinal tap or receiving spinal anesthesia (epidural), especially if for a genetic spinal defect, if they have a spinal catheter in place, if they have a history of spinal surgery or repeated spinal taps, or if they are also using other drugs that can affect blood clotting. This type of blood clot can lead to long-term or permanent paralysis.

The patient should get emergency medical help if they have *symptoms of a spinal cord blood clot* such as back pain, numbness, or muscle weakness in the lower body, or loss of bladder or bowel control.

Patients should not stop taking apixaban unless their doctor tells them to. Stopping suddenly can increase their risk of blood clot or stroke.

17.12.3 Before taking this medicine

- Patients should not take apixaban if they are allergic to it, or have active bleeding from a surgery, injury, or other cause.
- Apixaban may cause them to bleed more easily, especially if they have a bleeding disorder that is inherited or caused by disease.
- The patient should tell their doctor if they have an artificial heart valve, or have ever had:
- liver or kidney disease;
- are older than 80; or
- weigh less than 132 pounds (60 kg).

Apixaban can cause a very serious blood clot around the spinal cord if undergoing a spinal tap or receiving spinal anesthesia (epidural). This type of blood clot could cause long-term paralysis, and may be more likely to occur if:

- the patient has a spinal catheter in place or if a catheter has been recently removed;
- they have a history of spinal surgery or repeated spinal taps;
- they have recently had a spinal tap or epidural anesthesia;
- they are taking an NSAID—aspirin, ibuprofen (Advil, Motrin), naproxen (Aleve), diclofenac, indometacin, meloxicam, and others; or
- they are using other medicines to treat or prevent blood clots.

Taking apixaban may increase the risk of bleeding while pregnant or during delivery. Patients should tell their doctor if they are pregnant or plan to become pregnant.

- Patients should not breastfeed while using apixaban.

17.12.4 How should the patient take apixaban?

Follow all directions on the prescription label and read all medication guides or instruction sheets. The doctor may occasionally change the dose. Use the medicine exactly as directed.

- Apixaban can be taken with or without food.

Apixaban can make it easier for to bleed, even from a minor injury. The patient should seek medical attention if they have bleeding that will not stop.

If they need surgery or dental work, they should tell their doctor or dentist ahead of time if they have taken apixaban within the past 24 hours. They may need to stop taking apixaban for a short time.

They should not stop taking apixaban unless their doctor advises this. *Stopping suddenly can increase the risk of blood clot or stroke.*

If the patient stops taking apixaban for any reason, their doctor may prescribe another medication to prevent blood clots until they start taking apixaban again.

17.12.5 Apixaban side effects

Emergency medical help should be sought if the patient has *signs of an allergic reaction*: hives; chest pain, wheezing, difficult breathing; feeling light-headed; swelling of the face, lips, tongue, or throat.

They should also seek emergency medical attention if they have *symptoms of a spinal blood clot*: back pain, numbness or muscle weakness in the lower body, or loss of bladder or bowel control.

17.12.6 The patient should call their doctor at once if they have

- Easy bruising, unusual bleeding (nose, mouth, vagina, or rectum), bleeding from wounds or needle injections, any bleeding that will not stop;
- Heavy menstrual periods;
- Headache, dizziness, weakness, feeling like they might pass out;
- Urine that looks red, pink, or brown; or
- Black or bloody stools, coughing up blood or vomit that looks like coffee grounds.

17.12.7 What other drugs will affect apixaban?

Sometimes it is not safe to use certain medications at the same time. Some drugs can affect the blood levels of other drugs that are taken, which may increase side effects or make the medications less effective.

Many other drugs (including some over-the-counter medicines) can increase the risk of bleeding or blood clots, or the risk of developing blood clots around the brain or spinal cord during a spinal tap or epidural. *It is very important for the patient to tell their doctor about all medicines they have recently used*, especially:

- Any other medicines to treat or prevent blood clots;
- A blood thinner such as heparin or warfarin (Coumadin, Jantoven);
- An antidepressant; or
- An NSAID used long term.

17.12.8 Side effects

Some side effects can be serious. If the patient experiences any of these symptoms, they should call their doctor immediately or get emergency medical treatment:

- Bleeding gums;
- Nosebleeds;
- Heavy vaginal bleeding;
- Red, pink, or brown urine;
- Red or black, tarry stools;
- Coughing up or vomiting blood or material that looks like coffee grounds;
- Swelling or joint pain;
- Headache;
- Rash;
- Chest pain or tightness;

- Swelling of the face or tongue;
- Trouble breathing;
- Wheezing;
- Feeling dizzy or faint.

17.12.9 Direct thrombin inhibitors

17.12.9.1 Management of mild to moderate toxicity

Evaluate the need for anticoagulation and either decrease the dose or stop the infusion, whichever is clinically appropriate. Minor bleeding complications include epistaxis, mucous membrane bleeding, and poor clotting of minor cuts. Minor bleeding complications can usually be managed by simply stopping the medication. If the patient has blood loss that leads to cardiovascular instability, blood product transfusion should be considered. Patients with symptomatic anemia can be managed with red blood cell (RBC) transfusion alone. Allergic reactions should be treated with cessation of infusion, antihistamines, steroids, and intramuscular epinephrine for anaphylactic reactions.

Management of severe toxicity: stop the infusion; administer isotonic fluid for hypotension, packed RBCs for anemia. Patients with significant continued blood loss despite cessation of the direct thrombin inhibitor infusion should be given fresh frozen plasma (FFP) in an attempt to competitively antagonize the thrombin inhibition. Prothrombin complex concentrates or cryoprecipitate may be considered because they have a higher concentration of thrombin per volume when compared with FFP, however they carry a higher risk of thromboembolic complications.

17.13 Deep vein thrombosis/pulmonary embolism management with rivaroxaban (Xalerto)

Rivaroxaban is FDA approved for the acute treatment of DVT and PE and reduction of the risk of recurrence of DVT and PE.

17.13.1 Food and Drug Administration-approved indications

- NVAF, as an alternative to warfarin, for stroke prevention.
- Treatment of DVT and PE and reduction in risk of recurrence of DVT and PE.
- Prophylaxis of DVT following hip and knee replacement.
- About the drug:
- Mechanism of action: FXa inhibitor.
- Rapid absorption and peak activity in 2–4 hours.

- Half-life 5–9 hours, in the elderly half-life is 11–13 hours.
- The drug will accumulate in renal and liver impairment.
- There are a limited number of drugs that may affect rivaroxaban levels. Drug interactions include ketoconazole, itraconazole, ritonavir, carbamazepine, phenytoin, and rifampin (medications with combined P-gp and strong CYP3A4 interactions).
- Higher doses (15–20 mg tablets) should be taken with food, and lower doses (10 mg tablets) may be taken with or without food.

17.13.2 Contraindications

- Bleeding or fall risks.
- Renal impairment, contraindicated in:
- AF with CrCl <15 mL/minute (need lower dose in CrCl 15–50 mL/minute);
- DVT and PE with CrCl <30 mL/minute;
- DVT prophylaxis with CrCl <30 mL/minute;
- Liver impairment—Child–Pugh B and C.
- It has not been studied in pregnancy, but it does cross the placenta in animals and is excreted in rat milk.
- It has not been studied in valvular heart disease.

17.14 Precautions

- Antiplatelet agents increase the risk of hemorrhage; it is advised that antiplatelet agents be stopped when taking rivaroxaban, unless there is a good reason to continue them.
- It should not be used together with other anticoagulants, like heparin or LMWH.

17.15 Laboratory testing to be done prior to starting rivaroxaban

- Baseline labs prior to starting rivaroxaban: Complete Blood Count, platelets, liver function tests, Serum Creatinine, INR and aPTT—results within 1 month are acceptable.
- Routine monitoring is not required.
- Routine coagulation monitoring is not required. Liver and renal functions should be checked yearly, or more often, if clinically indicated.

17.15.1 Who is not an appropriate candidate for rivaroxaban?

- Renal impairment, contraindicated in:
- AF with CrCl <15 mL/minute (need a lower dose for CrCl 15–50 mL/minute);
- DVT and PE with CrCl <30 mL/minute;
- Liver impairment—contraindicated in Child–Pugh B and C.
- Patient concerns about higher drug costs. Most members will pay higher brand copay for rivaroxaban and lower generic copay for warfarin.
- If a patient is well controlled on warfarin, consider leaving them on warfarin.
- If the patient has significant risks for bleeding and anticoagulation is given on a trial basis the use of shorter acting reversible agents for initial therapy is recommended. This may require inpatient monitoring.

17.15.2 Side effects

- The risk of bleeding appears similar to other anticoagulants.
- DVT and PE treatment—any bleeding was similar (28.0%–28.3%). Clinically relevant nonmajor bleeding was similar (8.6%–8.7%). Major bleeding was numerically less with rivaroxaban (1.0% vs 1.7%).
- AF—major bleeding was similar (3.5–3.6 events per 100 patient-years).
- DVT prophylaxis following knee or hip replacement surgery—major bleeding was similar (0.2%–0.3%).
- Other adverse reactions include upper abdominal pain, dyspepsia, back pain, and oropharyngeal pain.

17.15.3 How rivaroxaban affects laboratory studies?

- Rivaroxaban prolongs PTT and INR and the antifactor Xa test is also influenced by the drug.
- The antifactor Xa test can be used to evaluate plasma levels, but appropriate calibration is not available to check for the presence of rivaroxaban or drug levels.
- Tests that will not be reliable on rivaroxaban include lupus anticoagulant testing, factor X chromogenic, protein S activity, and clotting factor activities.

17.16 Rivaroxaban: an oral direct FXa inhibitor

Rivaroxaban is a small-molecule, direct FXa inhibitor and may be a potentially attractive alternative to VKAs. Rivaroxaban is being investigated for the prevention and treatment of venous and arterial thrombosis. A broad search of Medline, clinical trials.gov, and the annual proceedings of the American Society of Hematology and the International Society on Thrombosis and Hemostasis was conducted. This review addresses the findings of this systematic search, including the need for new oral anticoagulants, the development and pharmacology of rivaroxaban, and the results of completed as well as ongoing trials with rivaroxaban. At present, the safety and efficacy of rivaroxaban for the prophylaxis and treatment of VTE have been evaluated in phase II and III trials involving 24,000 patients. Additionally, rivaroxaban is being evaluated for the treatment of PE, secondary prevention after acute coronary syndromes, and the prevention of stroke and noncentral nervous system embolism in patients with NVAF. The drug may have its greatest impact in providing a much-needed and attractive alternative to warfarin (Fig. 17.3).

Arterial thrombosis, venous thrombosis, and subsequent thromboembolism account for significant morbidity and mortality worldwide. In the United States, more than 200,000 patients develop VTE every year, and upto 30% of these patients die within 30 days. Despite the administration of current prophylaxis, 5%–20% of all hip replacement surgeries are complicated by VTE. Furthermore, deep venous thrombosis and PE are associated with a significant economic burden due to the costs of acute care as well as long-term costs associated with recurrent VTE and postthrombotic syndrome [379]. A problem growing at a faster rate is the increasing burden imposed by AF-associated thromboembolism. AF leads to stasis of blood in the atria and thrombus formation than can leave the heart and

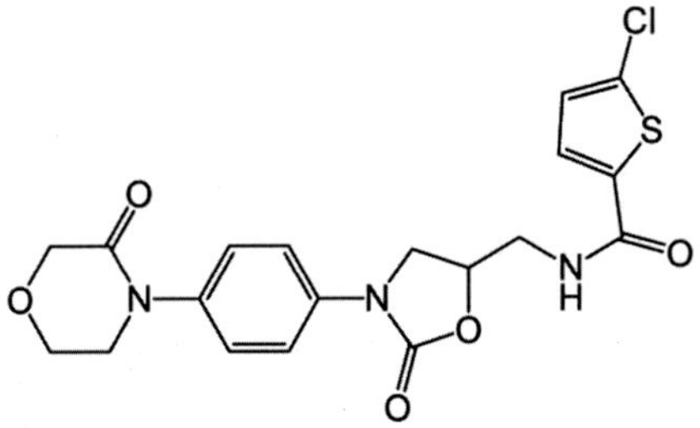

Figure 17.3 Molecular structure of Rivaroxaban.

embolize any vascular bed, most seriously the cerebral circulation, leading to stroke. At present, 2.3 million Americans have AF, including 10% of all patients 80 years and older. By 2050, an estimated 5.6 million Americans will have AF, increasing the risk of stroke fivefold, and ultimately accounting for 15%–20% of all strokes in the United States [380]. Furthermore, patients who have an AF-associated stroke are twice as likely to remain bedridden as other stroke victims.

17.16.1 Clinical pharmacology of rivaroxaban

Rivaroxaban is an oxazolidinone derivative which binds to the active site of FXa, leading to potent and selective inhibition of FXa. In animal models of both venous stasis and thrombosis, oral rivaroxaban inhibited FXa activity, leading to reduced thrombus formation and extension. Rivaroxaban inhibits FXa activity in a dose-dependent manner, accompanied by prolongation of the prothrombin time (PT). The pharmacokinetic profile of rivaroxaban is consistent with rapid oral absorption and 80% bioavailability [381]. The time to peak plasma concentration is between 2.5–4 hours and, after multiple doses, the drug half-life is 5–9 hours in healthy volunteers, and 9–13 hours in the elderly (mean age 65). There are no major active circulating metabolites of rivaroxaban. Rivaroxaban is excreted in the urine (one-third), and the remaining two-thirds is metabolized by the liver. Drug elimination demonstrates first-order kinetics and is impaired with advancing age, renal insufficiency, and in the presence of strong CYP3A4 inhibitors (such as ketoconazole, macrolide antibiotics such as clarithromycin, and many protease inhibitors) [382].

17.16.2 Pharmacodynamics interactions

Rivaroxaban absorption is improved when taken with food; however the drug can be administered to fasting patients. The pharmacokinetics of the drug are unaffected by ranitidine or alteration of gastric pH with antacids in healthy male volunteers. Many patients who will require oral factors documented coronary artery disease and thus require aspirin therapy. Additionally, rivaroxaban is being studied in patients with acute coronary syndromes, a patient population treated with aspirin in addition to other antiplatelet agents. In a randomized two-way crossover study, antiplatelet therapy with aspirin did not alter the pharmacokinetics or pharmacodynamics of rivaroxaban (as determined by bleeding and PTs). A 500 mg loading dose of aspirin, followed by 100 mg, was well tolerated by those subjects randomized to rivaroxaban [383]. A similar phase I, two-way crossover study demonstrated no clinically significant interaction between naproxen and rivaroxaban in healthy subjects. Therefore, based upon early phase I data, there is no evidence of a clinically important interaction between aspirin and rivaroxaban. Since rivaroxaban may have a role in stroke prophylaxis in AF patients, 20 mg of rivaroxaban was coadministered in 20 healthy male volunteers receiving 0.375 mg of digoxin.

Drug exposure was not significantly different between those patients receiving rivaroxaban alone and those who received rivaroxaban and digoxin. Based upon these results, there is no apparent interaction between rivaroxaban and digoxin, suggesting that they can be prescribed together. Bridging with LMWHs is common in patients receiving chronic oral anticoagulants. When given together with rivaroxaban, enoxaparin results in additive inhibition of FXa activity and prolongation of bleeding times; however, coadministration of LMWH and rivaroxaban is possible [384]. Nonetheless, given the short onset of action and half-life of rivaroxaban, the need for bridging with LMWH should be relatively infrequent in patients taking rivaroxaban. Overall, compared to the VKAs and other cardiovascular medications such as amiodarone, rivaroxaban has relatively low potential for substantial pharmacodynamics interactions, allowing for a wide range of concomitant pharmacotherapy.

17.16.3 Toxicity and adverse effects

Therapeutic anticoagulation always carries an attendant risk of bleeding, either due to errors in dosing and administration, occult pathology such as gastric ulceration, unrecognized bleeding diatheses, or urgent and emergency medical procedures. Therefore there is great interest and need for neutralizing agents in the event of significant bleeding. To address this concern, investigators explored the use of recombinant-activated factor VII (rFVIIa) as a partial reversal agent for rivaroxaban. In a rat model of mesenteric hemorrhage, rFVIIa was administered after high-dose rivaroxaban (2 mg/kg). In the presence of high-dose, supratherapeutic rivaroxaban, 400 μg/kg of rFVIIa partially reversed the prolongation of the PT and partially restored total thrombin activity, without affecting rivaroxaban-dependent FXa inhibition. Therefore rFVIIa may have a role as an IV antidote for major bleeding in patients taking rivaroxaban. As evidenced by the significant clinical and economic burden imposed by VTE, including deep venous thrombosis and PE, in both medical and surgical patients, the rising incidence of AF in the rapidly expanding elderly population and the global impact of ischemic heart disease, there are many potential patient populations and indications for novel, safe, and effective oral anticoagulants such as rivaroxaban [385]. Accordingly, broad networks of clinical trials have been designed to evaluate the safety and efficacy of rivaroxaban in patients at risk for arterial and venous thrombosis.

17.17 Rivaroxaban in acute coronary syndrome

The anti-Xa therapy to lower cardiovascular events in addition to aspirin with or without thienopyridine therapy in subjects with acute coronary syndrome [ATLAS ACS TIMI (Thrombolysis in Myocardial Infarction) 46] trial is a phase II placebo-controlled randomized study designed to evaluate the safety of rivaroxaban in patients with recent ACS. The trial will enroll patients between the ages of 18–75 who have symptoms suggestive of ACS, a diagnosis of ST (Spasmodic Torticollis) segment[7] elevation or non-ST segment elevation, myocardial infarction within the past 7 days and, at least one additional high-risk feature. Patients are being randomized to placebo, OD rivaroxaban, or twice-OD rivaroxaban. The trial will have two stages. The primary end point of the dose-escalation stage is significant TIMI bleeding [386]. In the subsequent dose confirmation phase, the primary end point will be a composite of major adverse cardiac events (including death, recurrent myocardial infarction, stroke, or recurrent ischemia requiring revascularization). Patient randomization will be stratified according to the presence or absence of thienopyridine treatment, in order to assess the risk/benefit of anti-FXa activity with mono- or dual antiplatelet therapy.

17.18 Rivaroxaban for stroke prophylaxis in nonvalvular atrial fibrillation

The rivaroxaban once-OD oral direct FXa inhibition compared with VKA for prevention of stroke and embolism trial in AF is a prospective, randomized, double-blind, double-dummy, parallel-group, multicenter, event-driven, noninferiority study comparing the efficacy and safety of once-OD oral rivaroxaban with adjusted-dose oral warfarin for the prevention of stroke and non-CNS (Central Nervous System) embolism in

[7] ST Segment represents the interval between ventricular depolarization and repolarization (ECG).

patients with nonvalvular AF. In ROCKET-AF, patients are being randomized to rivaroxaban 20 mg once OD (15 mg if creatinine clearance is 30–49 mL/minute) versus dose-adjusted warfarin [387]. Blinded treatment will be maintained through the use of a double-dummy system including sham INRs in the patients receiving rivaroxaban. The primary efficacy end point for the trial is a composite of stroke or non-CNS embolism. The objective of the primary efficacy analysis is to establish that rivaroxaban is not inferior to warfarin. The challenges of noninferiority trials against warfarin are well demonstrated. Given the low frequency of stroke in patients receiving dose-adjusted warfarin, a noninferiority trial will require a large sample size. There is a clinical need for new oral anticoagulants. FXa represents an attractive pharmacologic target for new agents. While efficacy is paramount, so too is safety, given the morbidity and mortality associated with bleeding, especially in the predominantly elderly population in whom oral anticoagulants are prescribed [388].

The drug development plan for rivaroxaban is aggressive, with simultaneous investigative programs spanning multiple indications rather than a sequential approach. At present, over 24,000 patients (12,000 are predominantly postoperative orthopedic patients) have been evaluated in completed phase II and III trials of rivaroxaban for thromboprophylaxis and treatment of DVT. By the time all the currently enrolling trials have concluded, more than 50,000 patients will have been evaluated in all randomized controlled trials of rivaroxaban. The advantages of rivaroxaban include the potential for once-OD dosing for all indications, no required dose adjustment for body weight, no unknown interactions with common cardiovascular medications, a relatively safe pharmacodynamics profile with respect to bleeding risk and hepatotoxicity, no clinically significant interaction with aspirin, and the ability to bridge with LMWH when necessary. On the other hand, rivaroxaban is partially renally cleared and will require dose adjustment in those with grade III chronic kidney disease (CKD), and is not being studied in patients with a creatinine clearance less than 30 mL per minute. Additionally, since rivaroxaban's metabolism is affected by potent CYP3A4 inhibitors, such as ketoconazole, clarithromycin, and protease inhibitors, its use will be restricted in some special populations. Nonetheless, after extensive phase II, and now emerging phase III trial data, it appears that rivaroxaban is effective in preventing and treating VTE with a bleeding risk comparable to other anticoagulants. The results of randomized trials evaluating rivaroxaban for the prevention of stroke and non-CNS embolism in AF and secondary prevention of acute coronary syndromes are currently ongoing.

17.19 Rivaroxaban: and then there were three

Rivaroxaban is a direct oral anticoagulant (DOAC) that prolongs blood clotting by preventing thrombin generation via inhibition of FXa. Since, August 1, 2018, rivaroxaban has been fully subsided without restriction. Currently, rivaroxaban is only subsided with special authority approval for a short period of prophylaxis following total hip or total knee replacement. The availability of a third fully subsided oral anticoagulant means that if treatment with a DOAC is preferable to warfarin, prescribers will be able to choose between dabigatran and rivaroxaban, depending on the patient's clinical circumstances and preference.

17.19.1 The "need to know" for prescribing

Rivaroxaban is available in 10, 15, and 20 mg tablets. Rivaroxaban tablets can be placed in blister packs if required. The dose of rivaroxaban is determined by indication and renal function. Dose adjustments are required in people with reduced renal function and an assessment of renal function is required in all patients before rivaroxaban is initiated. Dose adjustment for age alone is not routinely required when prescribing rivaroxaban to older patients. Rivaroxaban (15 and 20 mg tablets) should be taken with food to aid absorption. It is not necessary to take the 10 mg tablets with food. There is no listed interaction between rivaroxaban and grapefruit juice, but as rivaroxaban is metabolized by CYP3A4 and grapefruit inhibits this enzyme, it may be a theoretical risk. Rivaroxaban can be initiated immediately for the treatment of DVT and PE, without the need for prior parenteral anticoagulant treatment, for example, LMWH, as is required before treatment with dabigatran [389].

Modifiable risk factors for bleeding should be managed before treatment with any anticoagulant is initiated, for example, uncontrolled hypertension and alcohol intake greater than eight standard drinks per week. The HAS-BLED prediction tool can be used to assess bleeding risk in patients with AF. A medicine review should be conducted to determine if the patient is taking any medicines or supplements that may increase their risk of bleeding, for example, antiplatelets including aspirin, NSAIDs, or herbal extracts such as garlic, ginkgo, or ginseng [390].

17.19.2 Monitoring patients taking rivaroxaban

Routine testing of the anticoagulant effect is not required during treatment with rivaroxaban. Testing may, however, be required in certain clinical circumstances such as patients with moderate to severe renal dysfunction, prior to surgery, or in the event of bleeding. Annual testing of renal function is recommended for all patients taking rivaroxaban, as with all anticoagulants [3]. More frequent monitoring may be appropriate in patient with progressive kidney disease or in those with a dehydrating illness, hypervolemia, or if nephrotoxic medicines are initiated, for example, NSAIDs. Future dose reductions may be required for patients with declining renal function after they have begun taking rivaroxaban. If a patient develops acute kidney injury, consider withdrawing rivaroxaban until renal perfusion has been restored.

17.19.3 Bleeding risk may influence the treatment decisions

Rivaroxaban and dabigatran are associated with less risk of intracranial hemorrhage than warfarin, but a greater risk of GI bleeding. Intracranial bleeding is perhaps the most concerning adverse effect of any anticoagulant treatment and the lower risk associated with dabigatran and rivaroxaban may mean that a DOAC is preferred to warfarin [391]. If a patient has an elevated risk of GI bleeding, warfarin or reduced dose dabigatran, that is, 110 mg, twice OD, may be the preferred anticoagulants.

17.19.3.1 Rivaroxaban generally requires once-OD dosing

Once-OD dosing for rivaroxaban is sufficient for most indications, which may improve adherence to treatment in comparison to dabigatran. Treatment adherence to dabigatran is typically high in patients aged over 70 years but is substantially lower in patients aged under 50 years, therefore younger patients in particular may prefer the once-OD dosing of rivaroxaban. However, treatment adherence is not necessarily just related to frequency of dosing, and this pattern of lower adherence in younger people may also occur with other anticoagulants [392]. Rivaroxaban has a half-life of 5–9 hours in people with normal renal function, but as thrombin generation is inhibited for 24 hours, a single oral OD dose is sufficient for most indications. Twice-OD dosing of rivaroxaban is indicated for the first 21 days of treatment or prevention of DVT or PE. The consequences of missing a dose of once-OD rivaroxaban are theoretically greater than missing a dose of twice-OD dabigatran, as the patient would have declining anticoagulation for a 24-hour period as opposed to a 12-hour period. However, there is no evidence that this is an issue in practice [393].

17.19.4 Rivaroxaban may be preferred if renal dysfunction is present

Rivaroxaban is recommended in preference to dabigatran for patients with a creatinine clearance of 30–49 mL/minute. This is because rivaroxaban undergoes substantially less renal excretion (36%) than dabigatran (80%). In patients with moderate CKD, the plasma concentration of rivaroxaban is increased by approximately 50%, compared to approximately 210% in patients taking dabigatran. To account for declining renal function in older patients, a reduction from dabigatran, 150 mg, twice OD, to 110 mg, twice OD, is recommended in patients aged 80 years or older [394]. Dose adjustments based purely on age are not required for patients taking rivaroxaban, but dosing based on renal function is required when rivaroxaban is initiated, and regular monitoring of renal function is recommended.

17.19.5 Rivaroxaban may be better tolerated by patients with a history of dyspepsia

Dyspepsia is reported in approximately 11%–12% patients taking dabigatran, whereas dyspepsia associated with taking rivaroxaban is reported in 1%–10% of patients. Although there are direct comparative studies of dabigatran and rivaroxaban, this suggests that rivaroxaban may be better tolerated than dabigatran by patients with a history of dyspepsia. A large retrospective study found that treatment persistence was higher in patients taking rivaroxaban than in patients taking warfarin or dabigatran, possibly because INR testing was not needed, treatment was once OD, and the rate of dyspepsia may have been lower. After 2 years of treatment, 50% of patients were taking dabigatran and 27% of patients were taking warfarin. However, caution is advised when any DOAC is used in a patient with a history of esophagitis or GI ulceration. For other adverse effects, the profiles of rivaroxaban and dabigatran are broadly similar, with bleeding being the greatest clinical concern in patients taking either medicine. Nausea and diarrhea are common adverse effects for both rivaroxaban and dabigatran. Headache, dizziness, hypotension, and dermatological symptoms, for example, rash and pruritus, are more often reported in patients taking rivaroxaban, than in patients taking dabigatran.

17.19.6 Medication interactions with rivaroxaban and dabigatran

As with all anticoagulants, the risk of bleeding is increased by the concurrent use of antiplatelet medicines including aspirin, however, this may be necessary in some patients, for example, following an acute coronary syndrome [395]. Bleeding risk is also increased with concurrent use of other medicines that affect platelet function or coagulation such as NSAIDs, selective serotonin reuptake inhibitors, heparins, and some complementary products. Other medicines increase bleeding risk or reduce the clinical effect of dabigatran or rivaroxaban by increasing or decreasing their elimination.

17.19.7 Switching to rivaroxaban from warfarin

A switch from warfarin to rivaroxaban (or dabigatran) may be appropriate for patients with INR values that are often outside of the therapeutic range, for example [22]:

- Two INR values less than 1.5 in the previous 6 months;
- Two INR values greater than 5 or one value higher than 8 in the previous 6 months;
- Less than 65% of the time within the therapeutic range.

There is unlikely to be any clinical benefit in switching patients from warfarin to a DOAC if they spend >70% of the time within the therapeutic range, as their risk of stroke or systemic embolism is unlikely to be reduced. However, there may be additional reasons that the patient wishes to switch treatment, for example, inconvenience of INR testing. Intolerable adverse effects and medicine interactions may also influence the decision to switch from warfarin to a DOAC.

17.19.8 Switching to warfarin from rivaroxaban

Patients with declining renal function or those who have experienced persistent adverse effects with DOACs may benefit from switching treatment from rivaroxaban to warfarin. To ensure anticoagulation is adequate, warfarin and rivaroxaban should be taken concurrently, and rivaroxaban withdrawn when the patient's INR is >2.0. Warfarin initiation is recommended at a standard dose and, after 2 days, treatment should be guided by INR testing. However, rivaroxaban may continue to contribute to an elevated INR until 24 hours after the last dose [396].

17.19.9 Switching to rivaroxaban from dabigatran

Patients who develop intolerable adverse effects while taking dabigatran, for example, dyspepsia, or who find it difficult to adhere to the twice-OD dosing of dabigatran, may benefit from switching to rivaroxaban. It is recommended that patients take their first dose of rivaroxaban, 12 hours after their last dose of dabigatran to ensure that adequate anticoagulation is maintained.

B. Parenteral anticoagulants

1. Heparin

Source:

Biological origin. Derived from ox lung or pig intestine with a molecular weight of 10,000–20,000 (nonuniform mixture of straight-chain mucopolysaccharide). It is the strongest negatively charged organic acid in the body (unfractionated and may contain impurities).

Administration: SC or IV slow infusion

Anticoagulant action:

Indirectly activates plasma antithrombin factor-III (ATIII). This ATIII binds with the coagulation factors in the intrinsic and common pathways and selectively blocks FXa-mediated conversion of prothrombin to thrombin (F-II).

Bioavailability: only 20%–30%.

Antiplatelet: >bleeding time; <platelet aggregation.

Lipemia clearing: reduces the triglyceride level.

Dosage:

Available in vials of 5 mL (5000 U/mL).

5000–10,000 U bolus dose. Every 6 h; followed by 1000 U/h continuous infusion.

In low-dose therapy: 5000 U/8–12 hours before surgery and continued for 7–10 days.

Caution: do not mix heparin with penicillin or tetracycline or hydrocortisone.

Adverse reactions: Bleeding in 1%–3%; thrombcytopenia; transient alopecia; osteoporosis; hypersensitivity reactions.

Contraindications: Bleeding disorders; severe hypertension; SubAcute Bacterial Endocarditis; ocular and neurosurgery, LP:Ch alcoholism and liver disease; chronic renal failure; with aspirin and antiplatelet drugs.

2. LMWH

Highly purified fractionated form of heparin. Molecular weight 3000–7000.

Action:

Selective inhibition of Xa with little effect on IIa.

Effect change only in ATIII, and not by bringing together ATIII and thrombin.

Much smaller effect on aPTT and coagulation time.

Lower antiplatelet action and hence negligible thrombocytopenia.

Lower incidence of hemorrhage, compared to heparin.

No monitoring required as no interference with aPTT and clotting time.

Bioavailability: 70%–90%

Route of administration: SC, 12 hourly.

Indication:

Prophylaxis and treatment of VTE, Carotid Artery Stenting, prevention of thrombus in shunts and cannula.

Dosage: 20–40 mg 2 h before surgery.

*Enoxaparin, Nadroparin, Dalteparin.

3. Fondaparinux

A new paradigm in anticoagulation: the first antithrombotic in 20 years >50,000 patients.

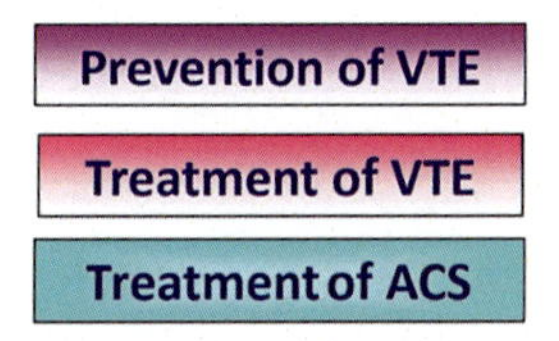

Fondaparinux (Table 17.5) acts exclusively by inhibiting FXa through the strong activation of ATIII, leading to potent and targeted physiologic control.

Fondaparinux approaches its binding site on ATIII and binds to ATIII with high affinity.

The binding of fondaparinux produces a conformational change in ATIII, exposing a loop that binds FXa.

Exposure of the arginine-containing loop greatly increases the affinity of ATIII for FXa.

Once ATIII binds FXa, a further conformational change releases fondaparinux from its binding site.

Table 17.5 Mechanism of action of Fondaparinux.

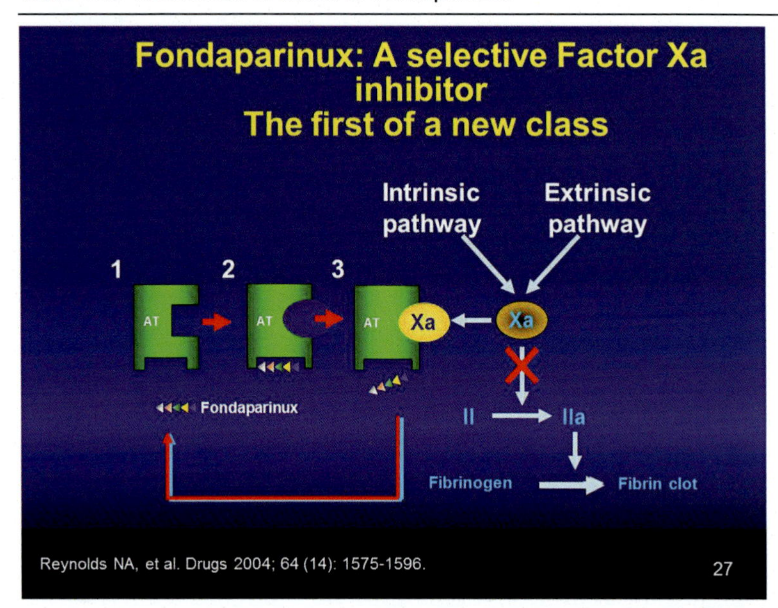

While the binding of ATIII to FXa is irreversible, once fondaparinux is released, it can go on to bind further ATIII molecules.

17.20 Fondaparinux

- Synthetic pentasaccharide;
- A good alternative to heparins;
- *Selective antithrombin-mediated inhibition of FXa;*
- Half-life longer than those of the LMWH;
- Dose: 5 mg (body weight <50 kg), 7.5 mg (50–100 kg), 10 mg (>100 kg);
- Complete (100%) bioavailability via the SC route;
- Peak concentration (C_{max}) reached in 2 hours;
- Rapid onset of action: $C_{max}/2$ reached within 25 minutes following subcutaneous administration;
- Excreted unchanged almost exclusively in the urine;
- Caution in patients with renal insufficiency. It is eliminated via the kidneys;
- *Terminal half-life = 17–21 hours* (allows for once-OD dosing) (Tables 17.6 and 17.7).

Table 17.6 Pharmacokinetic profile of Fondaparinux.

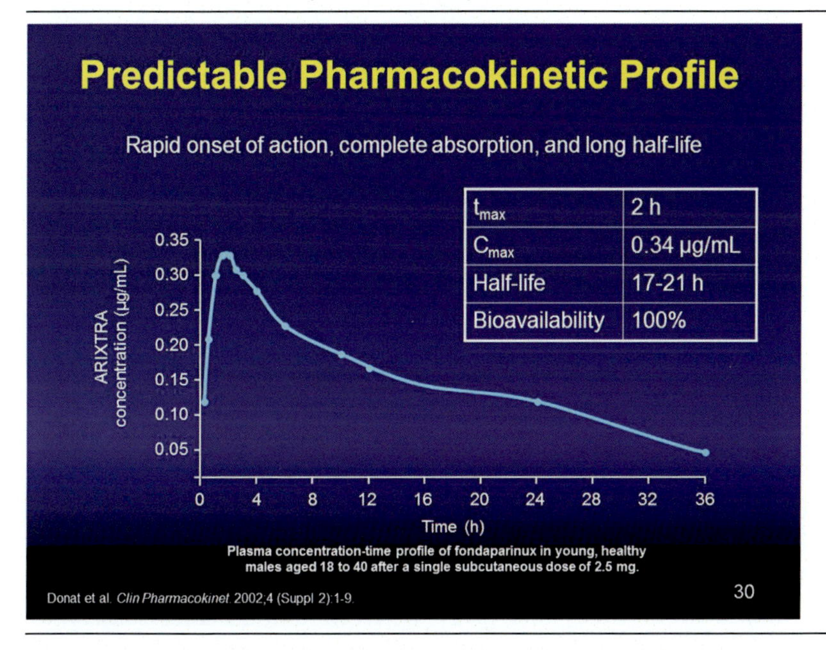

Table 17.7 Comparative study of Fondaparinux vs Enoxaparin.

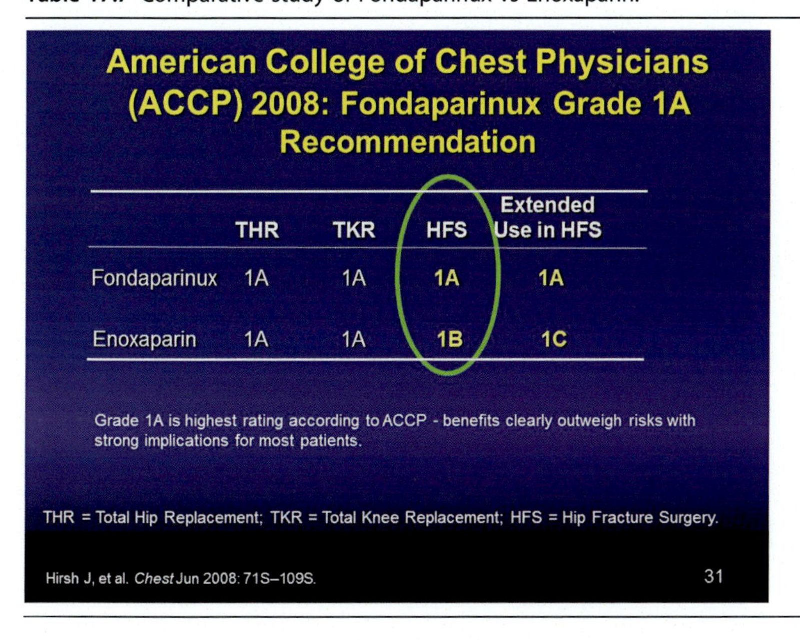

American College of Chest Physicians (ACCP) 2008: Fondaparinux Grade 1A Recommendation

	THR	TKR	HFS	Extended Use in HFS
Fondaparinux	1A	1A	1A	1A
Enoxaparin	1A	1A	1B	1C

Grade 1A is highest rating according to ACCP - benefits clearly outweigh risks with strong implications for most patients.

THR = Total Hip Replacement; TKR = Total Knee Replacement; HFS = Hip Fracture Surgery.

Hirsh J, et al. *Chest* Jun 2008: 71S–109S.

31

17.21 Fondaparinux: advantages

- Unique mechanism of action: selective FXa inhibition;
- Small, synthetic molecule;
- Broad spectrum of indications;
- Simple once OD dosage—not based on body weight;
- In NSTEMI, fondaparinux reduces major bleeding by 48% versus enoxaparin and also reduces mortality. Net clinical benefit superior to enoxaparin;
- In STEMI, fondaparinux reduces mortality and reinfarction without increasing the risk of bleeding;
- Highest recommendation by various guidelines in most patients with ACS;
- Efficacy superior to enoxaparin with a relative risk reduction >50% in VTE incidence;
- Comparable to enoxaparin in clinically relevant bleeding;
- Extended prophylaxis in hip fracture patients provides additional clinical benefit;
- Flexibility in initiating first dose—between 6 hours after surgery to next morning after surgery (Table 17.8).

Table 17.8 ACCP guideline for hip surgery.

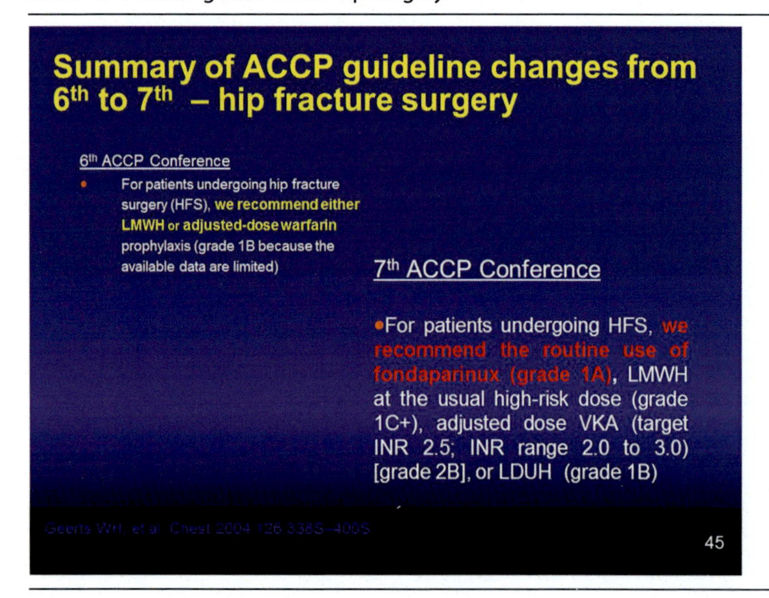

The evidence-based ACCP guidelines are updated every 3 years and the recent guidelines include changes to the treatment of VTE in Medical Outcomes Survey.

For patients undergoing hip fracture surgery, routine use of fondaparinux is now recommended. The use of LMWHs and VKAs are also still recommended for prophylaxis in elective hip arthroplasty.

The use of aspirin and low-dose UFH is *not* recommended in this situation.

NB:

Grade 1 = strong recommendation; the guideline developers are certain that the benefits do or do not outweigh the risks, burdens, and costs.

Grade 2 = weaker recommendation; the guideline developers are less certain of the magnitude of the benefits/risks and thus of the impact of the treatment.

Grade A = based on consistent results from RCTs.

Grade C+ = based on observational studies with very strong effects or secure generalizations from RCTs.

Grade B = based on inconsistent results from RCTs.

Grade C = based on inconsistent results from observational studies.

(Tables 17.9 and 17.10) [397].

Table 17.9 Efficacy analysis of Fondaparinux Vs LMWH in VTE incidents.

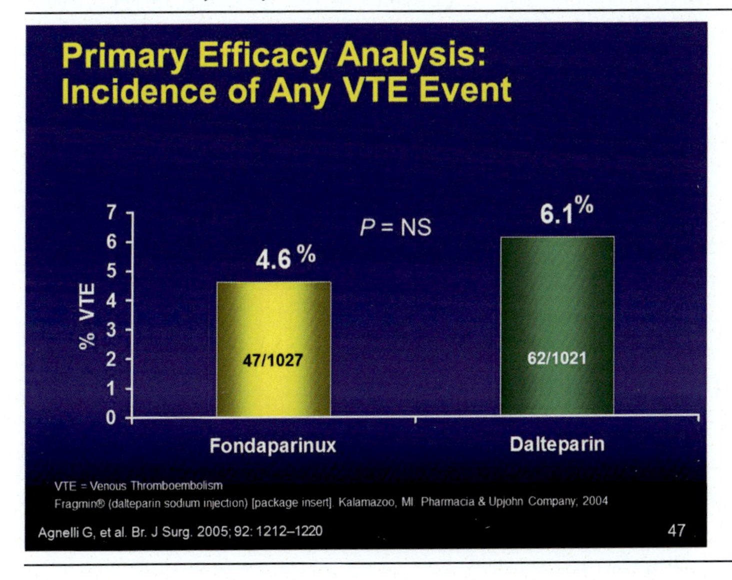

Table 17.10 Safety profile of Fondaparinux and LMWH.

Safety outcomes for treatment period

Fondaparinux and LMWH have equivalent safety profiles

As treated patients	Fondaparinux n=1,433 n (%)	Dalteparin n=1,425 n (%)	p value
Major bleeding	49 (3.4%)	34 (2.4%)	0.12
Minor bleeding	31 (2.2%)	23 (1.6%)	0.34
Death	15 (1.0%)	20 (1.4%)	0.40

Agnelli G, et al. Br. J Surg. 2005; 92: 1212–1220

49

17.22 Abdominal surgery

- Postoperative fondaparinux is as effective as preoperative dalteparin in the prevention of VTE after abdominal surgery;
- In cancer surgery patients, fondaparinux reduced VTE risk compared to dalteparin (odds ratio = 40.5%, $P = .02$);
- Fondaparinux and dalteparin have comparable safety in this clinical setting [398].

Key points:

- Fondaparinux is a synthetic pentasaccharide—selective FXa inhibitor.
- *Administration*: Fondaparinux is given subcutaneously as an OD dose.
- *Comparison with other agents*:

One potential advantage of fondaparinux over LMWH or UFH is that the risk for HIT is substantially lower.

Unlike direct FXa inhibitors, it mediates its effects indirectly through antithrombin III, but unlike heparin, it is selective for FXa.

Fondaparinux is similar to enoxaparin in reducing the risk of ischemic events at 9 days, but it substantially reduces major bleeding and improves long-term mortality and morbidity.

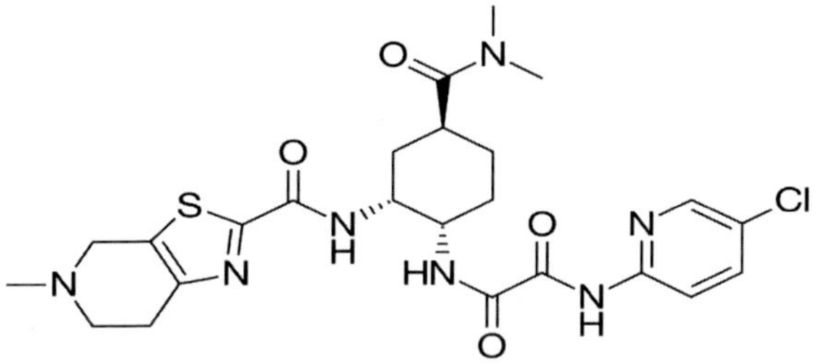

Figure 17.4 Molecular structure of Edoxaban.

17.23 Edoxaban

Edoxaban oral tablet is only available as a brandname drug. There is no generic version. Brand name: Savaysa.

Edoxaban comes only as a tablet to be taken by mouth.

Edoxaban is a blood-thinning drug. It is used to reduce the risk of stroke and blood clots in people with an irregular heart rate (NVAF). Edoxaban is also used to treat blood clots in the legs or lungs after treatment with an injectable blood thinner drug for 5–10 days (Fig. 17.4).

17.23.1 Food and Drug Administration warning

This drug has black box warnings. A black box warning is the most serious warning from the FDA. A black box warning alerts doctors and patients about drug effects that may be dangerous.

Decreased effectiveness in people with AF and good kidney function: The doctor should check how well the patient's kidneys are working before they start taking edoxaban. They carry out a test called creatinine clearance (CrCl). People with good kidney function (CrCl greater than 95 mL/minute) who have NVAF should not take this drug, because it may not work well in preventing a stroke.

Warning for stopping treatment early: patients should not stop taking edoxaban without talking to their doctor first. Stopping this drug before the treatment is complete will increase the risk of blood clots, which raises the risk of stroke. The doctor may have the patient stop taking this drug for a short time before a surgery or a medical or dental procedure. The doctor will advise when to start taking edoxaban again. If the patient has to stop taking edoxaban, the doctor may prescribe another drug to help prevent blood clots.

Spinal or epidural blood clots (hematoma) risk: some people who take edoxaban are at risk of forming a dangerous blood clot. This blood clot can cause long-term or permanent paralysis (loss of the ability to move). The patient is at risk of a blood clot if they have a thin tube called an epidural catheter placed in their back to give them medication. They are also at risk if they take NSAIDs or another medication to prevent blood from clotting, if they have a history of epidural or spinal punctures or problems with their spine, or if they have had spinal surgery.

If the patient has any of these risk factors, their doctor should watch closely for symptoms of spinal or epidural blood clots. The doctor should be informed immediately if the patient has back pain or tingling or numbness in the legs and feet. Also the doctor should be informed of incontinence (loss of control of the bowels or bladder) or muscle weakness, especially in the legs and feet.

17.23.2 Other warnings

Serious bleeding risk warning: Edoxaban can cause serious bleeding that can sometimes be fatal. This is because edoxaban is a blood-thinning drug that reduces blood clotting. While taking this drug, the patient may bruise more easily, and bleeding may take longer to stop. The patient should call their doctor or go to the emergency room immediately if they have any of the following symptoms of serious bleeding:

- Unexpected bleeding or bleeding that lasts a long time, such as:
- frequent nosebleeds;
- unusual bleeding from the gums;
- menstrual bleeding that is heavier than normal;
- bleeding that is severe or cannot control be controlled;
- red-, pink-, or brown-colored urine;
- bright red- or black-colored stools that look like tar;
- coughing up blood or blood clots;
- vomiting blood, or vomit that looks like coffee grounds;
- headaches, dizziness, or weakness.
- The patient may have a higher risk of bleeding if they take edoxaban and other drugs that increase the risk of bleeding, including:
- aspirin or products that contain aspirin;
- NSAIDs used long-term;
- other blood-thinning drugs used long-term, such as: warfarin sodium (Coumadin, Jantoven);any drugs that contains heparin;
- other drugs to prevent or treat blood clots.

17.23.3 What is edoxaban?

Edoxaban is a prescription drug. It comes as an oral tablet.

Edoxaban is only available as the brand-name drug Savaysa.

17.23.4 Why it is used?

Edoxaban is used to reduce the risk of stroke and blood clots in people with NVAF. This type of irregular heartbeat is not caused by a heart valve problem.

Edoxaban is also used to treat DVT (blood clots in the veins of the legs) or PEs (blood clots in the lungs) after treatment with an injectable blood-thinning medication for 5–10 days.

17.23.5 How it works?

Edoxaban belongs to a class of drugs called anticoagulants; specifically FXa inhibitors (blockers). A class of drugs is a group of medications that work in a similar way. These drugs are often used to treat similar conditions.

Edoxaban helps prevent blood clots from forming by blocking the substance FXa. This is a blood clotting factor that is needed for the blood to clot. When a drug like edoxaban blocks FXa, it decreases the amount of an enzyme called thrombin. Thrombin is a substance in the blood that is needed to form clots. Thrombin also makes platelets in the blood stick together, causing clots to form. When thrombin is decreased, this prevents a clot (thrombus) from forming [399].

With AF, part of the heart does not beat in the way that it should. This may lead to blood clots forming in the heart. These clots can travel to the brain, causing a stroke, or to other parts of the body. Edoxaban is a blood thinner that decreases the chances of having a stroke by helping to prevent clots from forming.

17.23.6 Edoxaban side effects

- Edoxaban oral tablet can cause certain side effects. The most common side effects that occur with edoxaban include:bleeding that takes longer to stop;
- bruising more easily;
- skin rash;
- reduced liver function;
- low RBC count (anemia).
- Symptoms may include:
- shortness of breath;

- feeling very tired;
- confusion;
- fast heart rate and palpitations;
- pale skin;
- trouble concentrating;
- headache;
- chest pain;
- cold hands and feet.

If these effects are mild, they may disappear within a few days or a couple of weeks. If they are more severe or do not go away, the patient should consult their doctor or pharmacist.

17.23.7 Serious side effects

Serious side effects and their symptoms can include the following:

- Serious bleeding, with symptoms that can include unexpected bleeding or bleeding that lasts a long time, such as:frequent nosebleeds;
- unusual bleeding from the gums;
- menstrual bleeding that is heavier than normal;
- bleeding that is severe or that cannot be controlled;
- red-, pink-, or brown-colored urine;
- bright red- or black-colored stools that look like tar;
- coughing up blood or blood clots;
- vomiting blood or vomit that looks like coffee grounds;
- headaches, dizziness, or weakness.
- Spinal or epidural blood clots (hematoma). If the patient is taking this drug and also receiving spinal anesthesia or has a spinal puncture, they are at risk for spinal or epidural blood clots that may cause paralysis. Symptoms can include:
- back pain;
- tingling or numbness in the legs and feet;
- muscle weakness, especially in the legs and feet;
- loss of control of the bowels or bladder (incontinence).

17.23.8 Edoxaban may interact with other medications

Edoxaban oral tablet can interact with other medications, vitamins, or herbs the patient may be taking. An interaction is when a substance changes the way a drug works. This can be harmful or prevent the drug from working well.

To help avoid interactions, the doctor should manage all of the patient's medications carefully. The patient should ensure they tell their doctor about all medications, vitamins, or herbs they are taking. To find out how this drug might interact with something else the patient is taking, they should talk to their doctor or pharmacist.

17.23.9 Examples of drugs that can cause interactions with edoxaban are listed below

- NSAIDs

Taking NSAIDs with this drug may increase the risk of bleeding, and caution should be applied when taking these drugs with edoxaban. Examples of these drugs include:

- Diclofenac;
- Etodolac;
- Fenoprofen;
- Flurbiprofen;
- Ibuprofen;
- Indometacin;
- Ketoprofen;
- Ketorolac;
- Meclofenamate;
- Mefenamic acid;
- Aspirin.

17.23.10 Antiplatelet drugs

Taking antiplatelet drugs with this drug may increase the risk of bleeding and caution should be applied when taking these drugs with edoxaban. Examples of these drugs include:

- Clopidogrel;
- Ticagrelor;
- Prasugrel;
- Ticlopidine;
- Blood thinners.

Edoxaban should not be taken with other blood thinners long term as this increases the chances of bleeding. It may be OK to use these

medications together briefly when switching from one to another. Examples of these drugs include:
- Warfarin (Coumadin, Jantoven);
- Heparin.

17.23.11 Drugs that affect how the body processes edoxaban

Edoxaban should not be taken with rifampin as it reduces the levels of edoxaban in the blood, which makes it less effective.

17.23.12 Edoxaban warnings

This drug comes with several warnings.

Warnings for people with certain health conditions include:
- *For people with bleeding problems*: If you currently have abnormal bleeding, you should not take edoxaban. Edoxaban is a blood thinner and may increase your risk for serious bleeding. Talk to your doctor if you have unusual bleeding, such as frequent nosebleeds, unusual bleeding from your gums, bleeding that is severe or that you cannot control, coughing up blood or blood clots, or vomiting blood.
- *For people with liver problems*: If you have liver problems, you may be prone to bleeding problems. Taking edoxaban may increase this risk even more. Edoxaban is not recommended in people with moderate to severe liver problems. Your doctor will do a blood test to see how well your liver is working and decide if this drug is safe for you to take.
- *For people with kidney problems*: You may not be able to take edoxaban or your doctor may give you a lower dose depending on how well your kidneys are working. If your kidneys are not working properly, your body will not be able to clear out the drug as well. This causes more of the drug to stay in your body, which may increase your risk for bleeding.
- *For people with mechanical heart valves*: If you have a mechanical heart valve, do not use edoxaban. It is not known if edoxaban will work or be safe for you to take.
- *For people with moderate to severe mitral stenosis*: If you have moderate to severe narrowing (stenosis) of your mitral valve, do not use edoxaban. It is not known if edoxaban will work or be safe for you to take.

17.23.13 Warnings for other groups

- For pregnant women: edoxaban is a category C pregnancy drug. That means two things:
- Research in animals has shown adverse effects to the fetus when the mother takes the drug.
- There have not been enough studies done in humans to be certain how the drug might affect the fetus.
- The patient should tell their doctor if they are pregnant or plan to become pregnant. Edoxaban should be used during pregnancy only if the potential benefit justifies the potential risk to the fetus.
- For women who are breastfeeding: it is not known if edoxaban passes through breast milk. If it does, it may cause serious effects in a breastfeeding child. The patient and doctor may need to decide if whether to take edoxaban or breastfeed.
- For children: The safety and effectiveness of edoxaban have not been established in people younger than 18 years old.

CHAPTER 18

Chronic venous ulcer

The most painful problem caused by chronic venous insufficiency is ulceration, which is erroneously described as *postphlebitic ulcer*. Slowly passing through a phase of six stages, the chronic venous disease enters the final phase of a chronic ulcer which fails to heal. This stage is generally misunderstood and is always subjected to improper treatment methodologies. The statistical data vary greatly for several reasons. Many cases are still being treated as a purely dermatological problem, as "chronic eczema," especially where there are no identifiable veins. The most apt terminology for chronic ulceration due to varicose veins would be venous stasis ulcer. When we consider postphlebitic ulceration, it is surprising as most patients have never had any deep vein thrombosis (DVT) or even superficial thrombophlebitis. It is also interesting to note that many patients with extensive full-length veins do not suffer from any stasis problem. They will not proceed beyond clinical, etiological, anatomical and pathophysiological (CEAP) Class II, even when they are old. Also, very young patients suffer from severe leg ulceration even before the age of 30, when they have no identifiable veins or DVT. The onset of ulceration has no relation to the age of the patient. The common concept that the chances of ulceration increases with advancing age is incorrect. The initiation of the pathological process by the type of vein and the site of involvement are the most important deciding criteria. The ulcerations always remain confined to around the ankle joint, with diminishing affinity toward the foot and lower two-third of the leg (refer to Chapter 19: "The Morbid Varicose Vein").

18.1 Etiopathogenesis

In venous hypertension, the pathophysiology of the microcirculation changes as follows: The increased venous pressure produces dilation of venules and capillaries, increased capillary permeability, an increase in tissue fluid formation and decrease in fluid absorption, progressive lymphatic failure; an increase in the size of the capillary pores allows blood cells to enter the tissue space and macromolecules to enter the circulation.

Genesis, Pathophysiology and Management of Venous and Lymphatic Disorders
DOI: https://doi.org/10.1016/B978-0-323-88433-4.00007-3

Increased venous pressure at the capillary level produces arteriolar constriction to reduce the input of blood, which as a consequence produces added ischemia in the tissues. There is pericapillary fibrin cuff formation on one side and on the other side blood cell trapping, adhesion, and activation of white blood cells. This produces impedance of the microcirculation, and release of free radicals, proteolytic enzymes, chemotactic agents, and cytokinins. Red blood cells disseminate the hemosiderin pigments, which produce discoloration and irritation. All these contribute to impairment of tissue perfusion and oxygenation with cellular vitality. Any minor trauma can trigger ulcer formation, which later helps to form eroding *chronic venous ulcers*.

The following list illustrates the changes at a tissue level and corroborates with the CEAP classification:

Stage I/C0: invisible veins/latent phase;

Stage II/C1: veins start appearing;

Stage III/C2: fully developed veins;

Stage IV/C3: stage of initiation of complications, excessive tissue fluid collection;

and edema appearing over the foot and ankle on long standing;

Stage V/C4: Skin changes (Pigmentation, irritation, itching, eczematous dermatitis, lipodermatosclerosis etc.);

Stage VI/C5: small ulcer formation with spontaneous healing; and with periodic recurrences;

Stage VII/C6: nonhealing ulcers.

From this it is easily seen that the venous hypertension progressively gets reflected at the tissue level and the odema develops at the most dependent part as the venous pressure at the ankle region is the highest. Overflooding at the tissue level progressively produces tissue damage and if not controlled it will pass into the various stages in succession resulting in nonhealing ulcers. This understanding is very important in treating all stages of this disease.

18.2 Harmful coexisting conditions

1. Diabetes mellitus
2. Diabetic peripheral neuropathy
3. Thromboangitis obliterans
4. Arterosclerosis
5. Vasculitis

6. Filariasis/lymphedema due to causes other than varicose vein
7. Nephropathy
8. Cirrhosis of the liver
9. Cardiac failure
10. Malignancies
11. Anemia
12. Arthropathies involving the lower limbs and spine
13. DVT
14. Old age

18.3 Management

The management of chronic venous disease depends on the stage of disease. Patients are divided into three categories:

Category I: patients included under C-1 and C-2;

Category II: C-3, C-4, and C-5 (venous edema with or without lymphatic decompensation);

Category IV: C-6 (the nonhealing ulcers).

18.3.1 Category I

These patients may be symptomatic or asymptomatic. Symptomatic patients usually present with discomfort in the leg, whereas asymptomatic patients present with any of the following. One of their close family members or friends suffered from venous disease or there was a family history or someone was disqualified during medical examination for recruitment into the defense services or another job, or purely for cosmetic reasons.

18.3.2 Category II

In this group, edema of the lower leg and foot remains the basic triggering problem. Hence the first thing to do is to correct the edema before any surgical correction.

18.3.3 Category III

This group is the most severe. The leg looks very ugly with awful ulcers. This may periodically become infected, when there is an offensive smell and discharge. Several cases have been seen to have maggots in the wound. Though very rarely turning malignant, a biopsy from the edge of the ulcer is always advised. In category III, the treatment involves seven phases:

1. After all relevant investigations, *control all coexisting problems* if any, *control infection* if present, and proper *wound care.*
2. Edema is the main predisposing factor of all complications. The most important thing to *control edema* is to keep the foot elevated for about 15–20 cm above the horizontal level with the knee fully extended. This is usually supplemented with some oral diuretics, preferably with a combination of frusemide and spironolactone. The main reason for the preference of this combination if diuretics is that frusemide is a potassium-losing drug, whereas spironolactone is a potassium-retaining drug. Moreover, these drugs may have to be continued for an indefinite period.
3. *Care and preparation of the wound* is the next step. Removal of slough and regular proper dressing are of vital importance. More than 50% of ulcers usually heal with proper care alone. However, at least one-third of patients will require skin grafting at a later date.
4. When the wound is clean and if there is no edema, the definitive *surgical procedure* can be undertaken. In DVT, only after assuring full recanalization by Doppler study and under cover of anticoaguant should treatment be offered.
5. Following surgical correction, if the ulcer looks healthy and requires *grafting*, it should be carried out.
6. *Aftercare* is very important, and attention should be focused on preventing edema. One cannot dictate how long the care of the leg is required, as this varies from patient to patient. Reversal of the stages of development of the ulcer from healing of the wound, progressive improvement of skin pigmentation and absence of edema even on normal standing conditions marks the complete regression.
7. *Follow-up* is equally important, and may usually require several years. Although the wound is healed and patient is obviously feeling healthy, they should be advised to appear for review on a planned schedule. It is advisable to have periodic checkups for at least 5 years. This will help to identify any opening up of veins or recanalization or neovacularization for which these veins are notorious.

18.4 Care of the ulcer

The management of venous ulcers differs from that of other types of ulcers. It is always amazing that most patients in the later stages of extensive chronic ulceration do not have as much pain as one would expect. However, the early small ulcers are very painful and distressing. The defective intricate body mechanism is corrected by the surgeon in the treatment process. The following principles may help the wound heal faster.

1. *Removal of slough:* Excise all the devitalized tissues and help the wound remain clean.
2. *Local applications: Never use povidone iodine* for dressing for two reasons. First, it is a wound-healing retardant by inhibiting granulation tissue formation. Second, it is an irritant to tender skin surrounding the ulcer. Instead, normal saline is more advisable. The use of Oxum for cleaning as well as dressing is very suitable. The efficacy of local applications of creams and ointments are always disputed. However, there is no harm in using it, as this will prevent the dressing material from sticking to the ulcer and help with less painful and easier removal. This can also help as an antibiotic barrier. Some advise silver creams, claiming that it has a specific effect on wound healing.
3. *Dressing material:* Never apply cotton directly on the wound. It should always be padded with surgical gauze. In weeping ulcers it is advisable to use a sterile polyvinyl sponge dipped in normal saline and squeezed out. The absorbing capacity of the sponge is 10 times more than that of normal cotton. The gauze acts only as a barrier to separate the ulcer from the dressing materials and to prevent them from sticking to the wound.
4. *Other considerations:* In addition to the above, the following require definite consideration:
   a. Honey
   b. Leech therapy
   c. Maggot therapy
   d. Collaspread and hydrogel
   e. Hydroheal
   f. Recombinant activated platelet factor
   g. Submucosa wound matrix
   h. Skin grafting

18.4.1 Honey

A quote from the Holy Quran (Surah Al-Nahal, verses 68 and 69) is thought provoking:

> *And your Lord revealed to the bee: Make hives in the mountains and in the trees and in what they build. Then eat of all the fruits and walk in the ways of your Lord submissively.* There comes forth from their bellies a beverage of many colors, in which there is healing for mankind. *Verily in this is a sign for those who give thought.*

Honey has been recognized for its medicinal properties since the Vedic Ages. It has been identified for its healing purposes in Ayurveda, the Bible, the Quran, and the Torah. It is mentioned in the Edwin Smith Papyrus, dating from the 17th century BCE, and was also highlighted by Hippocrates and Democritus in ancient Greece, Galen in ancient Rome, and Avicenna in medieval times. In the past century there have been various scientific reports of its use in the treatment of various wounds and infections. The wide antimicrobial effect of honey with its property of acceleration of the wound-healing process makes it a unique product in treating nonhealing ulcers. In recent years, there has been a resurgent interest in the use of honey. When directly applied to a wound with a dressing it is a good sealant, keeping the wound moist and free from contamination. It hyperosmotic in nature and hence draws out fluid from the wound, keeping the wound free of microbes and helping in the healing mechanism. On application on a wound it can be a mild irritant.

Honey is a natural product of bees of the genera *Apis* and *Meliponinae*. These bees collect nectar from flowering vegetation. The nectar is subjected to enzymatic processing in vivo in both the collecting bee and in a processing bee inside the hive. The processing bee then deposits the nectar into a wax cell in the hive, where due to the relative warmth and fanning, the water content is reduced by evaporation to 17%. The sugars in the nectar are converted enzymatically into glucose (35%), fructose (40%), and sucrose (5%). Glucose oxidase then converts the glucose into *gluconic acid and hydrogen peroxide*. The antimicrobial effects of honey are variably ascribed to the pH, the hydrogen peroxide content, the osmotic effect, and as-yet unidentified compounds, putatively described as *inhibines*. Various researchers have neutralized the hydrogen peroxide with catalase in vitro to exclude the activity of hydrogen peroxide, with varying results. For the bee's purposes, the antimicrobial effect is very useful; honey can feed a hive through a long winter, and likewise, has a shelf-life of many years for human consumption. Commercial processing involves heating of the honey to inactivate enzymes that may facilitate its crystallization, which makes it less attractive commercially. Honey can be purchased commercially in both unprocessed and processed forms. In Ayurveda, it is used traditionally for treating chronic ulcers with good results claimed. The older the age of the honey, the better the effect. Processed honey has not been found to be very effective clinically. The author had the privilege of using this for more than 500 patients with nonhealing venous ulcers with excellent results.

Honey has been used in various conditions with promising results, especially in necrotizing soft tissue infections, particularly in Fournier's gangrene. In various centers this has been successfully used for the treatment of peptic ulceration due to *Helicobacter pylori*, in burns, in chronic venous ulcers, in diabetic foot ulcers, and in decubitus ulcers. It is also reported to have a very good effect in pseudomonal infections and is very active against methicillin resistant *Staphylococcus aureus* (MRSA). The increased lymphocyte and phagocytic activity also stimulates monocytes in cell culture to release cytokines, tumor necrosis factor-alpha, and interleukin (IL)-1 and IL-6, which activate the immune response to infection.

Allergic reactions to honey are rare and have been attributed in some cases to a reaction to a specific pollen in the honey, which, when processed for use in wound care, is passed through fine filters making it almost pollen-free. A transient local stinging sensation was occasionally experienced in a very few cases and this is never severe enough to discontinue treatment, and with no adverse effects. A number of histological studies examining wound tissues also support the safe use of honey.

18.5 Leech therapy

The leech was an indispensable part of the practice of medicine until the 19th century and is very commonly used by Ayurvedic physicians in India. Scientific medical advances made it disappear for some time from the modern medical field. Currently it is seeing a renaissance in the area of modern plastic reconstructive surgery, particularly in microsurgery transplantation.

In the United States, medicinal leeches (*Hirudo medicinalis*) were cleared as a medical device in June 2004 by the Food and Drug Administration (FDA; shortly after maggots received clearance) and they are used today throughout the world as tools in skin grafts and reattachment microsurgery. The works of two Slovenian surgeons, in 1960, initiated the new increase in plastic surgery and, in 1985, Joseph Upton, a Harvard plastic surgeon, used leeches in the reattachment of an ear in a small child. Ears have been notoriously difficult to transplant successfully due to the clotting of minute blood vessels during the procedure. The use of leeches saved the boy's ear.

Leeches possess properties that make them uniquely able to assist with venous compromised tissues. Their saliva contains:

1. Hirudin, a direct thrombin inhibitor;
2. Hyaluronidase, which increases the local spread of leech saliva through human tissue at the site of the wound and also has antibiotic properties;
3. A histamine-like vasodilator that promotes local bleeding;
4. A local anesthetic.

18.6 Collaspread

These collagen sheets are available in different sizes. They act as a scaffold for the promotion of granulation tissue formation and faster healing. They should not be used on unhealthy ulcers with sloughing. They help relatively small ulcers of about 5 cm diameter to heal without skin grafting.

18.7 Hydrogel (colloidal silver gel)

Amorphous hydrogel wound dressing with colloidal silver is a unique clear odorless, transparent amorphous gel with colloidal silver that is used for wound management.

The mechanism of action is as follows:

1. It provides moisture to the wound bed, preventing dehydration of granulation tissue and also promotes autolytic debridement.
2. It helps with vital gaseous exchange across its surface but does not allow entry of microbes into the wound, thus forming an antiseptic barrier.
3. It is a soothening agent, through its cooling effect.
4. It provides slow, extended release of active ionic silver for broad antimicrobial activity and helps to prevent wound contamination.
5. It promotes wound healing by inhibition of matrix metalloproteinases.
6. It has extended antimicrobial spectrum involving *Pseudomonas* and MRSA.
7. It is biocompatible and can be washed away without disruption of granulation.

18.8 Recombinant activated platelet factor (Regranex, Plermin)

In complex wound-healing problems this helps to a greater extent. Platelet-derived growth factor (PDGF) is released by activated platelets to recruit inflammatory cells toward the wound bed. It has effects on promoting angiogenesis and granulation tissue formation. However, the effectiveness of topical PDGF on wound closure is variable.

18.9 Submucosa wound matrix

The native dermis is normally able to direct wound healing following damage but in chronic wounds, the dermal *extracellular matrix* (ECM) and cells within it are diseased and unable to provide the correct signals needed to stimulate and coordinate healing. In deep, chronic ulcers, the dermal ECM may be completely absent and wounds may not efficiently epithelialize because the wound-healing signals that are usually present in the dermis have been lost. Functional ECM is essential for wound healing. The healing process is remarkably retarded if functional ECM is absent. In these situations, therapeutic strategies must include the use of exogenous factors to act as a surrogate for the native dermis if healing is expected to occur.

Diabetic foot ulcers, venous leg ulcers, and pressure ulcers often fail to heal quickly, and ultimately become chronic as a result. They create many negative impacts in the quality of life of many patients. The standard of care is frequently insufficient to promote healing of these wounds and many require aggressive, active treatments to stimulate epithelialization.

Chronic wounds are characterized by high levels of matrix-degrading enzymes, proinflammatory cytokines, and senescent cells, which become unresponsive to their surrounding environment and are unable to effectively remodel their ECM. In these chronic wounds, the body fails to generate a functional ECM. A functional ECM is of prime importance in the process of wound healing, and acts as a scaffold for the migration of fibroblasts and other cells, regulates cell-to-cell communication, and directs the development of a biochemical environment that is conducive for healing. The absence of an ECM impedes granulation tissue formation and epithelialization. Strategies to replace the failing ECM may be beneficial in stimulating closure of chronic wounds.

The Small Intestinal Submucosa Wound Matrix (SISWM) is an active, intact, biologically derived ECM that recapitulates the complex structure and composition of the dermal matrix. It contains the major components present in the dermal ECM during normal wound healing. It also contains *bioactive growth factors*, which remain bound to the matrix during its isolation and processing for clinical use. Preclinical studies have confirmed capillary ingrowth into the matrix, as well as the migration, attachment, and proliferation of cells. Clinically, SISWM has been found to significantly enhance the healing of chronic wounds compared with standard care alone. As compared to other aggressive treatments available for chronic wounds, SISWM offers an effective combination of matrix and signals in a single weekly application while minimizing costs.

When treating high-risk wounds, aggressive, active interventions, such as SISWM are likely to be most beneficial if pursued early as part of a consistent treatment algorithm. Evidence indicates that the reduction in the area of the chronic wound during the first 4 weeks of treatment is a predictor of complete healing at 12 weeks for diabetic foot ulcers, and at 24 weeks for venous leg ulcers. Thus, for chronic wounds, a lack of response at 3–4 weeks to conventional therapy should prompt the reevaluation of the patient and consideration of advanced therapies. Clinical evidence shows that sharp debridement can shift a chronic wound toward a more acute state and is associated with better outcomes if performed prior to ECM replacement with SISWM. Rapid wound closure helps to minimize the detrimental effects of these chronic ulcers on the patients' daily lives and in some cases to avoid hospitalization, life-threatening complications, and amputations.

As a treatment that includes both matrix and signals in the form of growth factors, cell attachment factors, and matrix-binding sites, SISWM provides the elements needed to address the ECM deficiencies that characterize chronic wounds. Its clinical efficacy, reasonable cost, and ease of storage and use suggest that SISWM may be advantageous as an early and potentially cost-effective intervention for chronic wounds.

18.10 Maggot therapy

The presence of maggots in the wound can be really horrifying to the patient and others. The appearance of maggots in the wound can be rather sudden and surprising, especially when the patient is getting good care in a hospital under a specialist. Invariably this happens in an infected wound with a lot of slough. The patient and doctor will think alike; why have maggots appeared even when under the best care with antibiotics and daily dressing? It can be a quite embarassing situation. Immediately some turpentine is poured on the wound in the aggressive desire to eradicate the maggots. Some maggots are picked or pulled out. This continues for several days until the maggots are completely cleared. During this process some slough is also excised, causing pain to the patient. Actually, maggots are the defense sent by nature to protect the patient in a crisis. Though they look unsightly, they are angels! Their appearance is sudden, they eat off all the slough working earnestly, without rest, 24 hours a day, causing no pain to the patient and never disturbing the normal tissues. They help to reduce the expenses of surgical treatment and hospital stay. Once their purpose is served, they disappear spontaneously within 5–7 days. There are now some centers in the United States where maggots are cultured for this purpose. Their use also is approved by the US FDA.

18.11 Skin grafting

In very extensive ulcerations, when the wound becomes healthy, it is always advisable to graft (split skin graft). If only one leg is involved never take the graft from the thigh on the same side. The reason being that by applying a dressing for the donor site the bandaging will obstruct the venous as well as the lymphatic flow and there is the chance of edema formation on the leg which may interfere with smooth uptake of the graft. Even after successful grafting, one has to prevent edema occurring in the limb. Edema will also predispose to infection. Elevation of the foot and the application of a compression bandage or stockings are advised until complete recovery is obtained. The patient should be advised on follow-up as per the specified schedule to prevent the recurrence of varicose veins.

CHAPTER 19

The morbid varicose vein

19.1 Introduction

A journey through the history of varicose vein (VV) is very interesting. It can help us to understand the developments that have taken place over time, where we are now and who are our great experts on the subject. An unveiling of these facts will enable us to rethink the poor performance we have had in developing new areas in VV surgery over the last century. The history of VV surgery began about BCE 3500. All the best experts had very little surgical exposure; however, unfortunately we still blindly follow their footsteps without opening our minds to grasp new ideas in creating newer outlooks to improve the poor outcomes after VVs surgery. This chapter is directed toward a better understanding of the pathogenesis of VVs, especially with regard to pathological changes in the valve/vessel wall and their humoral and genetic associations, with an emphasis on the role of the microscopic venous valve and microcirculatory impediments.

19.2 Materials and methods

A total of 6350 patients were evaluated for the current study. All were examined evaluated, treated, and documented by the same surgeon for a period of 15 years. The majority of patients belonged to clinical, etiological, anatomical and pathophysiological (CEAP)-4, -5, and -6, and most of the patients in CEAP-6 presented after conventional surgical treatment, multiple ligation, or radiofrequency ablation/endovenous laser treatment.

The annual progressive curve of patients treated at this center is clear evidence of the acceptance of the patients to the principle of modality and an understanding of the treatment we have adopted.

Color Doppler scanning was carried out routinely in all cases with repeated assessments for *reflux* in various positions, including supine and erect.

Manometric venous pressure assessments were done at the ankle, knee, and upper thigh levels in many selected patients.

Genesis, Pathophysiology and Management of Venous and Lymphatic Disorders
DOI: https://doi.org/10.1016/B978-0-323-88433-4.00010-3

A genetic study was also carried out in collaboration with the Rajiv Gandhi Centre for Biotechnology, Thiruvananthapuram, Kerala, India, the most prestigious biotechnology institute in India.

From the routine detection and treatment in the surgical outpatient department of a general hospital, and treating less than 10 VV patients a year, it has now developed into a major research institute exclusively for venous diseases, with an average attendance of 100 outpatients per day, attracting patients from various places in India and abroad. The current number of patients undergoing surgical procedures at this center is about 250–350 per month (Tables 19.1–19.4).

19.3 Pathogenesis of varicose veins

Progressive venous hypertension (VHT) is the triggering factor for all pathologies and requires detailed understanding and analysis.

19.3.1 Anatomical considerations

The superficial venous (SVS) and the deep venous systems (DVS) mainly interconnect with each other through the two perforators in the thigh,

Table 19.1 Modified microfoam sclerotherapy (MMFST): patient acceptance graph.

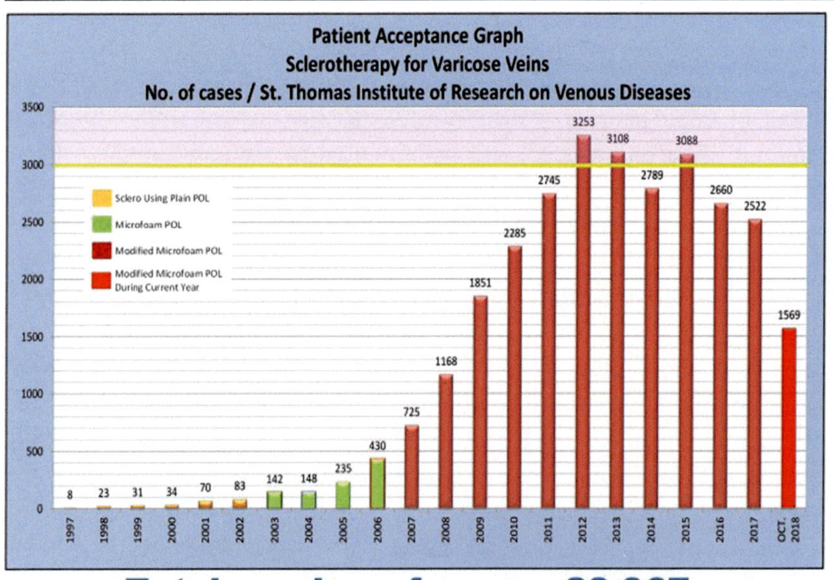

Table 19.2 Age related male-female ratio.

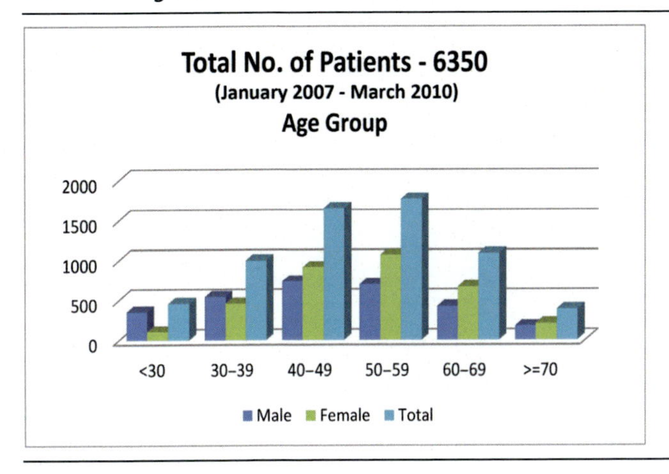

Table 19.3 Sex ratio.

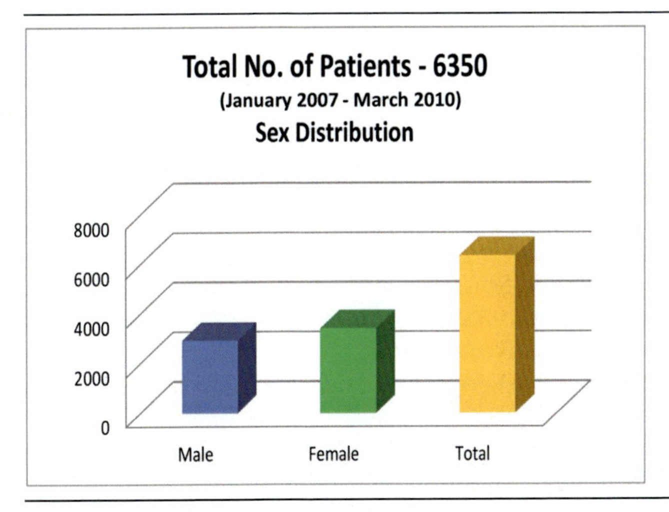

the four perforators in the leg, the saphenopopliteal (SP) valve behind the knee, and the saphenofemoral (SF) valve at the inguinal region. However, in addition to this, there are innumerable unnamed interconnections, the venous plexuses, venous sinuses, and tiny venous channels up to the tissue level. These are often overlooked by surgeons as most are not visible to the naked eye. It is this plethora of venous communications that helps one to sustain the normal venous return, even in adverse conditions.

Table 19.4 Presenting complaints of patients.

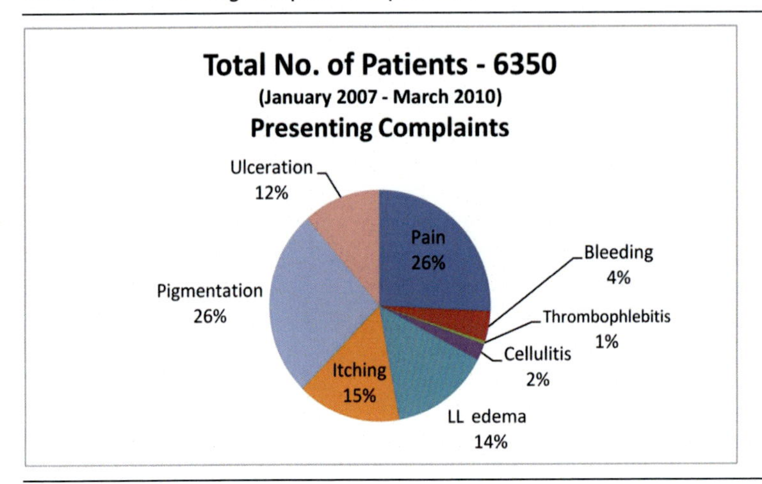

19.4 The venous valves and venous plethora

The normal efflux of venous blood is assisted by the nonreturn valves in the veins. The number of valves is higher in the smallest veins as compared to larger veins. Similarly, the number is increased and the distance between valves reduced toward the tissue level where the venous hydrostatic pressure is the highest. These valves prevent reflux during systole of the heart.

The two venous systems of the lower limb, the DVS and SVS, are not two entirely different venous compartments. At the tissue level they are very closely interconnected. Any obstruction to the smooth flow through the vein is surpassed through these interconnecting channels. The DVS passes through the muscular compartment and hence is strongly assisted in its upward flow by the muscular pump, the *peripheral heart*. However, the SVS remains outside the muscular compartment and under the skin, and hence depends entirely on valvular function. The veins of the lower limb are viewed as a river flowing from below upwards, climbing against gravity. The final destination of every vein is to reach the heart to maintain normal circulation (venous return). While the DVS, which is the mainstream and the SVS comprising of long saphenous vein and short saphenous vein are just major bypasses to surpass any block in the DVS. Each has its tributaries, and they join to form the trunk and at the tributaries and truncal level they interconnect. The truncal intercommunications are described as *perforators* which are named. The microperforators at the tributary level are innumerable and hence unnamed.

There are two types of veins functionally:

1. Type I, the tributaries;
2. Type II, the truncal veins.

The tributaries can be generally considered as veins with sizes of about 0.02–2 mm, starting from the venous end of the capillary to the beginning of the truncal system of the vein which can be more than 2 mm. While tributaries are the vital functioning element of the venous system, they can be qualified as the *collecting system* (*type I*). At the terminal end of the collecting system, the *conducting system* begins, with the truncal veins (type II). This is very important in the study of the morbidity of VVs. VVs, in association with the collecting system, have many informal names around the world depending on their appearance, such as reticular veins, spider veins, and intradermal VVs. Based on the physical architecture and functional nature it would be apt to describe them as *radicular veins* (type I).

19.5 The different types of varicosities

The time-old hypothesis that the trigging point of VVs is at the SF junction of the great saphenous vein (saphenocentric theory of Trendelenburg) is no longer accepted. In clinical practice there are different types of VVs of the leg:

1. With truncal involvement alone (type II);
2. With radicular vein involvement alone (type I);
3. The majority with types I and II, in varying proportions.

Depending of the clinical manifestations, VVs of the lower limb can be grouped into:

1. Group A—*the cosmetic group*: There are no clinical symptoms or manifestations, but they are cosmetically unattractive;
2. Group B—*the morbid group*: These present with clinical symptoms and later manifest with progressive complications of chronic venous disease (CVD).

The level of appearance of these veins carries great importance.

1. Level 1—These appear at, in, and around the ankle joint and on the foot;
2. Level 2—Above the ankle but below the knee;
3. Level 3—Above the knee, anywhere in the thigh area.

19.5.1 Type I veins

Formed by the tributaries, these are the most harmful veins, which show the greatest morbidity. At level 3 they remain practically harmless and usually remain unidentified and undetected, with the patient remaining totally ignorant of their presence at the front and back of the thighs.

From level 2, the morbidity intensity progressively increases, in proportion to the closeness to level 1.

At level 1, they are the most serious, causing patients severe pain while passing through the various stages of CVD, and resulting in chronic venous ulcer (CEAP 3–VI).

The pain is felt only in one area, that is, *in the area where socks are worn*, with the maximum impact around the ankle region. To sum up, where there is maximum venous pressure at a cellular level, morbidity develops. Age has absolutely no relation to the development of morbidity.

19.5.2 Type II veins

It is surprising to note that type 2 veins anywhere at the above-mentioned levels remain innocuous. They produce the following problems only:

1. Cosmetic;
2. Bleeding;
3. Thrombophlebitis.

19.6 Pathogenesis

To gain a deep knowledge of the morbidity of venous diseases, a detailed understanding of the pathogenesis of VV and a thorough knowledge of the pathophysiological changes of the microcirculation at a tissue level are very important.

19.7 Pathogenesis of varicose veins

The *saphenocentric theory of Trendelenburg* needs to be revised in terms of anatomy and physiology. Failure of venous blood to ascend the body is the main cause of VHT. The reflux is initiated at the lowermost level of valvular pathology. Therefore it is evident that the triggering point is not at the uppermost truncal end of the vein but somewhere far below, toward the tributary level or above. Impedance of the microcirculation occurs only if tributaries (type I veins) are affected, which will lead to pathological change at the tissue level.

19.7.1 The trigger point

Whether it is at the valvular level or at venous wall level is a dispute to be settled in the future by pathologists. For the clinician, whether it is at valvular or venous wall level is relatively unimportant as the functional changes remain the same.

19.7.2 The valvular theory

Based on venoscopic studies and histopathological analysis it can be seen that the valves are torn, thickened, maldeveloped, or absent. Endothelial changes are also seen. The dysfunctioning of the nonreturn valves develops a reflux which produces antegrade thrust. This thrust in turn expands the venal wall and further increases the gap at the valvular level. The leak gets progressively aggravated, with venous stagnation taking place accounting for progressive VHT and cyclical deterioration leading to changes in the microcirculation and clinical complications.

19.7.3 The venous wall theory

Many attribute the triggering of the whole process to the venous wall level. Changes in the extracellular matrix (ECM), as evidenced by ECM degradation leading to reduction in elastin, increase collagen, smooth muscle cell migration, changes in proteoglycans, and dysregulated apoptosis in association with alteration of matrix metalloproteinases, endogenous tissue inhibitors and endothelial changes with increases in different vascular endothelial growth factors, all suggest serious changes to the venous vessel wall. Any weakening of the vessel wall can expand the vessel both longitudinally and in a curricular fashion, which in turn increases the circumferential gap of the valve leading to incompetence and dysfunction.

Whether initiated at the valvular or wall level, the final impact is the same—valvular dysfunction—which will progress to VHT and thinning of the wall and pressure impact at the tissue level due to reflux. Once the process is initiated it will continue to be a progressive cyclical phenomenon.

Special mention is needed of the fact that the valves are in no way different from the vessel wall. The histological architecture is developed from the same structures, the endothelium, the basement membrane, and the inner longitudinal muscle layer of tunica media. Hence it can be considered as a general pathology initiated by a common factor—*a genetic factor which remains obscure.*

19.8 Pathophysiology at the level of the cellular microcirculation

The morbidity is decided by the following questions:

1. Where is the distal limit of valve/wall involvement?
2. What type of vein is involved? If type I, it is closer to the tissues and if type II it is involved distant from tissues.
3. Where is the level of involvement? At level 1 the morbidity is at its maximum.

If it is closer to the tissue level the morbidity is very high, and if it is more distant the morbidity is proportionately reduced. In type II level morbidity is extremely rare. Thinning of the wall leads to *bleeding* and stagnation of blood leads to *thrombophlebitis.*

If it is closer to tissues there is a direct impact of VHT at the tissue level. The capillaries are dilated, whereby the pore size is increased. In addition to excessive tissue fluid formation, there is also very poor tissue perfusion. Fluid accumulates at tissue spaces, the lymphatic channels fail to drain the massive quantity of tissue fluid, resulting in lower limb edema. Lymphedema marks the initiation of lymphatic failure and is the landmark of morbidity of VVs (CVD) and identified as CEAP 3. Progressive fluid accumulation leads to the impedance of tissue perfusion. Further expansion of capillary pores due to increasing VHT allows red blood cells to escape into the tissue spaces in addition to white blood cells and macrophages, which then become trapped before being destroyed, and hemosiderin pigments are released producing pigmentation and irritation. Dysregulated apoptosis at the tissue level, fibrin cuff formation at capillaries, decreased oxygen saturation, and release of oxidants, etc. contribute to further tissue destruction and finally ulcer formation.

If the trigger point is distant from tissues and the valves closer to the tissues are normally functioning, there is resistance to the impact at the tissue level, as the veins have more muscles and elastin toward the tissue area. Excessive VHT with the help of innumerable small venous connections is rerouted to the deep system which helps to protect the tissue and maintain normal venous return.

19.9 Conclusion

In summary, morbid VVs are the smallest veins, some visible and some not to the naked eye, centered on the ankle joint. These veins predispose to the development of CVDs. Maximal elimination of these veins in the lower end of the leg is the cardinal point in achieving very good success in the treatment of CVD, and not mere elimination of large innocent veins. The morbidity is decided by the number of type I veins at level 1, and whether or not it is associated with type II veins at any level.

CHAPTER 20

Photographic presentation of venous diseases

A visual aid is superior to any other teaching aid. Hence the following is a photographic representation of various manifestations of venous disease.

20.1 Congenital venous malformations

Congenital venous malformations are shown in Figs. 20.1–20.15.

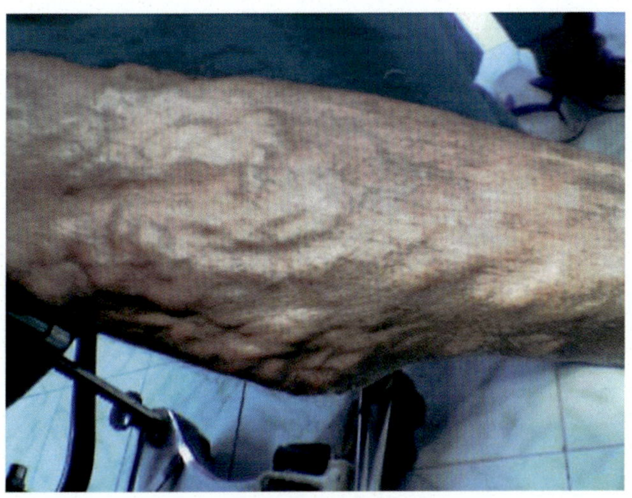

Figure 20.1 Congenital venous malformation of the leg.

Genesis, Pathophysiology and Management of Venous and Lymphatic Disorders
DOI: https://doi.org/10.1016/B978-0-323-88433-4.00014-0

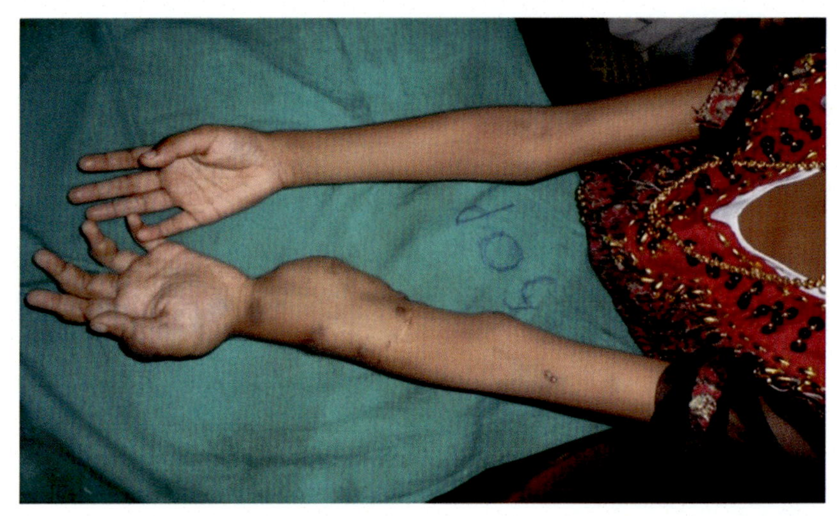

Figure 20.2 Congenital venous malformation of the forearm and hand in a child.

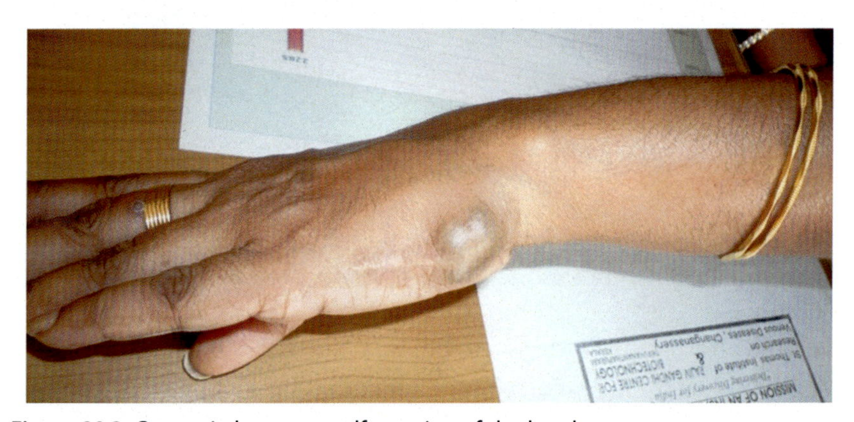

Figure 20.3 Congenital venous malformation of the hand.

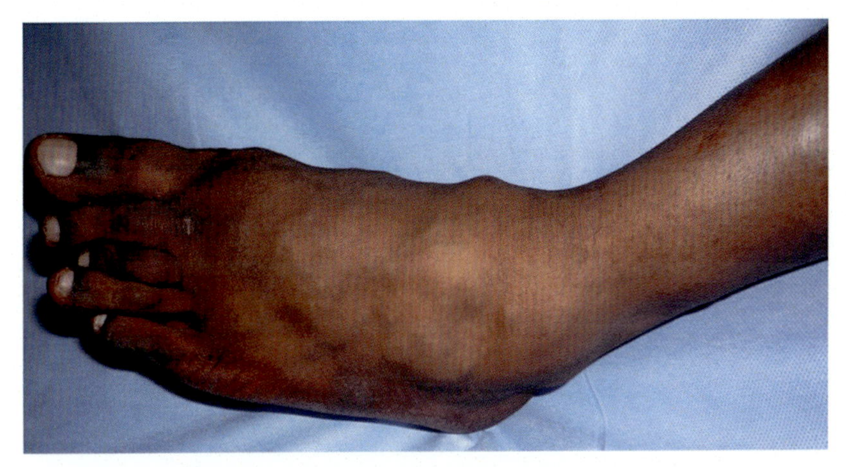

Figure 20.4 Congenital venous malformation of the leg.

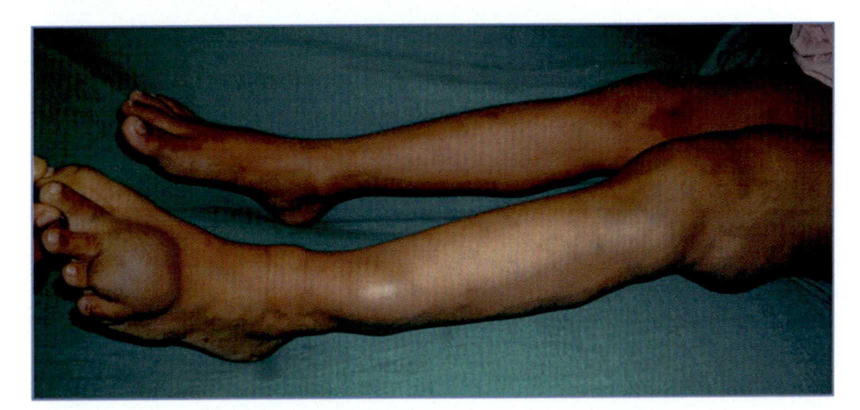

Figure 20.5 Congenital venous malformation of the leg in a child.

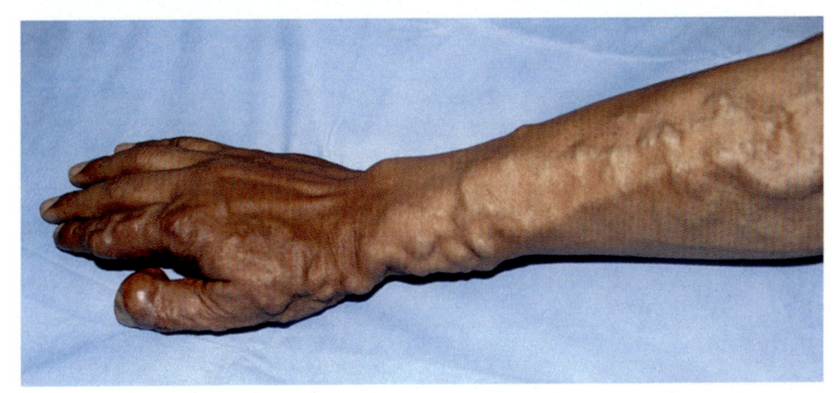

Figure 20.6 Congenital venous malformation just before sclerotherapy (April 12, 2012).

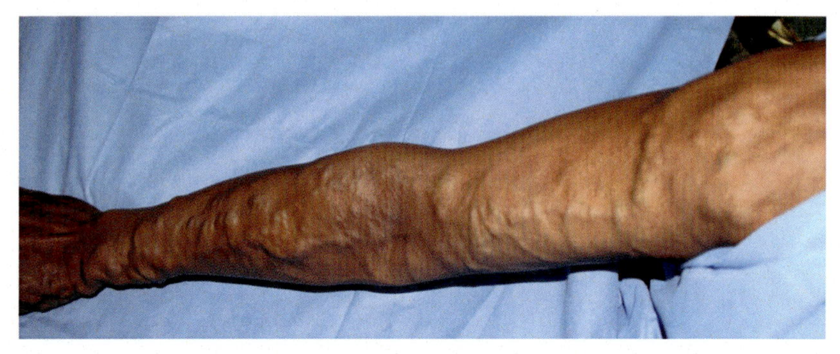

Figure 20.7 Congenital venous malformation just before sclerotherapy (April 12, 2012).

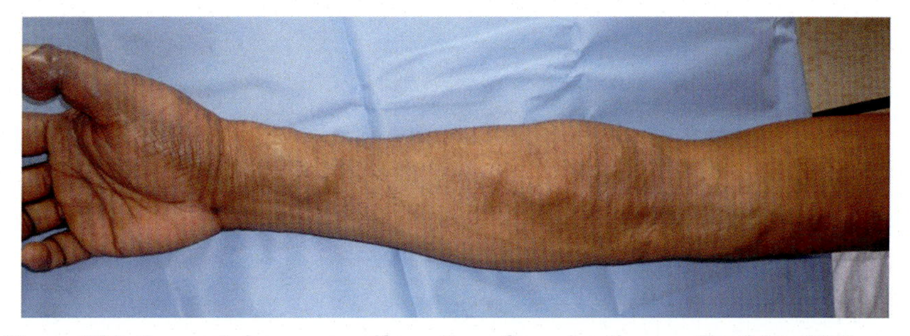

Figure 20.8 Congenital venous malformation after sclerotherapy (April 24, 2012).

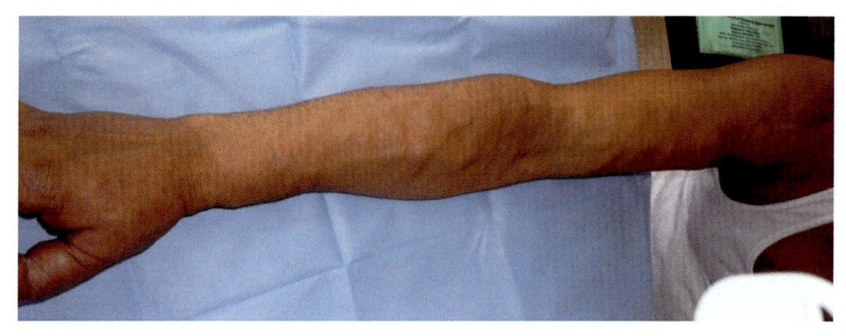

Figure 20.9 Congenital venous malformation after sclerotherapy (April 24, 2012).

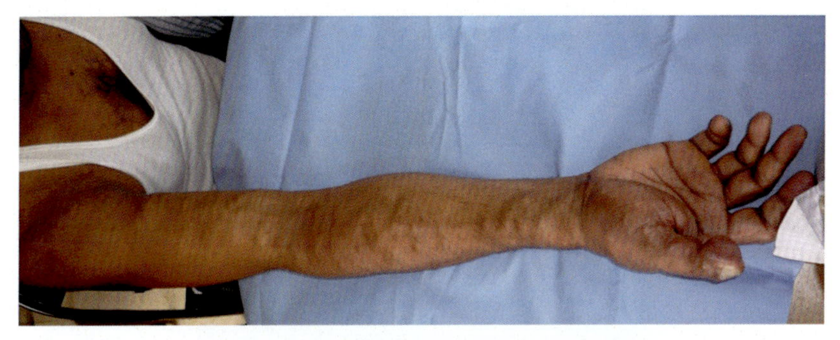

Figure 20.10 Congenital venous malformation after sclerotherapy (May 15, 2012).

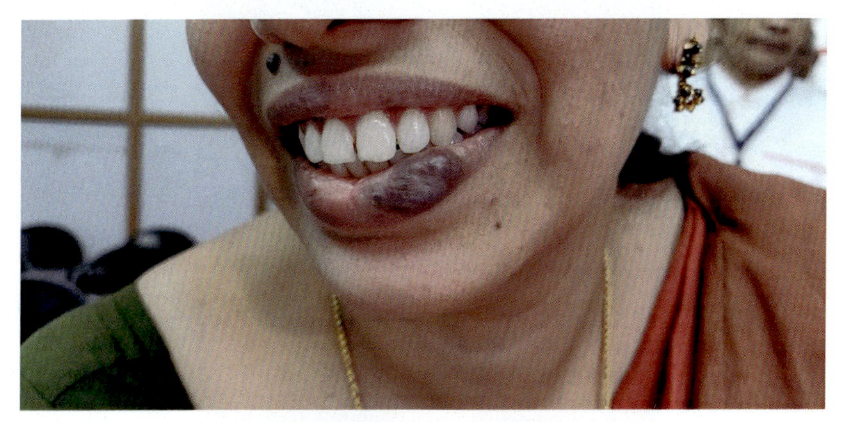

Figure 20.11 Congenital venous malformation on the lip.

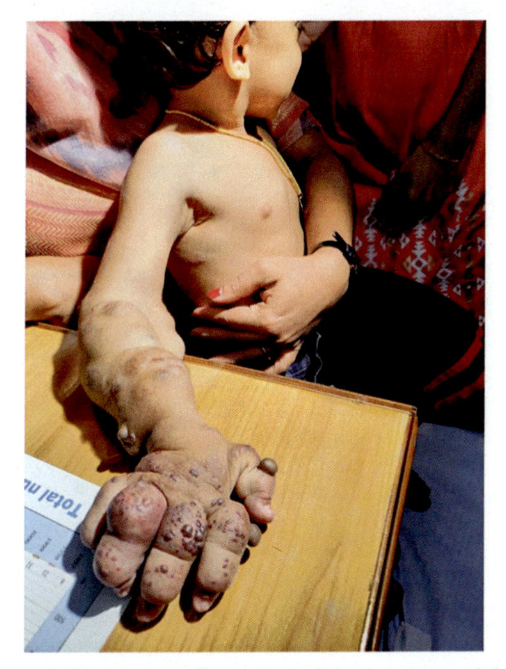

Figure 20.12 Congenital venous malformation of the hand in a child.

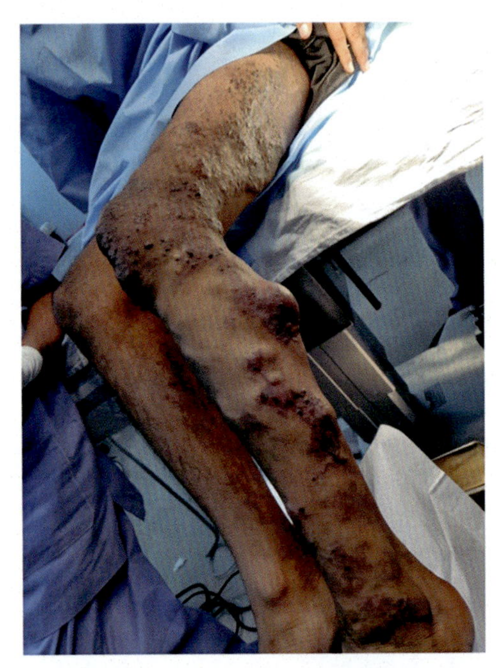

Figure 20.13 Congenital venous malformation of the leg.

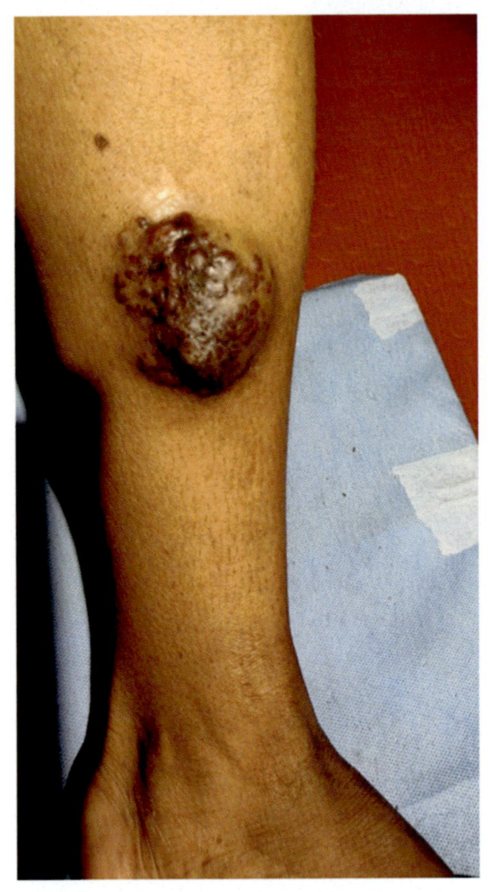

Figure 20.14 Congenital venous malformation of the leg.

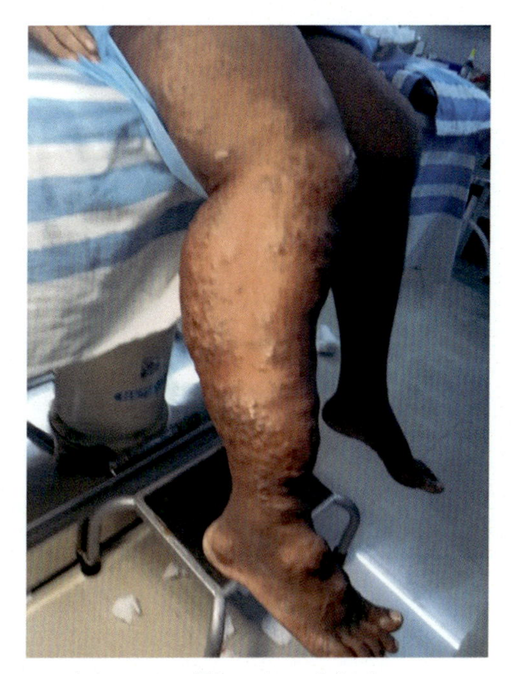

Figure 20.15 Congenital venous malformation of the leg.

20.2 Primary varicose veins

20.2.1 Uncomplicated

Description of Nomenclature [C used for Class (CEAP)]

Type 1 (T1): veins <3 mm size; type 2 (T2): veins >3 mm size

T1:C1

Type I: radicular veins (spider, reticular veins) are illustrated in Figs. 20.16–20.18.

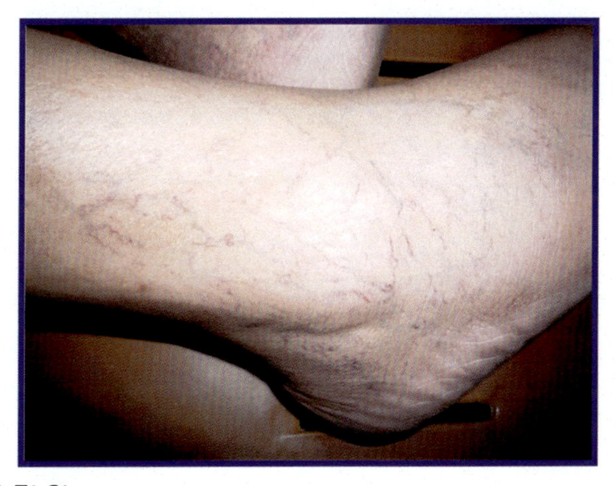

Figure 20.16 T1:C1.

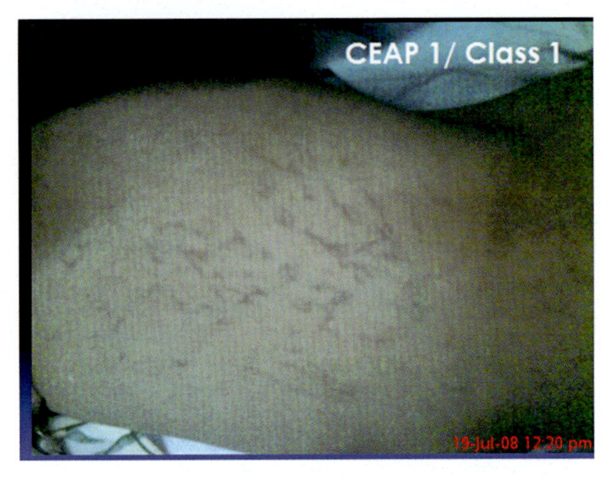

Figure 20.17 T1:C1.

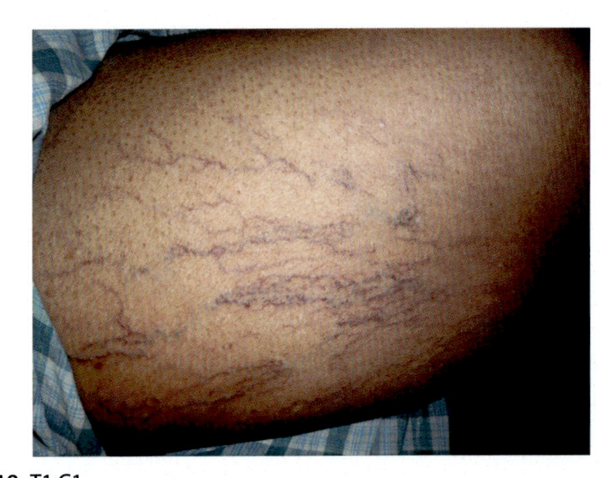

Figure 20.18 T1:C1.

T1:C2 is illustrated in Figs. 20.19–20.22.

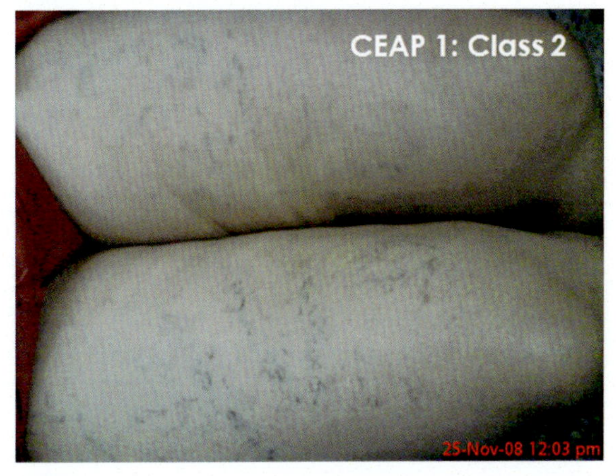

Figure 20.19 T1:C2.

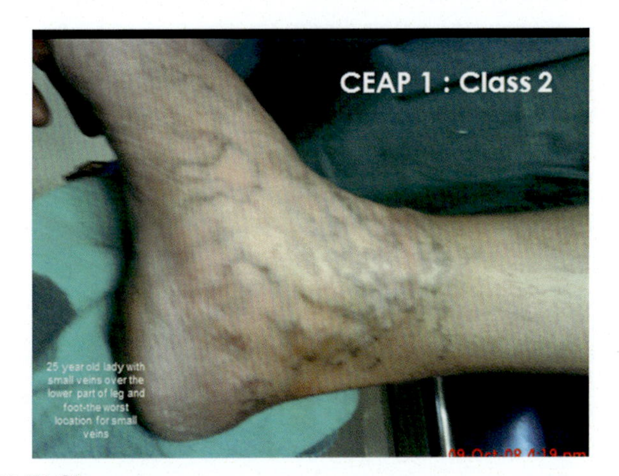

Figure 20.20 T1:C2.

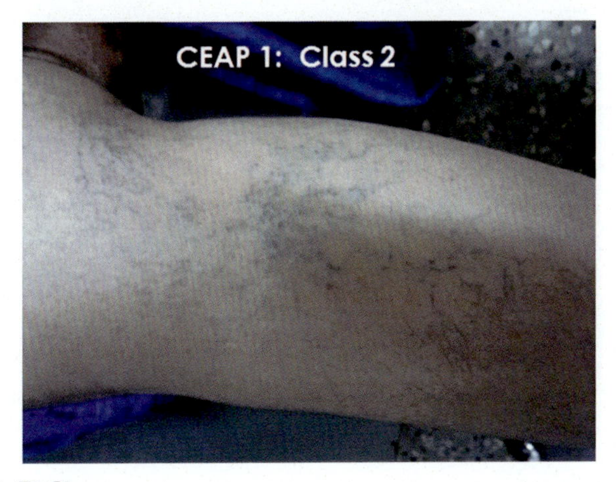

Figure 20.21 T1:C2.

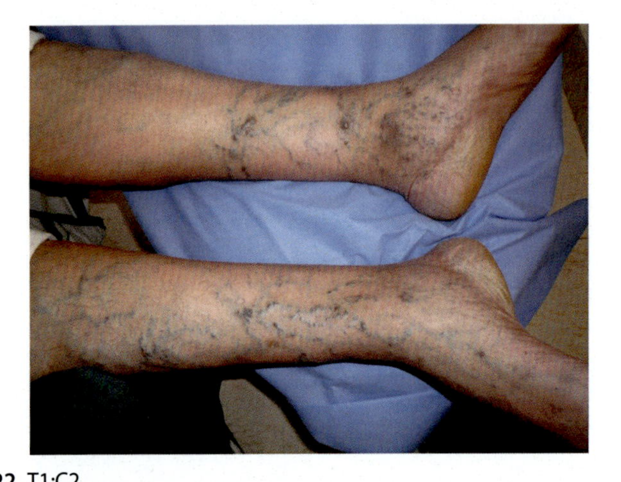

Figure 20.22 T1:C2.

T1:C1/C2 is illustrated in Figs. 20.23 and 20.24.

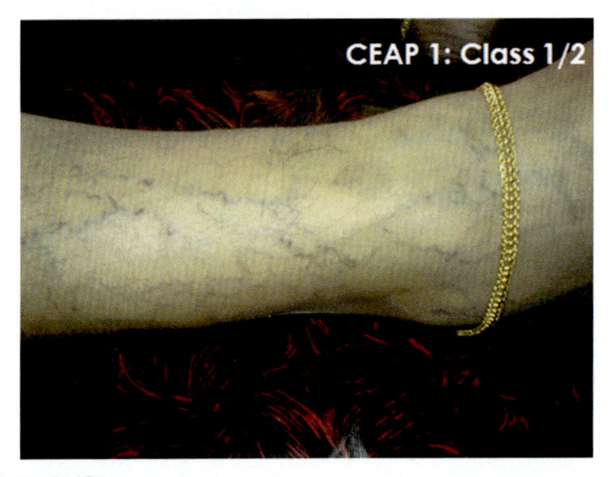

Figure 20.23 T1:C1/C2.

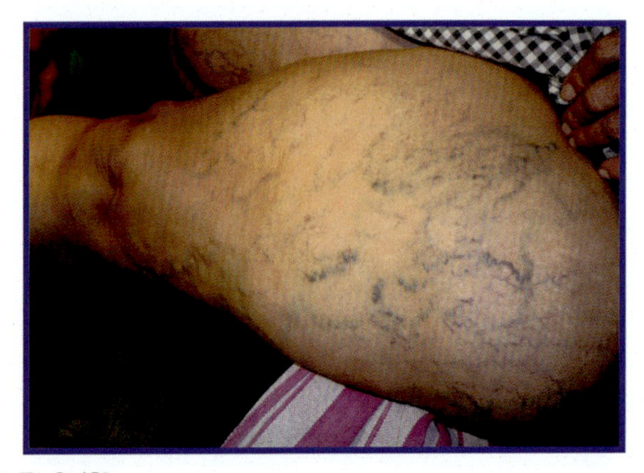

Figure 20.24 T1:C1/C2.

T1/T2:C2 is illustrated in Fig. 20.25.

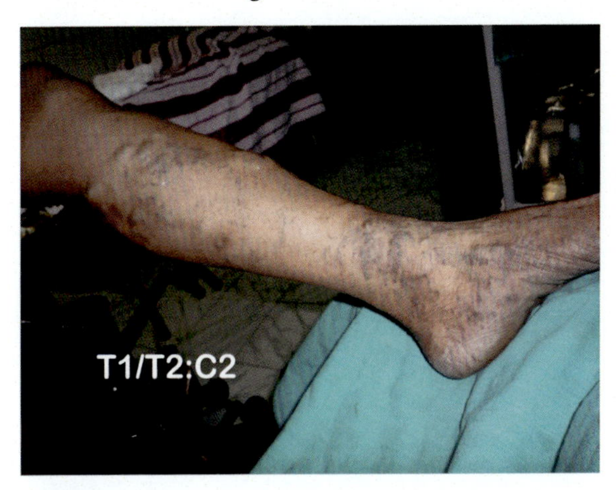

Figure 20.25 T1/T2:C2.

T2:C1 is illustrated in Figs. 20.26 and 20.27.

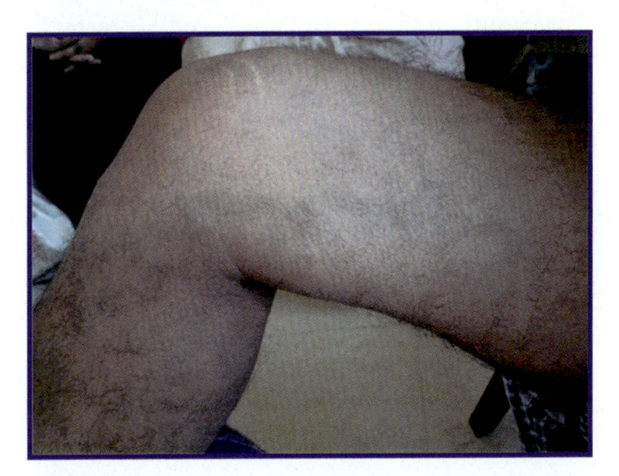

Figure 20.26 T2:C1.

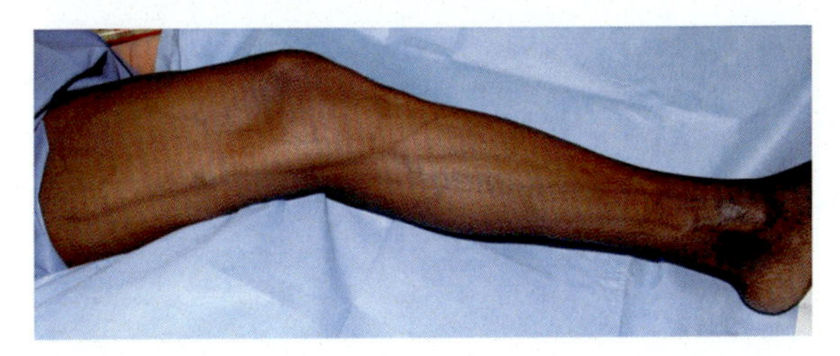

Figure 20.27 T2:C1.

T2:C2 is illustrated in Figs. 20.28–20.32.

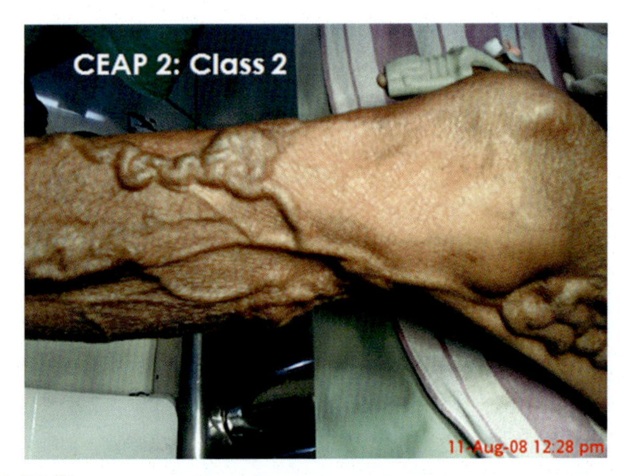

Figure 20.28 T2:C2.

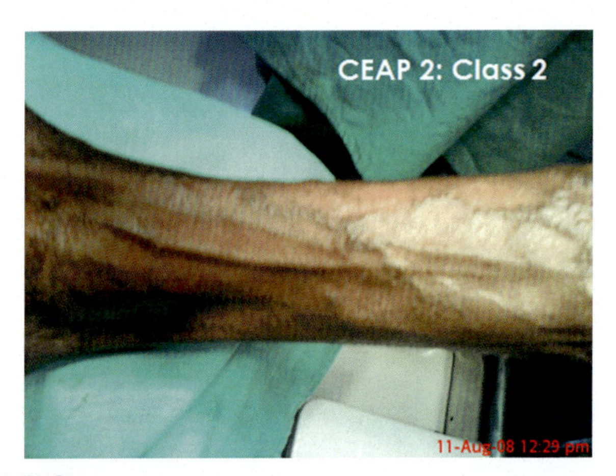

Figure 20.29 T2:C2.

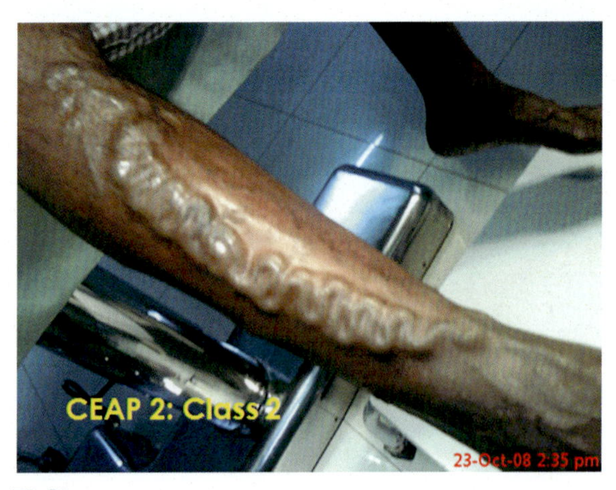

Figure 20.30 T2:C2.

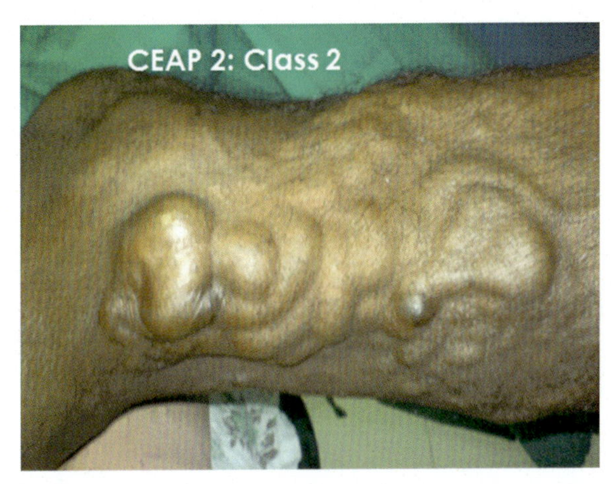

Figure 20.31 T2:C2.

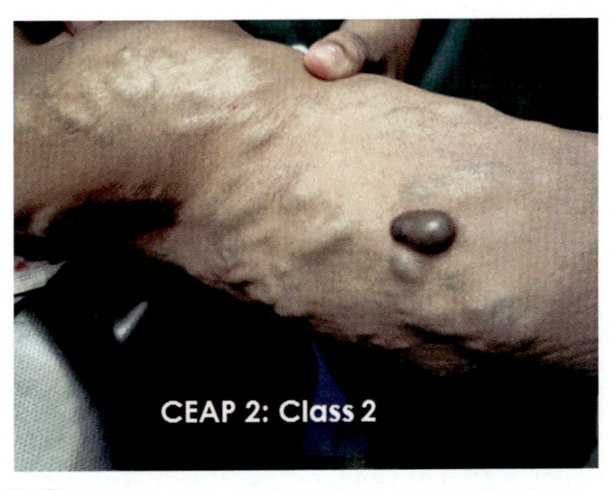

Figure 20.32 T2:C2.

20.2.2 Complicated

T1:C3 is illustrated in Figs. 20.33 and 20.34.

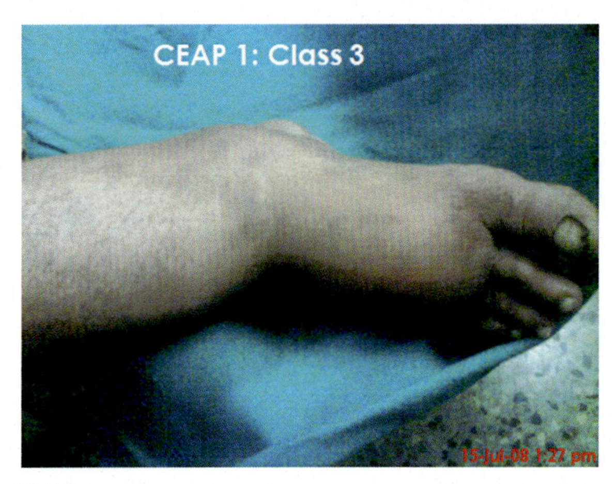

Figure 20.33 T1:C3.

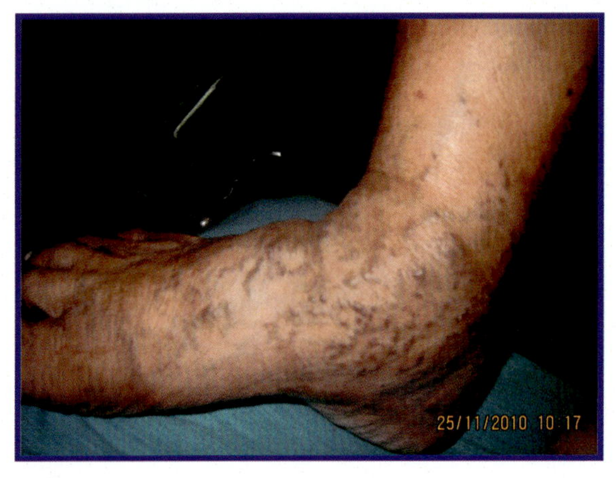

Figure 20.34 T1:C3.

T1:C4 is illustrated in Figs. 20.35–20.37.

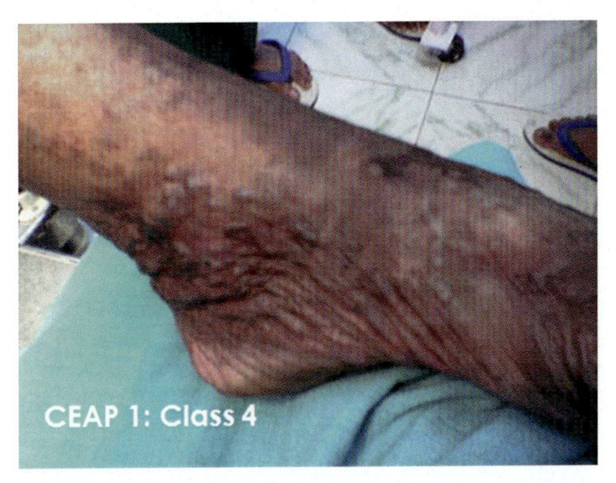

Figure 20.35 T1:C4.

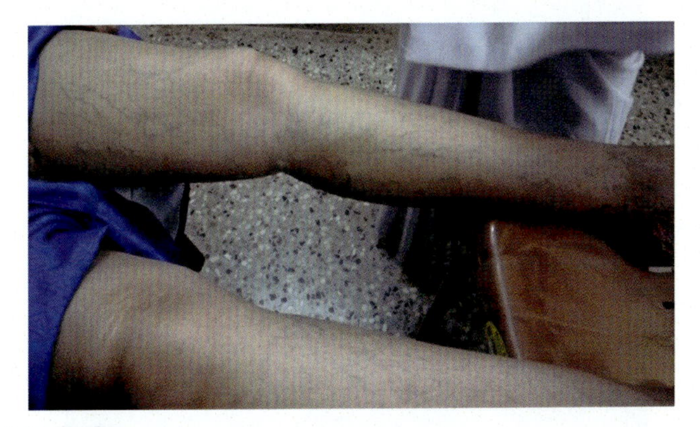

Figure 20.36 T1:C4.

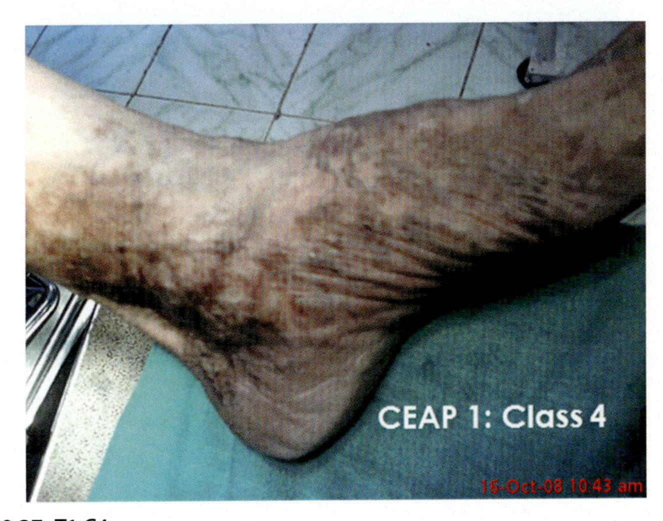

Figure 20.37 T1:C4.

T1/T2:C4 is illustrated in Figs. 20.38 and 20.39.

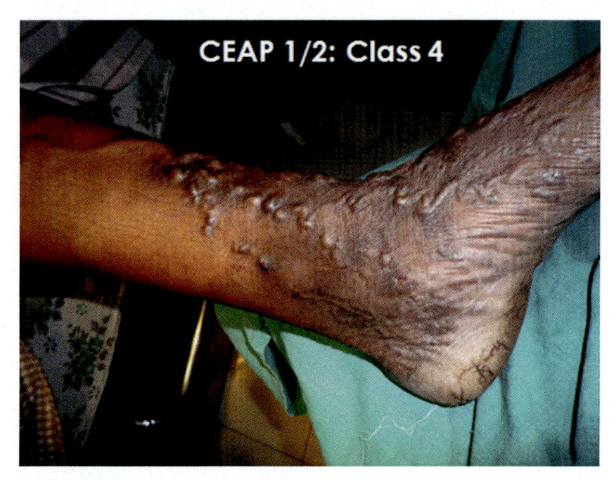

Figure 20.38 T1/T2:C4.

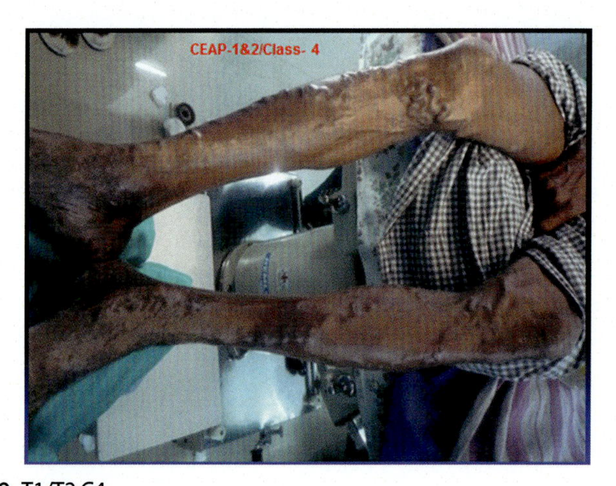

Figure 20.39 T1/T2:C4.

T1:C5 is illustrated in Figs. 20.40–20.42.

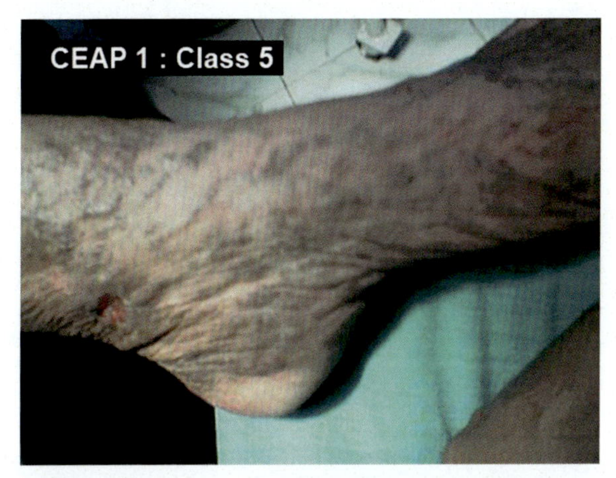

Figure 20.40 T1:C5.

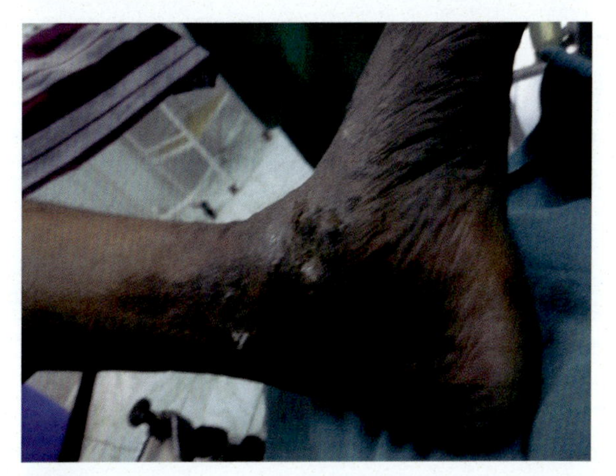

Figure 20.41 T1:C5.

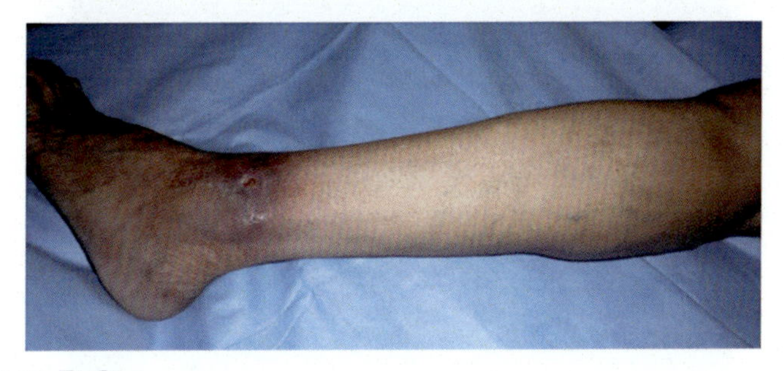

Figure 20.42 T1:C5.

T1/T2:C5 is illustrated in Figs. 20.43 and 20.44.

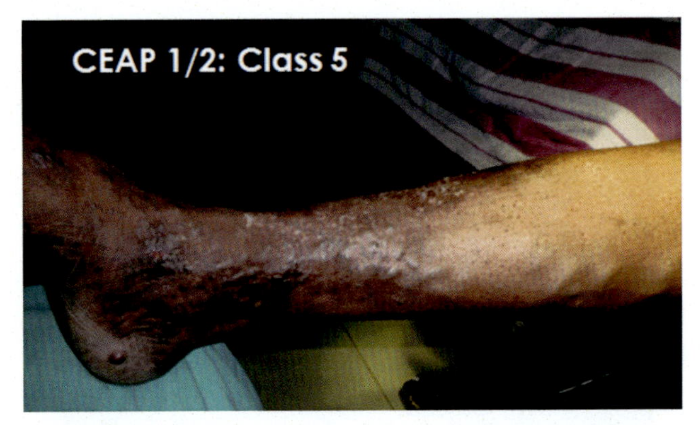

Figure 20.43 T1/T2:C5.

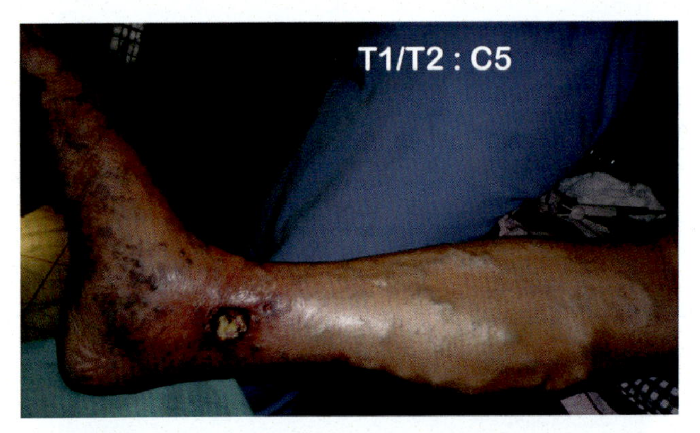

Figure 20.44 T1/T2:C5.

T1:C6 is illustrated in Figs. 20.45–20.50.

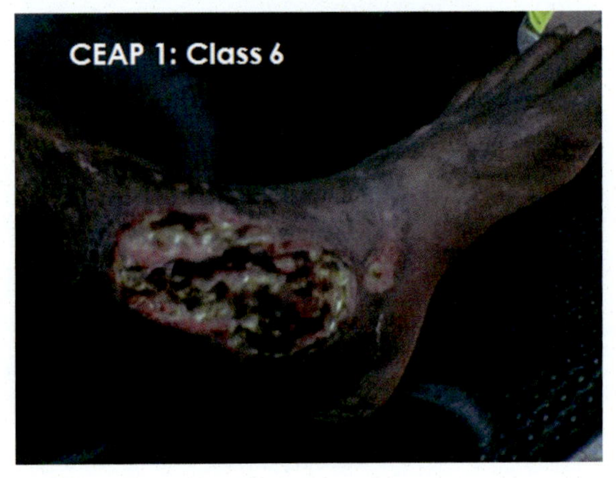

Figure 20.45 T1:C6.

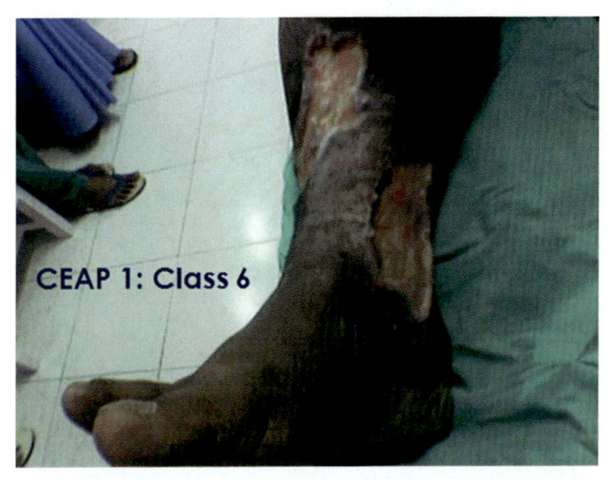

Figure 20.46 T1:C6.

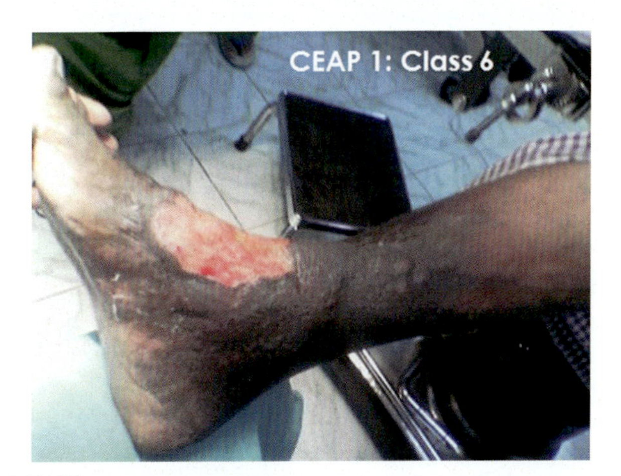

Figure 20.47 T1:C6.

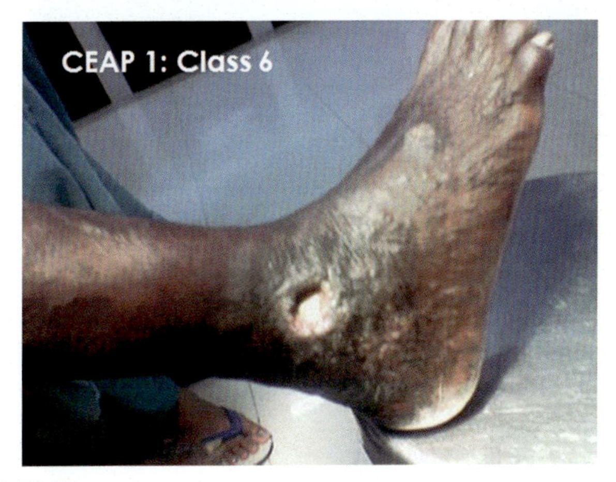

Figure 20.48 T1:C6.

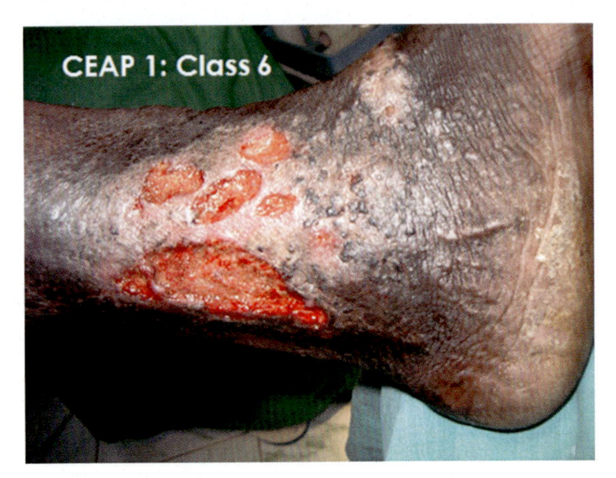

Figure 20.49 T1:C6.

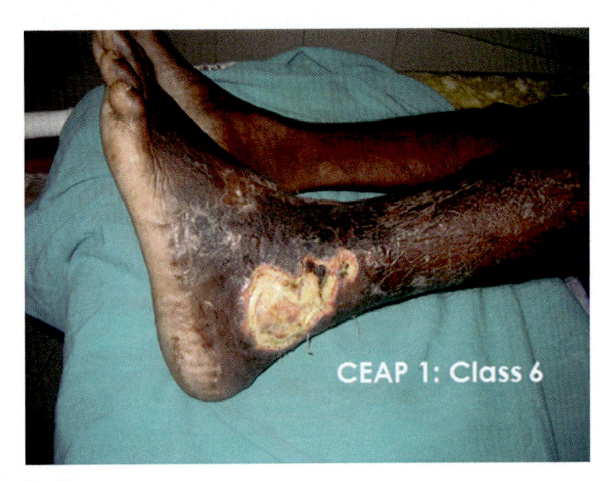

Figure 20.50 T1:C6.

T1/T2:C6 is illustrated in Figs. 20.51–20.57.

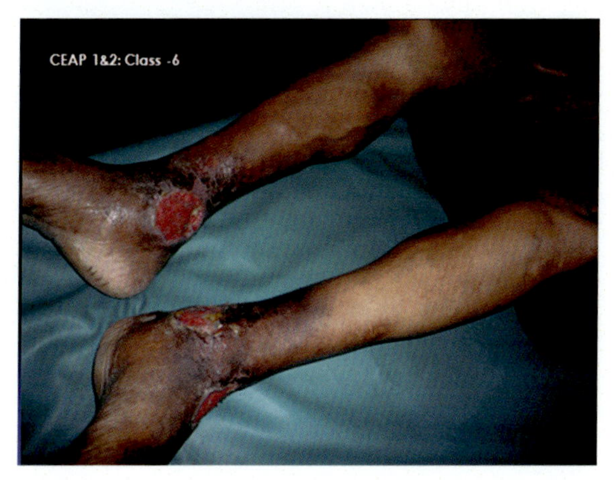

Figure 20.51 T1/T2:C6.

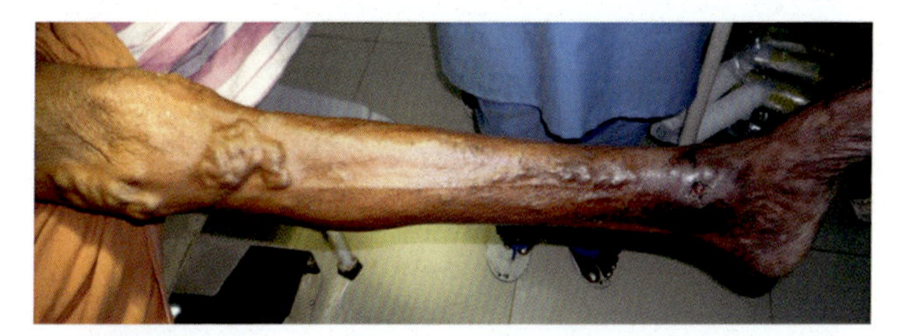

Figure 20.52 T1/T2:C6.

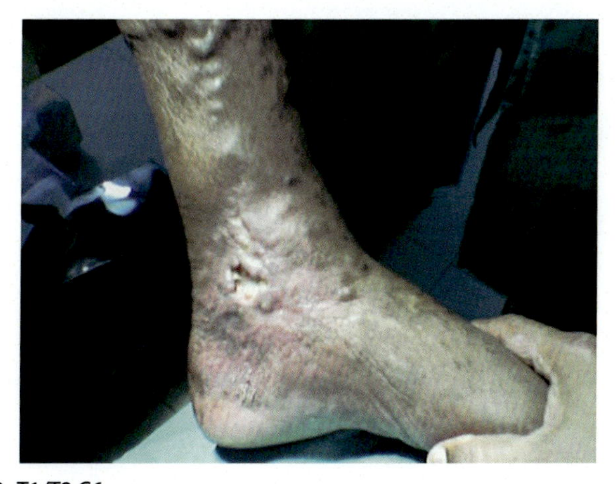

Figure 20.53 T1/T2:C6.

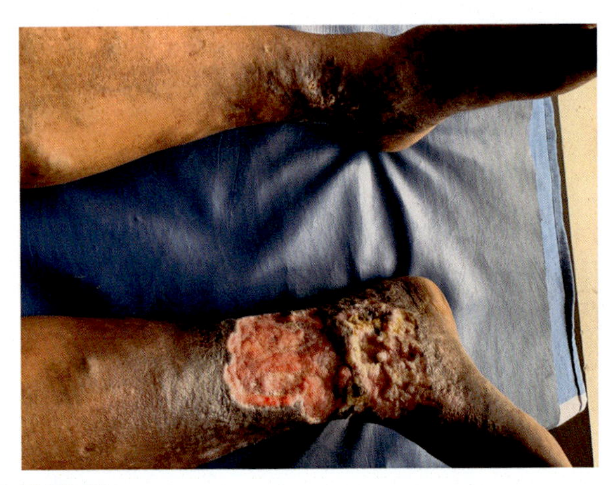

Figure 20.54 T1/T2:C6.

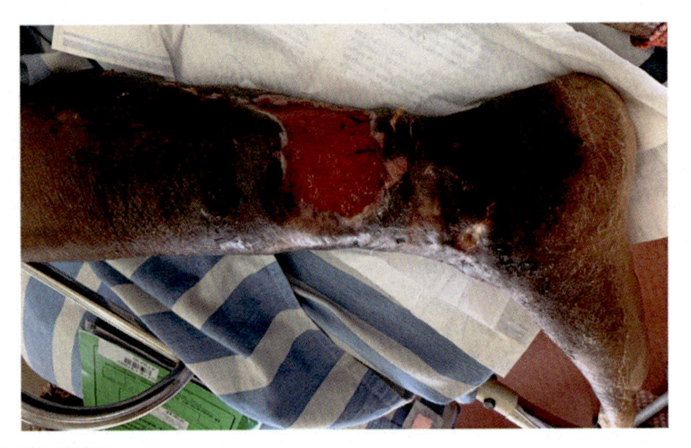

Figure 20.55 T1/T2:C6.

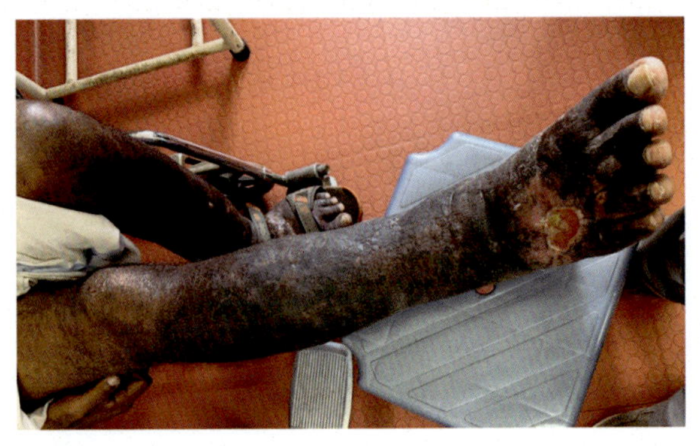

Figure 20.56 T1/T2:C6.

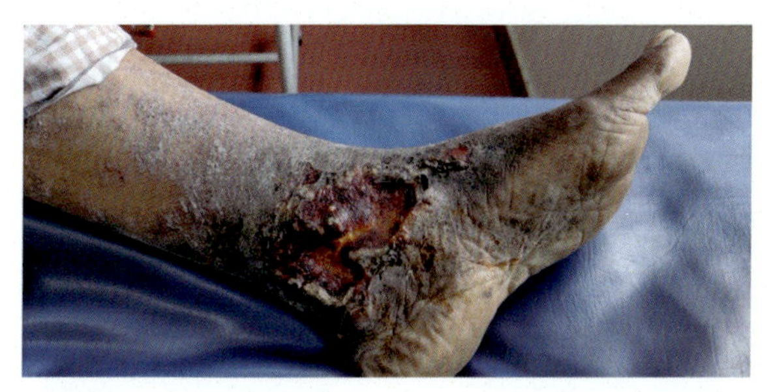

Figure 20.57 T1/T2:C6.

20.3 Maggots

Maggots are illustrated in Figs. 20.58–20.60.

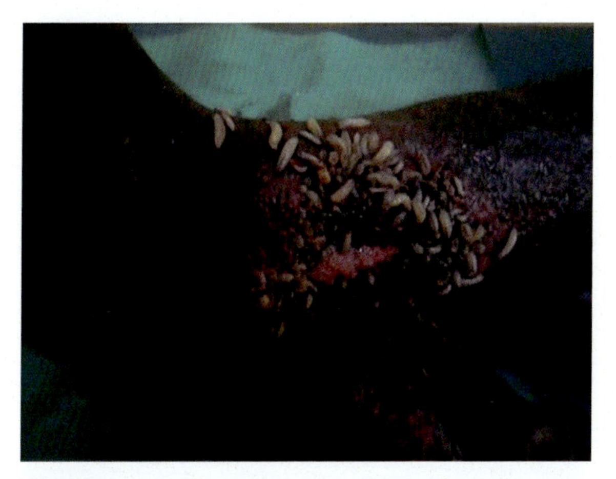

Figure 20.58 Maggots.

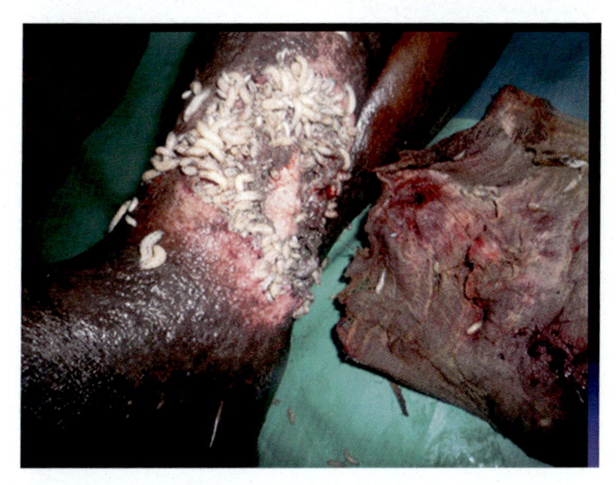

Figure 20.59 Maggots.

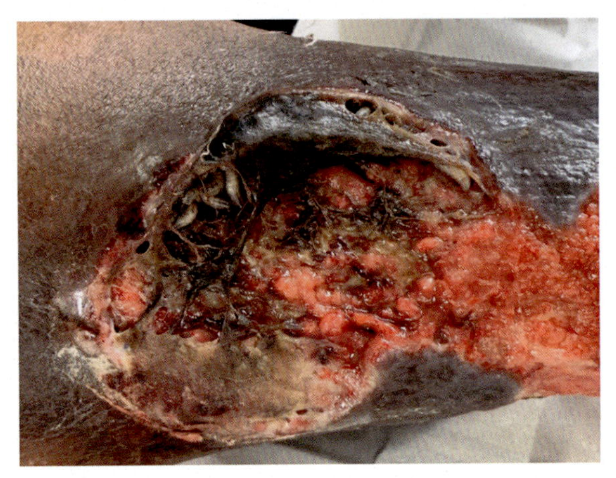

Figure 20.60 Maggots.

20.4 Recurrence after surgery

Images of recurrence after surgery are shown in Figs. 20.61–20.73.

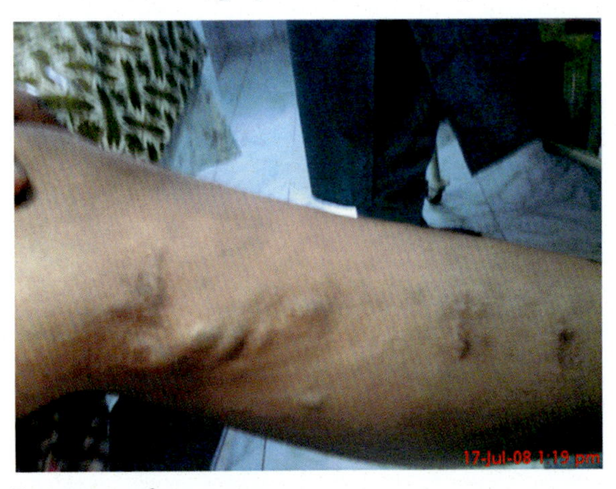

Figure 20.61 Recurrence after surgery.

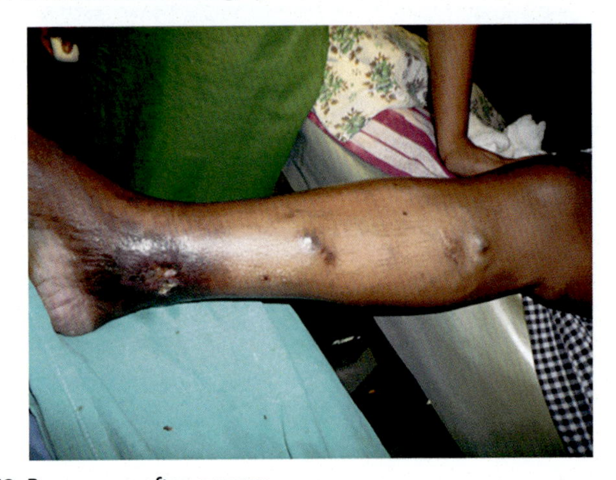

Figure 20.62 Recurrence after surgery.

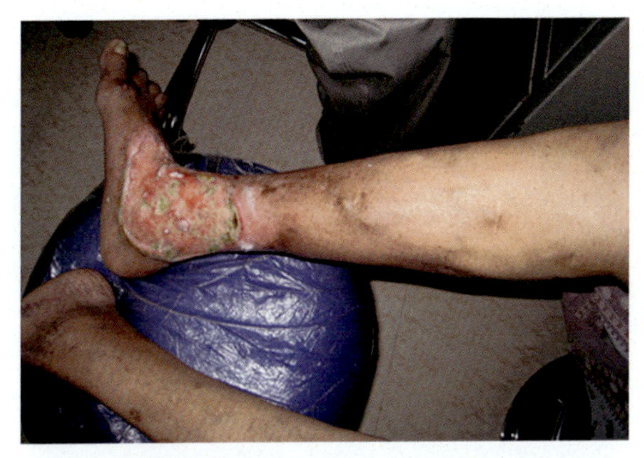

Figure 20.63 Recurrence after surgery.

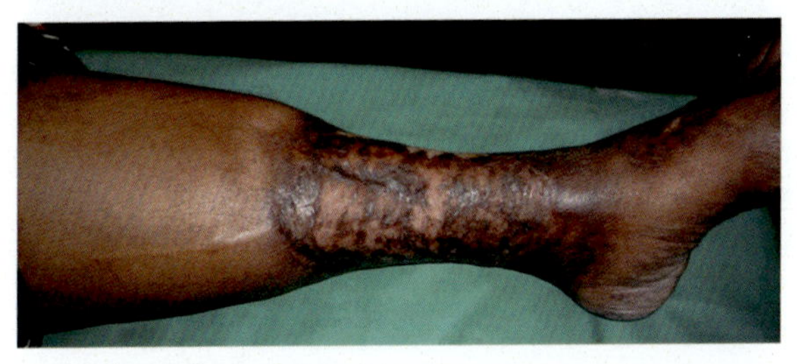

Figure 20.64 Recurrence after surgery.

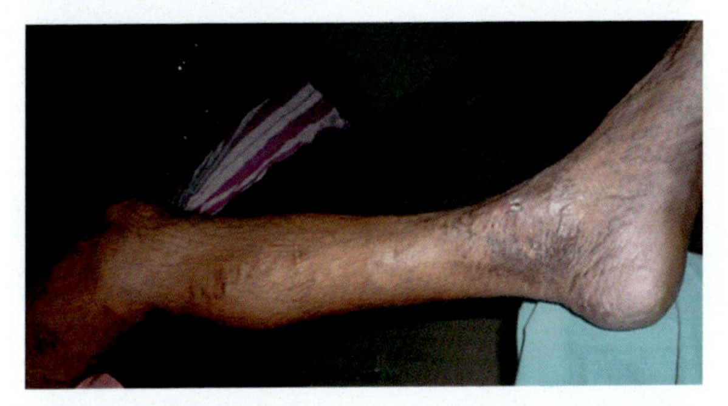

Figure 20.65 Recurrence after surgery.

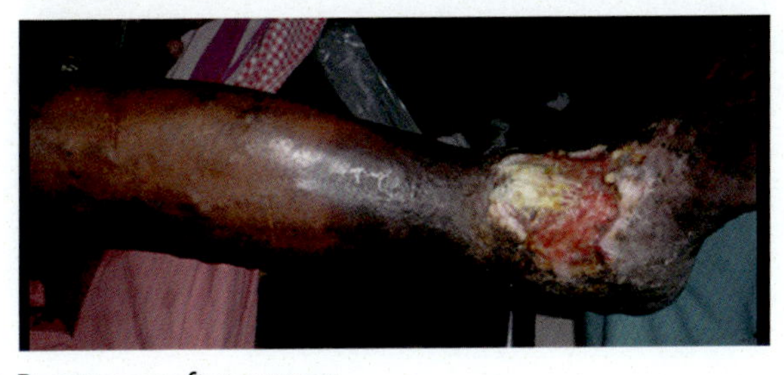

Figure 20.66 Recurrence after surgery.

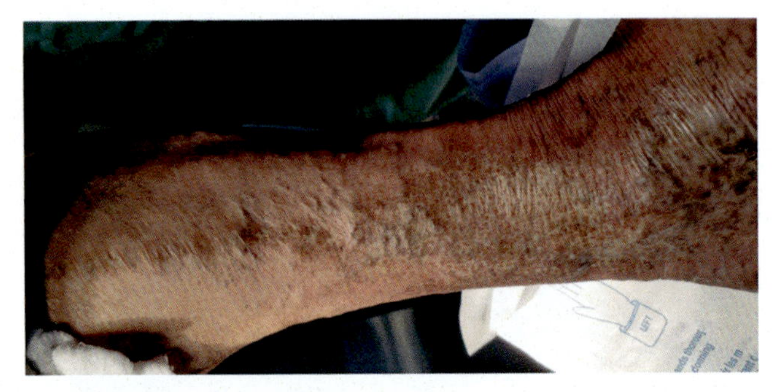

Figure 20.67 Recurrence after surgery.

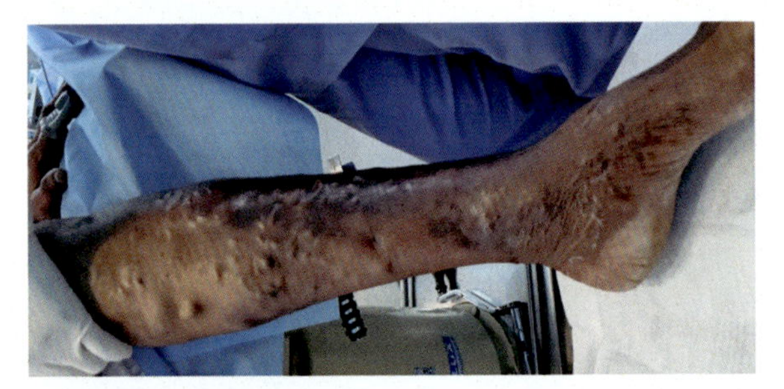

Figure 20.68 Recurrence after surgery.

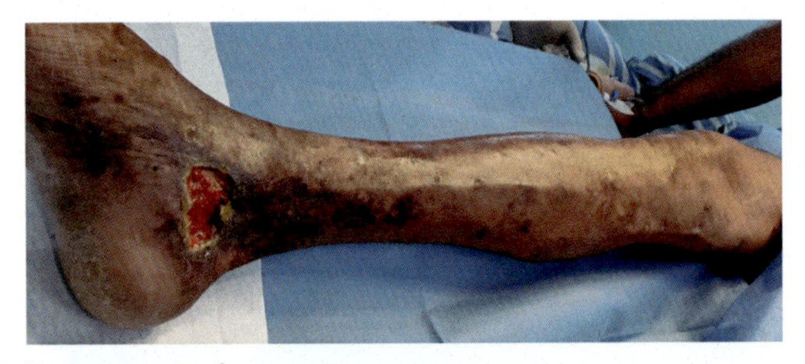

Figure 20.69 Recurrence after surgery.

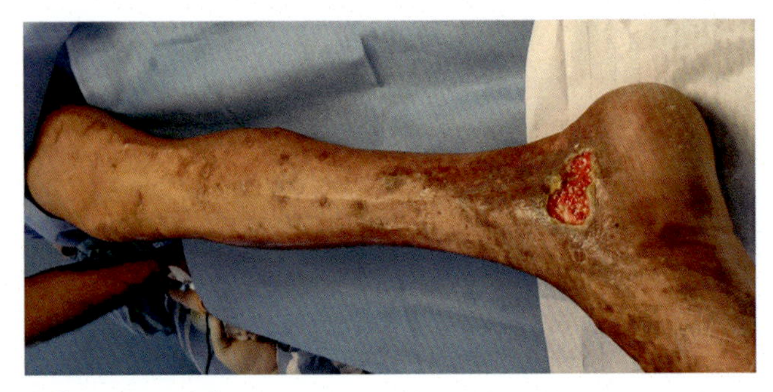

Figure 20.70 Recurrence after surgery.

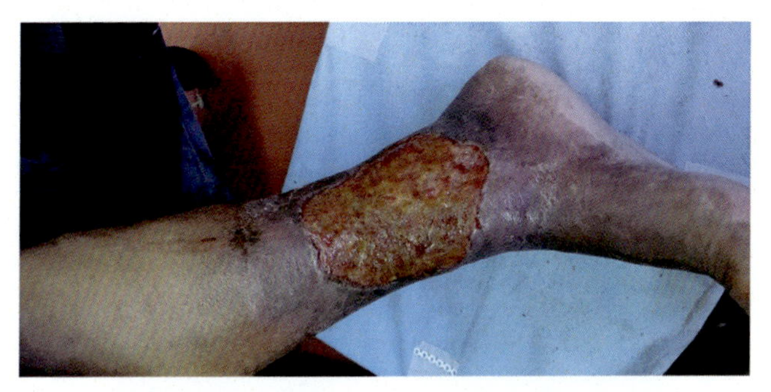

Figure 20.71 Recurrence after surgery.

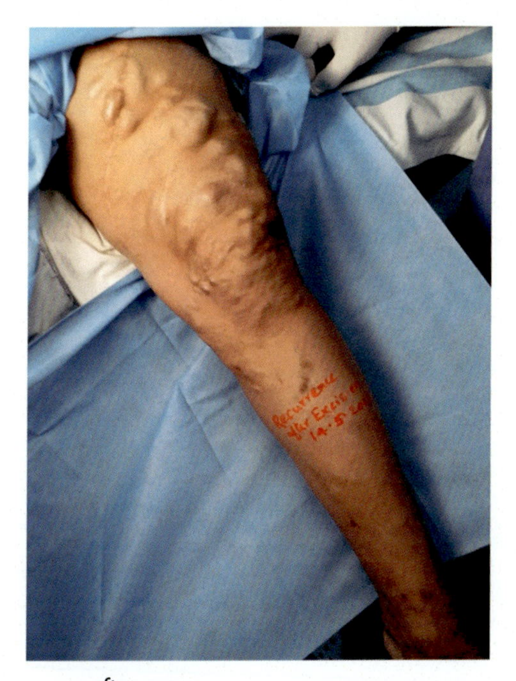

Figure 20.72 Recurrence after surgery.

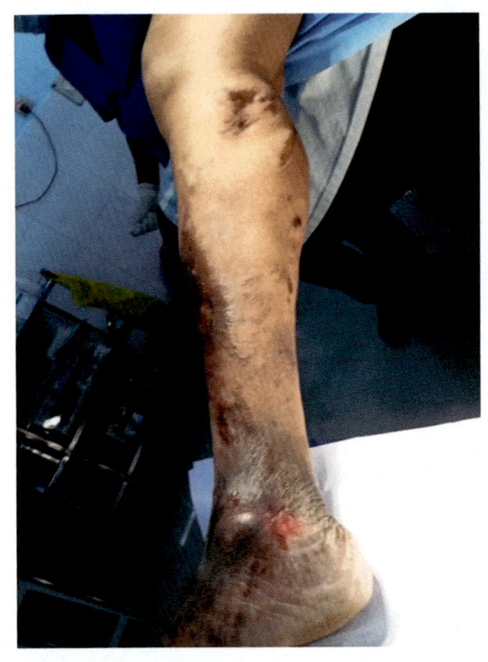

Figure 20.73 Recurrence after surgery.

20.5 Elephantoid change

Elephantoid change is illustrated in Figs. 20.74–20.79.

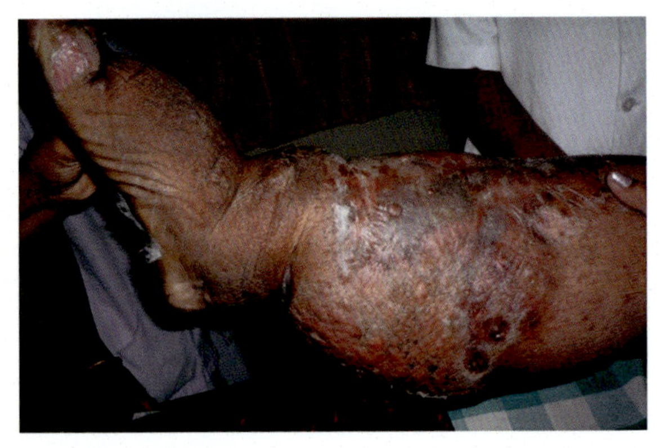

Figure 20.74 Elephantoid change.

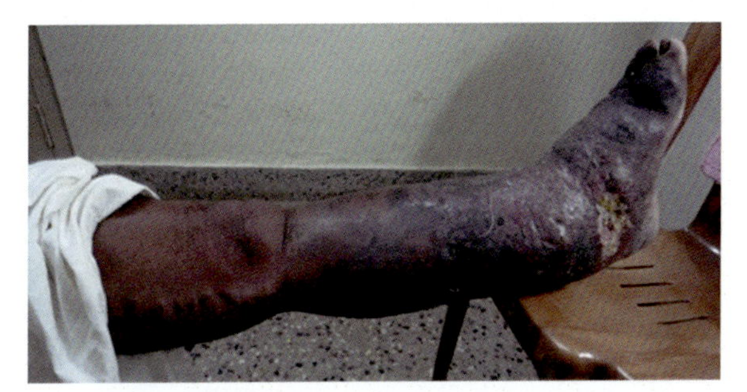

Figure 20.75 Elephantoid change.

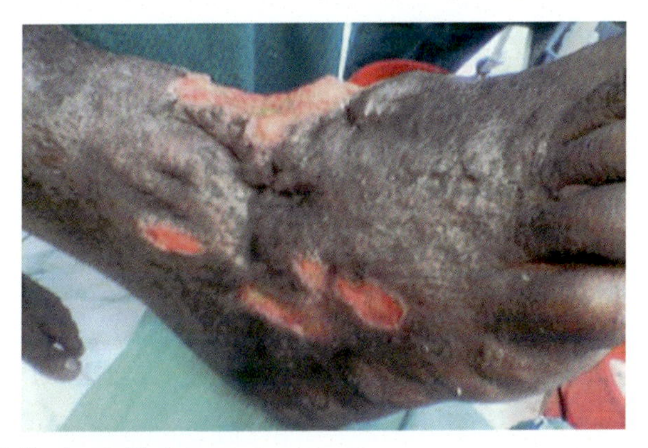

Figure 20.76 Elephantoid change.

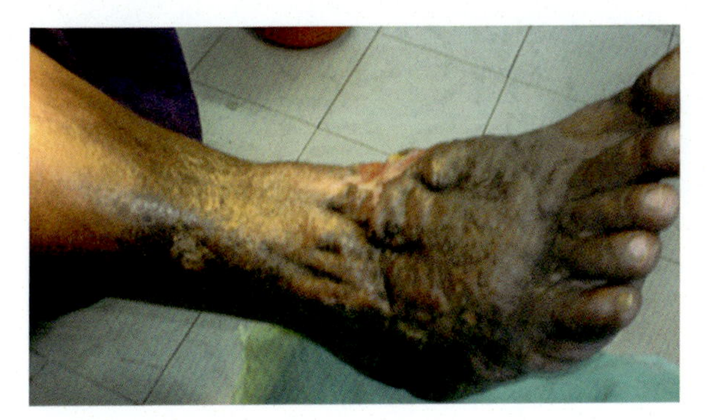

Figure 20.77 Elephantoid change.

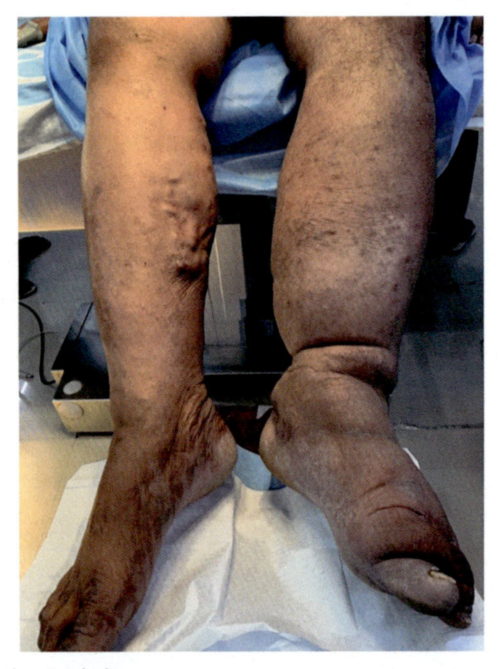

Figure 20.78 Elephantoid change.

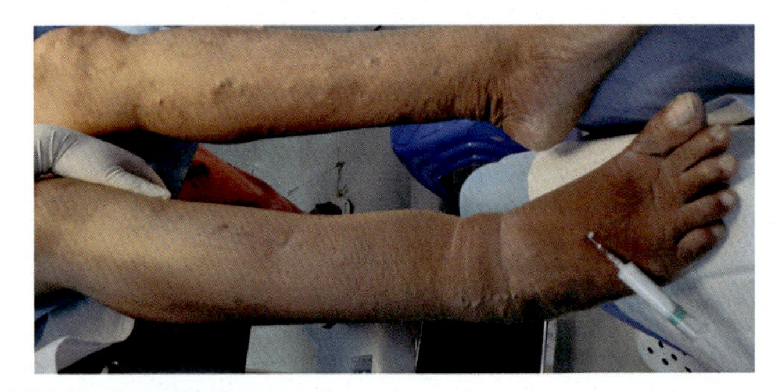

Figure 20.79 Elephantoid change.

20.6 Bleeding

Bleeding is illustrated in Figs. 20.80–20.84.

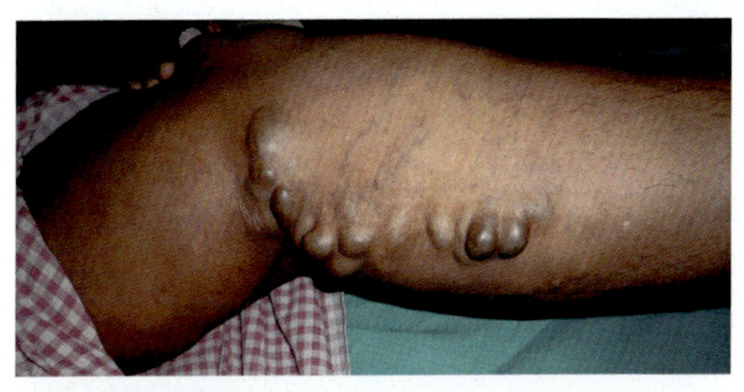

Figure 20.80 Threatening bleed.

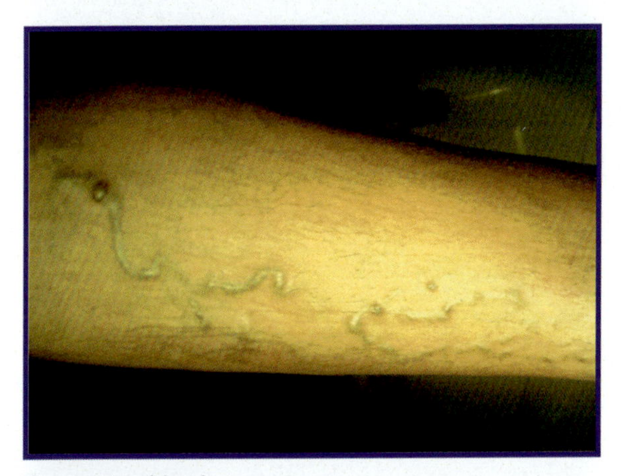

Figure 20.81 Threatening bleed.

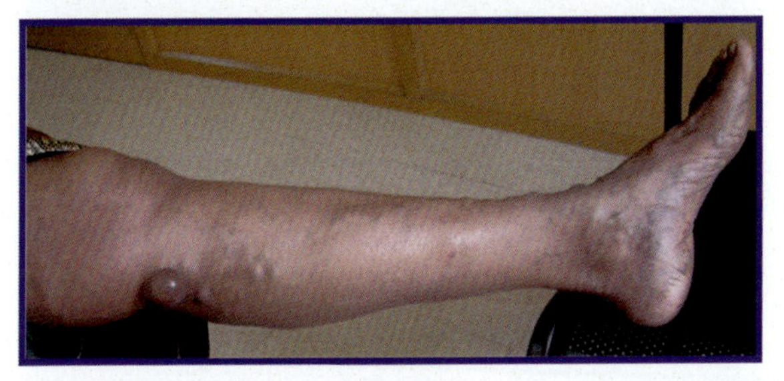

Figure 20.82 Threatening bleed.

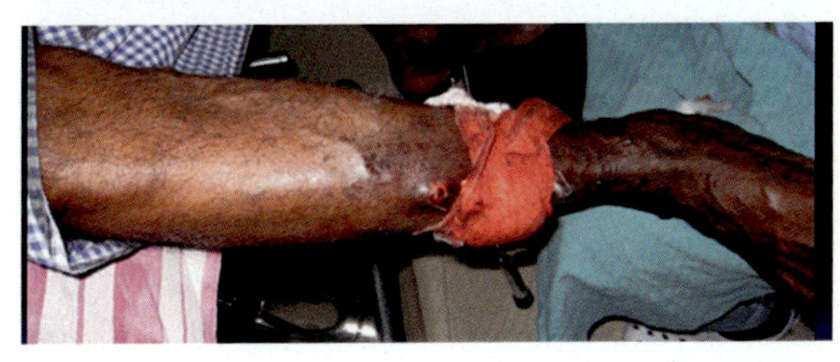

Figure 20.83 Bleeding.

Figure 20.84 Bleeding during consultation.

20.7 Thrombophlebitis

Thrombophlebitis is illustrated in Figs. 20.85 and 20.86.

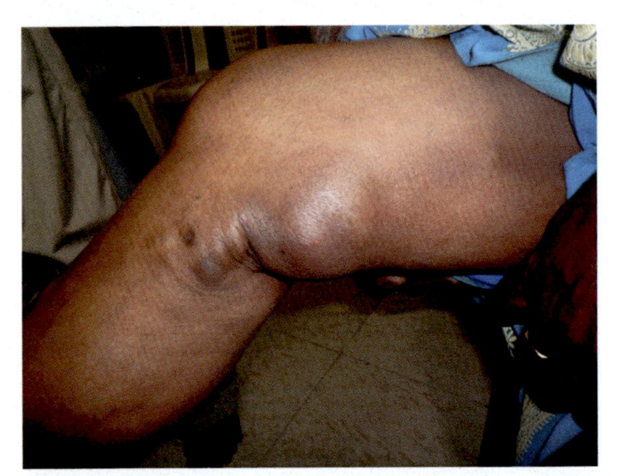

Figure 20.85 Thrombophlebitis.

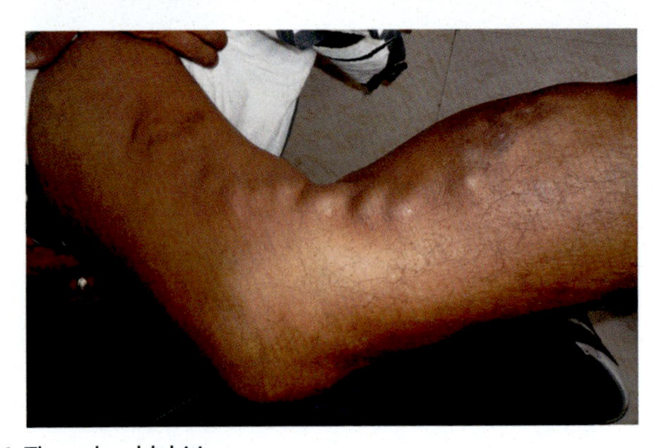

Figure 20.86 Thrombophlebitis.

20.8 Cellulitis

Cellulitis is illustrated in Figs. 20.87–20.89.

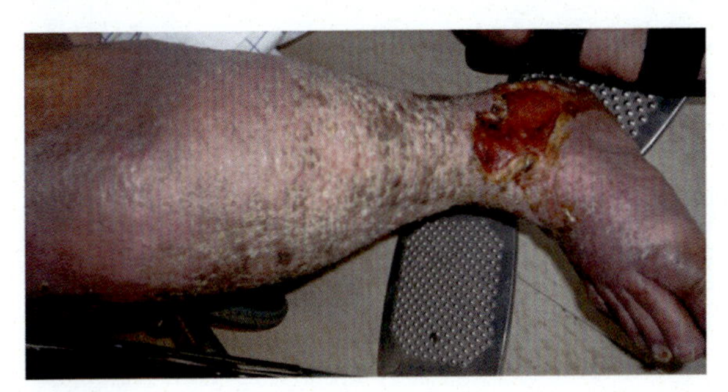

Figure 20.87 Cellulitis.

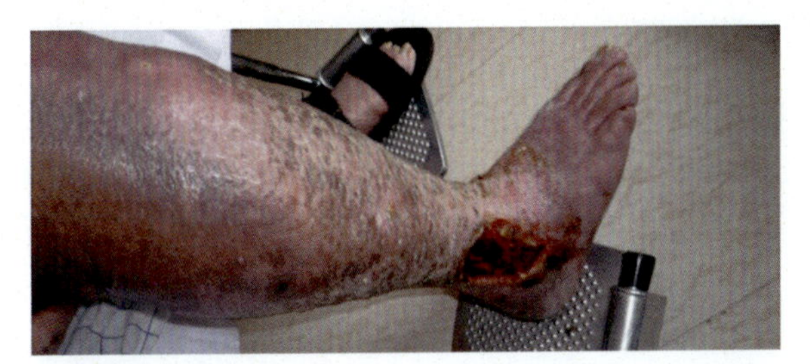

Figure 20.88 Cellulitis.

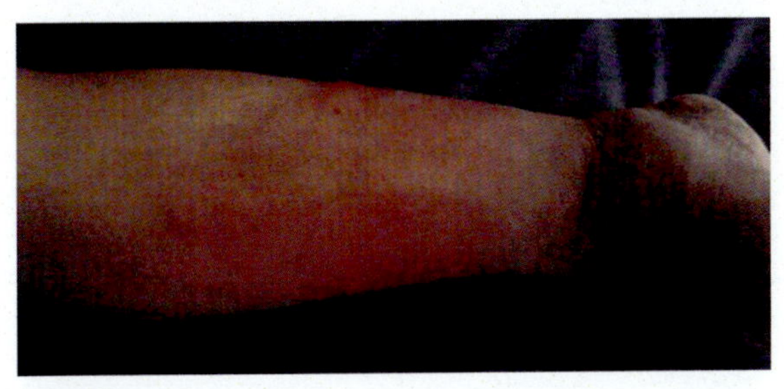

Figure 20.89 Cellulitis.

20.9 Postcellulitis

Postcellulitis is illustrated in Figs. 20.90–20.96.

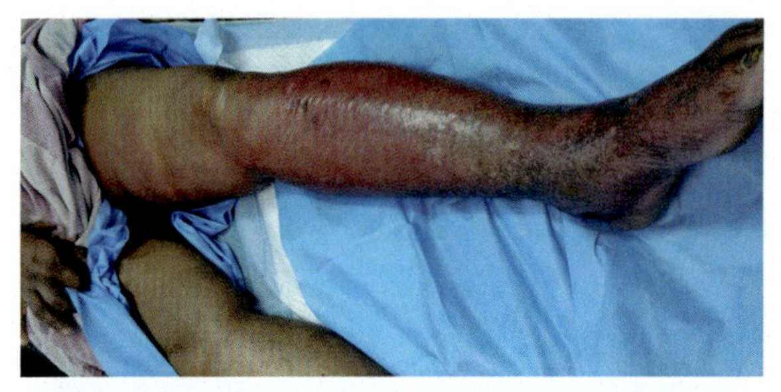

Figure 20.90 Postcellulitis.

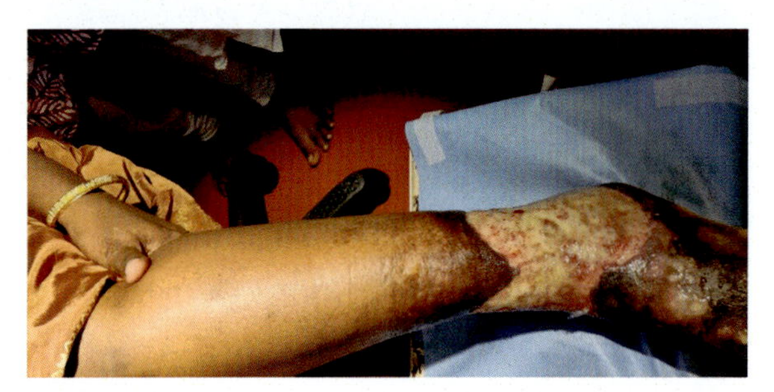

Figure 20.91 Postcellulitis.

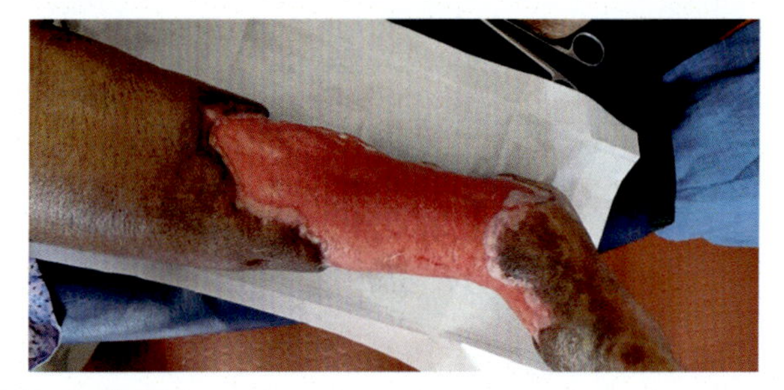

Figure 20.92 Postcellulitis.

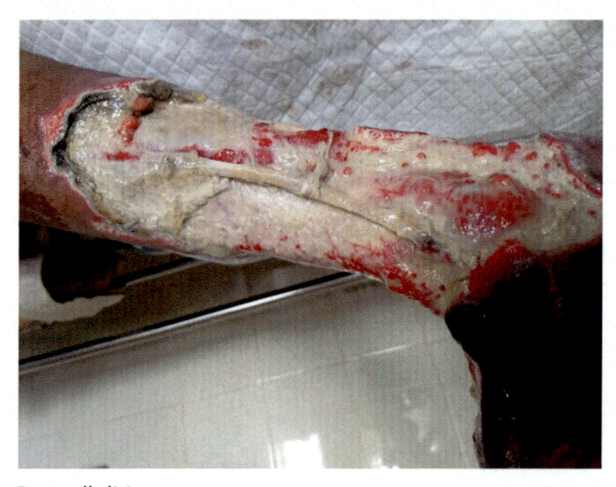

Figure 20.93 Postcellulitis.

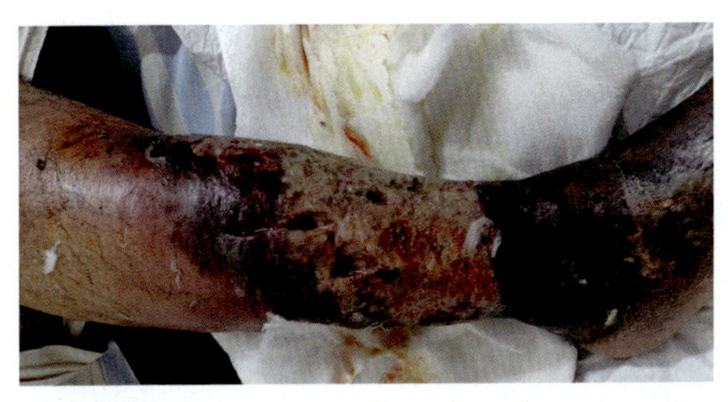

Figure 20.94 Postcellulitis.

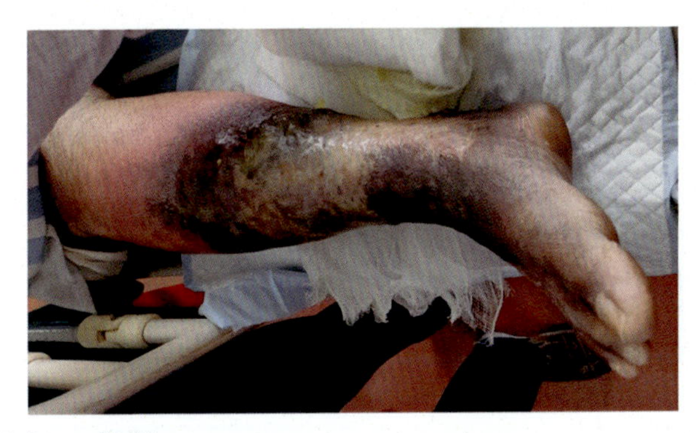

Figure 20.95 Postcellulitis.

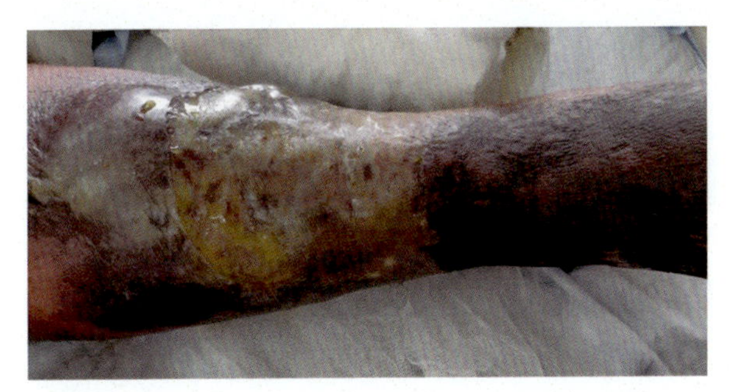

Figure 20.96 Postcellulitis.

20.10 Healing and healed ulceration

Healing and healed ulcerations are illustrated in Figs. 20.97–20.171.

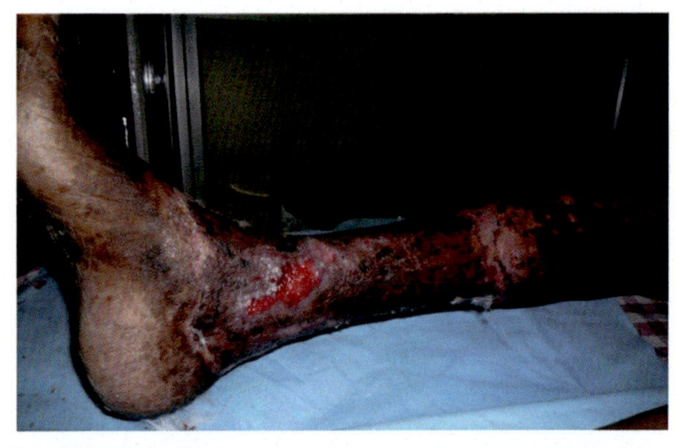

Figure 20.97 Healing ulceration.

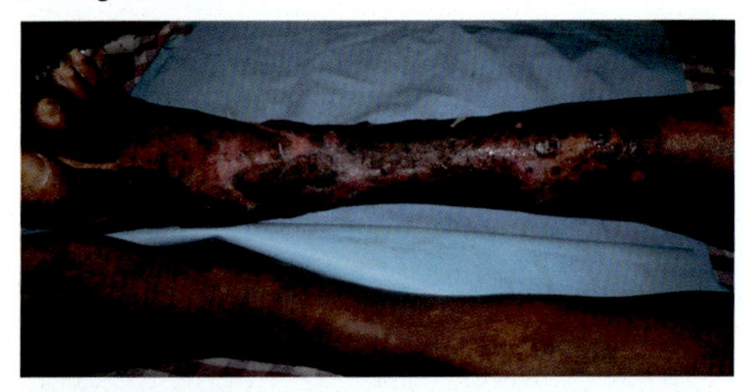

Figure 20.98 Healing ulceration.

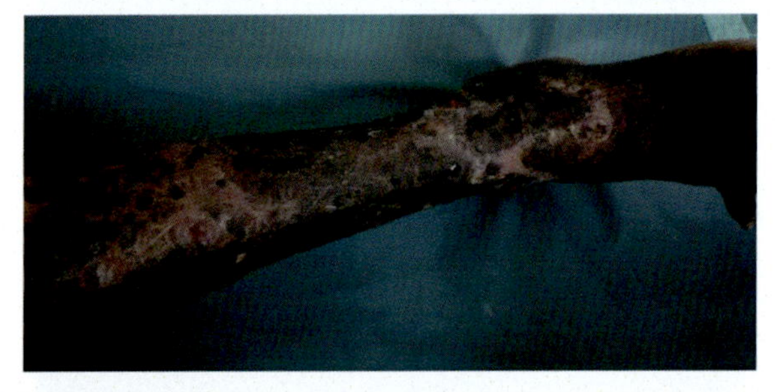

Figure 20.99 Healed ulceration.

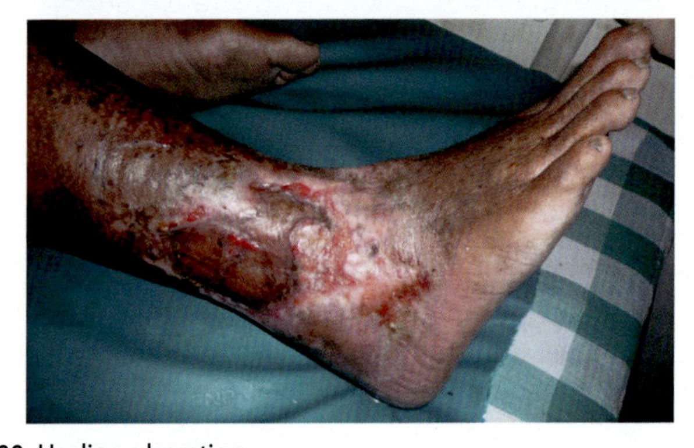

Figure 20.100 Healing ulceration.

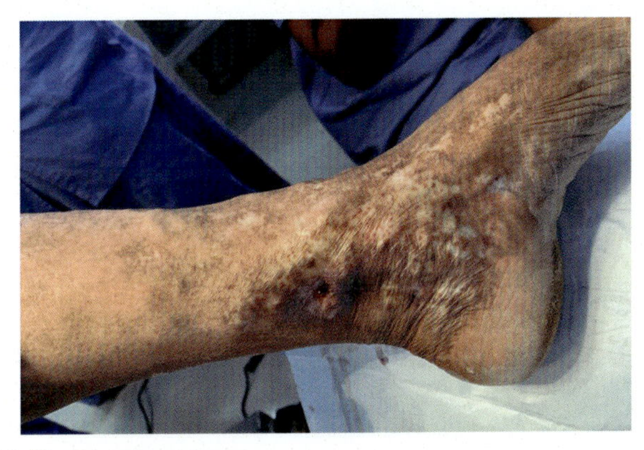

Figure 20.101 Healing ulceration.

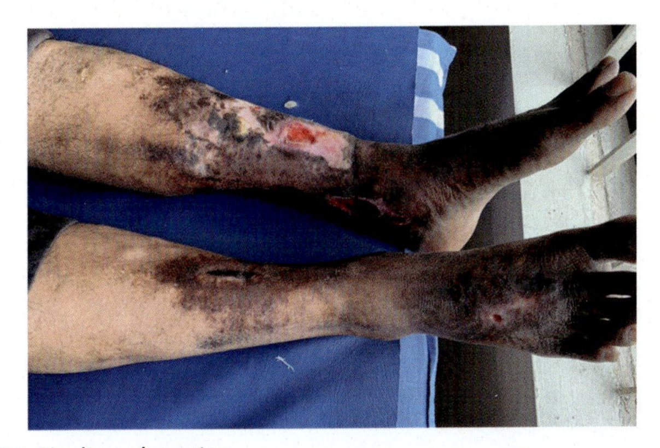

Figure 20.102 Healing ulceration.

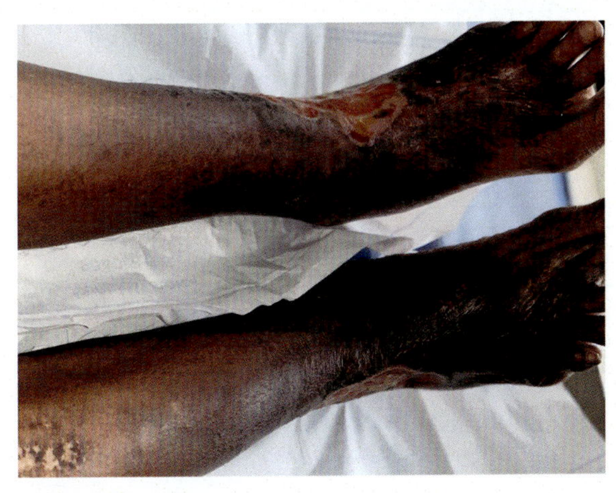

Figure 20.103 Healing ulceration.

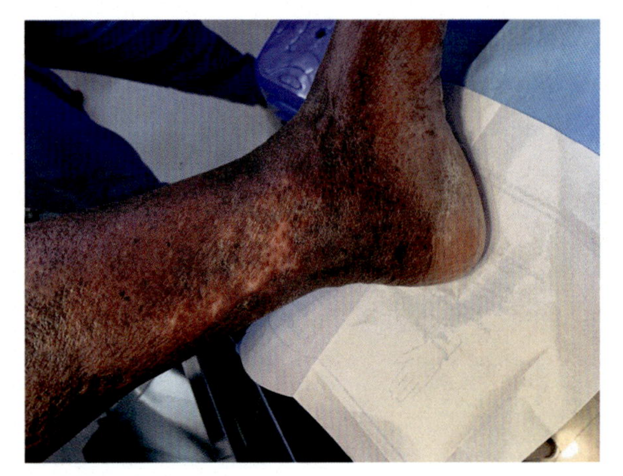

Figure 20.104 Healing ulceration.

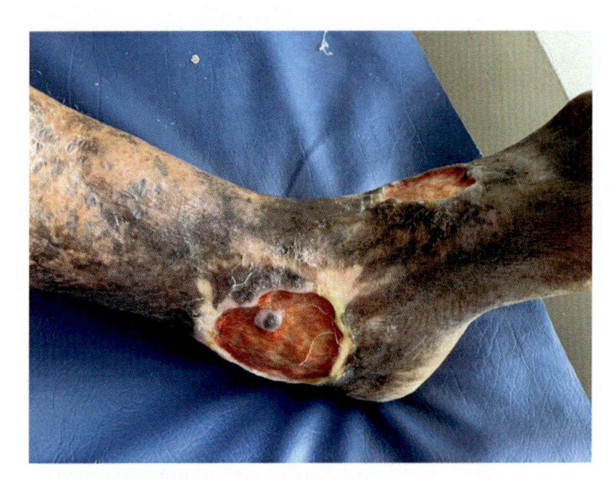

Figure 20.105 Healing ulceration.

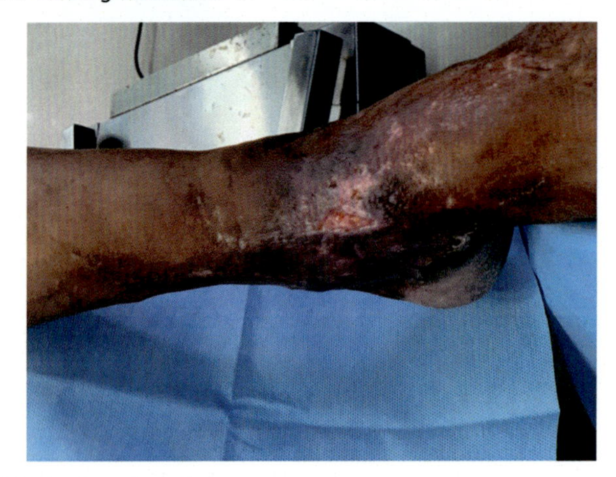

Figure 20.106 Healing ulceration.

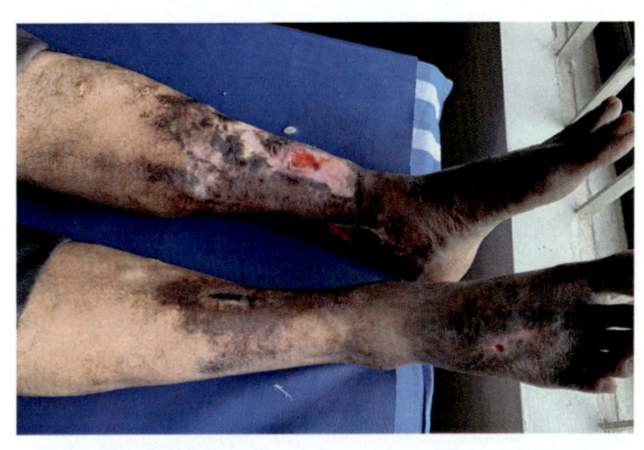

Figure 20.107 Healing ulceration.

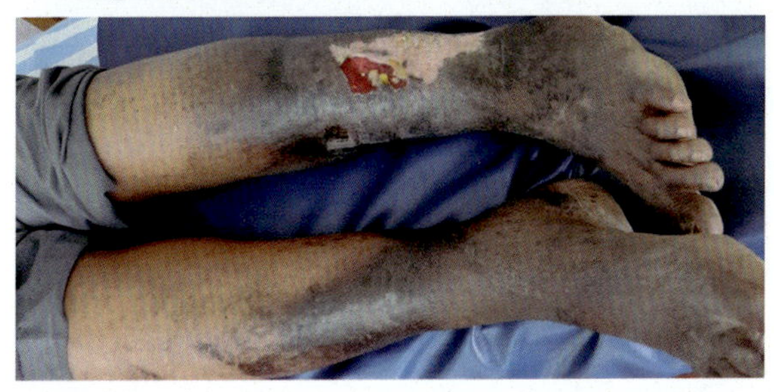

Figure 20.108 Healing ulceration.

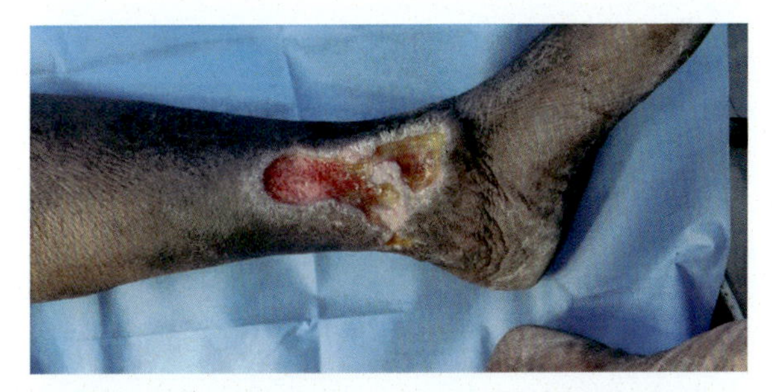

Figure 20.109 Healing ulceration.

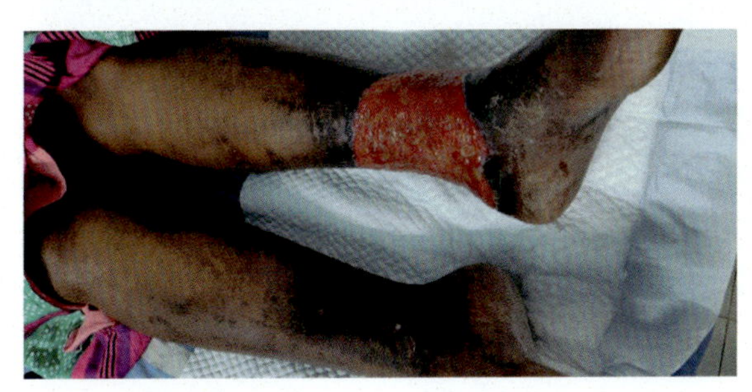

Figure 20.110 Healing ulceration.

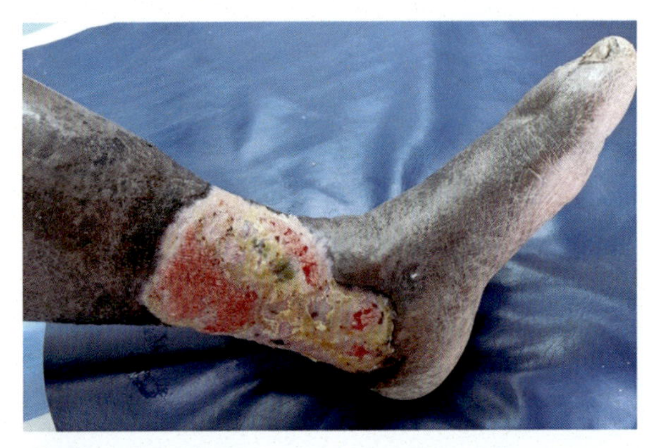

Figure 20.111 Healing ulceration.

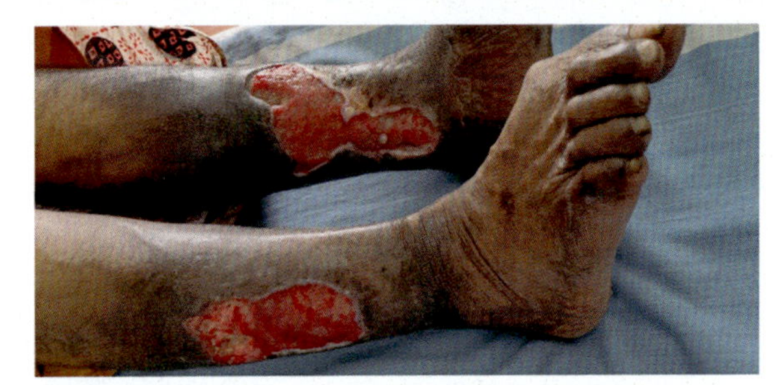

Figure 20.112 Healing ulceration.

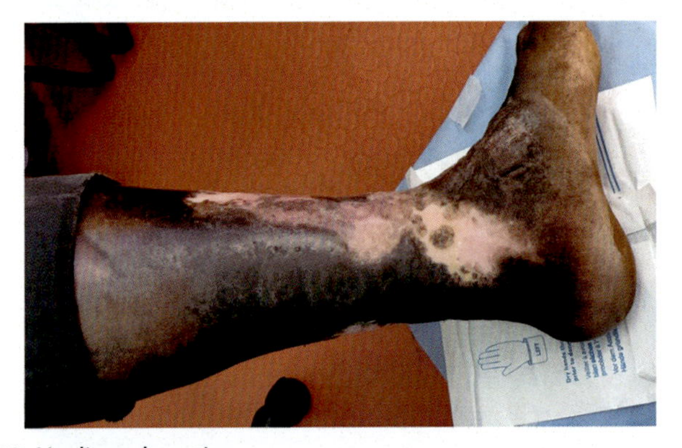

Figure 20.113 Healing ulceration.

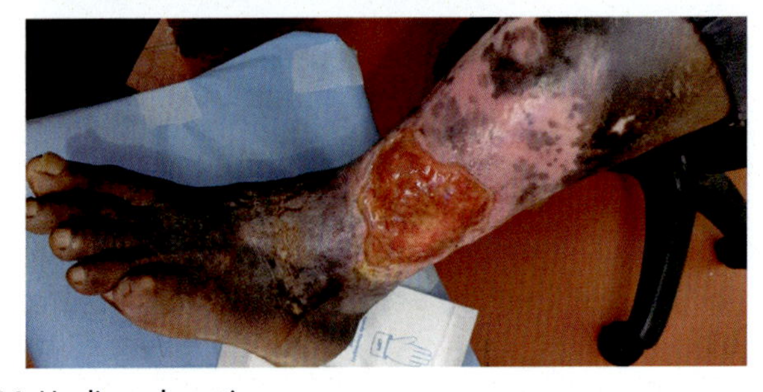

Figure 20.114 Healing ulceration.

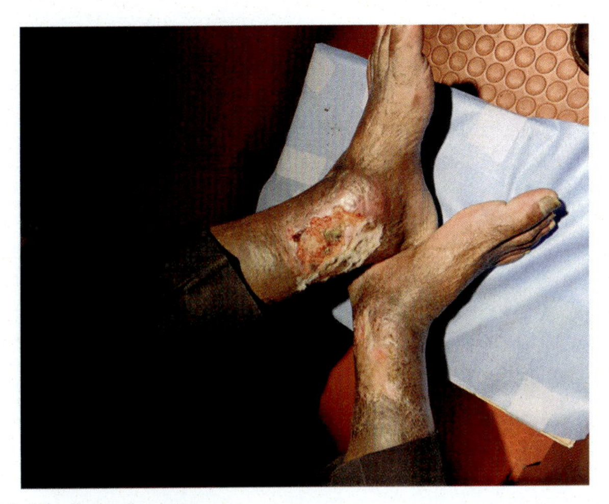

Figure 20.115 Healing ulceration.

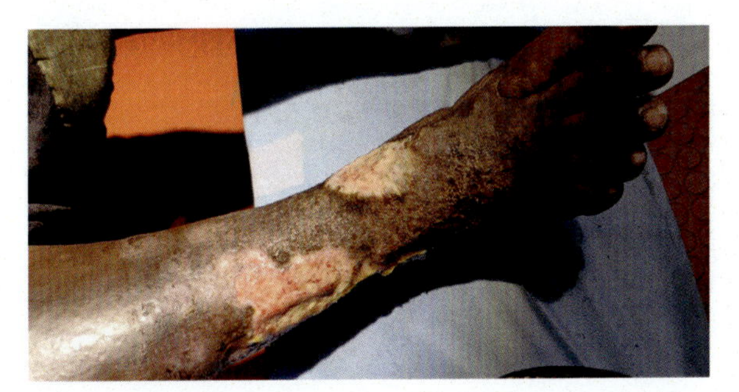

Figure 20.116 Healing ulceration.

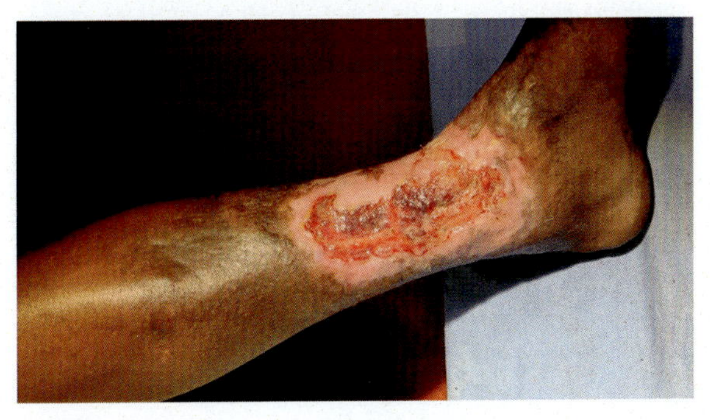

Figure 20.117 Healing ulceration.

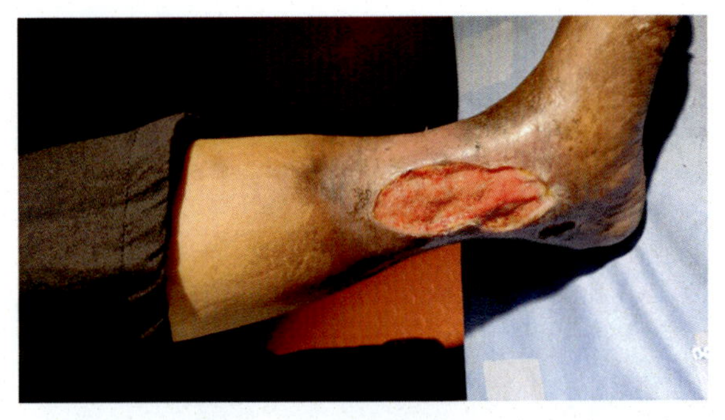

Figure 20.118 Healing ulceration.

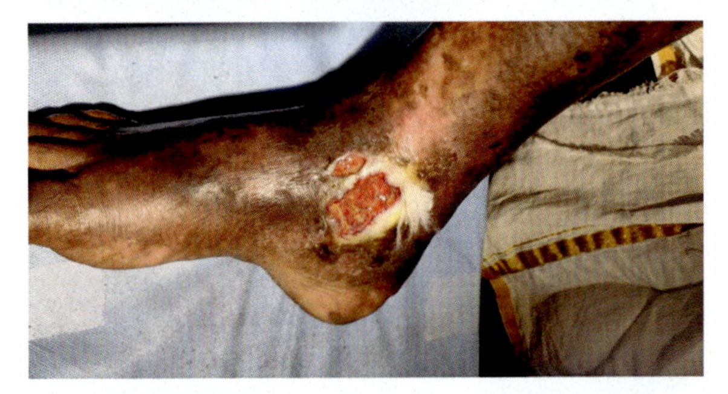

Figure 20.119 Healing ulceration.

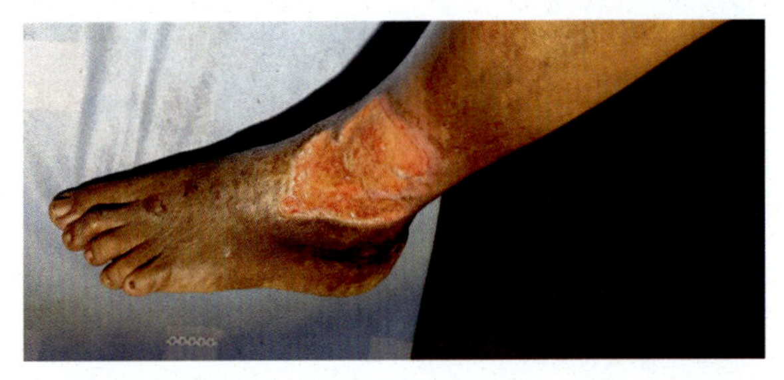

Figure 20.120 Healing ulceration.

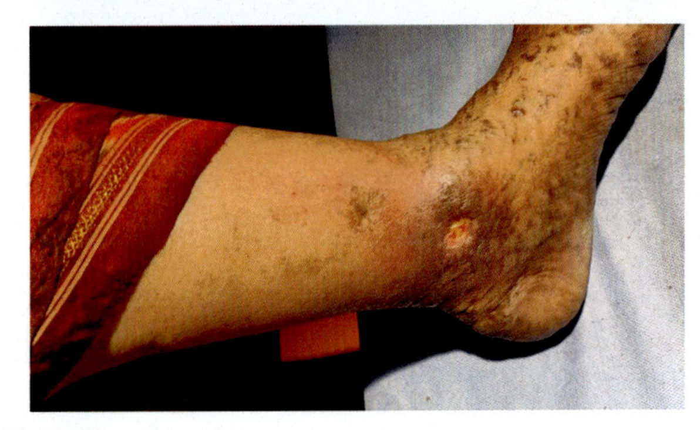

Figure 20.121 Healing ulceration.

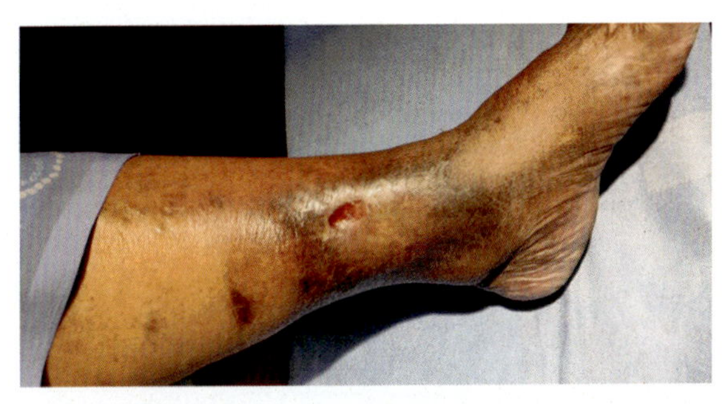

Figure 20.122 Healing ulceration.

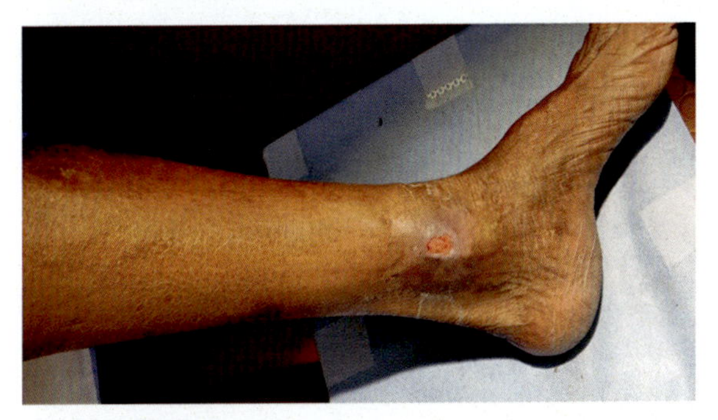

Figure 20.123 Healing ulceration.

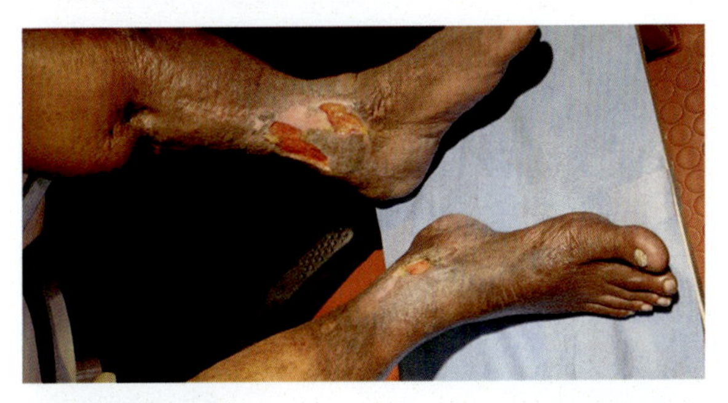

Figure 20.124 Healing ulceration.

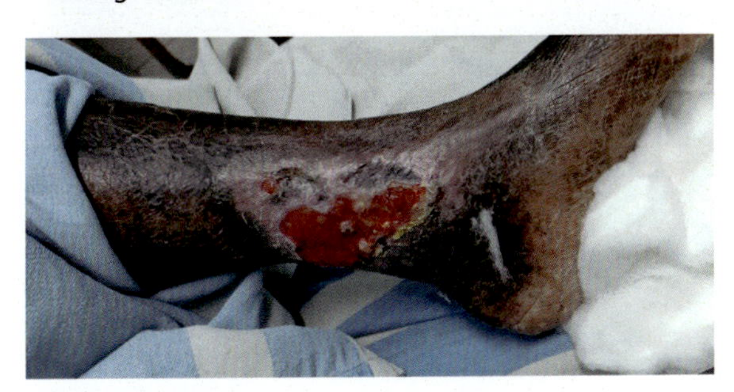

Figure 20.125 Healing ulceration.

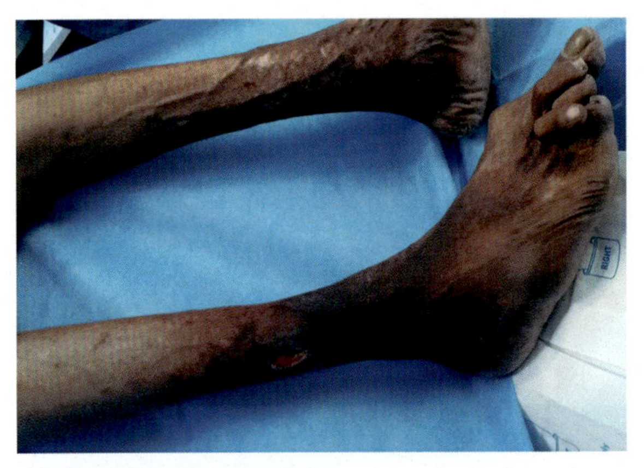

Figure 20.126 Healing ulceration.

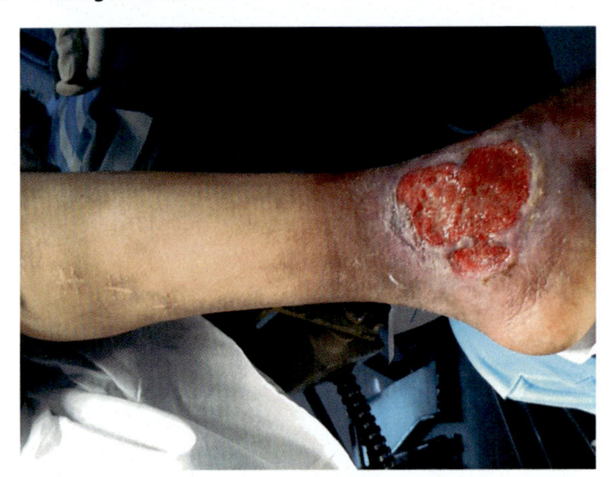

Figure 20.127 Healing ulceration.

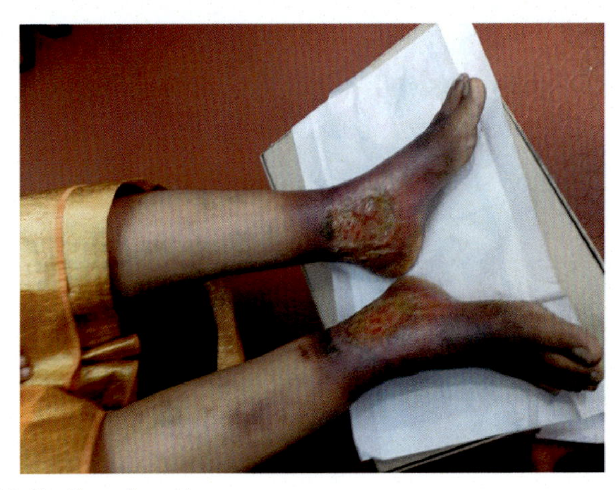

Figure 20.128 Healing ulceration.

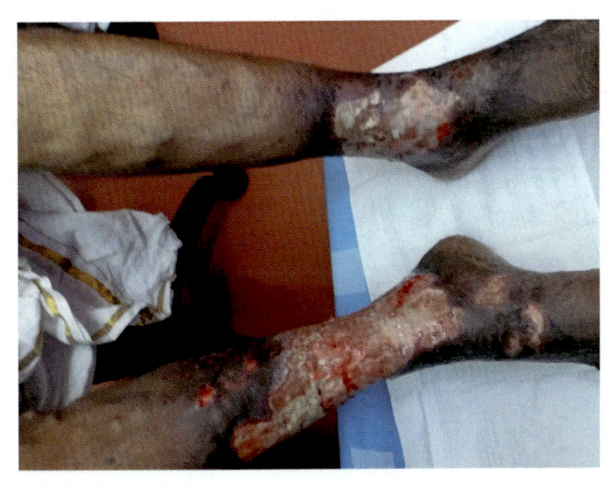

Figure 20.129 Healing ulceration.

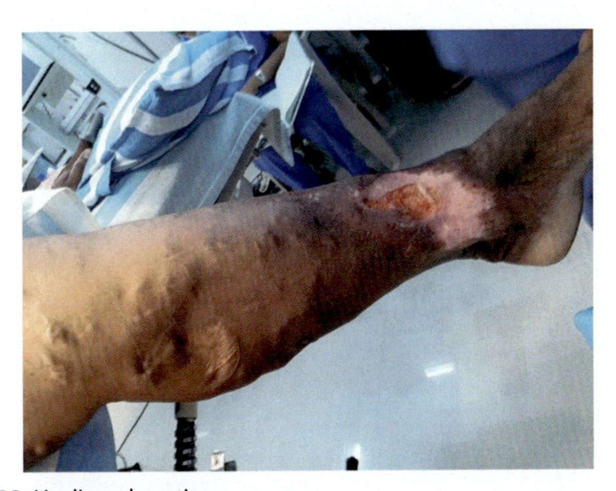

Figure 20.130 Healing ulceration.

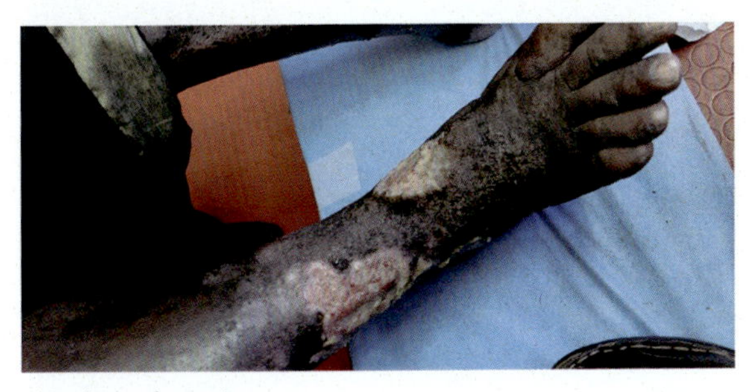

Figure 20.131 Healing ulceration.

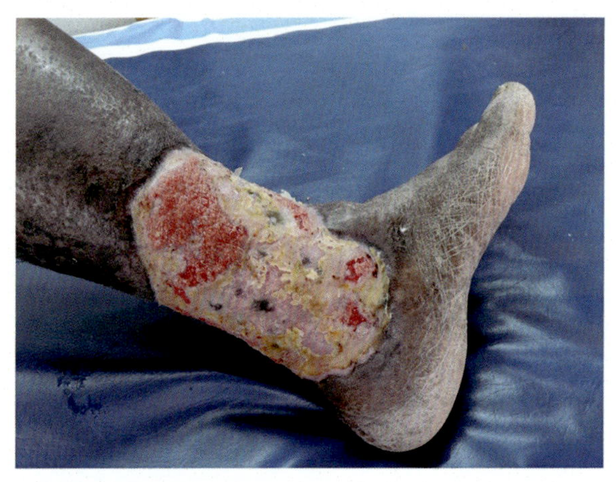

Figure 20.132 Healing ulceration.

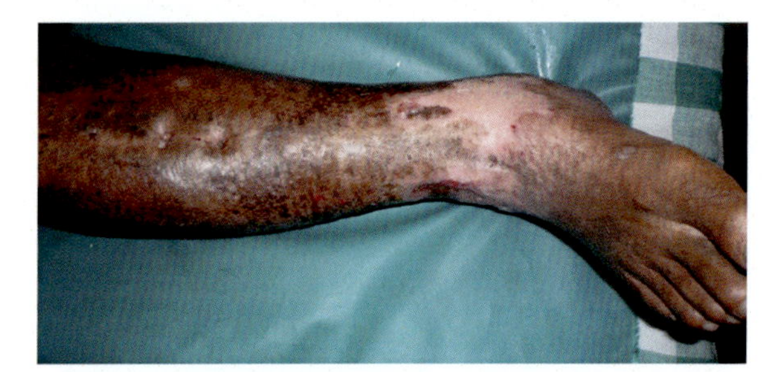

Figure 20.133 Healed ulceration.

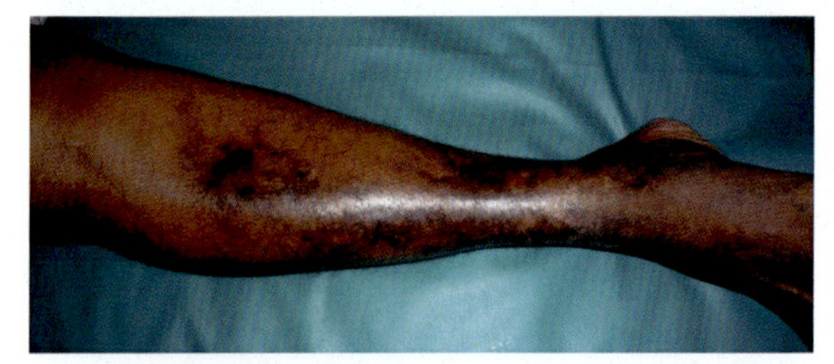

Figure 20.134 Healed ulceration.

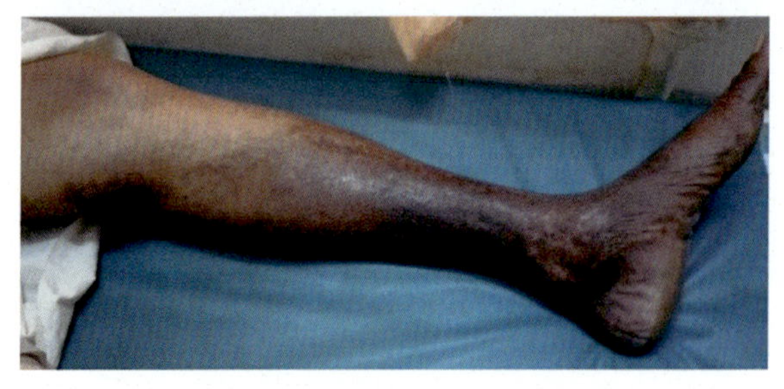

Figure 20.135 Healed ulceration.

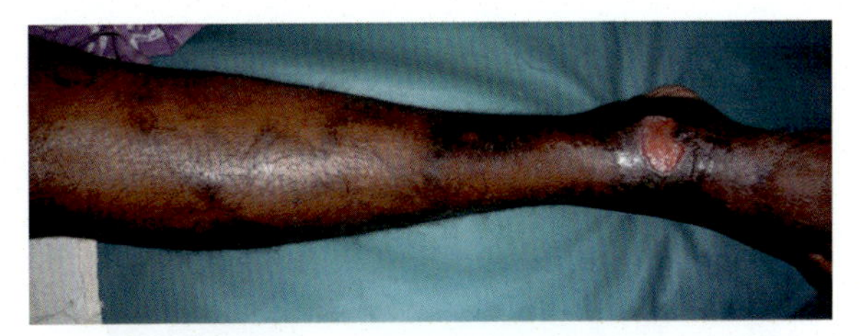

Figure 20.136 Healing ulceration.

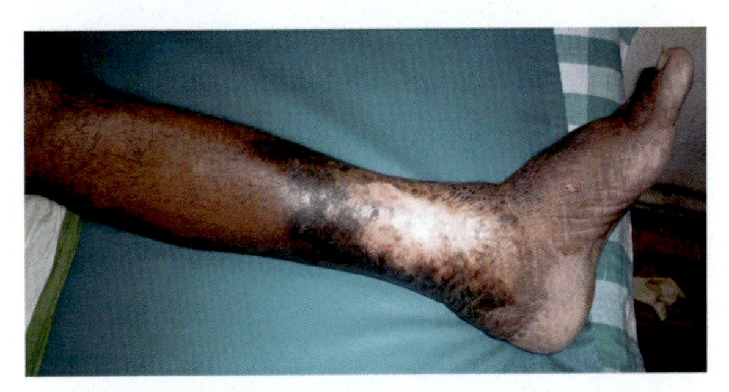

Figure 20.137 Healed ulceration.

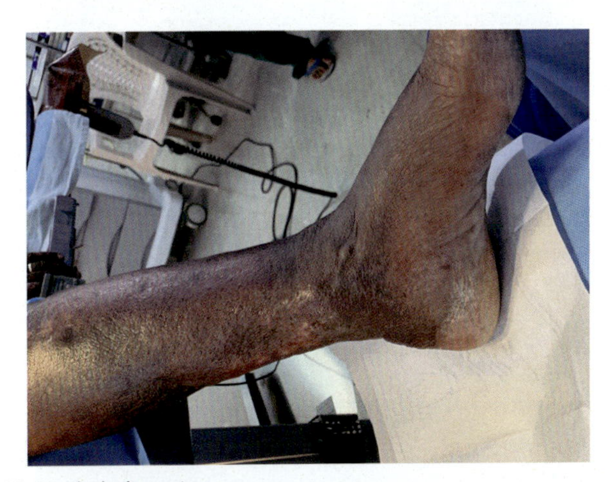

Figure 20.138 Healed ulceration.

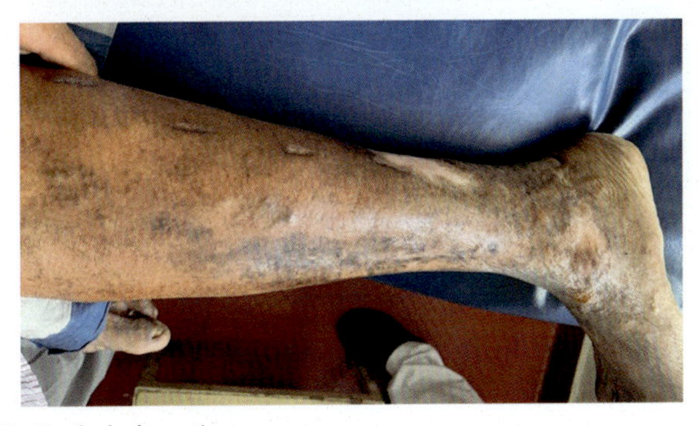

Figure 20.139 Healed ulceration.

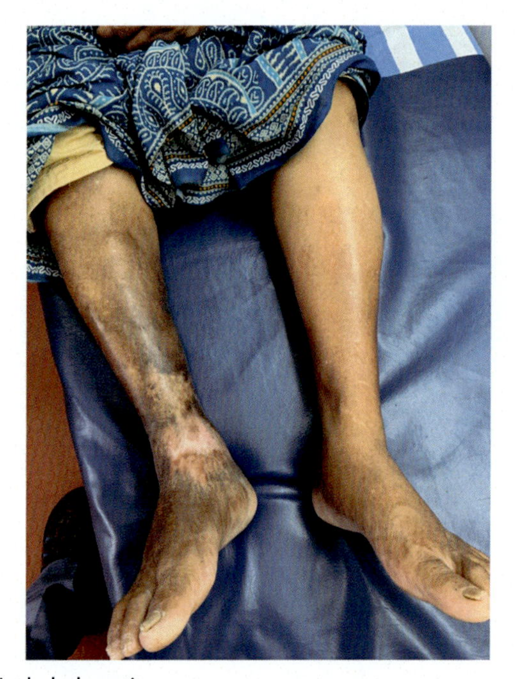

Figure 20.140 Healed ulceration.

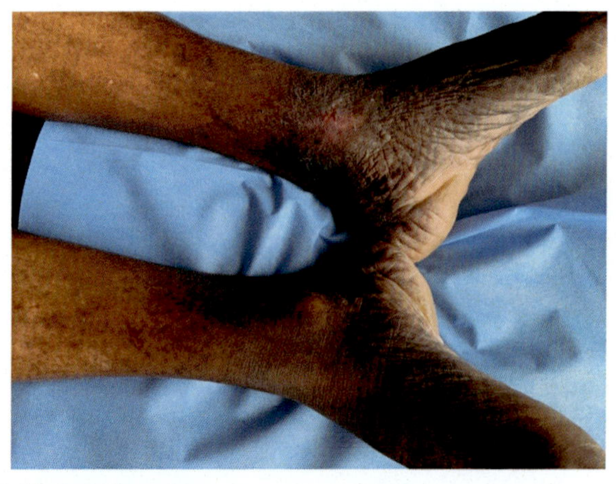

Figure 20.141 Healed ulceration.

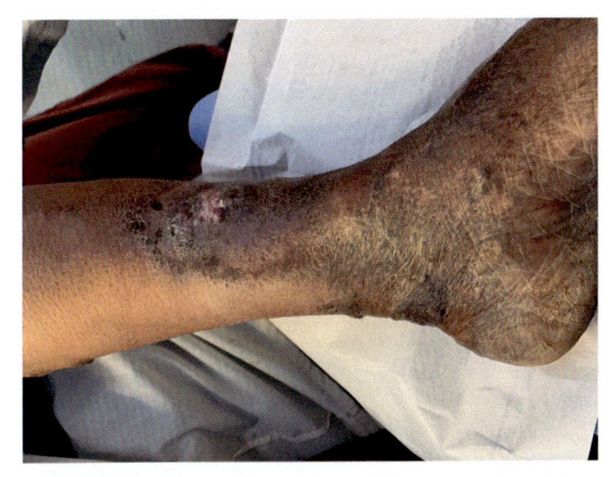

Figure 20.142 Healed ulceration.

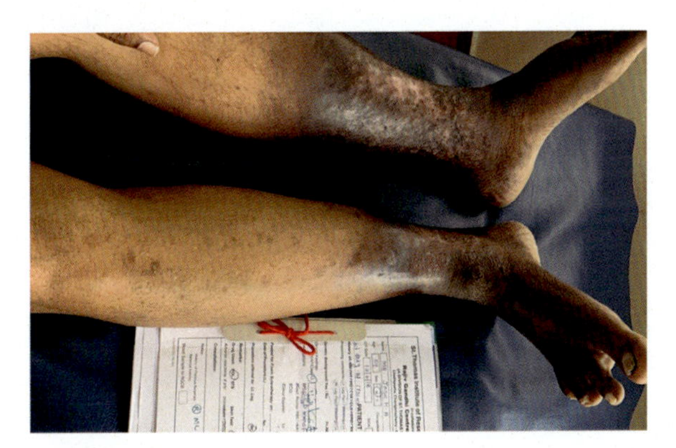

Figure 20.143 Healed ulceration.

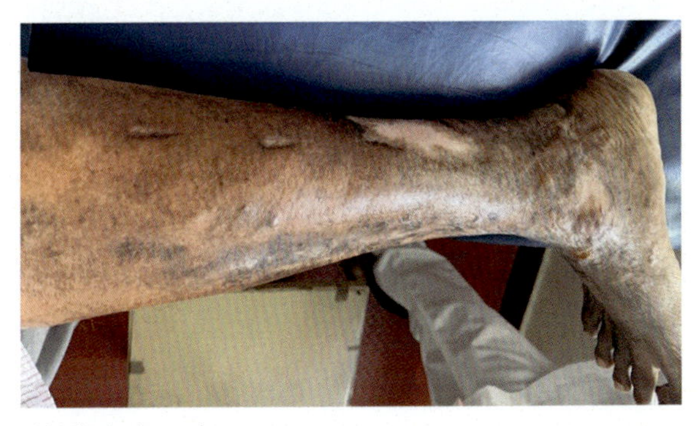

Figure 20.144 Healed ulceration.

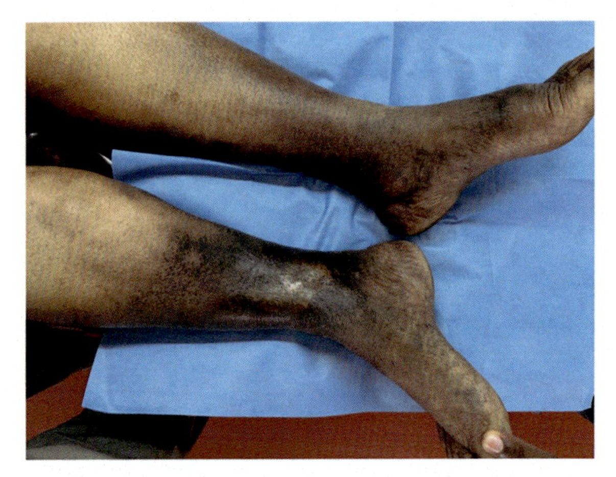

Figure 20.145 Healed ulceration.

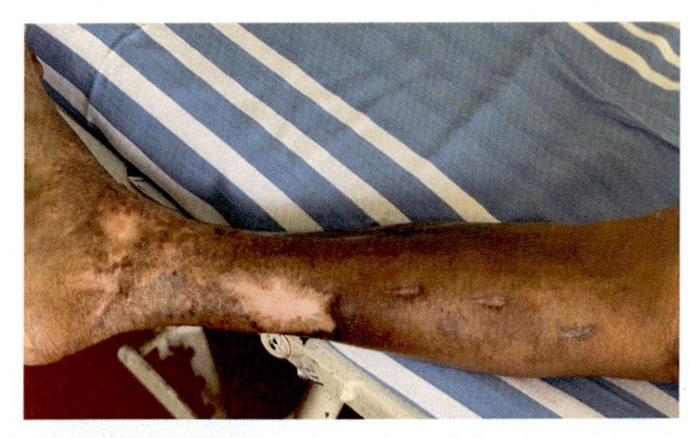

Figure 20.146 Healed ulceration.

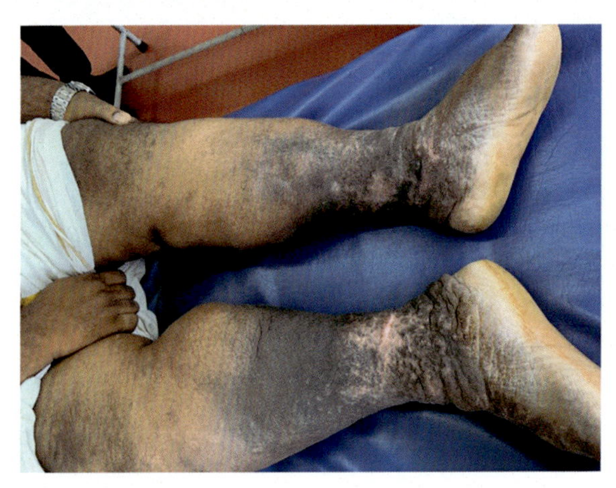

Figure 20.147 Healed ulceration.

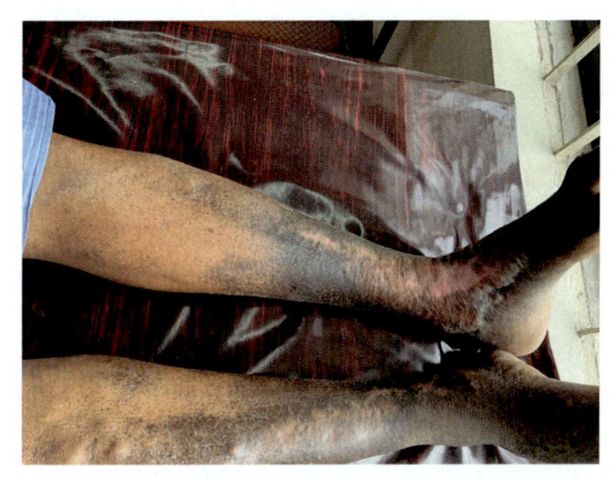

Figure 20.148 Healed ulceration.

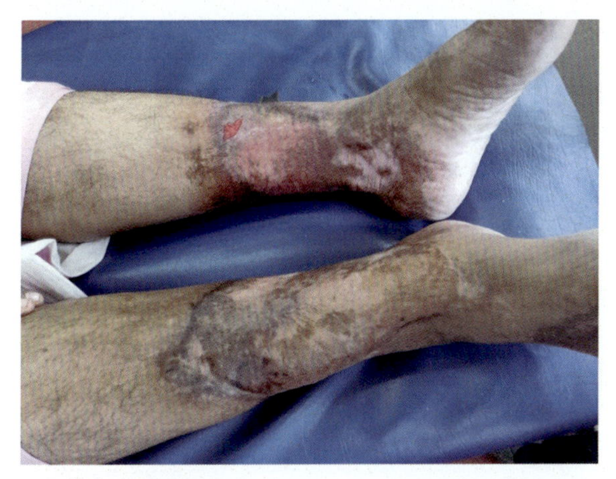

Figure 20.149 Healed ulceration.

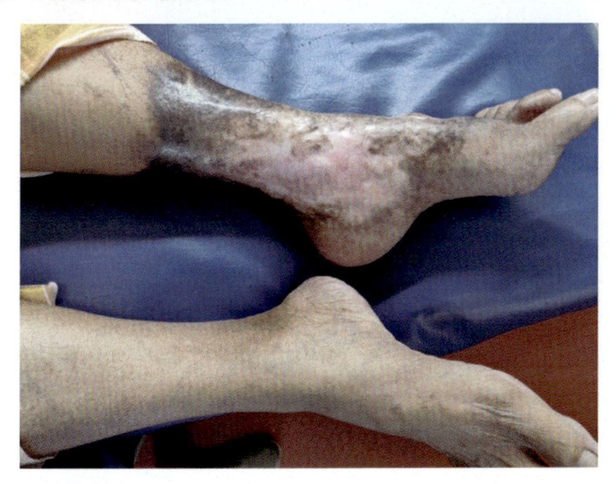

Figure 20.150 Healed ulceration.

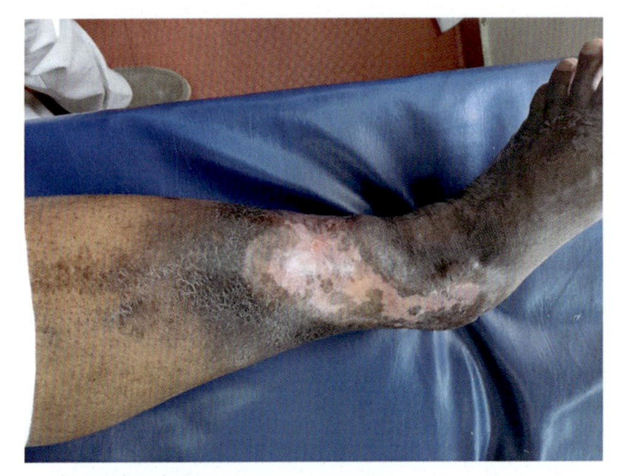

Figure 20.151 Healed ulceration.

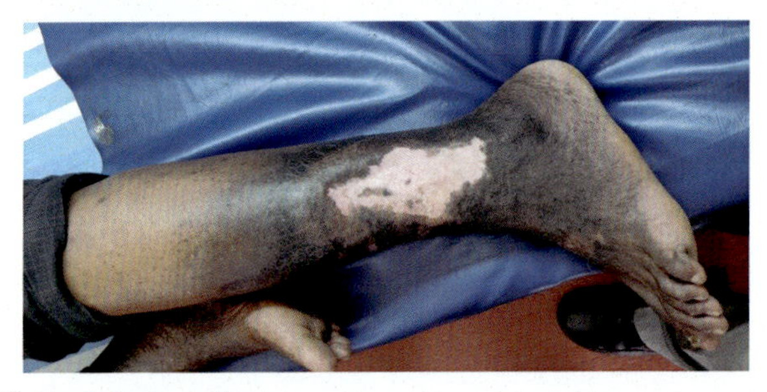

Figure 20.152 Healed ulceration.

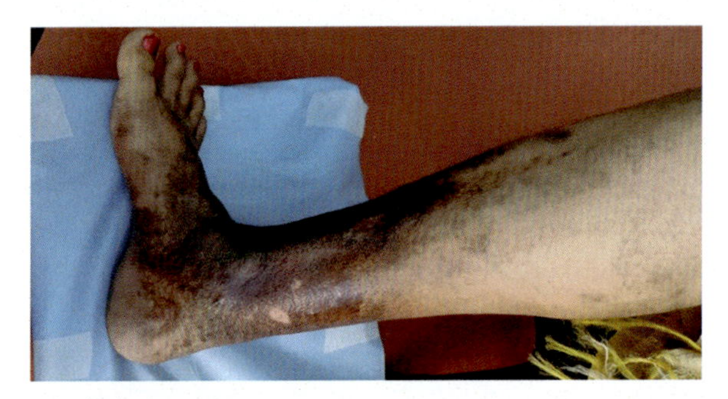

Figure 20.153 Healed ulceration.

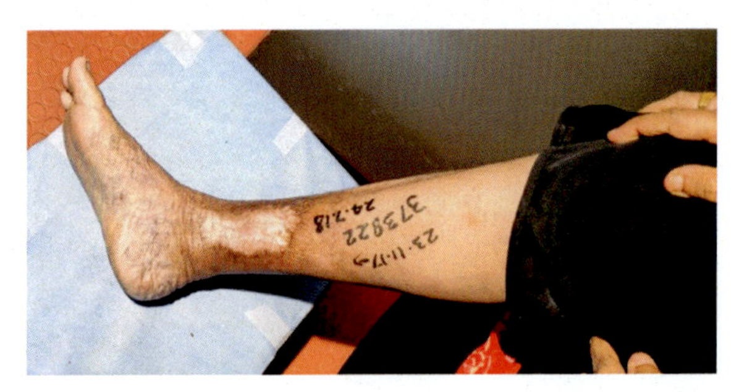

Figure 20.154 Healed ulceration.

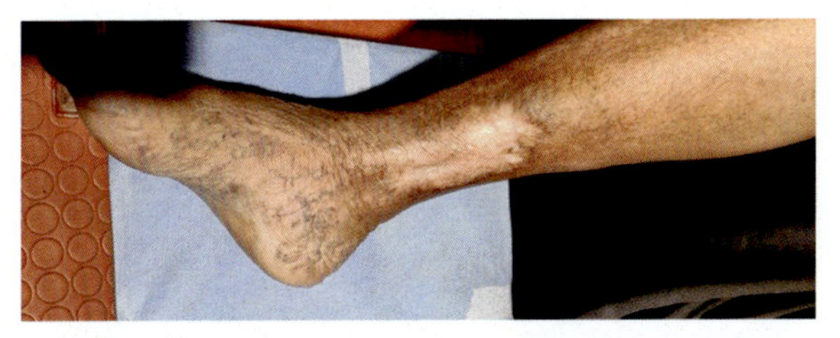

Figure 20.155 Healed ulceration.

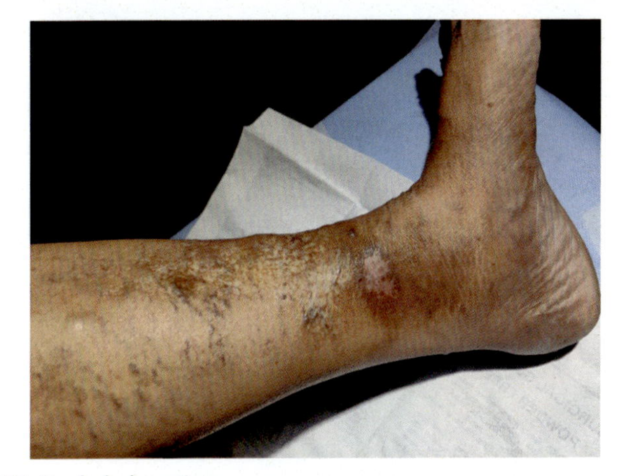

Figure 20.156 Healed ulceration.

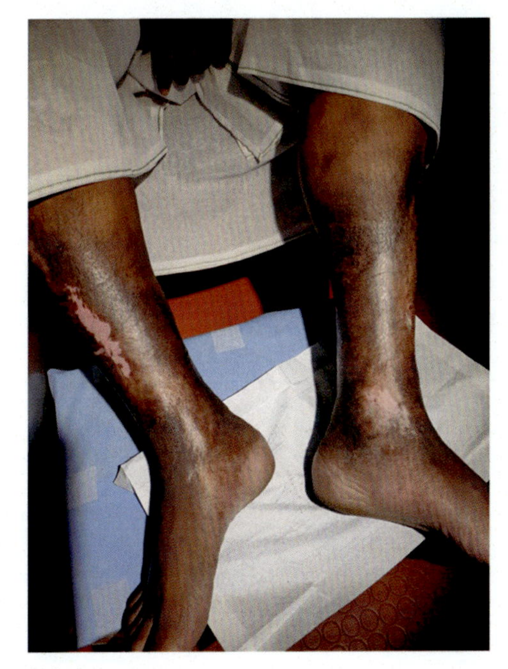

Figure 20.157 Healed ulceration.

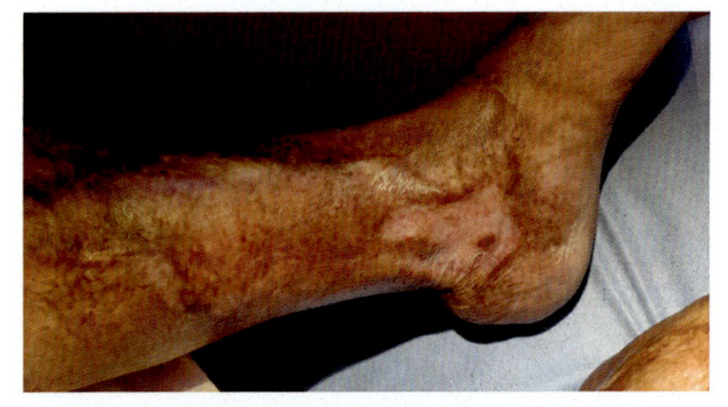

Figure 20.158 Healed ulceration.

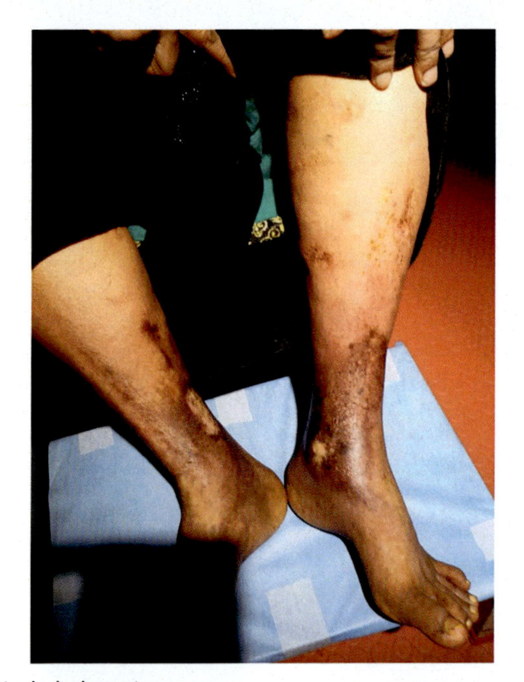

Figure 20.159 Healed ulceration.

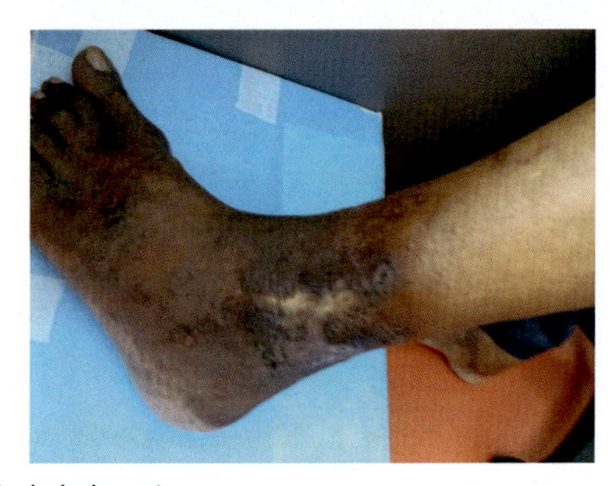

Figure 20.160 Healed ulceration.

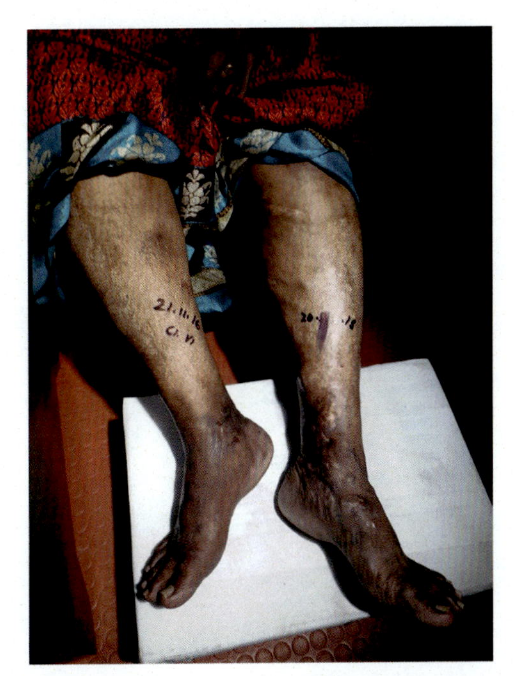

Figure 20.161 Healed ulceration.

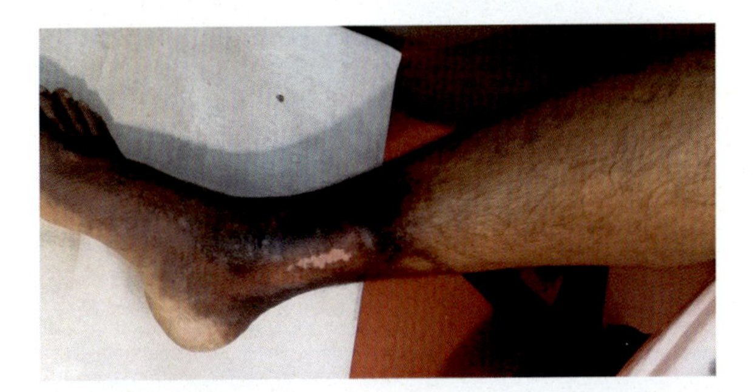

Figure 20.162 Healed ulceration.

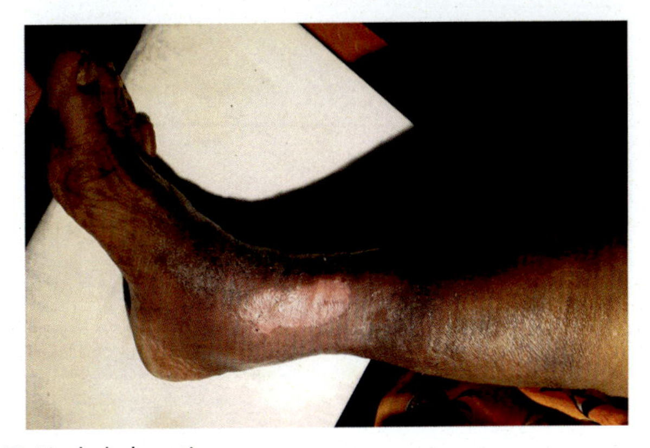

Figure 20.163 Healed ulceration.

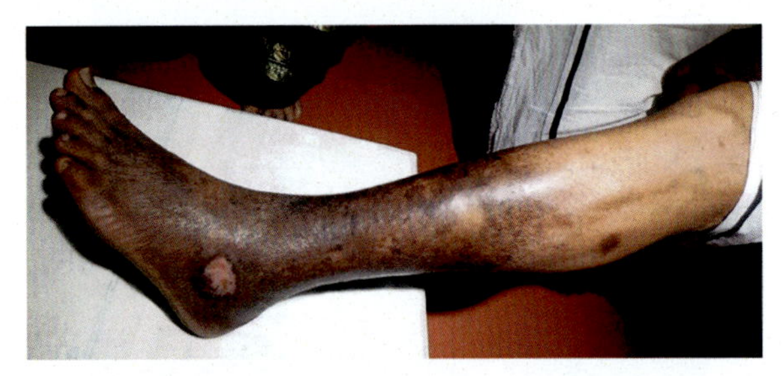

Figure 20.164 Healed ulceration.

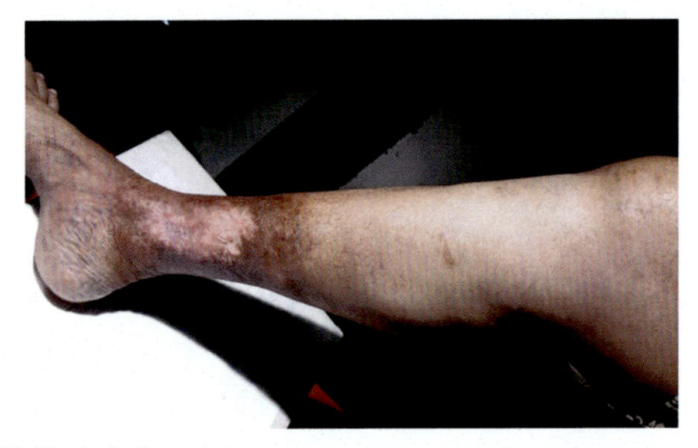

Figure 20.165 Healed ulceration.

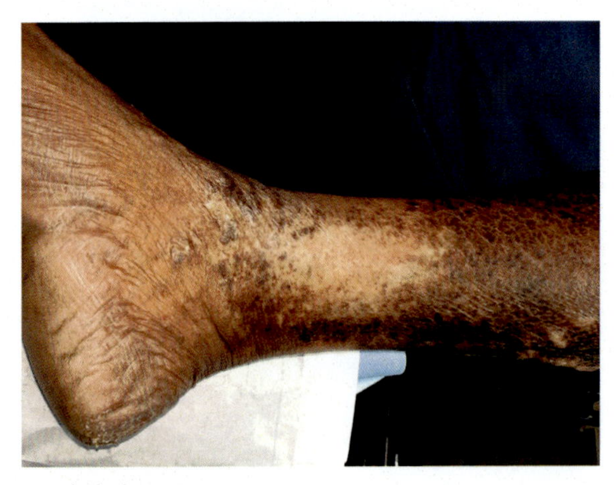

Figure 20.166 Healed ulceration.

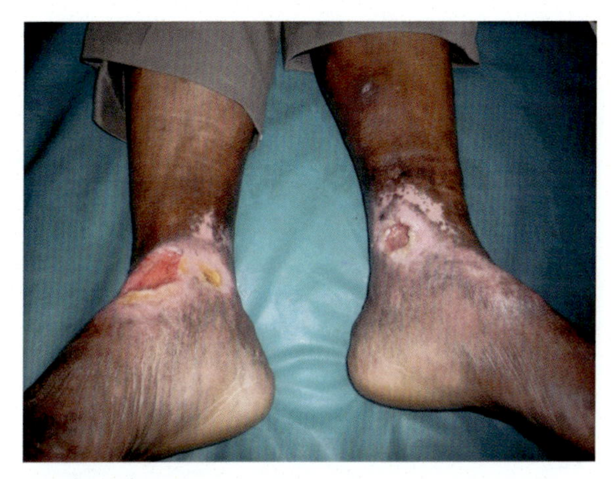

Figure 20.167 Healing ulceration.

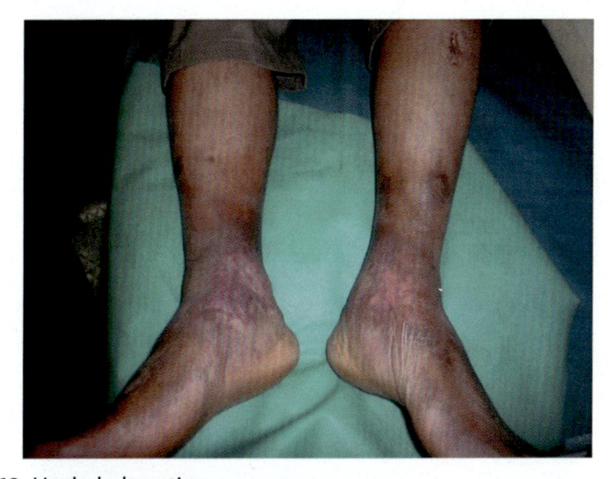

Figure 20.168 Healed ulceration.

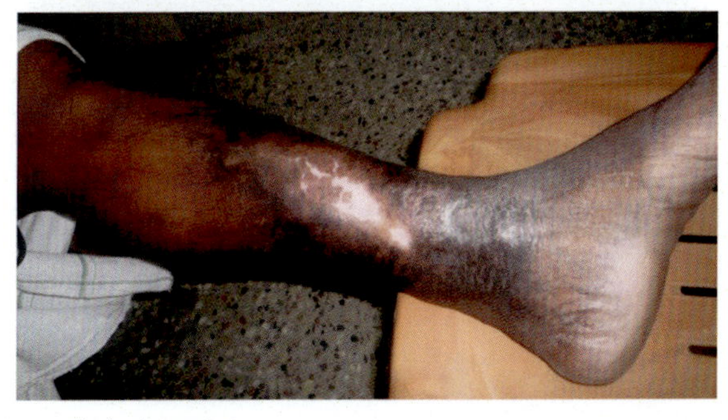

Figure 20.169 Healed ulceration.

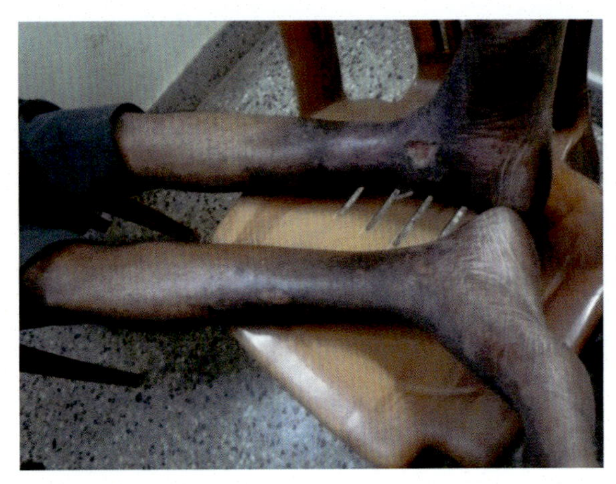

Figure 20.170 Healed ulceration.

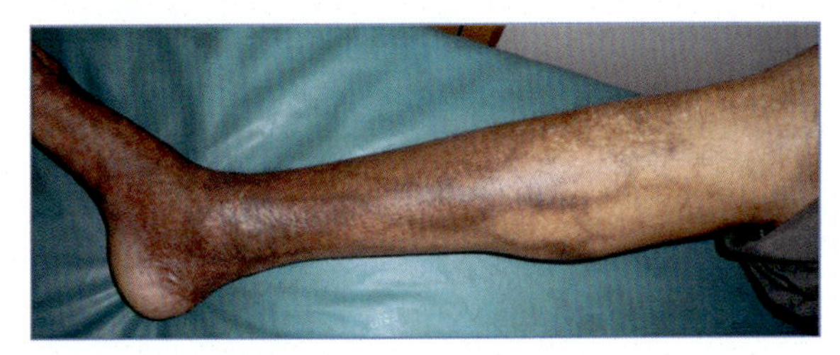

Figure 20.171 Healed ulceration.

20.11 Immediate effects of sclerotherapy

The immediate effect of sclerotherapy is illustrated in Figs. 20.172–20.175.

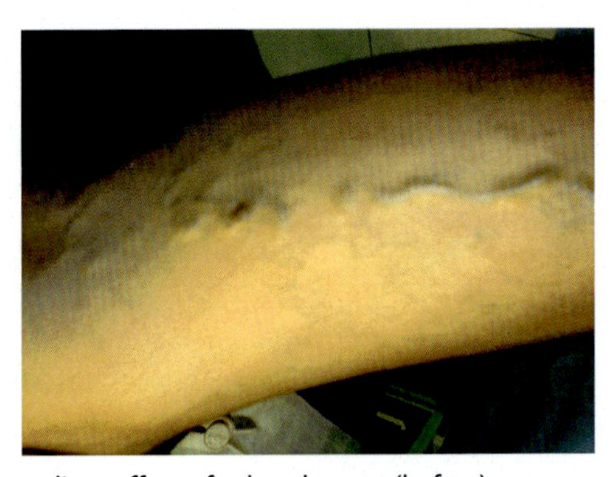

Figure 20.172 Immediate effect of sclerotherapy (before).

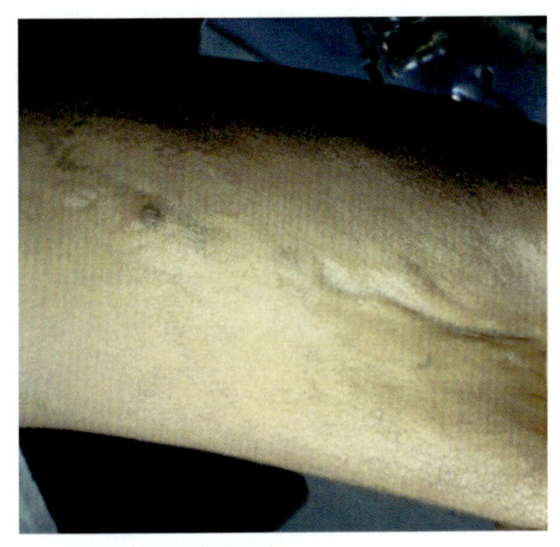

Figure 20.173 Immediate effect of sclerotherapy (after).

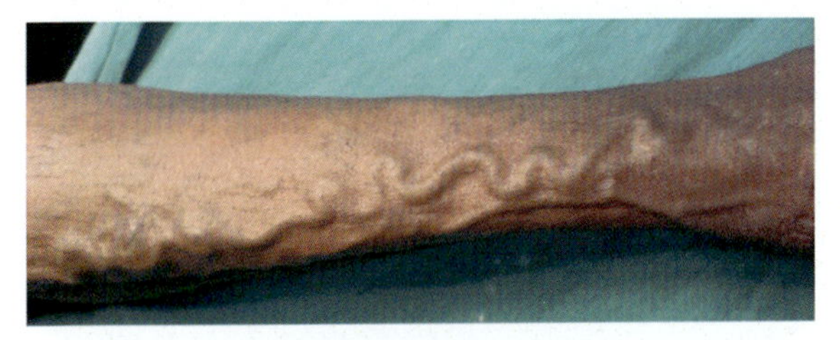

Figure 20.174 Immediate effect of sclerotherapy (before).

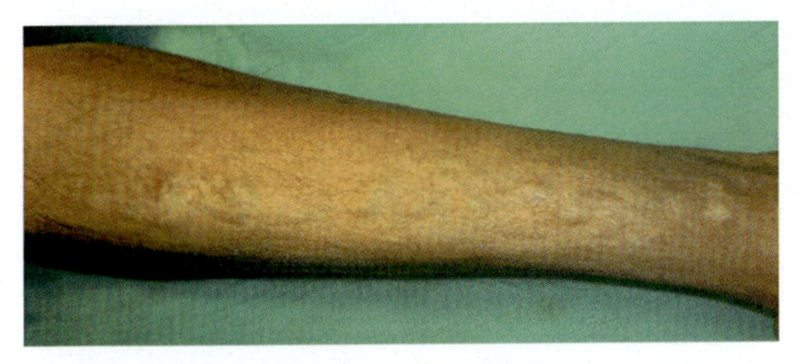

Figure 20.175 Immediate effect of sclerotherapy (after).

20.12 Late effect of sclerotherapy

The late effect of sclerotherapy on larger veins is illustrated in Fig. 20.176.

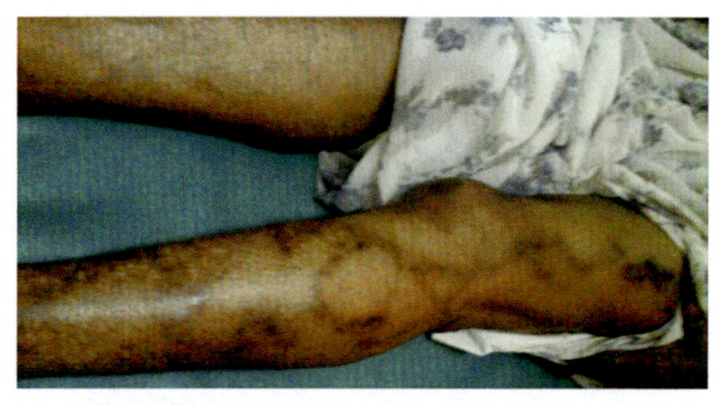

Figure 20.176 Late effect of sclerotherapy on larger veins.

20.13 Lipodermatosclerosis

Lipodermatosclerosis is illustrated in Figs. 20.177–20.180.

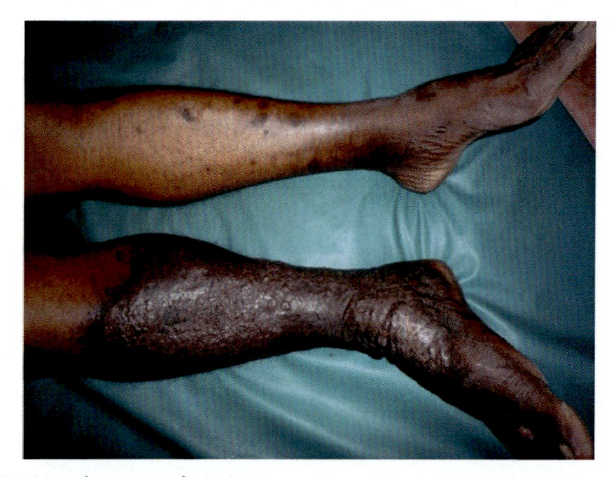

Figure 20.177 Lipodermatosclerosis.

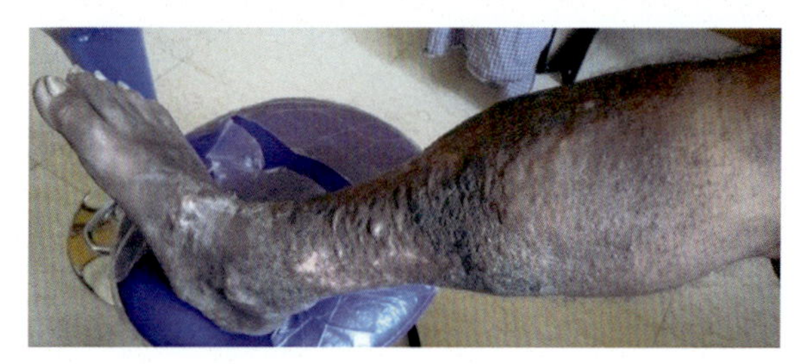

Figure 20.178 Lipodermatosclerosis.

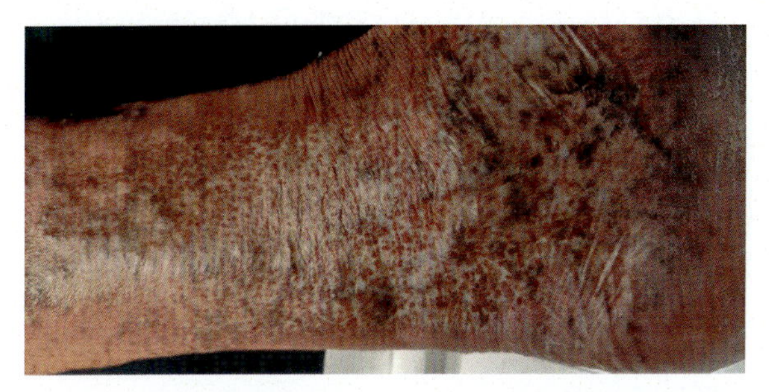

Figure 20.179 Lipodermatosclerosis.

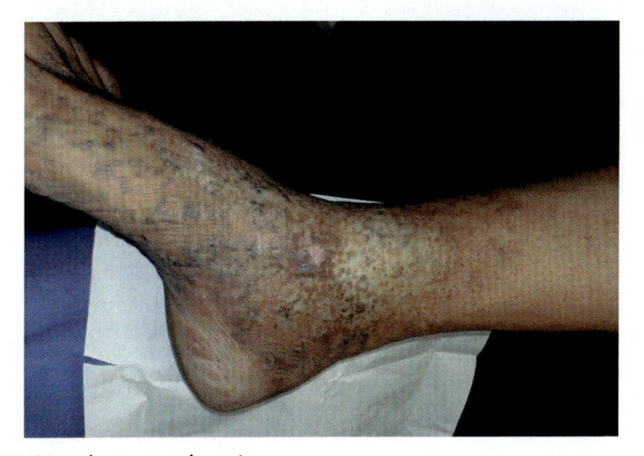

Figure 20.180 Lipodermatosclerosis.

20.14 Atrophie blanche (livedoid vasculopathy)

Atrophie blanche is illustrated in Figs. 20.181–20.183.

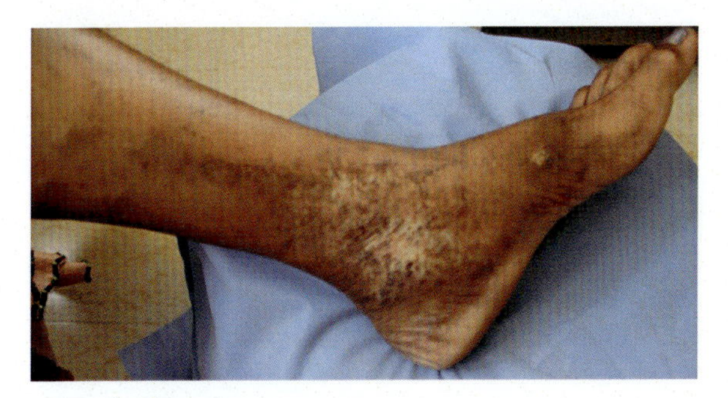

Figure 20.181 Atrophie blanche.

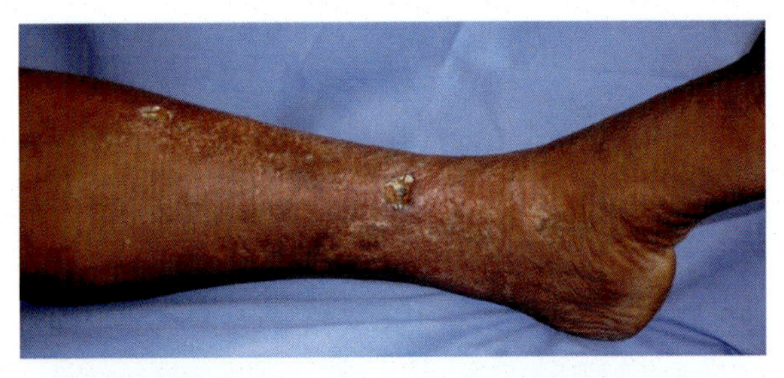

Figure 20.182 Atrophie blanche.

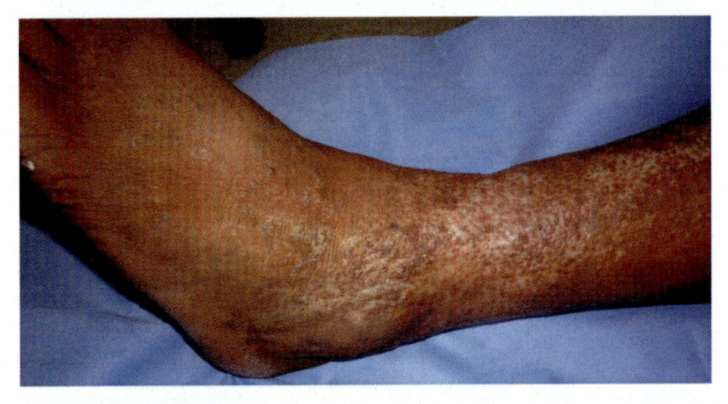

Figure 20.183 Atrophie blanche.

20.15 Hypertrichiasis

Hypertrichiasis as local hirsutism is illustrated in Figs. 20.184–20.187.

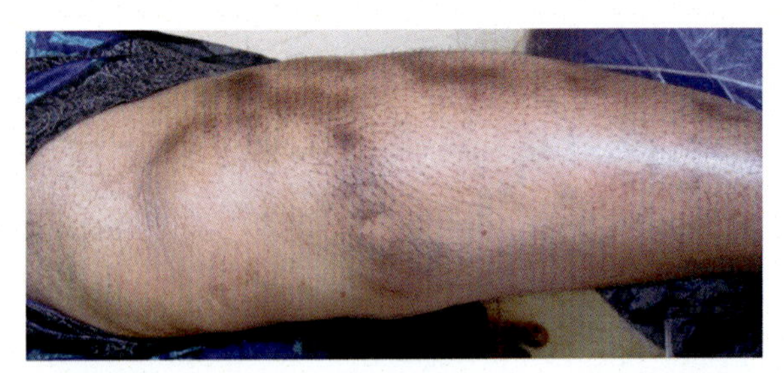

Figure 20.184 Localized hirsutism.

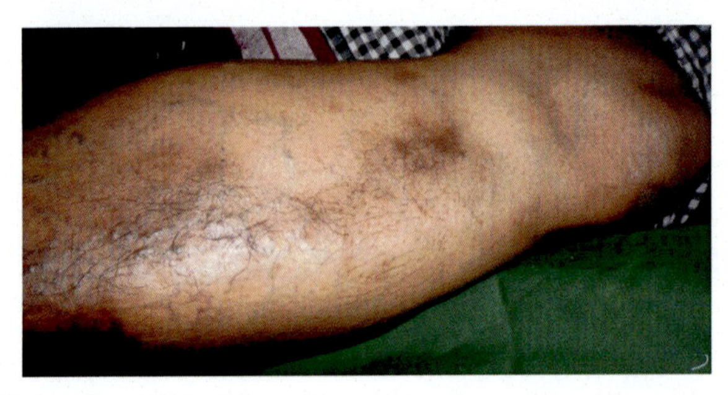

Figure 20.185 Localized hirsutism.

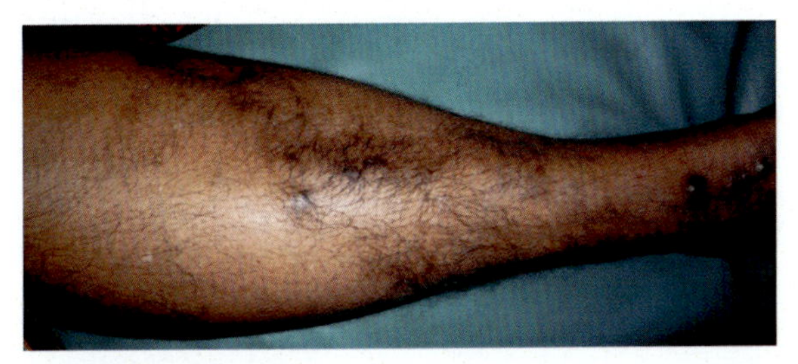

Figure 20.186 Localized hirsutism.

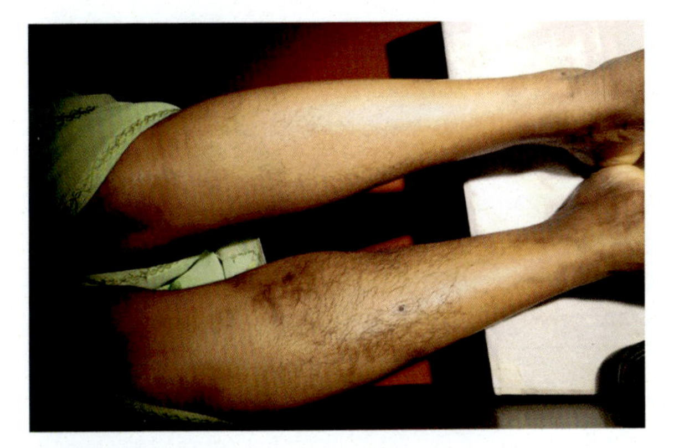

Figure 20.187 Localized hirsutism.

20.16 Varicose veins in the arm

Varicose veins in the hand are illustrated in Figs. 20.188 and 20.189.

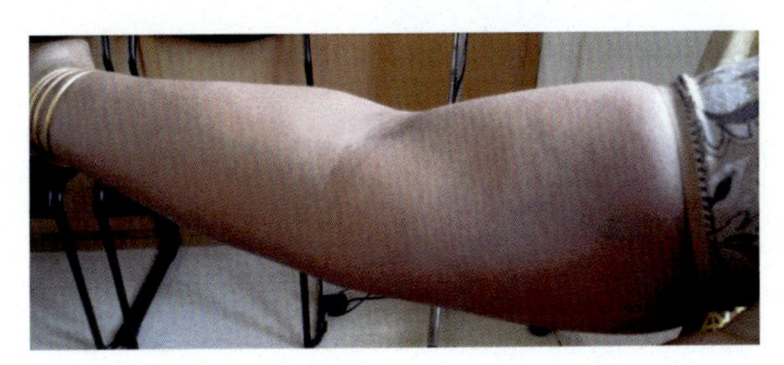

Figure 20.188 Varicose veins in the arm.

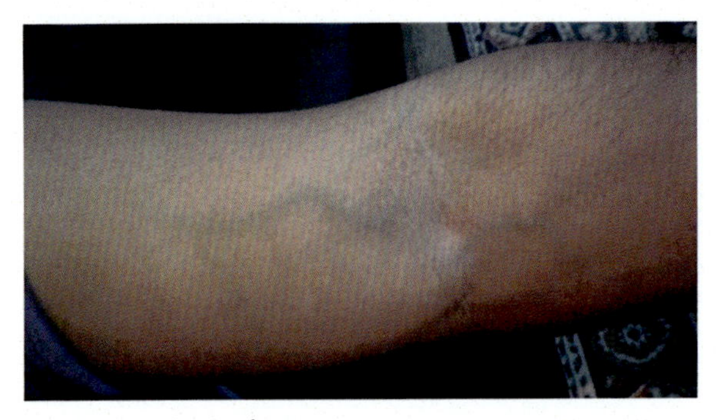

Figure 20.189 Varicose veins in the arm.

20.17 Morbid vein

Morbid veins are illustrated in Figs. 20.190–20.192.

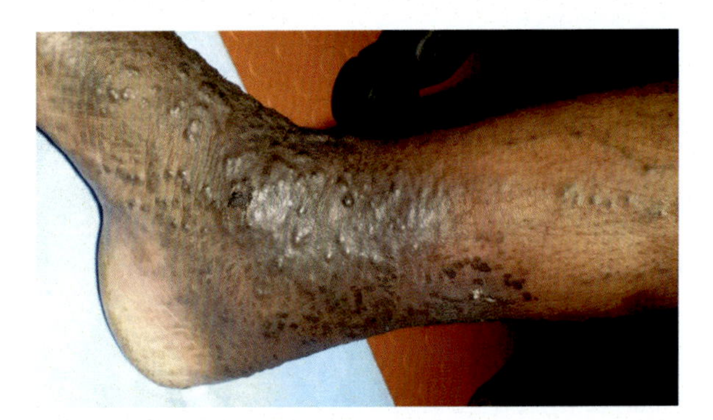

Figure 20.190 Morbid vein.

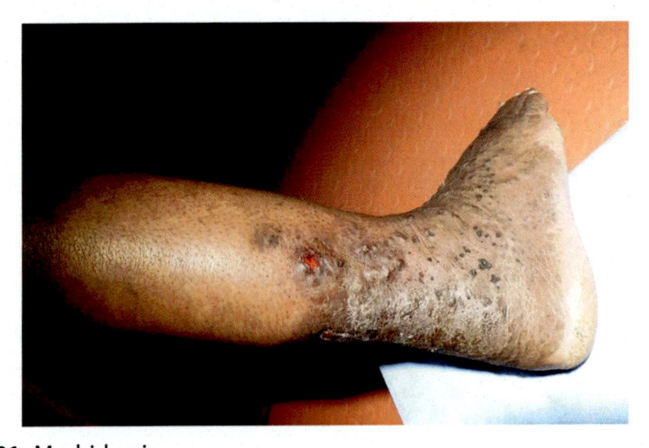

Figure 20.191 Morbid vein.

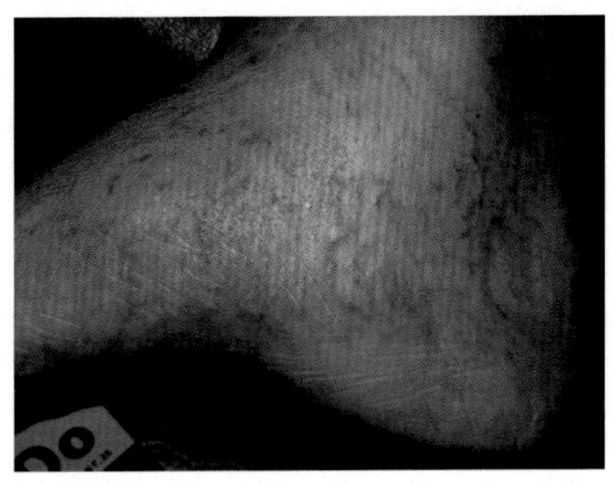

Figure 20.192 Morbid vein.

References

[1] Garcia MD, Larina IV. Vascular development and hemodynamic force in the mouse yolk sac. Front Physiol 2014;5:308.

[2] Majesky MW. Vascular development. Arterioscler Thromb Vasc Biol 2018;38(3): e17–24.

[3] Baron MH, Isern J, Fraser ST. The embryonic origins of erythropoiesis in mammals. Blood 2012;119(21):4828–37.

[4] Pathak NH, Barresi MJ. Zebrafish as a model to understand vertebrate development. The zebrafish in biomedical research. Elsevier; 2020. p. 559–91.

[5] Eriksson B, Löfberg J. Development of the hypochord and dorsal aorta in the zebrafish embryo (Danio rerio). J Morphol 2000;244:167–76.

[6] Sell S, Nicolini A, Ferrari P, Biava PM. Cancer: a problem of developmental biology; scientific evidence for reprogramming and differentiation therapy. Curr Drug Targets 2016;17(10):1103–10.

[7] Wei Y, Zhou F, Zhou H, Huang J, Yu D, Wu G. Endothelial progenitor cells contribute to neovascularization of non-small cell lung cancer via histone deacetylase 7-mediated cytoskeleton regulation and angiogenic genes transcription. Int J Cancer 2018;143(3):657–67.

[8] Donnem T, Reynolds AR, Kuczynski EA, Gatter K, Vermeulen PB, Kerbel RS, et al. Non-angiogenic tumours and their influence on cancer biology. Nat Rev Cancer 2018;18(5):323.

[9] Barminko J, Reinholt B, Baron MH. Development and differentiation of the erythroid lineage in mammals. Dev Comp Immunol 2016;58:18–29.

[10] Hofmann M, Heineke J. The impact of endothelial transcription factors in sprouting angiogenesis. In: Marmé D, editor. Tumor angiogenesis: a key target for cancer therapy. Cham: Springer International Publishing; 2018. p. 1–18.

[11] Abstracts by Author. Integr Comp Biol 2015;40(6):925–1273.

[12] Anschütz S, Schubert R. Modulation of the myogenic response by neurogenic influences in rat small arteries. Br J Pharmacol 2005;146(2):226–33.

[13] Rodriguez A, Csanyi G, Ranayhossaini D, Feck D, Blose K, Assatourian L, et al. MEF2B-Nox signaling is critical for stretch-induced phenotypic modulation of vascular smooth muscle cells. Arterioscler Thromb Vasc Biol 2014;35.

[14] Zeyer KA, Reinhardt DP. Engineered mutations in fibrillin-1 leading to Marfan syndrome act at the protein, cellular and organismal levels. Mutat Res Rev Mutat Res 2015;765:7–18.

[15] Sengle G, Ono RN, Sasaki T, Sakai LY. Prodomains of transforming growth factor beta (TGFbeta) superfamily members specify different functions: extracellular matrix interactions and growth factor bioavailability. J Biol Chem 2011;286(7):5087–99.

[16] Sinha S, Iyer D, Granata A. Embryonic origins of human vascular smooth muscle cells: implications for in vitro modeling and clinical application. Cell Mol Life Sci 2014;71(12):2271–88.

[17] Sawada H, Rateri DL, Moorleghen JJ, Majesky MW, Daugherty A. Smooth muscle cells derived from second heart field and cardiac neural crest reside in spatially distinct domains in the media of the ascending aorta-brief report. Arterioscler Thromb Vasc Biol 2017;37(9):1722–6.

[18] Majesky MW. Adventitia and perivascular cells. Arterioscler Thromb Vasc Biol 2015;35(8):e31–5.

[19] Lindsey ML, Bolli R, Canty Jr. JM, Du X-J, Frangogiannis NG, Frantz S, et al. Guidelines for experimental models of myocardial ischemia and infarction. Am J Physiol-Heart Circul Physiol 2018;314(4):H812–38.
[20] Zorn AM, Wells JM. Vertebrate endoderm development and organ formation. Ann Rev Cell Dev Biol 2009;25:221–51.
[21] Lécuyer E, Hoang T. SCL: from the origin of hematopoiesis to stem cells and leukemia. Exp Hematol 2004;32(1):11–24.
[22] Udan RS, Vadakkan TJ, Dickinson ME. Dynamic responses of endothelial cells to changes in blood flow during vascular remodeling of the mouse yolk sac. Development 2013;140(19):4041–50.
[23] Garcia M, Larina I. Corrigendum: vascular development and hemodynamic force in the mouse yolk sac. Front Physiol 2014;5:308.
[24] Teilmann SC, Christensen ST. Localization of the angiopoietin receptors Tie-1 and Tie-2 on the primary cilia in the female reproductive organs. Cell Biol Int 2005;29(5):340–6.
[25] Barton WA, Dalton AC, Seegar TCM, Himanen JP, Nikolov DB. Tie2 and Eph receptor tyrosine kinase activation and signaling. Cold Spring Harb Perspect Biol 2014;6(3):a009142.
[26] Jung B, Staudacher JJ, Beauchamp D. Transforming growth factor β superfamily signaling in development of colorectal cancer. Gastroenterology 2017;152(1):36–52.
[27] Zhang J, Hughes S. Role of the ephrin and Eph receptor tyrosine kinase families in angiogenesis and development of the cardiovascular system. J Pathol: A J Pathol Soc Gt Br Irel 2006;208(4):453–61.
[28] Gale NW, Yancopoulos GD. Growth factors acting via endothelial cell-specific receptor tyrosine kinases: VEGFs, angiopoietins, and ephrins in vascular development. Genes Dev 1999;13(9):1055–66.
[29] Kania A, Klein R. Mechanisms of ephrin—Eph signalling in development, physiology and disease. Nat Rev Mol Cell Biol 2016;17(4):240.
[30] Budi EH, Duan D, Derynck R. Transforming growth factor-β receptors and Smads: regulatory complexity and functional versatility. Trends Cell Biol 2017;27(9):658–72.
[31] Basu S, Barbur I, Calderon A, Banerjee S, Proweller A. Notch signaling regulates arterial vasoreactivity through opposing functions of Jagged1 and Dll4 in the vessel wall. Am J Physiol—Heart Circul Physiol 2018;315(6):H1835–50.
[32] Fang J, Hirschi K. Molecular regulation of arteriovenous endothelial cell specification. F1000Research 2019;8.
[33] Bolte C, Whitsett JA, Kalin TV, Kalinichenko VV. Transcription factors regulating embryonic development of pulmonary vasculature. Molecular and functional insights into the pulmonary vasculature. Springer; 2018. p. 1–20.
[34] Minami T, Muramatsu M, Kume T. Organ/tissue-specific vascular endothelial cell heterogeneity in health and disease. Biol Pharm Bull 2019;42(10):1609–19.
[35] Heldin CH, Lennartsson J, Westermark B. Involvement of platelet-derived growth factor ligands and receptors in tumorigenesis. J Intern Med 2018;283(1):16–44.
[36] Krishnan L, LaBelle SA, Ruehle MA, Weiss JA, Hoying JB, Guldberg RE. Mechanical Regulation of Microvascular Growth and Remodeling. In Holnthoner W, Banfi A, Kirkpatrick J, & Redl H, (Eds.), Vascularization for Tissue Engineering and Regenerative Medicine (pp. 1–45). Springer International Publishing; 2019. Available from: https://doi.org/10.1007/978-3-319-21056-8_19-1.
[37] Duchemin A-L, Vignes H, Vermot J, Chow R. Mechanotransduction in cardiovascular morphogenesis and tissue engineering. Curr OpGenet Dev 2019;57:106–16.
[38] Fu BM. Tumor metastasis in the microcirculation. Molecular, cellular, and tissue engineering of the vascular system. Springer; 2018. p. 201–18.
[39] Goumans M-J, ten Dijke P. TGF-β signaling in control of cardiovascular function. Cold Spring Harb Perspect Biol 2018;10(2):a022210.

[40] Talavera K, Startek JB, Alvarez-Collazo J, Boonen B, Alpizar YA, Sanchez A, et al. Mammalian transient receptor potential TRPA1 channels: from structure to disease. Physiol Rev 2019.
[41] Magalhães A, Dias S. Angiogenesis—vessels recruitment by tumor cells. Molecular and cell biology of cancer. Springer; 2019. p. 141–57.
[42] Whitcomb J, Gharibeh L, Nemer M. From embryogenesis to adulthood: critical role for GATA factors in heart development and function. IUBMB Life 2019.
[43] Hoog TG. Low Hemodynamic Loading Alters Heart Morphogensis in E8.5 to E9.5 Mouse Embryos; 2018. Available from: https://bearworks.missouristate.edu/cgi/viewcontent.cgi?article=4260&context=theses.
[44] Chao H, Hirschi KK. Hemato-vascular origins of endothelial progenitor cells? Microvasc Res 2010;79(3):169–73.
[45] Si R, Zhang Q, Tsuji-Hosokawa A, Watanabe M, Willson C, Lai N, et al. Overexpression of p53 due to excess protein *O*-GlcNAcylation is associated with coronary microvascular disease in type 2 diabetes. Cardiovasc Res 2019.
[46] Gammelsaeter R, Remmereit J. Cellular extracts. Google Pat 2018.
[47] Fashioning blood vessels by ROS signalling and metabolism. In: Santoro MM, editor. Seminars in cell and developmental biology. Elsevier; 2018.
[48] Costa C, Incio J. Angiogenesis and chronic inflammation: cause or consequence? Angiogenesis 2007;10:149–66.
[49] Rajabi M, Mousa SA. The role of angiogenesis in cancer treatment. Biomedicines 2017;5(2):34.
[50] De Spiegelaere W, Casteleyn C, Van den Broeck W, Plendl J, Bahramsoltani M, Simoens P, et al. Intussusceptive angiogenesis: a biologically relevant form of angiogenesis. J Vasc Res 2012;49(5):390–404.
[51] Vasudev NS, Reynolds AR. Anti-angiogenic therapy for cancer: current progress, unresolved questions and future directions. Angiogenesis 2014;17(3):471–94.
[52] Ucuzian AA, Gassman AA, East AT, Greisler HP. Molecular mediators of angiogenesis. J Burn Care Res 2010;31(1):158–75.
[53] Mentzer SJ, Konerding MA. Intussusceptive angiogenesis: expansion and remodeling of microvascular networks. Angiogenesis 2014;17(3):499–509.
[54] Makanya AN, Hlushchuk R, Djonov VG. Intussusceptive angiogenesis and its role in vascular morphogenesis, patterning, and remodeling. Angiogenesis 2009;12(2):113–23.
[55] Stockmann C, Schadendorf D, Klose R, Helfrich I. The impact of the immune system on tumor: angiogenesis and vascular remodeling. Front Oncol 2014;4(69).
[56] Karthik S, Djukic T, Kim J-D, Zuber B, Makanya A, Odriozola A, et al. Synergistic interaction of sprouting and intussusceptive angiogenesis during zebrafish caudal vein plexus development. Sci Rep 2018;8(1):9840.
[57] Udan RS, Culver JC, Dickinson ME. Understanding vascular development. Wiley Interdiscip Rev Dev Biol 2013;2(3):327–46.
[58] Djonov V, Baum O, Burri P. Vascular remodeling by intussusceptive angiogenesis. Cell Tissue Res 2003;314:107–17.
[59] Pepper M, Baetens D, Mandriota S, Sanza C, Oikemus S, Lane T, et al. Regulation of VEGF and VEGF receptor expression in the rodent mammary gland during pregnancy, lactation, and involution. Dev Dyn 2000;218:507–24.
[60] Martínez AK, Maroni L, Marzioni M, Ahmed ST, Milad M, Ray D, et al. Mouse models of liver fibrosis mimic human liver fibrosis of different etiologies. Curr Pathobiol Rep 2014;2(4):143–53.
[61] Toda N, Mukoyama M, Yanagita M, Yokoi H. CTGF in kidney fibrosis and glomerulonephritis. Inflamm Regen 2018;38(1):1–8.
[62] Conlon KC, Miljkovic MD, Waldmann TA. Cytokines in the treatment of cancer. J Interf Cytok Res 2019;39(1):6–21.

[63] Roberti A, Valdes AF, Torrecillas R, Fraga MF, Fernandez AF. Epigenetics in cancer therapy and nanomedicine. Clin Epigenet 2019;11(1):81.

[64] Rust R, Gantner C, Schwab ME. Pro- and antiangiogenic therapies: current status and clinical implications. FASEB J 2018;33(1):34–48.

[65] Qadura M, Terenzi DC, Verma S, Al-Omran M, Hess DA. Concise review: cell therapy for critical limb ischemia: an integrated review of preclinical and clinical studies. Stem Cell 2018;36(2):161–71.

[66] Smith AF, Nitzsche B, Maibier M, Pries AR, Secomb TW. Microvascular hemodynamics in the chick chorioallantoic membrane. Microcirculation 2016;23(7):512–22.

[67] Djonov V, Galli A, Burri P. Intussusceptive asborization contributes to vascular tree formation in the chick chorio-allantoic membrane. Anat Embryol 2000;202:347–57.

[68] Ribatti D, Crivellato E. "Sprouting angiogenesis", a reappraisal. Dev Biol 2012;372 (2):157–65.

[69] Culver JC, Dickinson ME. The effects of hemodynamic force on embryonic development. Microcirculation 2010;17(3):164–78.

[70] Barbarawi M, Zayed Y, Kheiri B, Gakhal I, Barbarawi O, Bala A, et al. The role of anticoagulation in venous thromboembolism primary prophylaxis in patients with malignancy: a systematic review and meta-analysis of randomized controlled trials. Thromb Res 2019;181:36–45.

[71] Zafar MI, Zheng J, Kong W, Ye X, Gou L, Regmi A, et al. The role of vascular endothelial growth factor-B in metabolic homoeostasis: current evidence. Biosci Rep 2017;37(4) BSR20171089.

[72] Bartlett CS, Jeansson M, Quaggin SE. Vascular growth factors and glomerular disease. Annu Rev Physiol 2016;78:437–61.

[73] De Spiegelaere W, Cornillie P, Erkens T, Van Loo D, Casteleyn C, Van Poucke M, et al. Expression and localization of angiogenic growth factors in developing porcine mesonephric glomeruli. J Histochem Cytochem 2010;58(12):1045–56.

[74] Murakami M, Simons M. Fibroblast growth factor regulation of neovascularization. Curr Opin Hematol 2008;15(3):215–20.

[75] Thijssen VL, Paulis YW, Nowak-Sliwinska P, Deumelandt KL, Hosaka K, Soetekouw PM, et al. Targeting PDGF-mediated recruitment of pericytes blocks vascular mimicry and tumor growth. J Pathol 2018;246(4):447–58.

[76] Pugliese SC, Poth JM, Fini MA, Olschewski A, El Kasmi KC, Stenmark KR. The role of inflammation in hypoxic pulmonary hypertension: from cellular mechanisms to clinical phenotypes. Am J Physiol Lung Cell Mol Physiol 2015;308(3):L229–52.

[77] Zimna A, Kurpisz M. Hypoxia-inducible factor-1 in physiological and pathophysiological angiogenesis: applications and therapies. BioMed Res Int 2015;2015:13.

[78] Lokman NA, Elder ASF, Ricciardelli C, Oehler MK. Chick chorioallantoic membrane (CAM) assay as an in vivo model to study the effect of newly identified molecules on ovarian cancer invasion and metastasis. Int J Mol Sci 2012;13(8):9959–70.

[79] Okamoto T, Usuda H, Tanaka T, Wada K, Shimaoka M. The functional implications of endothelial gap junctions and cellular mechanics in vascular angiogenesis. Cancers (Basel) 2019;11(2):237.

[80] Singh A, Winterbottom E, Daar IO. Eph/ephrin signaling in cell–cell and cell–substrate adhesion. Front Biosci (Landmark Ed) 2012;17:473–97.

[81] Kotini MP, Mae MA, Belting HG, Betsholtz C, Affolter M. Sprouting and anastomosis in the *Drosophila trachea* and the vertebrate vasculature: Similarities and differences in cell behaviour. Vasc Pharmacol 2019;112:8–16.

[82] Makanya A, Stauffer D, Ribatti D, Burri P, Djonov V. Microvascular growth, development, and remodeling in the embryonic avian kidney: the interplay between sprouting and intussusceptive angiogenic mechanisms. Microsc Res Tech 2005;66:275–88.

[83] Poole DC, Copp SW, Ferguson SK, Musch TI. Skeletal muscle capillary function: contemporary observations and novel hypotheses. Exp Physiol 2013;98(12):1645–58.
[84] Djonov VG, Kurz H, Burri PH. Optimality in the developing vascular system: branching remodeling by means of intussusception as an efficient adaptation mechanism. Dev Dyn 2002;224(4):391–402.
[85] Burri P, Hlushchuk R, Djonov V. Intussusceptive angiogenesis: its emergence, its characteristics, and its significance. Dev Dyn: Off Publ Am Assoc Anatom 2004;231:474–88.
[86] Ricard N, Simons M. When it is better to regress: dynamics of vascular pruning. PLoS Biol 2015;13(5) e1002148-e.
[87] Fraser HM. Regulation of the ovarian follicular vasculature. Reprod Biol Endocrinol 2006;4:18 -.
[88] Bisht M, Dhasmana DC, Bist SS. Angiogenesis: future of pharmacological modulation. Indian J Pharmacol 2010;42(1):2–8.
[89] Logsdon EA, Finley SD, Popel AS, Mac Gabhann F. A systems biology view of blood vessel growth and remodelling. J Cell Mol Med 2014;18(8):1491–508.
[90] Davies PF. Hemodynamic shear stress and the endothelium in cardiovascular pathophysiology. Nat Clin Pract Cardiovasc Med 2009;6(1):16–26.
[91] Tressel S, Huang R-P, Tomsen N, Jo H. Laminar shear inhibits tubule formation and migration of endothelial cells by an angiopoietin-2-dependent mechanism. Arterioscler Thromb Vasc Biol 2007;27:2150–6.
[92] Park S, Sorenson CM, Sheibani N. PECAM-1 isoforms, eNOS and endoglin axis in regulation of angiogenesis. Clin Sci (Lond) 2015;129(3):217–34.
[93] Niu G, Chen X. Vascular endothelial growth factor as an anti-angiogenic target for cancer therapy. Curr Drug Targets 2010;11(8):1000–17.
[94] Jo DH, Kim JH, Heo JI, Kim JH, Cho CH. Interaction between pericytes and endothelial cells leads to formation of tight junction in hyaloid vessels. Mol Cell 2013;36(5):465–71.
[95] Andrae J, Gallini R, Betsholtz C. Role of platelet-derived growth factors in physiology and medicine. Genes Dev 2008;22(10):1276–312.
[96] Thurston G. Complementary actions of VEGF and angiopoietin-1 on blood vessel growth and leakage. J Anat 2002;200:575–80.
[97] Roux Q, Gavard J. Endothelial cell–cell junctions in tumor angiogenesis. In: Marmé D, editor. Tumor angiogenesis: a key target for cancer therapy. Cham: Springer International Publishing; 2019. p. 91–119.
[98] Blanco R, Gerhardt H. VEGF and Notch in tip and stalk cell selection. Cold Spring Harb Perspect Med 2013;3(1) a006569-a.
[99] Cheung E, Wudali R, Halasz G, Wei Y, Economides A, Lin H, et al. The DII4/notch pathway controls postangiogenic blood vessel remodeling and regression by modulating vasoconstriction and blood flow. Blood 2011;117:6728–37.
[100] Siemerink MJ, Klaassen I, Van Noorden CJF, Schlingemann RO. Endothelial tip cells in ocular angiogenesis: potential target for anti-angiogenesis therapy. J Histochem Cytochem 2013;61(2):101–15.
[101] Escudero CA, Herlitz K, Troncoso F, Guevara K, Acurio J, Aguayo C, et al. Pro-angiogenic role of insulin: from physiology to pathology. Front Physiol 2017;8(204).
[102] Chappell JC, Wiley DM, Bautch VL. Regulation of blood vessel sprouting. Stem Cell Dev Biol 2011;22(9):1005–11.
[103] Khorolskaya J, Aleksandrova O, Samusenko I, Mikhailova N, Lobov I, Yudintceva N, et al. The effect of soluble recombinant protein Dll4-Fc on the functional activity of endothelial cells in vitro and vascularization in vivo. Cell Tissue Biol 2019;13(4):276–82.

[104] Chappell JC, Mouillesseaux KP, Bautch VL. Flt-1 (vascular endothelial growth factor receptor-1) is essential for the vascular endothelial growth factor—Notch feedback loop during angiogenesis. Arterioscler Thromb Vasc Biol 2013;33(8):1952–9.
[105] Kuhnert F, Kirshner JR, Thurston G. Dll4-Notch signaling as a therapeutic target in tumor angiogenesis. Vasc Cell 2011;3(1):20.
[106] Thomas J-L, Baker K, Han J, Calvo C, Nurmi H, Eichmann AC, et al. Interactions between VEGFR and notch signaling pathways in endothelial and neural cells. Cell Mol Life Sci 2013;70(10):1779–92.
[107] Herbert SP, Stainier DYR. Molecular control of endothelial cell behaviour during blood vessel morphogenesis. Nat Rev Mol Cell Biol 2011;12(9):551–64.
[108] Holderfield MT, Hughes CCW. Crosstalk between vascular endothelial growth factor, notch, and transforming growth factor-β in vascular morphogenesis. Circul Res 2008;102(6):637–52.
[109] Worzfeld T, Schwaninger M. Apicobasal polarity of brain endothelial cells. J Cereb Blood Flow Metab 2016;36(2):340–62.
[110] Gandalovičová A, Vomastek T, Rosel D, Brábek J. Cell polarity signaling in the plasticity of cancer cell invasiveness. Oncotarget 2016;7(18):25022–49.
[111] Hirakow R, Hiruma T. Scanning electron microscopic study on the development of primitive blood vessels in chick embryos at the early somite-stage. Anat Embryol 1981;163:299–306.
[112] Lenard A, Daetwyler S, Betz C, Ellertsdottir E, Belting H-G, Huisken J, et al. Endothelial cell self-fusion during vascular pruning. PLoS Biol 2015;13(4):e1002126.
[113] Stenmark KR, Meyrick B, Galie N, Mooi WJ, McMurtry IF. Animal models of pulmonary arterial hypertension: the hope for etiological discovery and pharmacological cure. Am J Physiol—Lung Cell Mol Physiol 2009;297(6):L1013–32.
[114] Gjini E, Hekking LH, Küchler A, Saharinen P, Wienholds E, Post J-A, et al. Zebrafish Tie-2 shares a redundant role with Tie-1 in heart development and regulates vessel integrity. Dis Model Mechan 2011;4(1):57–66.
[115] Benjamin LE, Hemo I, Keshet E. A plasticity window for blood vessel remodeling is defined by pericyte coverage of the preformed endothelial network and is regulated by PDGF-B and VEGF. Development 1998;125:1591–8.
[116] Eklund L, Kangas J, Saharinen P. Angiopoietin–Tie signalling in the cardiovascular and lymphatic systems. Clin Sci (Lond) 2017;131(1):87–103.
[117] Jones D, Min W. An overview of lymphatic vessels and their emerging role in cardiovascular disease. J Cardiovasc Dis Res 2011;2(3):141–52.
[118] Nutman TB. Insights into the pathogenesis of disease in human lymphatic filariasis. Lymphat Res Biol 2013;11(3):144–8.
[119] Breslin JW, Yang Y, Scallan JP, Sweat RS, Adderley SP, Murfee WL. Lymphatic vessel network structure and physiology. Compr Physiol 2018;9(1):207–99.
[120] Huxley VH, Scallan J. Lymphatic fluid: exchange mechanisms and regulation. J Physiol 2011;589(Pt 12):2935–43.
[121] Scallan JP, Zawieja SD, Castorena-Gonzalez JA, Davis MJ. Lymphatic pumping: mechanics, mechanisms and malfunction. J Physiol 2016;594(20):5749–68.
[122] Ikomi F, Kawai Y, Ohhashi T. Recent advance in lymph dynamic analysis in lymphatics and lymph nodes. Ann Vasc Dis 2012;5(3):258–68.
[123] Scavelli C, Weber E, Aglianò M, Cirulli T, Nico B, Vacca A, et al. Lymphatics at the crossroads of angiogenesis and lymphangiogenesis. J Anat 2004;204(6):433–49.
[124] Breslin JW. Mechanical forces and lymphatic transport. Microvasc Res 2014;96:46–54.
[125] Yao L-C, Baluk P, Srinivasan RS, Oliver G, McDonald DM. Plasticity of button-like junctions in the endothelium of airway lymphatics in development and inflammation. Am J Pathol 2012;180(6):2561–75.

[126] Bridenbaugh EA, Nizamutdinova IT, Jupiter D, Nagai T, Thangaswamy S, Chatterjee V, et al. Lymphatic muscle cells in rat mesenteric lymphatic vessels of various ages. Lymphat Res Biol 2013;11(1):35−42.

[127] Moore Jr. JE, Bertram CD. Lymphatic system flows. Annu Rev Fluid Mech 2018;50:459−82.

[128] Johnson NC, Dillard ME, Baluk P, McDonald DM, Harvey NL, Frase SL, et al. Lymphatic endothelial cell identity is reversible and its maintenance requires Prox1 activity. Genes Dev 2008;22(23):3282−91.

[129] Pazgal I, Boycov O, Shpilberg O, Okon E, Bairey O. Expression of VEGF-C, VEGF-D and their receptor VEGFR-3 in diffuse large B-cell lymphomas. Leuk Lymphoma 2007;48(11):2213−20.

[130] Kvietys PR, Granger DN. Role of intestinal lymphatics in interstitial volume regulation and transmucosal water transport. Ann N Y Acad Sci 2010;1207(Suppl 1):E29−43.

[131] Kohan AB, Yoder SM, Tso P. Using the lymphatics to study nutrient absorption and the secretion of gastrointestinal hormones. Physiol Behav 2011;105(1):82−8.

[132] Davis RB, Ding S, Nielsen NR, Pawlak JB, Blakeney ES, Caron KM. Calcitonin-receptor-like receptor signaling governs intestinal lymphatic innervation and lipid uptake. ACS Pharmacol Transl Sci 2019;2(2):114−21.

[133] Cifarelli V, Eichmann A. The intestinal lymphatic system: functions and metabolic implications. Cell Mol Gastroenterol Hepatol 2019;7(3):503−13.

[134] Escobedo N, Oliver G. The lymphatic vasculature: its role in adipose metabolism and obesity. Cell Metab 2017;26(4):598−609.

[135] Forchielli ML, Walker WA. The role of gut-associated lymphoid tissues and mucosal defence. Br J Nutr 2005;93(Suppl 1):S41−8.

[136] Papadopoulou PL, Patsikas MN, Charitanti A, Kazakos GM, Papazoglou LG, Karayannopoulou M, et al. The lymph drainage pattern of the mammary glands in the cat: a lymphographic and computerized tomography lymphographic study. Anat Histol Embryol 2009;38(4):292−9.

[137] Suami H, Pan W-R, Mann GB, Taylor GI. The lymphatic anatomy of the breast and its implications for sentinel lymph node biopsy: a human cadaver study. Ann Surg Oncol 2008;15(3):863−71.

[138] Betterman KL, Harvey NL. Histological and morphological characterization of developing dermal lymphatic vessels. Methods Mol Biol 2018;1846:19−35.

[139] Zawieja DC. Contractile physiology of lymphatics. Lymphat Res Biol 2009;7(2):87−96.

[140] Abdallah F, Mijouin L, Pichon C. Skin immune landscape: inside and outside the organism. Mediators Inflamm 2017;2017:5095293.

[141] Richmond JM, Harris JE. Immunology and skin in health and disease. Cold Spring Harb Perspect Med 2014;4(12) a015339-a.

[142] Wang Y, Jin Y, Mäe MA, Zhang Y, Ortsäter H, Betsholtz C, et al. Smooth muscle cell recruitment to lymphatic vessels requires PDGFB and impacts vessel size but not identity. Development 2017;144(19):3590−601.

[143] Wilting J, Papoutsi M, Othman-Hassan K, Rodriguez-Niedenführ M, Pröls F, Tomarev S, et al. Development of the avian lymphatic system. Microsc Res Tech 2001;55:81−91.

[144] Taya Y, Sato K, Shirako Y, Soeno Y. Migration of lymphatic endothelial cells and lymphatic vascular development in the craniofacial region of embryonic mice. Int J Dev Biol 2018;62(4-5):293−301.

[145] Yang Y, Oliver G. Development of the mammalian lymphatic vasculature. J Clin Invest 2014;124(3):888−97.

[146] Ma W, Oliver G. Lymphatic endothelial cell plasticity in development and disease. Physiol (Bethesda) 2017;32(6):444−52.

[147] Yang Y, García-Verdugo JM, Soriano-Navarro M, Srinivasan RS, Scallan JP, Singh MK, et al. Lymphatic endothelial progenitors bud from the cardinal vein and intersomitic vessels in mammalian embryos. Blood 2012;120(11):2340–8.

[148] Huang SS, Li YW, Wu JL, Johnson FE, Huang JS. Development of the LYVE-1 gene with an acidic-amino-acid-rich (AAAR) domain in evolution is associated with acquisition of lymph nodes and efficient adaptive immunity. J Cell Physiol 2018;233(4):2681–92.

[149] Srinivasan RS, Escobedo N, Yang Y, Interiano A, Dillard ME, Finkelstein D, et al. The Prox1-Vegfr3 feedback loop maintains the identity and the number of lymphatic endothelial cell progenitors. Genes Dev 2014;28(19):2175–87.

[150] Vaahtomeri K, Karaman S, Mäkinen T, Alitalo K. Lymphangiogenesis guidance by paracrine and pericellular factors. Genes Dev 2017;31(16):1615–34.

[151] Alderfer L, Wei A, Hanjaya-Putra D. Lymphatic tissue engineering and regeneration. J Biol Eng 2018;12(1):32.

[152] Eelen G, de Zeeuw P, Treps L, Harjes U, Wong BW, Carmeliet P. Endothelial cell metabolism. Physiol Rev 2018;98(1):3–58.

[153] Fatima A, Culver A, Culver F, Liu T, Dietz WH, Thomson BR, et al. Murine Notch1 is required for lymphatic vascular morphogenesis during development. Dev Dyn: Off Publ Am Assoc Anatom 2014;243(7):957–64.

[154] James J, Nalbandian A, Mukouyama Y-S. TGF signaling is required for sprouting lymphangiogenesis during lymphatic network development in the skin. Development 2013;140.

[155] Ohtani O, Ohtani Y. Recent developments in morphology of lymphatic vessels and lymph nodes. Ann Vasc Dis 2012;5(2):145–50.

[156] Jiang X, Nicolls MR, Tian W, Rockson SG. Lymphatic dysfunction, leukotrienes, and lymphedema. Annu Rev Physiol 2018;80:49–70.

[157] Blancas AA, Balaoing LR, Acosta FM, Grande-Allen KJ. Identifying behavioral phenotypes and heterogeneity in heart valve surface endothelium. Cell Tissues Organs 2016;201(4):268–76.

[158] Bautch VL, Caron KM. Blood and lymphatic vessel formation. Cold Spring Harb Perspect Biol 2015;7(3) a008268-a.

[159] Kume T. Lymphatic vessel development: fluid flow and valve-forming cells. J Clin Invest 2015;125(8):2924–6.

[160] Xia L. Platelet CLEC-2: a molecule with 2 faces. Blood 2017;130(20):2158–60.

[161] Boulaftali Y, Hess PR, Kahn ML, Bergmeier W. Platelet immunoreceptor tyrosine-based activation motif (ITAM) signaling and vascular integrity. Circul Res 2014;114(7):1174–84.

[162] Janardhan HP, Milstone ZJ, Shin M, Lawson ND, Keaney Jr. JF, Trivedi CM. Hdac3 regulates lymphovenous and lymphatic valve formation. J Clin Invest 2017;127(11):4193–206.

[163] Fatima A, Wang Y, Uchida Y, Norden P, Liu T, Culver A, et al. Foxc1 and Foxc2 deletion causes abnormal lymphangiogenesis and correlates with ERK hyperactivation. J Clin Invest 2016;126.

[164] Brouillard P, Boon L, Vikkula M. Genetics of lymphatic anomalies. J Clin Invest 2014;124(3):898–904.

[165] Geng X, Cha B, Mahamud MR, Srinivasan RS. Intraluminal valves: development, function and disease. Dis Model Mech 2017;10(11):1273–87.

[166] Tatin F, Taddei A, Weston A, Fuchs E, Devenport D, Tissir F, et al. Planar cell polarity protein celsr1 regulates endothelial adherens junctions and directed cell rearrangements during valve morphogenesis. Dev Cell 2013;26.

[167] Padera TP, Meijer EFJ, Munn LL. The lymphatic system in disease processes and cancer progression. Annu Rev Biomed Eng 2016;18:125–58.

[168] Santambrogio L. Immunology of the lymphatic system. New York: Springer; 2011. p. 177. Available from: https://doi.org/10.1007/978-1-4614-3235-7.
[169] Hashimoto T, Tsuneki M, Foster TR, Santana JM, Bai H, Wang M, et al. Membrane-mediated regulation of vascular identity. Birth Defects Res C Embryo Today 2016;108(1):65–84.
[170] Su S-A, Xie Y, Zhang Y, Xi Y, Cheng J, Xiang M. Essential roles of EphrinB2 in mammalian heart: from development to diseases. Cell Commun Signal 2019;17(1):29.
[171] Corliss BA, Azimi MS, Munson JM, Peirce SM, Murfee WL. Macrophages: an inflammatory link between angiogenesis and lymphangiogenesis. Microcirculation 2016;23(2):95–121.
[172] Grada AA, Phillips TJ. Lymphedema: pathophysiology and clinical manifestations. J Am Acad Dermatol 2017;77(6):1009–20.
[173] Hantusch B. Morphological and functional characteristics of blood and lymphatic vessels. Fundamentals of vascular biology. Springer; 2019. p. 1–43.
[174] Connell F, Brice G, Jeffery S, Keeley V, Mortimer P, Mansour S. A new classification system for primary lymphatic dysplasias based on phenotype. Clin Genet 2010;77(5):438–52.
[175] Mortimer PS, Rockson SG. New developments in clinical aspects of lymphatic disease. J Clin Invest 2014;124(3):915–21.
[176] Frueh FS, Gousopoulos E, Rezaeian F, Menger MD, Lindenblatt N, Giovanoli P. Animal models in surgical lymphedema research—a systematic review. J Surg Res 2016;200(1):208–20.
[177] Hruby A, Hu FB. The epidemiology of obesity: a big picture. PharmacoEconomics 2015;33(7):673–89.
[178] Mehrara BJ, Greene AK. Lymphedema and obesity: is there a link? Plast Reconstr Surg 2014;134(1):154e–160ee.
[179] Rubin DC, Shaker A, Levin MS. Chronic intestinal inflammation: inflammatory bowel disease and colitis-associated colon cancer. Front Immunol 2012;3:107.
[180] Alexander JS, Ganta VC, Jordan PA, Witte MH. Gastrointestinal lymphatics in health and disease. Pathophysiology 2010;17(4):315–35.
[181] Maas SLN, Breakefield XO, Weaver AM. Extracellular vesicles: unique intercellular delivery vehicles. Trends Cell Biol 2017;27(3):172–88.
[182] Park RJ, Hong YJ, Wu Y, Kim PM, Hong Y-K. Exosomes as a communication tool between the lymphatic system and bladder cancer. Int Neurourol J 2018;22(3):220–4.
[183] Mansilha A, Sousa J. Pathophysiological mechanisms of chronic venous disease and implications for venoactive drug therapy. Int J Mol Sci 2018;19(6):1669.
[184] Kakkos SK, Nicolaides AN. Efficacy of micronized purified flavonoid fraction (Daflon(R)) on improving individual symptoms, signs and quality of life in patients with chronic venous disease: a systematic review and meta-analysis of randomized double-blind placebo-controlled trials. Int Angiol 2018;37(2):143–54.
[185] Galkina E, Ley K. Immune and inflammatory mechanisms of atherosclerosis. Annu Rev Immunol 2009;27:165–97.
[186] Labropoulos N, Wierks C, Golts E, Volteas S, Leon M, Volteas N, et al. Microcirculatory changes parallel the clinical deterioration of chronic venous insufficiency. Phlebology 2004;19:81–6.
[187] Labropoulos N. How does chronic venous disease progress from the first symptoms to the advanced stages? A review. Adv Ther 2019;36(Suppl 1):13–19.
[188] Dahm KT, Myrhaug HT, Strømme H, Fure B, Brurberg KG. Effects of preventive use of compression stockings for elderly with chronic venous insufficiency and swollen legs: a systematic review and meta-analysis. BMC Geriatr 2019;19 (1):76.

[189] Serralheiro P, Soares A, Costa Almeida CM, Verde I. TGF-β1 in vascular wall pathology: unraveling chronic venous insufficiency pathophysiology. Int J Mol Sci 2017;18(12):2534.
[190] Raffetto J, Khalil RA. Mechanisms of varicose vein formation: valve dysfunction and wall dilation. Phlebol/Venous Forum R Soc Med 2008;23:85−98.
[191] Hinck AP, Mueller TD, Springer TA. Structural biology and evolution of the TGF-β family. Cold Spring Harb Perspect Biol 2016;8(12):a022103.
[192] Pardali E, Ten Dijke P. TGFβ signaling and cardiovascular diseases. Int J Biol Sci 2012;8(2):195−213.
[193] Serralheiro P, Soares A, Almeida CMC, Verde I. TGF-β1 in vascularwall pathology: unraveling chronic venous insufficiency pathophysiology. Int J Mol Sci 2017;18.
[194] Zhang Y-H, He M, Wang Y, Liao A-H. Modulators of the balance between M1 and M2 macrophages during pregnancy. Front Immunol 2017;8(120).
[195] Costanza B, Umelo IA, Bellier J, Castronovo V, Turtoi A. Stromal modulators of TGF-β in cancer. J Clin Med 2017;6(1):7.
[196] Del Amo-Maestro L, Marino-Puertas L, Goulas T, Gomis-Rüth FX. Recombinant production, purification, crystallization, and structure analysis of human transforming growth factor β2 in a new conformation. Sci Rep 2019;9(1):8660.
[197] Hata A, Chen Y-G. TGF-β signaling from receptors to smads. Cold Spring Harb Perspect Biol 2016;8:a022061.
[198] Liu T, Feng X-H. Regulation of TGF-beta signalling by protein phosphatases. Biochemical J 2010;430(2):191−8.
[199] Miyazawa K, Miyazono K. Regulation of TGF-β family signaling by inhibitory smads. Cold Spring Harb Perspect Biol 2017;9(3):a022095.
[200] Wu M, Chen G, Li Y-P. TGF-β and BMP signaling in osteoblast, skeletal development, and bone formation, homeostasis and disease. Bone Res 2016;4(1):16009.
[201] Lu D, Kassab GS. Role of shear stress and stretch in vascular mechanobiology. JR Soc Interf 2011;8(63):1379−85.
[202] Hortells L, Sur St. S, Hilaire C. Cell phenotype transitions in cardiovascular calcification. Front Cardiovasc Med 2018;5(27).
[203] Qi Y-X, Jiang J, Jiang X-H, Wang X-D, Ji S-Y, Han Y, et al. PDGF-BB and TGF-{beta}1 on cross-talk between endothelial and smooth muscle cells in vascular remodeling induced by low shear stress. Proc Natl Acad Sci USA 2011;108(5):1908−13.
[204] Prandi F, Piola M, Soncini M, Colussi C, D'Alessandra Y, Penza E, et al. Adventitial vessel growth and progenitor cells activation in an ex vivo culture system mimicking human saphenous vein wall strain after coronary artery bypass grafting. PLoS One 2015;10(2) e0117409-e.
[205] Lee SB, Kalluri R. Mechanistic connection between inflammation and fibrosis. Kidney Int Suppl 2010;119:S22−6.
[206] Barrett EJ, Liu Z, Khamaisi M, King GL, Klein R, Klein BEK, et al. Diabetic microvascular disease: an endocrine society scientific statement. J Clin Endocrinol Metab 2017;102(12):4343−410.
[207] Liu R-M, Desai LP. Reciprocal regulation of TGF-β and reactive oxygen species: a perverse cycle for fibrosis. Redox Biol 2015;6:565−77.
[208] Trionfini P, Benigni A. MicroRNAs as master regulators of glomerular function in health and disease. J Am Soc Nephrol 2017;28(6):1686−96.
[209] Chen Y, Peng W, Raffetto JD, Khalil RA. Matrix metalloproteinases in remodeling of lower extremity veins and chronic venous disease. Prog Mol Biol Transl Sci 2017;147:267−99.
[210] Małkowski A, Kowalewski R, Gacko M, Sobolewski K. Influence of thrombophlebitis on TGF-β1 and its signaling pathway in the vein wall. Folia Histochem Cytobiol 2011;48.

[211] Sayer G, Smith P. Immunocytochemical characterisation of the inflammatory cell infiltrate of varicose veins. Eur J Vasc Endovasc Surg: Off J Eur Soc Vasc Surg 2004;28:479–83.
[212] Hamilos M, Petousis S, Parthenakis F. Interaction between platelets and endothelium: from pathophysiology to new therapeutic options. Cardiovas Diagn Ther 2018;8(5):568–80.
[213] Wang X, Khalil RA. Matrix metalloproteinases, vascular remodeling, and vascular disease. Adv Pharmacol 2018;81:241–330.
[214] Pappas P, Lal B, Ohara N, Saito S, Zapiach L, Durán W. Regulation of matrix contraction in chronic venous disease. Eur J Vasc Endovasc Surg: Off J Eur Soc Vasc Surg 2009;38:518–29.
[215] Walton KL, Johnson KE, Harrison CA. Targeting TGF-β mediated SMAD signaling for the prevention of fibrosis. Front Pharmacol 2017;8(461).
[216] Ortega MA, Asunsolo A, Romero B, Alvarez-Rocha MJ, Sainz F, Leal J, et al. Unravelling the role of MAPKs (ERK1/2) in venous reflux in patients with chronic venous disorder. Cell Tissues Organs 2018;206(4-5):272–82.
[217] Dhanarak N, Kanchanabat B. Comparative histopathological study of the venous wall of chronic venous insufficiency and varicose disease. Phlebology 2016;31 (9):649–53.
[218] Lohr JM, Bush RL. Venous disease in women: epidemiology, manifestations, and treatment. J Vasc Surg 2013;57(4):37S–45S Supplement.
[219] Gui T, Sun Y, Shimokado A, Muragaki Y. The roles of mitogen-activated protein kinase pathways in TGF-β-induced epithelial–mesenchymal transition. J Signal Transduct 2012;2012:289243.
[220] Lyons O, Saha P, Seet C, Kuchta A, Arnold A, Grover S, et al. Human venous valve disease caused by mutations in FOXC2 and GJC2. J Exp Med 2017;214 (8):2437–52. Available from: https://doi.org/10.1084/jem.20160875.
[221] Munger SJ, Geng X, Srinivasan RS, Witte MH, Paul DL, Simon AM. Segregated Foxc2, NFATc1 and Connexin expression at normal developing venous valves, and Connexin-specific differences in the valve phenotypes of Cx37, Cx43, and Cx47 knockout mice. Dev Biol 2016;412(2):173–90.
[222] Basu R, Bose A, Thomas D, Das Sarma J. Microtubule-assisted altered trafficking of astrocytic gap junction protein connexin 43 is associated with depletion of connexin 47 during mouse hepatitis virus infection. J Biol Chem 2017;292(36):14747–63.
[223] Pandey R, Botros MA, Nacev BA, Albig AR. Cyclosporin a disrupts notch signaling and vascular lumen maintenance. PLoS One 2015;10(3) e0119279-e.
[224] Olbina G, Eckhart W. Mutations in the second extracellular region of connexin 43 prevent localization to the plasma membrane, but do not affect its ability to suppress cell growth. Mol Cancer Res 2003;1(9):690–700.
[225] Kamalapurkar G, Kartha CC, Surendran S, Girijamma A, Radhakrishnan N, Ramegowda KS, et al. Forkhead box C2 promoter variant c.-512C > T is associated with increased susceptibility to chronic venous diseases. PLoS One 2014;9:e90682.
[226] Zolotukhin IA, Seliverstov EI, Shevtsov YN, Avakiants IP, Nikishkov AS, Tatarintsev AM, et al. Prevalence and risk factors for chronic venous disease in the general Russian population. Eur J Vasc Endovasc Surg 2017;54(6):752–8.
[227] Petrie JR, Guzik TJ, Touyz RM. Diabetes, hypertension, and cardiovascular disease: clinical insights and vascular mechanisms. Can J Cardiol 2018;34(5):575–84.
[228] Rajendran P, Rengarajan T, Thangavel J, Nishigaki Y, Sakthisekaran D, Sethi G, et al. The vascular endothelium and human diseases. Int J Biol Sci 2013;9(10):1057–69.
[229] Gomes LR, Terra LF, Wailemann RA, Labriola L, Sogayar MC. TGF-β1 modulates the homeostasis between MMPs and MMP inhibitors through p38 MAPK and ERK1/2 in highly invasive breast cancer cells. BMC Cancer 2012;12:26.

[230] Mansilha A, Sousa J. Pathophysiological mechanisms of chronic venous disease and implications for venoactive drug therapy. Int J Mol Sci 2018;19:1–21.
[231] Zeng Y. Endothelial glycocalyx as a critical signalling platform integrating the extracellular haemodynamic forces and chemical signalling. J Cell Mol Med 2017;21 (8):1457–62.
[232] Tisato V, Zauli G, Voltan R, Gianesini S, Iasio M, Volpi I, et al. Endothelial cells obtained from patients affected by chronic venous disease exhibit a pro-inflammatory phenotype. PLoS One 2012;7:e39543.
[233] Bergan JJ, Pascarella L, Schmid-Schönbein GW. Pathogenesis of primary chronic venous disease: insights from animal models of venous hypertension. J Vasc Surg 2008;47(1):183–92.
[234] Barratt SL, Flower VA, Pauling JD, Millar AB. VEGF (vascular endothelial growth factor) and fibrotic lung disease. Int J Mol Sci 2018;19(5):1269.
[235] Broszczak DA, Sydes ER, Wallace D, Parker TJ. Molecular aspects of wound healing and the rise of venous leg ulceration: omics approaches to enhance knowledge and aid diagnostic discovery. Clin Biochem Rev 2017;38(1):35–55.
[236] Lee B-B, Nicolaides A, Myers K, Meissner M, Kalodiki E, Allegra C, et al. Venous hemodynamic changes in lower limb venous disease: the UIP consensus according to scientific evidence. Int Angiol: J Int Union Angiol 2016;35.
[237] Kowalewski R, Malkowski A, Sobolewski K, Gacko M. Evaluation of aFGF/bFGF and FGF signaling pathway in the wall of varicose veins. J Surg Res 2009;155 (1):165–72.
[238] Markovic JN, Shortell CK. Genomics of varicose veins and chronic venous insufficiency. Semin Vasc Surg 2013;26(1):2–13.
[239] Pereira C, Queirós S, Galaghar A, Sousa H, Pimentel-Nunes P, Brandão C, et al. Genetic variability in key genes in prostaglandin E2 pathway (COX-2, HPGD, ABCC4 and SLCO2A1) and their involvement in colorectal cancer development. PLoS One 2014;9(4):e92000-e.
[240] Shadrina AS, Sharapov SZ, Shashkova TI, Tsepilov YA. Varicose veins of lower extremities: insights from the first large-scale genetic study. PLoS Genet 2019;15(4): e1008110.
[241] Claesson-Welsh L. Vascular permeability–the essentials. Ups J Med Sci 2015;120 (3):135–43.
[242] Swirski FK, Nahrendorf M. Leukocyte behavior in atherosclerosis, myocardial infarction, and heart failure. Science (NY) 2013;339(6116):161–6.
[243] Weyl A, Vanscheidt W, Weiss JM, Peschen M, Schopf E, Simon J. Expression of the adhesion molecules ICAM-1, VCAM-1, and E-selectin and their ligands VLA-4 and LFA-1 in chronic venous leg ulcers. J Am Acad Dermatol 1996;34(3):418–23.
[244] Charles CA, Tomic-Canic M, Vincek V, Nassiri M, Stojadinovic O, Eaglstein WH, et al. A gene signature of nonhealing venous ulcers: potential diagnostic markers. J Am Acad Dermatol 2008;59(5):758–71.
[245] Tönjes M, Barbus S, Park YJ, Wang W, Schlotter M, Lindroth AM, et al. BCAT1 promotes cell proliferation through amino acid catabolism in gliomas carrying wild-type IDH1. Nat Med 2013;19(7):901–8.
[246] Dong N, Xu B, Xu J. EGF-mediated overexpression of Myc attenuates miR-26b by recruiting HDAC3 to induce epithelial-mesenchymal transition of lens epithelial cells. Biomed Res Int 2018;2018:7148023.
[247] Zamboni P, Tognazzo S, Izzo M, Pancaldi F, Scapoli G, Liboni A, et al. Hemochromatosis C282Y gene mutation increases the risk of venous leg ulceration. J Vasc Surg: Off Pub Soc Vasc Surg [and] Int Soc Cardiovasc Surg North Am Chapter 2005;42:309–14.

[248] Gemmati D, Federici F, Campo G, Tognazzo S, Serino ML, De Mattei M, et al. Factor XIIIA-V34L and factor XIIIB-H95R gene variants: effects on survival in myocardial infarction patients. Mol Med 2007;13(1-2):112–20.

[249] Eming SA, Martin P, Tomic-Canic M. Wound repair and regeneration: mechanisms, signaling, and translation. Sci Transl Med 2014;6(265):265sr6-sr6.

[250] Landén NX, Li D, Ståhle M. Transition from inflammation to proliferation: a critical step during wound healing. Cell Mol Life Sci 2016;73(20):3861–85.

[251] Martínez-Zapata M, Moreno R, Gich I, Urrútia G, Bonfill X. A randomized, double-blind multicentre clinical trial comparing the efficacy of calcium dobesilate with placebo in the treatment of chronic venous disease. Eur J Vasc Endovasc Surg: Off J Eur Soc Vasc Surg 2008;35:358–65.

[252] Muller WA. Getting leukocytes to the site of inflammation. Vet Pathol 2013;50 (1):7–22.

[253] Shoab SS, Porter J, Scurr J, Coleridge-Smith P. Effect of oral micronized purified flavonoid fraction treatment on leukocyte adhesion molecule expression in patients with chronic venous disease: a pilot study. J Vasc Surg: Off Pub, Soc Vasc Surg [and] Int Soc Cardiovasc Surg North Am Chapter 2000;31:456–61.

[254] Kucukguven A, Khalil RA. Matrix metalloproteinases as potential targets in the venous dilation associated with varicose veins. Curr Drug Targets 2013;14(3):287–324.

[255] Rabe E, Agus G, Roztocil K. Analysis of the effects of micronized purified flavonoid fraction versus placebo on symptoms and quality of life in patients suffering from chronic venous disease: from a prospective randomized trial. Int Angiol: J Int Union Angiol 2015;34.

[256] Katsenis K. Micronized purified flavonoid fraction (MPFF): a review of its pharmacological effects, therapeutic efficacy and benefits in the management of chronic venous insufficiency. Curr Vasc Pharmacol 2005;3:1–9.

[257] Ivetic A, Hoskins Green HL, Hart SJ. L-selectin: a major regulator of leukocyte adhesion, migration and signaling. Front Immunol 2019;10(1068).

[258] das Gracas CdSM, Cyrino FZ, de Carvalho JJ, Blanc-Guillemaud V, Bouskela E. Protective effects of micronized purified flavonoid fraction (MPFF) on a novel experimental model of chronic venous hypertension. Eur J Vasc Endovasc Surg 2018;55(5):694–702.

[259] Paysant J, Sansilvestri-Morel P, Bouskela E, Verbeuren TJ. Different flavonoids present in the micronized purified flavonoid fraction (Daflon® 500 mg) contribute to its anti-hyperpermeability effect in the hamster cheek pouch microcirculation. Int Angiol: J Int Union Angiol 2008;27:81–5.

[260] Kauss T, Moynet D, Rambert J, Al-Kharrat A, Brajot S, Thiolat D, et al. Rutoside decreases human macrophage-derived inflammatory mediators and improves clinical signs in adjuvant-induced arthritis. Arthritis Res Ther 2008;10(1):R19.

[261] Li T, Liu X, Zhao Z, Ni L, Liu C. Sulodexide recovers endothelial function through reconstructing glycocalyx in the balloon-injury rat carotid artery model. Oncotarget 2017;8(53):91350–61.

[262] Ligi D, Mosti G, Croce L, Raffetto J, Mannello F. Chronic venous disease. Part I. Inflammatory biomarkers in wound healing. Biochim Biophys Acta (BBA)—Mol Basis Dis 2016;1862.

[263] Coccheri S, Scondotto G, Agnelli G, Aloisi D, Palazzini E, Zamboni V. Randomised, double blind, multicentre, placebo controlled study of sulodexide in the treatment of venous leg ulcers. Thromb Haemost 2002;87:947–52.

[264] Thergaonkar R, Bhardwaj S, Sinha A, Dinda A, Kumar R, Bagga A, et al. Posttransplant lymphoproliferative disorder: Experience from a pediatric nephrology unit in North India. Indian Dermatol Online J 2014;5(3):374–7.

[265] Radhakrishnan N. A treatise on venous diseases. Jaypee Brothers Medical Publishers (P) Ltd; 2014. Available from: https://doi.org/10.5005/jp/books/12365.

[266] Rautio T, Ohinmaa A, Perälä J, Ohtonen P, Heikkinen T, Wiik H, et al. Endovenous obliteration versus conventional stripping operation in the treatment of primary varicose veins: a randomized controlled trial with comparison of the costs. J Vasc Surg 2002;35(5):958–65.

[267] Korkmaz K, Yener AÜ, Gedık HS, Budak AB, Yener Ö, Genç SB, et al. Tumescentless endovenous radiofrequency ablation with local hypothermia and compression technique. Cardiovasc J Afr 2013;24(8):313–17.

[268] Rautio TT, Perälä JM, Wiik HT, Juvonen TS, Haukipuro KA. Endovenous obliteration with radiofrequency-resistive heating for greater saphenous vein insufficiency: a feasibility study. J Vasc Intervent Radiol 2002;13(6):569–75.

[269] Recek C. Significance of reflux abolition at the saphenofemoral junction in connection with stripping and ablative methods. Int J Angiol 2015;24(4):249–61.

[270] Sarma N. Guidelines and recommendation on surgery for venous incompetence and leg ulcer. Indian Dermatol Online J 2014;5(3):390–5.

[271] van Rij AM, Jones GT, Hill GB, Jiang P. Neovascularization and recurrent varicose veins: more histologic and ultrasound evidence. J Vasc Surg 2004;40(2):296–302.

[272] Abd El-Mabood E-S, Salama R. Efficacy of endovenous laser ablation (endovenous laser ablation) versus conventional stripping in the treatment of great saphenous vein reflux. Egypt J Surg 2017;36(3):222–32.

[273] Poluektova AA, Malskat WS, van Gemert MJ, Vuylsteke ME, Bruijninckx CM, Neumann HA, et al. Some controversies in endovenous laser ablation of varicose veins addressed by optical–thermal mathematical modeling. Lasers Med Sci 2014;29(2):441–52.

[274] Ashpitel HF, Dabbs EB, Salguero FJ, Nemchand JL, La Ragione RM, Whiteley MS. Histopathologic differences in the endovenous laser ablation between jacketed and radial fibers, in an ex vivo dominant extrafascial tributary of the great saphenous vein in an in vitro model, using histology and immunohistochemistry. J Vasc Surg Venous Lymphat Disord 2019;7(2):234–45.

[275] Hirokawa M, Kurihara N. Comparison of bare-tip and radial fiber in endovenous laser ablation with 1470 nm diode laser. Ann Vasc Dis 2014;7(3):239–45.

[276] Foster KR, Ziskin MC, Balzano Q. Thermal response of human skin to microwave energy: a critical review. Health Phys 2016;111(6):528–41.

[277] Kurihara N, Hirokawa M, Yamamoto T. Postoperative venous thromboembolism in patients undergoing endovenous laser and radiofrequency ablation of the saphenous vein. Ann Vasc Dis 2016;9(4):259–66.

[278] Malskat WSJ, Poluektova AA, van der Geld CWM, Neumann HAM, Weiss RA, Bruijninckx CMA, et al. Endovenous laser ablation (EVLA): a review of mechanisms, modeling outcomes, and issues for debate. Lasers Med Sci 2014;29(2):393–403.

[279] Al Wahbi AM. Evaluation of pain during endovenous laser ablation of the great saphenous vein with ultrasound-guided femoral nerve block. Vasc Health Risk Manag 2017;13:305–9.

[280] Fan CM, Anderson R. Endovenous laser ablation: mechanism of action. Phlebology/Venous Forum R Soc Med 2008;23:206–13.

[281] Yu D-Y, Chen H-C, Chang S-Y, Hsiao Y-C, Chang C-J. Comparing the effectiveness of 1064 vs. 810 nm wavelength endovascular laser for chronic venous insufficiency (varicose veins). Laser Ther 2013;22(4):247–53.

[282] Rohringer S, Holnthoner W, Chaudary S, Slezak P, Priglinger E, Strassl M, et al. The impact of wavelengths of LED light-therapy on endothelial cells. Sci Rep 2017;7(1):10700.

[283] Uchino IJ. Endovenous laser closure of the perforating vein of the leg. Phlebology 2007;22(2):80–2.

[284] Tseng S-H, Bargo P, Durkin A, Kollias N. Chromophore concentrations, absorption and scattering properties of human skin in-vivo. Opt Express 2009;17 (17):14599–617.

[285] Mahapatra S, Ramakrishna P, Gupta B, Anusha A, Para MA. Correlation of obesity and comorbid conditions with chronic venous insufficiency: results of a single-centre study. Indian J Med Res 2018;147(5):471–6.

[286] Fitzpatrick T, Perrier L, Shakik S, Cairncross Z, Tricco AC, Lix L, et al. Assessment of long-term follow-up of randomized trial participants by linkage to routinely collected data: a scoping review and analysis. JAMA Netw Open 2018;1(8):e186019.

[287] Kayssi A, Pope M, Vucemilo I, Werneck C. Endovenous radiofrequency ablation for the treatment of varicose veins. Can J Surg 2015;58(2):85–6.

[288] Sanioglu S, Yerebakan H, Farsak MB. Effects of two current great saphenous vein thermal ablation methods on visual analog scale and quality of life. BioMed Res Int 2017;2017:6.

[289] Zhang Y-J, Chen M-S. Role of radiofrequency ablation in the treatment of small hepatocellular carcinoma. World J Hepatol 2010;2(4):146–50.

[290] Ahmed M, Solbiati L, Brace CL, Breen DJ, Callstrom MR, Charboneau JW, et al. Image-guided tumor ablation: standardization of terminology and reporting criteria--a 10-year update. Radiology 2014;273(1):241–60.

[291] Tang TY, Kam JW, Gaunt ME. ClariVein®—early results from a large single-centre series of mechanochemical endovenous ablation for varicose veins. Phlebology 2017;32(1):6–12.

[292] Attaran RR. Latest innovations in the treatment of venous disease. J Clin Med 2018;7(4):77.

[293] Mueller RL, Raines JK. ClariVein mechanochemical ablation: background and procedural details. Vasc Endovasc Surg 2013;47(3):195–206.

[294] Witte ME, Reijnen MM, de Vries JP, Zeebregts CJ. Mechanochemical endovenous occlusion of varicose veins using the ClariVein(R) device. Surg Technol Int 2015;26:219–25.

[295] Vun SV, Rashid ST, Blest NC, Spark JI. Lower pain and faster treatment with mechanico-chemical endovenous ablation using ClariVein®. Phlebology 2015;30 (10):688–92. Available from: https://doi.org/10.1177/0268355514553693.

[296] Bootun R, Lane TRA, Dharmarajah B, Lim CS, Najem M, Renton S, et al. Intra-procedural pain score in a randomised controlled trial comparing mechanochemical ablation to radiofrequency ablation: the Multicentre Venefit™ versus ClariVein® for varicose veins trial. Phlebology 2016;31(1):61–5. Available from: https://doi.org/10.1177/0268355514551085.

[297] Lin F, Zhang S, Sun Y, Ren S, Liu P. The management of varicose veins. Int Surg 2015;100(1):185–9.

[298] Körner C. Plant adaptation to cold climates. F1000Research 2016;5 F1000 Faculty Rev-2769.

[299] Osman O, El-Heeny AA, El-Razeq MM. Management of primary uncomplicated varicose veins, endovenous laser ablation with sclerotherapy versus traditional surgery: which is the best option? Egypt J Surg 2019;38(2):319–27.

[300] Niedzwiecki G. Endovenous thermal ablation of the saphenous vein. Semin Intervent Radiol 2005;22(3):204–8.

[301] Morrison N, Gibson K, McEnroe S, Goldman M, King T, Weiss R, et al. Randomized trial comparing cyanoacrylate embolization and radiofrequency ablation for incompetent great saphenous veins (VeClose). J Vasc Surg 2015;61:985–94. Available from: https://doi.org/10.1016/j.jvs.2014.11.071.

[302] McHugh SM, Leahy AL. What next after thermal ablation for varicose veins: non-thermal ablation? Surgeon 2014;12(5):237–8.

[303] Allen RC, Tawes RL, Wetter LA, Fogarty TJ. Endoscopic perforator vein surgery: creation of a subfascial space. In: Gloviczki P, Bergan JJ, editors. Atlas of endoscopic perforator vein surgery. London: Springer London; 1998. p. 153–63.
[304] Tawes RL, Wetter LA, Hermann GD, Fogarty TJ. Endoscopic technique for subfascial perforating vein interruption. J Endovasc Ther 1996;3(4):414–20.
[305] Ombrellino M, Kabnick LS. Varicose vein surgery. Semin Intervent Radiol 2005;22(3):185–94.
[306] Sahoo MR, Misra L, Deshpande S, Mohanty SK, Mohanty SK. Subfascial endoscopic perforator surgery: a safe and novel minimal invasive procedure in treating varicose veins in 2(nd) trimester of pregnancy for below knee perforator incompetence. J Minim Access Surg 2018;14(3):208–12.
[307] Eckmann DM. Polidocanol for endovenous microfoam sclerosant therapy. Expert Opin Invest Drugs 2009;18(12):1919–27.
[308] Hajar R. The air of history (part II) medicine in the middle ages. Heart Views 2012;13(4):158–62.
[309] Almeida JI, Raines JK. FDA-approved sodium tetradecyl sulfate (STS) versus compounded STS for venous sclerotherapy. Dermatol Surg 2007;33(9):1037–44 discussion 44.
[310] Kaul S, Gulati N, Verma D, Mukherjee S, Nagaich U. Role of nanotechnology in cosmeceuticals: a review of recent advances. J Pharm (Cairo) 2018;2018:3420204.
[311] Lugus JJ, Ngoh GA, Bachschmid MM, Walsh K. Mitofusins are required for angiogenic function and modulate different signaling pathways in cultured endothelial cells. J Mol Cell Cardiol 2011;51(6):885–93.
[312] Sternebring O, Christensen JK, Bjørnsdottir I. Pharmacokinetics, tissue distribution, excretion, and metabolite profiling of PEGylated rFIX (nonacog beta pegol, N9-GP) in rats. Eur J Pharm Sci 2016;92:163–72.
[313] Xu J, Wang Y-F, Chen A-W, Wang T, Liu S-H. A modified Tessari method for producing more foam. Springerplus 2016;5:129.
[314] Peterson JD, Goldman MP. An investigation into the influence of various gases and concentrations of sclerosants on foam stability. Dermatol Surg 2011;37(1):12–17.
[315] Ravishankar K, Chakravarty A, Chowdhury D, Shukla R, Singh S. Guidelines on the diagnosis and the current management of headache and related disorders. Ann Indian Acad Neurol 2011;14(Suppl 1):S40–59.
[316] Crous-Bou M, Harrington LB, Kabrhel C. Environmental and genetic risk factors associated with venous thromboembolism. Semin Thromb Hemost 2016;42 (8):808–20.
[317] O'Donnell M, Weitz JI. Thromboprophylaxis in surgical patients. Can J Surg 2003;46(2):129–35.
[318] Yau JW, Teoh H, Verma S. Endothelial cell control of thrombosis. BMC Cardiovasc Disord 2015;15:130.
[319] Devis P, Knuttinen MG. Deep venous thrombosis in pregnancy: incidence, pathogenesis and endovascular management. Cardiovascular diagnosis and therapy. AME Publishing Company; 2017. Available from: https://doi.org/10.21037/cdt.2017.10.08.
[320] Galanaud J-P, Kahn SR. 18—Postthrombotic syndrome. In: Kitchens CS, Kessler CM, Konkle BA, Streiff MB, Garcia DA, editors. Consultative hemostasis and thrombosis. 4th ed. Philadelphia: Content Repository Only!;; 2019. p. 338–45.
[321] Kahn SR, Galanaud J-P, Vedantham S, Ginsberg JS. Guidance for the prevention and treatment of the post-thrombotic syndrome. J Thromb Thrombol 2016;41 (1):144–53.
[322] Fleck D, Albadawi H, Shamoun F, Knuttinen G, Naidu S, Oklu R. Catheter-directed thrombolysis of deep vein thrombosis: literature review and practice considerations. Cardiovasc Diagn Ther 2017;7(Suppl 3):S228–37.

[323] Wang KL, Chu PH, Lee CH, Pai PY, Lin PY, Shyu KG, et al. Management of venous thromboembolisms. Part I. The consensus for deep vein thrombosis. Acta Cardiol Sin 2016;32(1):1–22. Available from: https://doi.org/10.6515/ACS20151228A.

[324] Vedantham S, Piazza G, Sista AK, Goldenberg NA. Guidance for the use of thrombolytic therapy for the treatment of venous thromboembolism. J Thromb Thrombol 2016;41(1):68–80. Available from: https://doi.org/10.1007/s11239-015-1318-z.

[325] Yurdakok M. Fetal and neonatal effects of anticoagulants used in pregnancy: a review. Turk J Pediatr 2012;54(3):207–15.

[326] Bates SM, Middeldorp S, Rodger M, James AH, Greer I. Guidance for the treatment and prevention of obstetric-associated venous thromboembolism. J Thromb Thrombol 2016;41(1):92–128.

[327] Yang G, Staercke C, De, Hooper WC. The effects of obesity on venous thromboembolism: a review. Open J Prev Med 2012;02(04):499–509. Available from: https://doi.org/10.4236/ojpm.2012.24069.

[328] Hughes S, Szeki I, Nash MJ, Thachil J. Anticoagulation in chronic kidney disease patients—the practical aspects. Clin Kidney J 2014;7(5):442–9.

[329] Ratib S, Walker AJ, Card TR, Grainge MJ. Risk of venous thromboembolism in hospitalised cancer patients in England—a cohort study. J Hematol Oncol 2016;9(1):60.

[330] Cohen AT, Bauersachs R. Rivaroxaban and the Einstein clinical trial programme. Blood coagulation and fibrinolysis. Lippincott Williams and Wilkins; 2019. Available from: https://doi.org/10.1097/MBC.0000000000000800.

[331] Behrouzi R, Punter M. Diagnosis and management of cerebral venous thrombosis. Clin Med, J R Coll Phys Lond 2018;18(1):75–9. Available from: https://doi.org/10.7861/clinmedicine.18-1-75.

[332] Hmoud B, Singal AK, Kamath PS. Mesenteric venous thrombosis. J Clin Exp Hepatol 2014;4(3):257–63.

[333] Di Nisio M, Carrier M. Incidental venous thromboembolism: is anticoagulation indicated? Hematol Am Soc Hematol Educ Program 2017;2017(1):121–7.

[334] Ahmed I, Majeed A, Powell R. Heparin induced thrombocytopenia: Diagnosis and management update. Postgrad Med J. BMJ Publishing Group; 2007. Available from: https://doi.org/10.1136/pgmj.2007.059188.

[335] Warkentin TE, Greinacher A. Management of heparin-induced thrombocytopenia. Curr Opin Hematol. Lippincott Williams and Wilkins; 2016. Available from: https://doi.org/10.1097/MOH.0000000000000273.

[336] Kumaresan A, Bose S. Delayed seroconversion in a patient with heparin-induced thrombocytopenia. J Cardiothorac Vasc Anesth 2018;32(2):e37–8.

[337] Ahmed I, Majeed A, Powell R. Heparin induced thrombocytopenia: diagnosis and management update. Postgrad Med J 2007;83(983):575–82.

[338] Holcomb JB, Tilley BC, Baraniuk S, Fox EE, Wade CE, Podbielski JM, et al. Transfusion of plasma, platelets, and red blood cells in a 1:1:1 vs a 1:1:2 ratio and mortality in patients with severe trauma: the PROPPR randomized clinical trial. JAMA 2015;313(5):471–82.

[339] Kozlowski D, Budrejko S, Raczak G, Rysz J, Banach M. Anticoagulant prevention in patients with atrial fibrillation: alternatives to vitamin K antagonists. Curr Pharm Des 2013;19(21):3816–26.

[340] Blommel ML, Blommel AL. Dabigatran etexilate: a novel oral direct thrombin inhibitor. Am J Health Syst Pharm 2011;68(16):1506–19.

[341] Stangier J, Rathgen K, Stähle H, Gansser D, Roth W. The pharmacokinetics, pharmacodynamics and tolerability of dabigatran etexilate, a new oral direct thrombin inhibitor, in healthy male subjects. Br J Clin Pharmacol 2007;64(3):292–303.

[342] Reilly PA, Lehr T, Haertter S, Connolly SJ, Yusuf S, Eikelboom JW, et al. The effect of dabigatran plasma concentrations and patient characteristics on the frequency of ischemic stroke and major bleeding in atrial fibrillation patients: the RE-LY trial (randomized evaluation of long-term anticoagulation therapy). J Am Coll Cardiol 2014;63(4):321–8.
[343] Laizure SC, Parker RB, Herring VL, Hu Z-Y. Identification of carboxylesterase-dependent dabigatran etexilate hydrolysis. Drug Metab Dispos 2014;42(2):201–6.
[344] Lutz J, Jurk K, Schinzel H. Direct oral anticoagulants in patients with chronic kidney disease: patient selection and special considerations. Int J Nephrol Renovasc Dis 2017;10:135–43.
[345] Posner J. Optimizing the dose of dabigatran etexilate. Br J Clin Pharmacol 2012;74(5):741–3.
[346] Stangier J, Rathgen K, Stahle H, Reseski K, Kornicke T, Roth W. Coadministration of dabigatran etexilate and atorvastatin: assessment of potential impact on pharmacokinetics and pharmacodynamics. Am J Cardiovasc Drugs 2009;9(1):59–68.
[347] Ellis CR, Kaiser DW. The clinical efficacy of dabigatran etexilate for preventing stroke in atrial fibrillation patients. Vasc Health Risk Manag 2013;9:341–52.
[348] Camm AJ, Fox KAA, Peterson E. Challenges in comparing the non-vitamin K antagonist oral anticoagulants for atrial fibrillation-related stroke prevention. EP Europace 2017;20(1):1–11.
[349] O'Dea D, Whetteckey J, Ting N. A prospective, randomized, open-label study to evaluate two management strategies for gastrointestinal symptoms in patients newly on treatment with dabigatran. Cardiol Ther 2016;5(2):187–201.
[350] Korenstra J, Wijtvliet EPJ, Veeger NJGM, Geluk CA, Bartels GL, Posma JL, et al. Effectiveness and safety of dabigatran versus acenocoumarol in 'real-world' patients with atrial fibrillation. EP Europace 2016;18(9):1319–27.
[351] Tsivgoulis G, Krogias C, Sands KA, Sharma VK, Katsanos AH, Vadikolias K, et al. Dabigatran etexilate for secondary stroke prevention: the first year experience from a multicenter short-term registry. Ther Adv Neurol Disord 2014;7(3):155–61.
[352] Jain N, Reilly RF. Clinical pharmacology of oral anticoagulants in patients with kidney disease. Clin J Am Soc Nephrol 2019;14(2):278–87.
[353] Garnock-Jones KP. Dabigatran etexilate: a review of its use in the prevention of stroke and systemic embolism in patients with atrial fibrillation. Am J Cardiovasc Drugs 2011;11(1):57–72.
[354] Topcuoglu MA, Liu L, Kim D-E, Gurol ME. Updates on prevention of cardioembolic strokes. J Stroke 2018;20(2):180–96.
[355] Chapter 1: Definition and classification of CKD. Kidney Int Suppl (2011) 2013;3(1):19–62.
[356] Steffel J, Verhamme P, Potpara TS, Albaladejo P, Antz M, Desteghe L, et al. The 2018 European Heart Rhythm Association Practical Guide on the use of non-vitamin K antagonist oral anticoagulants in patients with atrial fibrillation. Eur Heart J 2018;39(16):1330–93.
[357] Ezekowitz MD, Eikelboom J, Oldgren J, Reilly PA, Brueckmann M, Kent AP, et al. Long-term evaluation of dabigatran 150 vs. 110 mg twice a day in patients with non-valvular atrial fibrillation. EP Europace 2016;18(7):973–8.
[358] Mekaj YH, Mekaj AY, Duci SB, Miftari EI. New oral anticoagulants: their advantages and disadvantages compared with vitamin K antagonists in the prevention and treatment of patients with thromboembolic events. Ther Clin Risk Manag 2015;11:967–77.
[359] Hori M, Connolly S, Zhu J, Liu L, Lau C-P, Pais P, et al. Dabigatran versus warfarin effects on ischemic and hemorrhagic strokes and bleeding in Asians and non-Asians with atrial fibrillation. Stroke: J Cereb Circul 2013;44.

[360] Eltringham-Smith LJ, Yu R, Qadri SM, Wang Y, Bhakta V, Pryzdial EL, et al. Prothrombin, alone or in complex concentrates or plasma, reduces bleeding in a mouse model of blood exchange-induced coagulopathy. Sci Rep 2019;9(1):13029.
[361] Pirlog A-M, Pirlog CD, Maghiar MA. DOACs vs vitamin K antagonists: a comparison of phase III clinical trials and a prescriber support tool. Open Access Maced J Med Sci 2019;7(7):1226−32.
[362] Blair HA, Keating GM, Dabigatran. Etexilate: a review in nonvalvular atrial fibrillation. Drugs 2017;77(3):331−44.
[363] Ganetsky M, Babu KM, Salhanick SD, Brown RS, Boyer EW. Dabigatran: review of pharmacology and management of bleeding complications of this novel oral anticoagulant. J Med Toxicol 2011;7(4):281−7.
[364] Ezekowitz MD, Nagarakanti R, Noack H, Brueckmann M, Litherland C, Jacobs M, et al. Comparison of dabigatran and warfarin in patients with atrial fibrillation and valvular heart disease: the RE-LY trial (randomized evaluation of long-term anticoagulant therapy). Circulation 2016;134(8):589−98.
[365] Vranckx P, Valgimigli M, Heidbuchel H. The significance of drug−drug and drug−food interactions of oral anticoagulation. Arrhythm Electrophysiol Rev 2018;7(1):55−61.
[366] Ezekowitz MD, Connolly S, Parekh A, Reilly PA, Varrone J, Wang S, et al. Rationale and design of RE-LY: randomized evaluation of long-term anticoagulant therapy, warfarin, compared with dabigatran. Am Heart J 2009;157(5):805−10.
[367] Uchiyama S, Atarashi H, Inoue H, Kitazono T, Yamashita T, Shimizu W, et al. Primary and secondary prevention of stroke and systemic embolism with rivaroxaban in patients with non-valvular atrial fibrillation: sub-analysis of the EXPAND study. Heart Vessel 2019;34(1):141−50.
[368] Milling TJ, Fromm C, Ganetsky M, Pallin DJ, Cong J, Singer AJ. Management of major bleeding events in patients treated with dabigatran for nonvalvular atrial fibrillation: a retrospective, multicenter review. Ann Emerg Med 2017;69(5):531−40.
[369] Roskell NS, Samuel M, Noack H, Monz BU. Major bleeding in patients with atrial fibrillation receiving vitamin K antagonists: a systematic review of randomized and observational studies. Europace 2013;15(6):787−97.
[370] Foody JM. Reducing the risk of stroke in elderly patients with non-valvular atrial fibrillation: a practical guide for clinicians. Clin Interv Aging 2017;12:175−87.
[371] D'Andrea G, Chetta M, Margaglione M. Inherited platelet disorders: thrombocytopenias and thrombocytopathies. Blood Transfus 2009;7(4):278−92.
[372] Yu Y-B, Liu J, Fu G-H, Fang R-Y, Gao F, Chu H-M. Comparison of dabigatran and warfarin used in patients with non-valvular atrial fibrillation: meta-analysis of random control trial. Medicine 2018;97(46):e12841-e.
[373] Ribic C, Crowther M. Thrombosis and anticoagulation in the setting of renal or liver disease. Hematol Am Soc Hematol Educ Program 2016;2016(1):188−95.
[374] Nardi F, Gulizia MM, Colivicchi F, Abrignani MG, Di Fusco SA, Di Lenarda A, et al. ANMCO position paper: direct oral anticoagulants for stroke prevention in atrial fibrillation: clinical scenarios and future perspectives. Eur Heart J Suppl 2017;19(Suppl D):D70−88.
[375] Hanley JP. Warfarin reversal. J Clin Pathol 2004;57(11):1132−9.
[376] McKeage K. Dabigatran etexilate: a pharmacoeconomic review of its use in the prevention of stroke and systemic embolism in patients with atrial fibrillation. PharmacoEconomics 2012;30(9):841−55.
[377] Chan Y-H, See L-C, Tu H-T, Yeh Y-H, Chang S-H, Wu L-S, et al. Efficacy and safety of apixaban, dabigatran, rivaroxaban, and warfarin in Asians with nonvalvular atrial fibrillation. J Am Heart Assoc 2018;7(8):e008150.

[378] Acanfora D, Ciccone MM, Scicchitano P, Ricci G, Acanfora C, Uguccioni M, et al. Efficacy and safety of direct oral anticoagulants in patients with atrial fibrillation and high thromboembolic risk. Syst Rev Front Pharmacol 2019;10:1048.
[379] Buesing KL, Mullapudi B, Flowers KA. Deep venous thrombosis and venous thromboembolism prophylaxis. Surg Clin North Am 2015;95(2):285−300.
[380] Violi F, Pastori D, Pignatelli P. Mechanisms and management of thromboembolism in atrial fibrillation. J Atr Fibrillation 2014;7(3):1112.
[381] Parry TJ, Huang Z, Chen C, Connelly MA, Perzborn E, Andrade-Gordon P, et al. Arterial antithrombotic activity of rivaroxaban, an orally active factor Xa inhibitor, in a rat electrolytic carotid artery injury model of thrombosis. Blood Coagul Fibrinol 2011;22(8):720−6.
[382] Mueck W, Stampfuss J, Kubitza D, Becka M. Clinical pharmacokinetic and pharmacodynamic profile of rivaroxaban. Clin Pharmacokinet 2014;53(1):1−16.
[383] Mueck W, Schwers S, Stampfuss J. Rivaroxaban and other novel oral anticoagulants: pharmacokinetics in healthy subjects, specific patient populations and relevance of coagulation monitoring. Thromb J 2013;11(1):10.
[384] Vimalesvaran K, Dockrill SJ, Gorog DA. Role of rivaroxaban in the management of atrial fibrillation: insights from clinical practice. Vasc Health Risk Manag 2018;14:13−21.
[385] Ahmed Z, Hassan S, Salzman GA. Novel oral anticoagulants for venous thromboembolism with special emphasis on risk of hemorrhagic complications and reversal agents. Curr Drug ther 2016;11(1):3−20.
[386] Ibanez B, James S, Agewall S, Antunes MJ, Bucciarelli-Ducci C, Bueno H, et al. ESC guidelines for the management of acute myocardial infarction in patients presenting with ST-segment elevation: the Task Force for the management of acute myocardial infarction in patients presenting with ST-segment elevation of the European Society of Cardiology (ESC). Eur Heart J 2017;39(2):119−77.
[387] Bansilal S, Bloomgarden Z, Halperin JL, Hellkamp AS, Lokhnygina Y, Patel MR, et al. Efficacy and safety of rivaroxaban in patients with diabetes and nonvalvular atrial fibrillation: the rivaroxaban once-daily, oral, direct factor Xa inhibition compared with vitamin K antagonism for prevention of stroke and embolism trial in atrial fibrillation (ROCKET AF trial). Am Heart J 2015;170(4):675−82 e8.
[388] Hanley CM, Kowey PR. Are the novel anticoagulants better than warfarin for patients with atrial fibrillation? J Thorac Dis 2015;7(2):165−71.
[389] Trujillo T, Dobesh PP. Clinical use of rivaroxaban: pharmacokinetic and pharmacodynamic rationale for dosing regimens in different indications. Drugs 2014;74 (14):1587−603.
[390] Williams B, Mancia G, Spiering W, Agabiti Rosei E, Azizi M, Burnier M, et al. ESC/ESH Guidelines for the management of arterial hypertension: the Task Force for the management of arterial hypertension of the European Society of Cardiology (ESC) and the European Society of Hypertension (ESH). Eur Heart J 2018;39 (33):3021−104.
[391] Adeboyeje G, Sylwestrzak G, Barron JJ, White J, Rosenberg A, Abarca J, et al. Major bleeding risk during anticoagulation with warfarin, dabigatran, apixaban, or rivaroxaban in patients with nonvalvular atrial fibrillation. J Manag Care Spec Pharm 2017;23(9):968−78.
[392] Brown JD, Shewale AR, Talbert JC. Adherence to rivaroxaban, dabigatran, and apixaban for stroke prevention in incident, treatment-naïve nonvalvular atrial fibrillation. J Manag Care Specialty Pharm 2016;22(11):1319−29.
[393] Ageno W, Beyer-Westendorf J, Rubboli A. Once- versus twice-daily direct oral anticoagulants in non-valvular atrial fibrillation. Expert Opin Pharmacother 2017;18 (13):1325−32.

[394] Heine GH, Brandenburg V, Schirmer SH. Oral anticoagulation in chronic kidney disease and atrial fibrillation. Dtsch Arztebl Int 2018;115(17):287–94.
[395] So CH, Eckman MH. Combined aspirin and anticoagulant therapy in patients with atrial fibrillation. J Thromb Thrombol 2017;43(1):7–17.
[396] Moore KT, Byra W, Vaidyanathan S, Natarajan J, Ariyawansa J, Salih H, et al. Switching from rivaroxaban to warfarin: an open label pharmacodynamic study in healthy subjects. Br J Clin Pharmacol 2015;79(6):907–17.
[397] Kim JH, Lim K-M, Gwak HS. New anticoagulants for the prevention and treatment of venous thromboembolism. Biomol Ther (Seoul) 2017;25(5):461–70.

Index

T

U

V

W

Y

Z

Printed in the United States
by Baker & Taylor Publisher Services